Low-Voltage Digital BiCMOS Circuits:
Circuit Design, Comparative Study, and Sensitivity Analysis

Samir S. Rofail

Kiat-Seng Yeo

Prentice Hall PTR,
Upper Saddle River, NJ 07458
http://www.phptr.com

ISBN 0-13-011380-8

Library of Congress Cataloging-in-Publication Data

```
Low-voltage, low-power digital BiCMOS circuits : circuit design,
    comparative study, and sensitivity analysis / by Samir S. Rofail and
    Yeo Kiat Seng.
        p.   cm.
    Includes bibliographical references and index.
    ISBN 0-13-011380-8
    1. Metal oxide semiconductors, Complementary.  2. Bipolar
integrated circuits.  3. Digital integrated circuits.  4. Low
voltage integrated circuits.  I. Rofail, Samir S., 1948-      .
II. Yeo, Kiat Seng, 1964-       .
TK7871.99.M44L69    1999
621.39'732--dc21                                          99-11804
                                                              CIP
```

Editorial/production supervision: *Joanne Anzalone*
Cover designer: *Anthony Gemmellaro*
Cover design director: *Jerry Votta*
Manufacturing manager: *Alan Fischer*
Marketing manager: *Lisa Konzelmann*
Acquisitions editor: *Bernard Goodwin*
Editorial assistant: *Diane Spina*
Book design and layout: *Aurelia Scharnhorst, Vanessa Moore*

© 2000 by Prentice Hall PTR
　Prentice Hall, Inc.
　Upper Saddle River, New Jersey 07458

Prentice Hall books are widely used by corporations and government agencies for training, marketing, and resale.
The publisher offers discounts on this book when ordered in bulk quantities.
For more information, contact Corporate Sales Department. Phone: 800-382-3419;
Fax: 201-236-7141; E-mail: corpsales@prenhall.com
Or write: Prentice Hall PTR, Corp. Sales Dept., One Lake Street, Upper Saddle River, NJ 07458

Product names mentioned herein are the trademarks or registered trademarks of their respective owners.

All rights reserved. No part of this book may be reproduced, in any form or by any means, without permission in writing
　　from the publisher.

Printed in the United States of America
　　10　9　8　7　6　5　4　3　2　1

ISBN 0-13-011380-8

Prentice-Hall International (UK) Limited, *London*
Prentice-Hall of Australia Pty. Limited, *Sydney*
Prentice-Hall Canada Inc., *Toronto*
Prentice-Hall Hispanoamericana, S.A., *Mexico*
Prentice-Hall of India Private Limited, *New Delhi*
Prentice-Hall of Japan, Inc., *Tokyo*
Simon & Schuster Asia Pte. Ltd., *Singapore*
Editora Prentice-Hall do Brasil, Ltda., *Rio de Janeiro*

Table of Contents

Preface	ix
Acknowledgments	xi
Nomenclature	xiii
1 Introduction	**1**
1.1 Why Low-Voltage, Low-Power?	2
1.2 Why BiCMOS Technology?	4
1.3 Applications of BiCMOS	7
1.3.1 Random Access Memories (RAMs)	7
1.3.2 Digital Signal Processing (DSP)	9
1.3.3 Telecommunications Applications	9
1.4 Low-Voltage Low-Power Design	10
1.4.1 Low-Voltage Low-Power Design Limitations	10
1.4.2 Power Dissipation of a Single Logic Gate	13
1.4.3 Power Dissipation of LSIs	14
1.4.4 Techniques of Reducing the Power Dissipation	15
1.4.5 Future BiCMOS Directions	17
1.5 Conclusions	21
1.6 References	21

2 BiCMOS Process Technology — 27

2.1 Introduction — 27
2.2 Bipolar and CMOS Processes Convergence — 28
2.3 CMOS Processing Issues of the Twin-Well BiCMOS Process — 30
    2.3.1 Various Source/Drain Structures and the Channel Profiles — 30
    2.3.2 Threshold Voltage Issues — 33
2.4 Bipolar Process Techniques — 33
    2.4.1 Base Design Techniques — 36
    2.4.2 Collector Design Techniques — 37
    2.4.3 Emitter Design Techniques — 38
2.5 BiCMOS Isolation Issues — 39
    2.5.1 Latch-up Phenomenon — 39
    2.5.2 Trench Isolation — 40
    2.5.3 Epitaxial Layer — 42
    2.5.4 Buried Layers — 43
    2.5.5 Active Device Isolation — 43
2.6 BiCMOS Interconnect Issues — 44
    2.6.1 Silicidation — 44
    2.6.2 Local Interconnect — 46
    2.6.3 Metallization and Planarization — 48
2.7 Classification of BiCMOS Technologies — 50
    2.7.1 Process Flow of a High-Performance 5 V, 0.8 µm Digital BiCMOS — 50
    2.7.2 Process Changes for the 0.5 µm Digital BiCMOS Technology — 56
    2.7.3 An Analog/Digital BiCMOS Process Flow — 60
2.8 A New Low-Power Ultra Low-Capacitance BiCMOS Process — 65
    2.8.1 Process Description — 66
    2.8.2 Shallow Trench Isolation (STI) — 66
2.9 Manufacturing Considerations — 68
2.10 Future Trends in BiCMOS Technology — 68
    2.10.1 Bipolar Device Structure Improvements — 69
    2.10.2 Silicon-On-Insulator (SOI) Technology — 72
2.11 Conclusions — 73
2.12 References — 75

3 MOS Device Modeling 83

3.1	The Threshold Voltage Models	83
3.2	The MOSFET Current Models	87
3.3	The MOSFET in a Hybrid Mode Environment	96
	3.3.1 Surface P-Channel for Sub-Half Micron Devices	97
	3.3.2 Device Fabrication	100
	3.3.3 Model Parameters Extraction	100
	3.3.4 Sub-Half Micron DC Model Formulation	102
3.4	Concluding Remarks	124
3.5	Summary	125
3.6	References	126

4 Low-Voltage BiCMOS Digital Circuits 131

4.1	Introduction	131
4.2	Source-Well Tie and Quasi-Reduction of Bipolar Turn-On Voltage Techniques	132
	4.2.1 pMOS/NPN Pull-Down Technique With and Without Source-Well Tie	132
	4.2.2 Quasi-Reduction of Bipolar Turn-On Voltage Technique	136
4.3	Full-Swing BiCMOS Logic Circuits with Complementary Emitter-Follower Driver Configuration	140
	4.3.1 Introduction	140
	4.3.2 BiCMOS Driver Configurations	147
	4.3.3 Full-Swing Techniques	149
	4.3.4 Comparison Between the Three Different Driver Configurations	150
	4.3.5 Full-Swing Complementary MOS/Bipolar Logic (FS-CMBL) Circuits	153
	4.3.6 Experimental Results and Analysis	155
	4.3.7 Circuit Variations of the FS-CMBL with Feedback	156
	4.3.8 Experimental Results and Analysis	160
4.4	Merged BiCMOS (MBiCMOS) Logic Gates	166
	4.4.1 Introduction	166
	4.4.2 MBiCMOS Gate	167
	4.4.3 Circuit Performance and Comparison	169
	4.4.4 Experimental Tests and Analysis	172

	4.4.5 Simulation Results and Analysis	175
	4.4.6 Full-Swing MBiCMOS Gate	177
4.5	Full Voltage Swing Multi-Drain/Multi-Collector Complementary BiCMOS Buffers	177
	4.5.1 Introduction	177
	4.5.2 Conventional BiCMOS and CBiCMOS Buffers	178
	4.5.3 Multi-Drain/Multi-Collector BiCMOS Buffers	179
	4.5.4 Simulation Results and Discussion	181
	4.5.5 Full-Swing Multi-Drain/Multi-Collector BiCMOS Buffers	183
	4.5.6 Circuit Implementation and Operation	184
	4.5.7 Simulation Results and Discussion	185
4.6	Quasi-Complementary BiCMOS Logic Circuits	189
	4.6.1 Introduction	189
	4.6.2 Circuit Concept of the QC-BiCMOS and Its Advantages	190
	4.6.3 Circuit Performance Comparison and Analysis	193
	4.6.4 Delay Analysis	201
	4.6.5 Design Issues	204
	4.6.6 Experimental Results	204
	4.6.7 Appendix	206
4.7	Full-Swing Schottky BiCMOS/BiNMOS Logic Circuits	208
	4.7.1 Introduction	208
	4.7.2 Basic Concept	208
	4.7.3 Circuit Implementations	210
	4.7.4 Circuit Simulation and Discussion	212
	4.7.5 Scaling the Supply Voltage	217
	4.7.6 Delay Dependence on the Frequency of Operation	218
	4.7.7 Power Dissipation	219
	4.7.8 Area Comparison	221
	4.7.9 Crossover Capacitance	222
4.8	Feedback-Type BiCMOS Logic Circuits	223
	4.8.1 Introduction	223
	4.8.2 Conventional BiCMOS Circuits and Their Limitations	223
	4.8.3 R + N Type and Feedback-Type (FB) BiCMOS Logic Gates	224
	4.8.4 BiCMOS Full-Swing Circuits Utilizing a Positive Capacitively Coupled Feedback Technique	230
	4.8.5 1.2 V Complementary Feedback BiCMOS Logic Gates	244
	4.8.6 1.5 V BiCMOS Dynamic Logic Circuit	252

4.9	High-Beta BiCMOS (Hβ-BiCMOS) Logic Circuits	258
	4.9.1 Introduction	258
	4.9.2 Hβ-BiCMOS Logic Circuit	258
	4.9.3 Circuit Operation of the Hβ-BiCMOS Circuit	261
	4.9.4 Circuit Simulations and Discussion	262
	4.9.5 A BiCMOS Charge Pump with Hβ-BiCMOS	264
4.10	Transiently Saturated Full-Swing BiCMOS (TS-FS-BiCMOS) Logic Circuits	269
	4.10.1 Introduction	269
	4.10.2 Circuit Concept and Operation	269
	4.10.3 Circuit Performance Comparison and Analysis	272
	4.10.4 Design Issues	276
	4.10.5 Experimental Results and Analysis	278
4.11	Bootstrapped BiCMOS Logic Circuits	279
	4.11.1 Introduction	279
	4.11.2 1.5 V Bootstrapped BiCMOS Logic Gate (BS-BiCMOS)	281
	4.11.3 Bootstrapped Full-Swing BiCMOS/BiNMOS Inverter	283
	4.11.4 Double Bootstrapped BiCMOS Logic Gates (DB-BiCMOS)	297
4.12	References	305

5 Delay Time and Power Dissipation Sensitivity Analyses of Multi-Generation BiCMOS Digital Circuits — 311

5.1	Introduction	311
5.2	Relationship Between Key BJT/MOS Process and Device/Circuit Parameters	312
5.3	Sensitivity Analysis of the Conventional BiCMOS Circuit	313
	5.3.1 Delay Time Sensitivity to Key Device Parameters	313
	5.3.2 The Effects of the Load Capacitance on the Circuit Delay Sensitivity	317
	5.3.3 The Effects of the Quality of the BJT on the Circuit Delay Sensitivity	319
	5.3.4 The Effects of Scaling the Technology on the Circuit Delay Sensitivity	320
	5.3.5 HSPICE Simulations and Analysis	322
5.4	Sensitivity Evaluation and Comparison of Low-Voltage, Low-Power BiCMOS Circuits	326

5.4.1 The Effects of Loading on the Delay and Power
　　　　　　Dissipation Sensitivities　　　　　　　　　　　　　331
　　　5.4.2 The Effects of the Quality of the BJT on the Delay and
　　　　　　Power Dissipation Sensitivities　　　　　　　　　　336
　　　5.4.3 The Effects of Scaling the Technology on the Delay and
　　　　　　Power Dissipation Sensitivities　　　　　　　　　　338
　　5.5　Delay Sensitivity Upper and Lower Bounds:
　　　　　A Worst Case Scenario　　　　　　　　　　　　　　　338
　　5.6　Conclusions　　　　　　　　　　　　　　　　　　　　339
　　5.7　References　　　　　　　　　　　　　　　　　　　　341
　　5.8　Appendix　　　　　　　　　　　　　　　　　　　　　342

The Authors　　　　　　　　　　　　　　　　　　　　　　**343**

Index　　　　　　　　　　　　　　　　　　　　　　　　　**345**

Preface

The quest for excellence in integrated circuit and system design for Very Large-Scale Integration (VLSI) has been geared in recent years towards developing innovative techniques and methodologies to achieve low-voltage, minimum power circuits (designs). This, together with the uncompromising constraint of high chip performance, has shaped a new arena in which the Bipolar compatible Complementary Metal Oxide Semiconductor (BiCMOS) technology has found itself gaining growing significance and attracting keen interest. The high momentum developed in the direction of low-voltage low-power has created a trend sustained and fuelled by the strong demand from the field of portable electronics in general, and the mobile communications market in particular. This trend has led many business leaders and market visionaries to predict that the best is yet to come, and the established trend will stay with us for quite a few years.

This book addresses the design of low-voltage, low-power VLSI digital circuits with a focus on the BiCMOS technology. Such focus has been conceived by the need, greatly felt by the authors, to provide an in-depth coverage of the BiCMOS circuit design ideas, the detailed analyses, and the design methodologies backed by experimental data. This focus, in turn, should help students, designers, and researchers gain deeper insights into the circuit design aspects of BiCMOS integrated circuit and further their understanding of the subject.

This book begins with an introductory chapter on the low-voltage, low-power design where light is being shed on some historical facts, the semiconductor process technology, and its evolution throughout the years. Chapter 2 presents a comprehensive description of the BiCMOS process technology. It highlights the various process requirements and techniques for realizing a high-performance BiCMOS technology. A typical submicron BiCMOS process flow, a new low-power ultra low capacitance BiCMOS process and the future trends in the BiCMOS process technology are illustrated in the same chapter.

Chapter 3 presents and evaluates the main models of the MOS device, highlights the assumptions made in developing each model, and accurately describes its performance in a scaled technology environment. Much coverage is given in this chapter to the analytical and experimental characterization of the sub-half micron MOS devices, the modeling of lateral pnp bipolar transistors in the pMOSFET structures, and the trends and general features of the various device/process parameters of scaled pMOSFETs operating in a hybrid-mode environment. A key highlight of Chapter 3 is the presentation of a methodology to show how to construct a device model, for a given wafer, with the aid of device characterization tools and successfully retrieved experimental data.

Chapter 4 presents an in-depth analysis and the development of a new generation of BiCMOS circuits for present and future VLSI requirements. Circuit design ideas and techniques are explained in detail to show how they are implemented to improve the circuit performance or to propose a novel circuit structure. Much attention is devoted to full-swing BiCMOS circuits and the inherent techniques such as bootstrapping and transient saturation. The Quasi-Complementary BiCMOS, the Schottky BiCMOS/BiNMOS, the high-β BiCMOS, and the feedback BiCMOS circuits are also covered in-depth. This chapter also covers a comparative evaluation of the various BiCMOS circuits and how these circuits are being applied in a low-voltage low-power environment.

Chapter 5 deals with the sensitivity of the BiCMOS circuit delay time and power dissipation to changes in key device and circuit parameters. This is especially crucial for small technologies and high-performance BiCMOS circuits whereby the circuit speed in particular can be easily affected by adverse effects caused by the reduced dimensions and an increase in the device and process parameters tolerances. This chapter demonstrates a method to calculate the worst case speed degradation for a given set of tolerances for the device and process parameters.

We hope that this book appeals to students, instructors, circuit designers, engineers, and researchers in the area of BiCMOS circuit design, enhances their knowledge of the subject, and stimulates their ambitions to advance the subject to higher elevations.

Acknowledgments

We are indebted to those who helped us in bringing this book through the entire process, from idea to reality. Samir Rofail is grateful to his family Mona, Sherry, Chrissy, and Patrick. Their continuous support and encouragement were instrumental in the successful completion of this book. Yeo Kiat Seng is thankful to his family Shannen, Shun Yuan, and Shun Yi. In particular to Shannen for her encouragement, patience, and tolerance of all those late hours.

We express our gratitude to Professor Er Meng Hwa, the Dean of the School of Electrical and Electronic Engineering at Nanyang Technological University, and Professor Do Manh Anh, the Head of the Division of Circuits and Systems. Their enthusiasm and wisdom have created an environment in which one cannot help but to persevere.

Finally, the success of this book would not have been possible without the kind assistance provided by our research students: Johnny Chew Kok Wai, Law Chong Fatt, Shridhar Mubaraq Mishra and Steven Yeo Boon Hwang.

Nomenclature

α	Space-charge current model parameter
α_1	Coefficient 1 in the effective hole mobility model
α_2	Coefficient 2 in the effective hole mobility model
α_3	Coefficient 3 in the effective hole mobility model
β_f	Current gain
χ	Short-channel effect factor in the drain current equation of the Level-2 MOSFET model
δ	Narrow width factor for adjusting the threshold voltage
ε_{ox}	Permittivity of silicon dioxide
ε_{si}	Permittivity of silicon
ϕ_f	Potential difference between the Fermi level and the intrinsic Fermi level
ϕ_s	Surface potential of silicon
φ	Empirical parameter in defining E_{eff}
γ	Forward source-body effect factor
γ_B	Body effect coefficient
η	Drain-Induced Barrier Lowering (DIBL) parameter
η'	DIBL and channel length modulation effects factor
λ	Source barrier height lowering fitting parameter
μ	Carrier mobility
μ_{eff}	Effective carrier mobility
μ_o	Low field mobility
μ_n	Electron mobility
μ_p	Hole mobility

μ_{peff}	Effective hole mobility
μ_1	Empirical parameter 1 in the effective hole mobility model
μ_2	Empirical parameter 2 in the effective hole mobility model
μ_3	Empirical parameter 3 in the effective hole mobility model
v_{sat}	Saturation velocity of the carriers
θ	Charge-sharing factor
ρ_s	Resistivity of the well
σ	Channel length modulation effect factor
τ_f	Forward transit time
ξ	Mobility degradation factor
ζ	Bipolar current model parameter
$\zeta_1(V_{SB})$	Source-body bias dependent bipolar current parameter
$\zeta_2(T)$	Temperature dependent bipolar current parameter
ΔE_g	Bandgap narrowing between source and body
ΔL	Channel length shortening
ΔV_T	Threshold voltage shift due to short-channel effects
ψ_{Bh}	Injection barrier for holes in hybrid-mode lateral BJT
Λ	LDD region fitting parameter
g_{m1}	Transconductance associated with the variation of the source-gate voltage
g_{m2}	Transconductance associated with the variation of the source-body voltage
g_{sd}	Output conductance
h	Planck constant
k	Boltzmann constant
l_c	Characteristic length
m_e^*	Density-of-state effective mass for electrons
m_h^*	Density-of-state effective mass for holes
m_o	Electron mass at rest
n_f	Ideality factor
n_i	Intrinsic doping concentration of silicon
q	Electronic charge
$q\chi_s'$	Electron affinity of a semiconductor
$q\phi_B$	Energy difference between the Fermi level and the intrinsic Fermi level
$q\phi_o'$	Energy difference between the conduction band edge and the Fermi level
r_j	LDD p-implant junction depth
$A_{bipolar}$	Cross-sectional area for the bipolar current emission
A_{bulk}	Parameter modeling the bulk charge effect
A_{space}	Cross-sectional area for the space-charge current emission
C_d	Depletion capacitance

Nomenclature

C_{ox}	Gate oxide capacitance per unit area
D_i	Threshold adjust implant dosage per unit area
D_p	Hole diffusion constant
E_c	Conduction band energy level
E_{crit}	Critical electric field
E_{eff}	Effective transverse electric field
E_g	Bandgap energy
E_F	Fermi level of extrinsic semiconductor
E_{Fi}	Fermi level of intrinsic semiconductor
E_p	Lateral electric field at the pinch-off point
E_v	Valence band energy level
F_N	Correction factor for narrow-channel MOSFETs
F_S	Correction factor for short-channel effects in short-channel MOSFETs
$I_{bipolar}$	Lateral bipolar current
I_{body}	Body current
I_{DS}	Drain-source current of MOSFET
I_K	Knee current
I_{mos}	MOS current
I_S	Forward-bias saturation current of a p-n junction
I_{SD}	Source-drain current of MOSFET
I_{space}	Space-charge limiting current
K_1	Nonuniform substrate doping effect factor 1
K_2	Nonuniform substrate doping effect factor 2
K_p	Saturation field factor
L	Device drawn channel length
L_d	Electrical channel length for the MOS current
L_e	Electrical channel length
L_{eff}	Device effective channel length
L_{effav}	Average effective channel length
L_g	Gate length of MOSFETs
L_p	Length of the LDD p-region
N_a	Source/Drain concentration
N_{a-}	LDD P-concentration
N_c	Effective density of states in conduction band
N_{eff}	Fitting parameter for the channel length shortening factor
N_{FS}	Number of fast superficial states
N_{LX}	Lateral nonuniform doping parameter
N_{sub}	Channel doping concentration

N_v	Effective density of states in valence band
P_e	Hole concentration at the edge of the source(emitter)/n-well junction
Q_{dep}	Depletion carrier density
Q_{inv}	Inversion carrier density
R_b	Base resistance
R_C	Collector resistance
R_e	Emitter resistance
R_{well}	Average well resistance
T	Temperature at which measurements are carried out
TC_1	Temperature compensation factor 1
TC_2	Temperature compensation factor 2
T_{nom}	Ambient room temperature
T_{ox}	Gate oxide thickness
U_a	First-order mobility degradation coefficient
U_b	Second-order mobility degradation coefficient
U_c	Third-order mobility degradation coefficient
U_1	Contribution of the drain voltage to the gate-to-channel field
U_2	Exponential fitting parameter in the mobility model
V_A	Early voltage
V_{ASCBE}	Early voltage due to substrate current induced body effect
V_{bi}	Built-in potential of the LDD P-/n-well regions
V_{BS}	Body-source voltage
V_{bseff}	Upper bound for the body bias
V_{DD}	Supply voltage
V_{Dsat}	Saturation voltage
V_{DS}	Drain-source voltage
V_{DSeff}	Effective drain-source voltage
V_{FB}	Flatband voltage
V_{GB}	Gate-body voltage
V_{GS}	Gate-source voltage
V_{GSteff}	Effective gate-source voltage
V_{j1}	Junction potential between the source and the body
V_{j2}	Junction potential between the drain and the body
V_{off}	Offset voltage
V_{ON}	Threshold potential above which the device enters the strong inversion region and below which the device operates in the weak inversion region
V_{SB}	Source-body voltage
V_{SBeff}	Effective source-body voltage

Nomenclature

V_{SD}	Source-drain voltage
V_{SDeff}	Effective source-drain voltage
V_{SG}	Source-gate voltage
V_t	Thermal voltage
V_T	Effective threshold voltage
V_{TO}	MOSFET threshold voltage at V_{SD} and $V_{SB} = 0V$
W	Device channel width
W_{eff}	Effective device channel width
W_c	Depletion-layer width of a cylindrical junction
W_D	Depletion-layer width beneath the gate
X_{dep}	Depletion layer thickness
X_{dm}	Depletion width at the drain's surface
X_j	P+ implant junction depth
X_s	Depletion width at the source
X_d	Depletion width at the drain

CHAPTER 1

Introduction

The history of semiconductor devices begins in the 1930s, when Lilienfeld and Heil [1,2] first proposed the Metal Oxide Semiconductor (MOS) field-effect transistor. However, it took 30 years before this idea was to be applied to functioning devices for use in practical applications [3], and, up to the late 1970s, bipolar devices were the mainstream digital technology. This trend took a turn around 1980, when MOS technology caught up and there was a crossover between bipolar and MOS shares. Complementary-MOS (CMOS) was finding more widespread use due to its low power dissipation, high packing density and simple design, such that by 1990, CMOS covered more than 90% of the total MOS sales and the relation between MOS and bipolar sales was two to one.

In digital circuit applications, there was a performance gap between CMOS and bipolar logic. The existence of this gap implied that neither CMOS nor bipolar had the flexibility required to cover the full delay-power space. This was achieved by the emergence of bipolar compatible CMOS (BiCMOS) technology, the objective of which is to combine bipolar and CMOS so as to exploit the advantages of both at the circuit and system levels.

In 1983, a bipolar compatible process based on CMOS technology was developed, and BiCMOS technology — with both the MOS and bipolar devices fabricated on the same chip — was developed and studied [4–6]. The principal BiCMOS circuit in the early days was the BiCMOS totem-pole gate [7] as shown in Figure 1–1. This circuit was proposed by Lin et al. and is one of the earliest versions to be used in practice. It is commonly referred to as the conventional BiCMOS circuit. Since 1985, BiCMOS technologies developed beyond initial experimentation to become widespread production processes. The state-of-the-art bipolar and CMOS structures have been converging. Today, BiCMOS has become one of the dominant technologies used for high speed, low-power and highly functional VLSI circuits [8–11], especially when the BiCMOS process has been enhanced and integrated into the CMOS process without any addi-

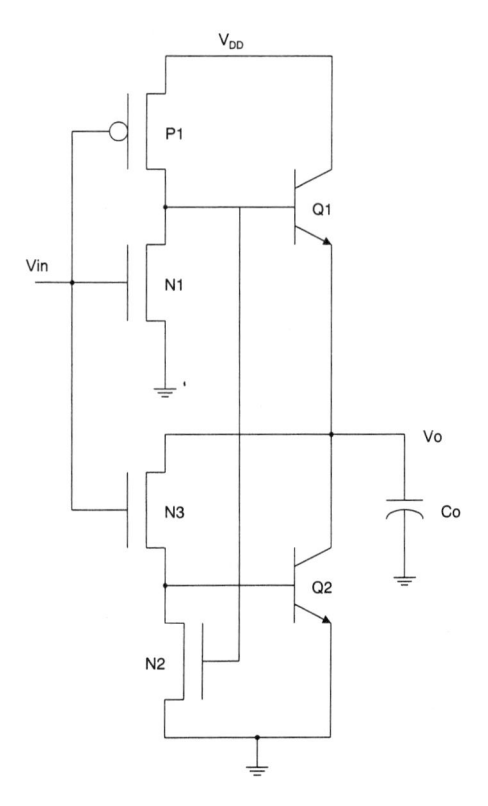

Figure 1-1 Conventional BiCMOS inverter.

tional steps [12]. The similarity between the process steps required for both devices, CMOS and bipolar, means that these steps can be shared for both of them.

Since the early 1970s, the concern of power dissipation has been included as part of the design process, but it was not considered to be of paramount importance. Instead, high speed operation and designing with minimum area, especially in memories, were the main design constraints. As the state of the art was driven towards higher speed and smaller die size, CAD developers and IC designers were all geared towards achieving these two goals. As a result, power dissipation was considered only as a figure of merit in most applications. However, with the advent in the area of portable electronics and the area of high-performance chips exceeding power dissipation limits, power consumption has become the main design criteria in many applications.

1.1 Why Low-Voltage, Low-Power?

In the past, due to a high degree of process complexity and the exorbitant costs involved, low-power circuit design and applications involving CMOS and BiCMOS technologies were used only in applications where very low-power dissipation was absolutely essential, such as in wrist watches, pocket calculators, pacemakers, and some integrated sensors. However, low-power design is becoming the norm for all high-performance applications, as power is the most important single design constraint. Although designers have different reasons for lowering the power consumption, depending on the target application, minimizing the overall power system has become a high priority.

One of the most important reasons for this trend is the advent of portable systems. As the "on the move with anyone, anytime, and anywhere" era becomes reality, portability becomes an essential feature of the electronic systems interfacing with nonelectronic systems, emphasizing efficient use of energy as a major design objective. Low-voltage operation is necessary in order to decrease the power consumption, so as to extend the operating life of the battery in systems such as laptop and notebook computers as well as in handheld communication devices such as cellular phones. Such portability will enable users to carry these systems easily for long periods of time. The low-power requirement for such portable equipment arises from the need to reduce the weight of the battery and also to extend its life before recharging. Many portable electronics use the rechargeable Nickel Cadmium (NiCd) batteries. Although the battery industry has been making efforts to develop batteries with higher energy capacity than that of NiCd, there seems to be no significant improvement in the near future, and expected improvement is only 40% by the turn of the century [13]. Even with advanced battery technologies such as Nickel-Metal Hydride (Ni-MH), which provide large energy density characteristics, the lifetime of the battery is still low. Since battery technology offers only limited improvement in battery life, low-power design techniques are essential for portable devices.

Performance improvement causes a drastic increase in power dissipation due to increased operating frequency and increased chip size. This is evident especially in high-performance systems with high clock frequencies. With high-speed products such as microprocessors [14–18], the cost associated with packaging, cooling and fans required to remove the heat increases significantly. Table 1–1 shows the power consumption of various microprocessors that operate in the frequency of 133 to 300 MHz. Note that at higher frequencies, power dissipation is very high. There is a need, therefore, to reduce the cooling requirements of complex integrated circuits. It will definitely be more cost effective and efficient to design a low-power high performance chip than to invest in expensive fancy packaging or air-cooling or liquid-cooling techniques to the chip.

Also, with high power dissipation, the reliability of the chip becomes a major concern. Due to the generation of high temperatures on the chip, failure mechanisms such as silicon interconnect fatigue, package-related failure, electrical parameter shift, electromigration, and junction fatigue are provoked [19]. This would imply a much shorter lifetime for these systems and less integration (which translates to lower performance and higher costs). Hence, in order to reduce the power consumption without sacrificing the operating frequency, it is essential to scale the power supply voltage.

Table 1–1 Power dissipation of microprocessors. [13]

Processor	Clock (MHz)	Technology (μm)	V_{DD} (V)	Power Dissipation (W)	Reference
PowerPC 620	133	0.50	3.3	30	[14, 15]
UltraSparc	167	0.45	3.3	30	[14, 16]
MIPS R10000	200	0.50	3.3	30	[17]
DEC Alpha 21164	300	0.50	3.3	50	[18]

As the demand for high-speed, low-power consumption and high-packing density continues to grow, there is a need to scale the device to become smaller and smaller. A component with small dimensions is subjected to breakdown at high voltages, compelling the use of low-supply voltages. The power dissipation issues and the devices' reliability problems, when they are scaled down to 0.5 μm and below, have motivated the electronics industry to adopt a supply voltage lower than the old standard of 5 V. The new industry standard for Integrated Circuit (IC) operating voltage is 3.3 V (±10%) [13]. The effect of decreasing the voltage to much lower values is impressive in terms of power saving.

1.2 Why BiCMOS Technology?

The BiCMOS process combines bipolar and CMOS to give the best balance between available output current and power consumption. Historically, BiCMOS technology has been developed as either a "CMOS speed enhancer" or a "bipolar power miser". Thus early BiCMOS processes were based either on CMOS [20,21] or bipolar [22] process flows, resulting in suboptimal performance for either the bipolar or MOSFET devices. However, today's BiCMOS technology offers both high-performance CMOS and bipolar devices on the same chip in an uncompromised manner. Therefore, by retaining the benefits of bipolar and CMOS, BiCMOS is able to achieve VLSI circuits with speed-power-density performance previously unattainable with either technology individually.

CMOS technology maintains an advantage over bipolar in power dissipation, noise margins, packing density, and the ability to integrate large complex functions with high-speed yields. One of the greatest advantages of CMOS technologies is their high packing density. The reason is that MOS devices of the same type can be placed in the same well such that distances between the devices are very short and are therefore determined only by the field isolation pitch, while bipolar transistors must be positioned in separate n-wells, which decreases the packing density [23].

The power required by CMOS gates originates from the need to charge the subsequent load capacitance and is therefore a linear function of the frequency. On the other hand, bipolar

Why BiCMOS Technology?

circuits in a given design always consume the same amount of power, regardless of the processing speed (see Figure 1–2) [23]. For duty factors, s (see Figure 1–2), larger than 0.3, such as in fast data-transmission systems, bipolar is faster than CMOS at the same power level, and even consumes less power than CMOS at the same speed. However, it is true that CMOS is advantageous over bipolar with respect to the power dissipation for small duty factors (or switching activities) of 0.1 and less, which occur in most of the standard logic circuits within the low to medium performance areas.

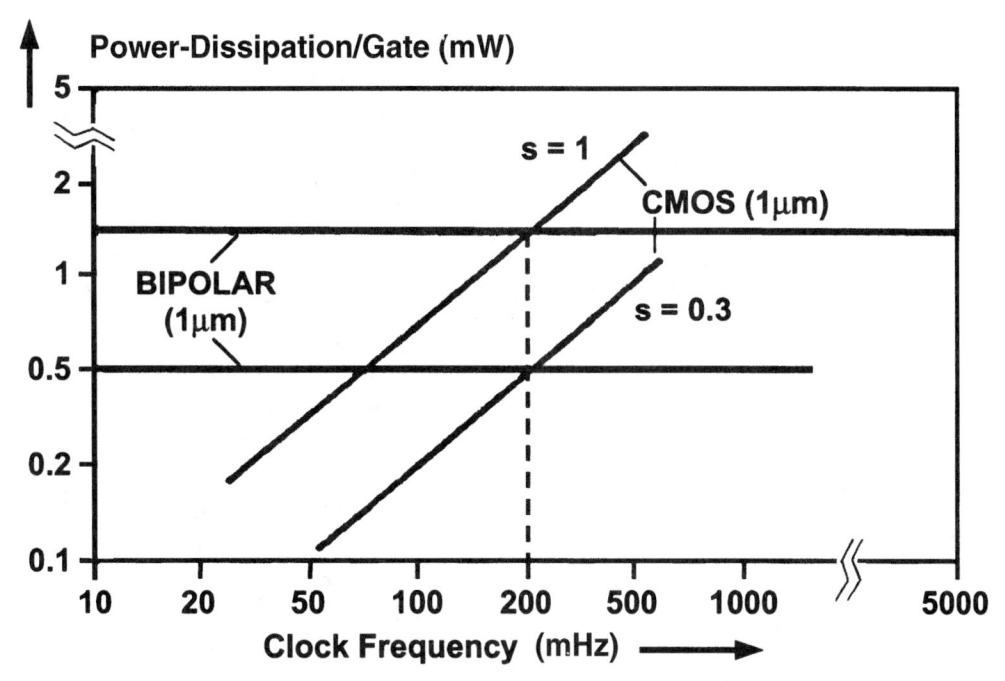

Figure 1–2 Comparison of the power dissipation/gate vs. clock frequency for bipolar and CMOS circuits.

The bipolar technology has advantages over CMOS in the switching speed, the current drive per unit area, the noise performance, and the Input/Output (I/O) speed. Bipolar transistors have an exponential current-voltage characteristic, while the characteristics of unipolar devices have an approximately quadratic function. The voltage swing required to switch a given current can be much smaller at the input of bipolar transistors than at MOS transistors: The transconductance of the bipolar transistors is considerably higher than that of a MOS transistor when the same current is applied. This means that the charging process takes place approximately ten times faster in Emitter-Coupled Logic (ECL) circuits than in CMOS, and bipolar is very much ahead of CMOS in terms of speed.

Although there are solutions to compensate for the disadvantages of both CMOS and bipolar, they generally require a large silicon area space which leads to an undesirable increase in the overall cost. An all-bipolar high-speed Very Large Scale Integration (VLSI) circuits are usually very difficult to implement and require very elaborate heat-sink arrangements.

Thus, BiCMOS technology makes it possible to combine the "best of both worlds" and is able to provide both high speed and high complexity: A BiCMOS VLSI provides the benefits of the low-power dissipation of CMOS and the high-output drive capability of bipolar devices. The advantages of BiCMOS can be summarized as follows [24]:

1. Improved speed performance over CMOS technology
2. Lower power dissipation than bipolar technology
3. Lower delay sensitivity to fan-out and output capacitive loading
4. Low dependence on process and temperature variation
5. Flexibility in input/output (I/O) interface, e.g., Transistor-Transistor Logic (TTL), Emitter-Coupled Logic (ECL) and other CMOS-compatible logic
6. Reduced clock skew
7. Improved internal gate delay
8. Reduced scaling-down as compared to CMOS, e.g., one to two micron BiCMOS offers circuit speed as high as submicron CMOS

The BiCMOS technology also has the practical advantage of being less susceptible to latch-up. Latch-up is the condition in which inherent pnp and npn parasitic bipolar transistors in the CMOS structure are forward biased by certain conditions: terminal over-voltage stress, transient displacement currents or ionizing radiation, lateral currents in the well and substrate, which drives the CMOS structure into a low impedance, high current state. Latch-up in the CMOS can result in the momentary or permanent loss of circuit function, and with the trend of scaling CMOS devices, better parasitic bipolar transistors are formed. In BiCMOS, there is a reduction of substrate and n-well resistances, due to the presence of highly doped P+ and N+ buried layers, which leads to higher latch-up initiation current. Also the N+ buried layer significantly reduces the current gain of the parasitic vertical pnp transistor. Therefore, BiCMOS is less likely to have latch-up.

The main disadvantages of the BiCMOS technology are the higher costs and longer fabrication cycle time due to the increase in complexity of the fabrication process. However, many conventional CMOS process steps can be utilized to fabricate bipolar transistor structures concurrently with MOS transistors. The BiCMOS fabrication requires only three to four masks in addition to well-established CMOS processes [25].

BiCMOS technology is the most suitable technology for low-power applications because it can keep power manageable by focusing it when required to optimize the speed-power tradeoff, with bipolar transistors used only in critical speed paths and in sense circuits. BiCMOS will play an important role in devices with a reduced power supply environment.

1.3 Applications of BiCMOS

In general, BiCMOS fills the market niche between the very high-speed (~200 ps/gate), power-consuming (~2 mW/gate) bipolar ECL and the very high-density (~100,000 gates/mm^2), medium-speed (~700 ps/gate) CMOS [26]. In a way, BiCMOS technology can be viewed as the "AND" function of high speed and high density. BiCMOS technology is used in two areas: in pure bipolar technology in order to replace some bipolar circuits with BiCMOS circuits to obtain lower power consumption; and in pure CMOS technology, in order to replace CMOS circuits with high-load capacitance to obtain higher speeds. When the power budget is unconstrained, a bipolar technology optimized for speed will almost always be faster than a BiCMOS technology, but in applications where a finite power budget exists, the ability to focus power where it is required usually allows BiCMOS speed performance to surpass that of bipolar [27]. In other words, in typical BiCMOS applications, CMOS is used in the circuit to provide low power and high integration, while the bipolar structures are used to play fast logic roles to ensure that BiCMOS gates work well in driving highly capacitive loads. BiCMOS technology is used in the following areas:

1.3.1 Random Access Memories (RAMs)

One of the most successful applications of BiCMOS technology was in the design of SRAMs [28]. By combining the high-density and low-power dissipation of CMOS with the extremely fast speed of bipolar ECL techniques, BiCMOS SRAMs achieved speeds very close to that of bipolar SRAMs at power levels close to that of CMOS SRAMs [29]. The rapid performance progress in computer systems, such as engineering workstations (EWS) and super computers, has been achieved by the development of high-speed memories as well as Central Processing Units (CPUs). Recently developed pipelined microprocessors with over 100 MHz clock frequencies strongly require high-speed and large-capacity memories [30]. BiCMOS memories have been developed as promising high-performance SRAMs, and at present, submicron BiCMOS SRAMs are widely used in the design of cache memories for high-performance EWS and mainframes. Submicron BiCMOS devices for high-speed SRAMs with more than 256-kb memory capacity have been developed to achieve high transistor performance and minimized process complexity, as well as high reliability.

Figure 1–3 shows the development trends in memory cell sizes and chip sizes for CMOS and BiCMOS SRAMs with 16-kb to 16-Mb memory capacities. The latest generation BiCMOS SRAM is required for transfer of data to/from super high-speed Reduced Instruction Set Computer (RISC) microprocessors synchronously with more than 100 MHz clock frequencies. The main application is for second-level cache memory systems in high-performance EWS. One such SRAM is the recently developed 16-Mb Gate-Transistor Logic (GTL) I/O SRAM, fabricated using the 0.4 µm process [31]. Its key features are a 3.3 V supply voltage, pMOS thin film transistors (TFT), triple-well structure, and BiNMOS gates. The super high-speed synchronous BiCMOS SRAM used is composed of a fully static SRAM core and input/output flip-flop (F/F) registers, in which data are latched synchronously with the external clock. Since bipolar circuits

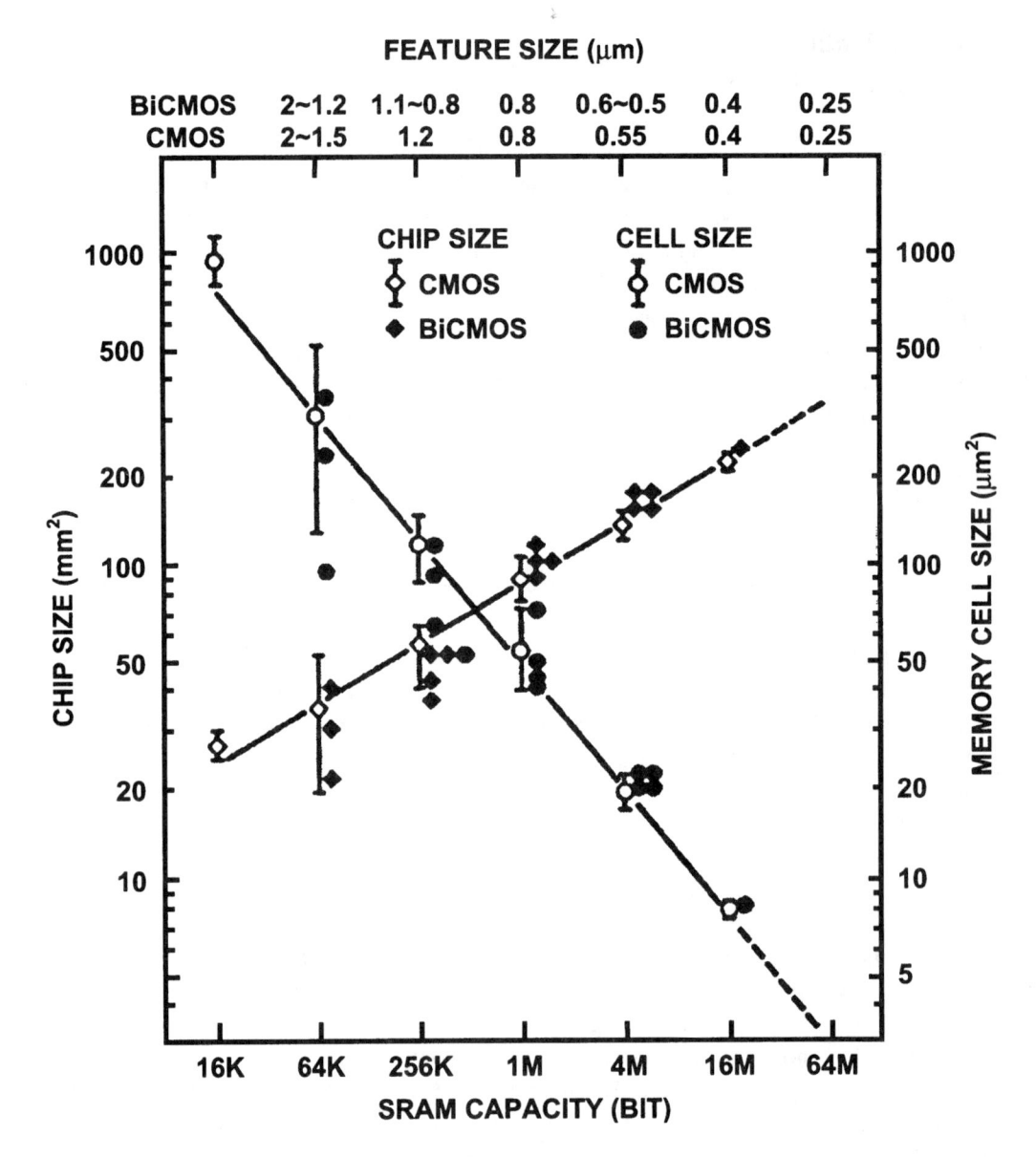

Figure 1–3 Progresses in memory cell sizes and changes in chip sizes for CMOS and BiCMOS SRAMs.

have higher speeds and less operational delay dependence on load capacitance than CMOS circuits, BiNMOS logic circuits and bipolar sense amplifiers are superior to CMOS circuits for high-frequency clocked operation.

Applications of BiCMOS 9

BiCMOS technologies are also used in the design of high-speed, mega-byte DRAMs [32,33]. However, their use is limited due to the increased costs associated with the additional process steps and increased chip size. Nevertheless, BiCMOS DRAMs are capable of covering the medium performance area of pure CMOS SRAMs.

Future BiCMOS SRAMs will still be confined to high-speed and large-capacity cache memories for super high-performance microprocessors with over 100 MHz clock frequency. However, for the submicron era, BiCMOS SRAMs must meet the twin requirements of low-voltage operation (at 2.5 V power supply voltage and below) and reduced process complexity. So far, the design rules for the BiCMOS process is lagging behind the CMOS process by about two years. Hence, in order to fully utilize the BiCMOS advantage, design rules for the BiCMOS, memories in particular, should be developed concurrently with the CMOS.

1.3.2 Digital Signal Processing (DSP)

Digital Signal Processing (DSP) is the technology at the heart of the next generation of personal communication systems. In the past, most of the research and design efforts have been focused on increasing the speed and throughput of DSPs. As a result, present technologies possess computing capacities that allow the realization of computationally intensive computing tasks such as speech recognition and real time digital video. However, the demand for high-performance portable systems incorporating multimedia capabilities has elevated the design for low power to the forefront of design requirements in order to maintain reliability and provide longer hours of operation. High-performance video codecs capable of efficient motion picture coding and decoding are needed for multimedia systems [34–36].

In 1993, a very fast video codec, a 300 MHz Video Signal Parallel Pipeline Processor (VSP3) [37] using BiCMOS technology, was announced. It was able to carry out an encoding algorithm at a rate of 30 frames/sec with a resolution of 352×288 lines per frame. BiCMOS will be at the forefront of this technology as the active mode power dissipation in video codecs needs to be reduced to the level of about 1.0 W to 1.5 W for inexpensive plastic packages and to well under 1 W for battery back-up portable systems. The power dissipation in the standby mode also needs to be small for portable multimedia systems using batteries, with these systems using low voltages such as 0.9 V [38].

1.3.3 Telecommunications Applications

Today's cellular telephones require a gigahertz level of operation. Therefore, coupling very high speed with very low voltage is a challenge to Ultra Large Scale Integration (ULSI) design, especially for circuit design. The BiCMOS technology has proven to be an excellent workhorse for telecommunication applications [39]. This is due to its excellent digital and analog capabilities as well as the variety of I/Os it offers, which make the design process, using BiCMOS technology, a very flexible one. For the I/O intensive telecommunications applications, BiCMOS is the technology of choice, as it allows the designer to concentrate the power dissipation where it is needed most in high-speed blocks, while using the low-power moderate speed CMOS in other blocks.

Lately, due to the ever increasing complexity of telecommunication switches, the need has arisen for novel circuits with low on-chip power consumption. This is particularly important in two classes of circuits that conventionally consume the most power: signal translation circuits and I/O circuits. The DSP portions of telecommunication chips are usually implemented with CMOS logic, while the front-end of the transceiver is usually realized in BiCMOS to take advantage of the characteristics of bipolar devices.

1.4 Low-Voltage Low-Power Design

Until recently, the two main design challenges faced by circuit and system designers were designing with minimum chip area and maximum speed constraints. However, with the advent of personal data processing and communication, coupled with device miniaturization, battery-based systems have been pushing toward low-voltage operation in order to decrease the power consumption and hence increase the battery operation time. There are two main issues facing device designers in the field of low-power design: the reduction of the supply voltage to as low as possible to reduce operating power, as well as the reduction of the threshold voltage to increase the performance of the chip.

1.4.1 Low-Voltage Low-Power Design Limitations

1.4.1.1 Power Supply Voltage

From the device designer's viewpoint, "The lower the supply voltage, the better." While the dynamic power is largely dependent on the supply voltage, stray capacitance, and frequency of operation, the overall supply voltage has the largest effect. Therefore, with overall supply voltage lowered, the power dissipation of the circuits can be largely reduced, without compromising the frequency of operation, or in other words, the speed performance. However, there are various problems associated with lowering the voltage: In CMOS circuitry, the drivability of MOSFET's will decrease, signals will become smaller, and the threshold voltage variations will become more limiting. As shown in Figure 1–4, the increase of the gate delay time is serious when the operating voltage is reduced to 2 V or less, even with scaling down the device dimensions. The supply voltage scaling in BiCMOS circuits puts even more serious constraints on the circuit performance. Although BiCMOS ULSI systems realize the benefits of the low-power dissipation of CMOS and high-output drive capability of bipolar devices, under low-power supply voltage conditions, the gate delay time significantly increases. This is because the effective voltage applied to MOS devices is dropped by the inherent built-in voltage ($V_{BE} \sim 0.7$ V) of the bipolar devices in the conventional totem-pole type circuit. New methods, therefore, must be devised in order to overcome these obstacles to lowering the supply voltage.

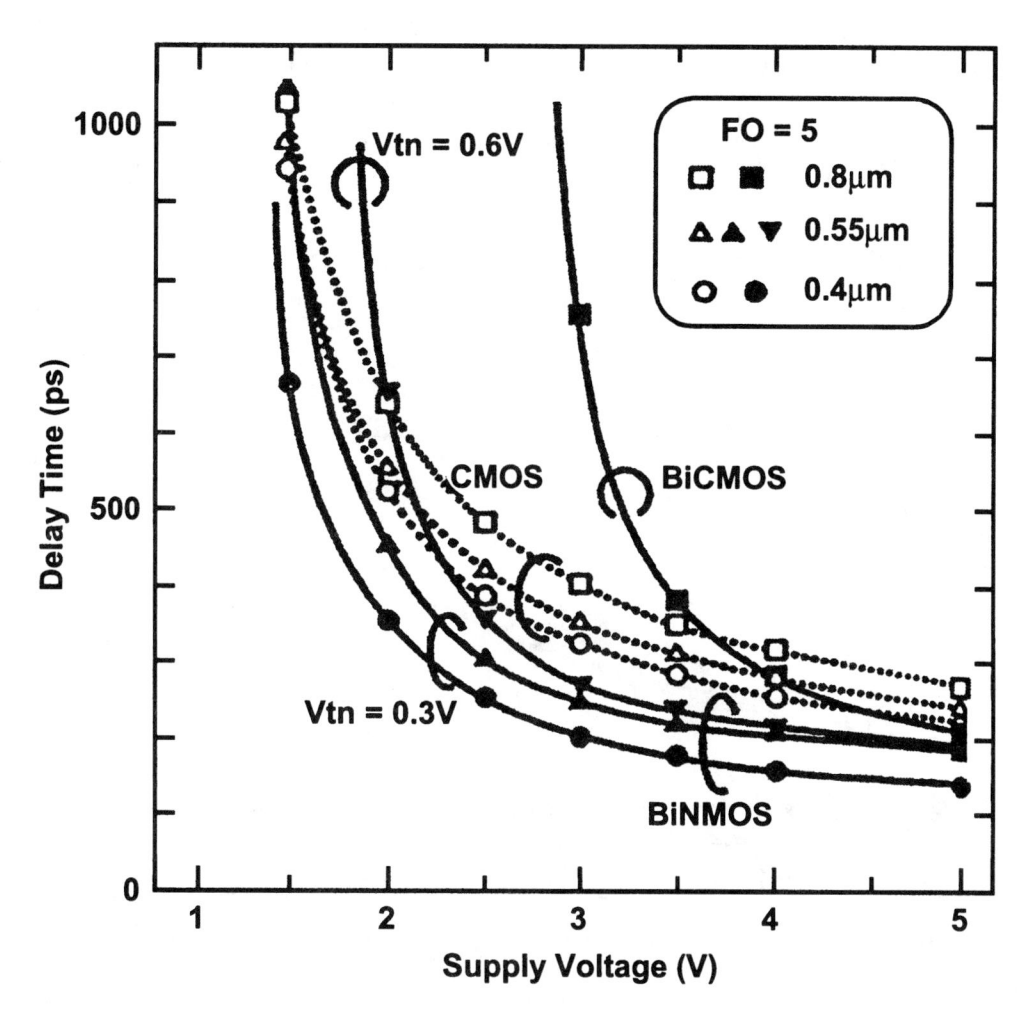

Figure 1-4 Inverter time vs. supply voltage. [30]

1.4.1.2 Threshold Voltage

Another key issue of scaling the power-supply voltage is the threshold voltage restriction. At a low-power supply voltage, a low-threshold voltage is preferable to maintain the performance trend. However, since the reduction of the threshold voltage causes a drastic increase in the cutoff current, the lower limit of the threshold voltage should be considered carefully by taking into account the stability of the circuit operation and the power dissipation. Furthermore, the threshold voltage dispersion must be suppressed proportional to the power supply. The dispersion of threshold voltage affects the noise margin, the standby power dissipation, and the transient power dissipation. Since the LSI performance is restricted by the worst case critical path, it is influenced by the threshold voltage dispersion. Therefore, suppressing the threshold voltage is strongly required for low power LSI from the process control and the circuit design point of view [40].

Figure 1–5 shows the V_{th}/V_{DD} dependence of the gate delay time of the CMOS inverter [41]. When the threshold voltage approaches $V_{DD}/2$, the delay time increases rapidly causing a

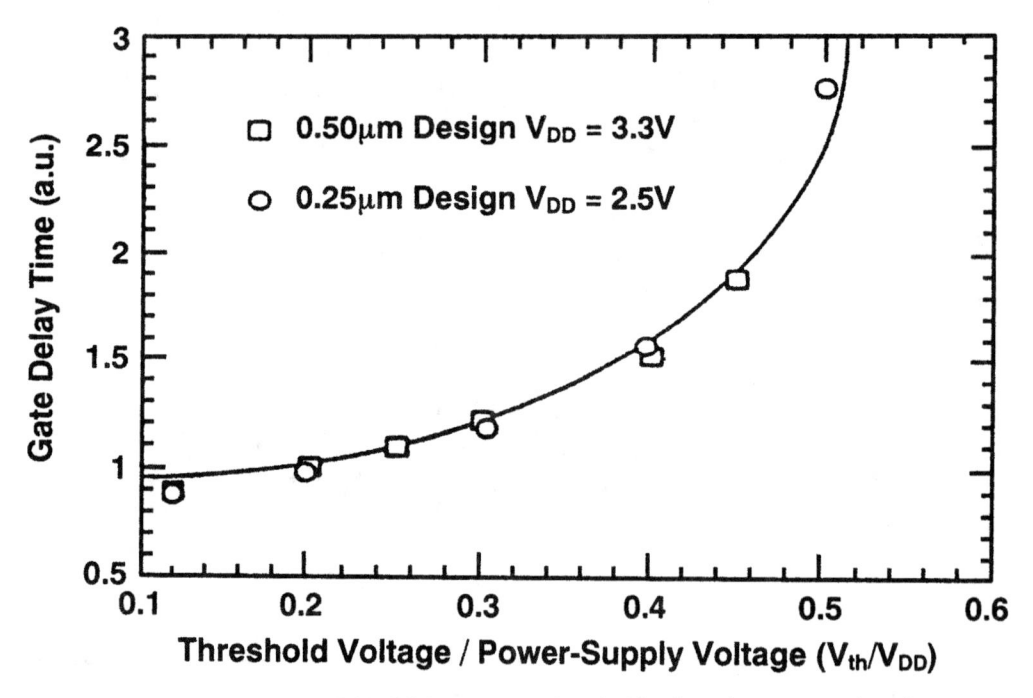

Figure 1–5 Gate delay time of CMOS inverter vs. threshold voltage/power supply voltage.

drastic reduction of the MOSFET current and a corresponding increase in CMOS inverter threshold. On the contrary, lowering the threshold voltage drastically improves the gate delay time. Therefore a V_{th}/V_{DD} ratio of 0.2 and below is required for high-speed operation, and it is necessary to reduce the threshold voltage to as low as possible when lowering the power supply voltage. However, since the subthreshold swing is almost constant in any device generation, reduction of the threshold voltage sharply increases the MOSFET cutoff current and degrades its On/Off ratio. Moreover, the threshold voltage reduction increases the power dissipation due to the switching transient current. At high-threshold voltages, the transient power dissipation is negligible compared to the total power dissipation. On the other hand, at low-threshold voltage, the transient power greatly increases with the transient current. Thus a compromise needs to be found for the V_{th}/V_{DD} ratio in order to have both low-power and high-speed operation.

1.4.1.3 Scaling

As the demand for high-speed, low-power consumption and high-packing density continues to grow each year, there is a need to scale the device to smaller dimensions. As the market trend moves towards a greater scale of integration, the move towards a reduced supply voltage also has the advantage of improving the reliability of IC components of ever-reducing dimensions. This can be easily understood as IC components of smaller dimensions have a higher tendency of breaking down at high voltages. It has already been accepted that scaled down CMOS devices even at 2.5 V do not sacrifice device performance with maintaining device reliability [42].

Scaling brings about the following benefits:

1. Improved device characteristics for low-voltage operation, due to the improvement in the current driving capabilities
2. Reduced capacitance through small geometries and junction capacitances
3. Improved interconnect technology
4. Availability of multiple and variable threshold devices, which results in good management of active and standby power trade-off
5. Higher density of integration (It has been shown that the integration of a whole system, into a single chip, provides orders of magnitude in power savings.)

However, during the scaling process, the supply voltage would have to decrease to limit the field strength in the insulator of the CMOS, and relax the electric field from the reliability point of view. This leads to a tremendous increase in the propagation delay of the BiCMOS gates, especially if the supply voltage is scaled below 3 V [43]. Also, scaling down the supply voltage causes the output voltage swing of the BiCMOS circuits to decrease [44,45]. Moreover, external noise does not scale down as the device features size reduces, giving rise to adverse effects on the circuit performance and reliability.

The major device problem associated with the simple scaling lies in the increase of the threshold voltage and the decrease of the carrier surface mobility, when the substrate doping concentration is increased in order to prevent punch-through. To sustain the low-threshold voltage with a high-carrier surface mobility and a high immunity to punch-through simultaneously, substrate engineering will be a prerequisite.

1.4.2 Power Dissipation of a Single Logic Gate

In order to reduce the power dissipation of a logic gate, it is necessary to determine the sources of power dissipation. The total power dissipation (P_t) of a single logic gate can be classified into power dissipation in the active mode (P_{at}), and power dissipation in the standby mode (P_{st}).

P_{at} consists of two components and is given by:

$$P_{at} = P_{cd} + P_{dp} \qquad 1.1$$

where P_{cd} is the power dissipation due to the charging and discharging of the load capacitance of a logic gate. It is given by:

$$P_{cd} = f_c \times C_L \times V_{DD}^2 \qquad 1.2$$

where f_c is the switching frequency of the input, C_L is the load capacitance of logic gate, and V_{DD} is the supply voltage.

P_{dp} is the power dissipation due to a transient short circuit current flowing through the direct path of a pMOSFET and an nMOSFET. It is given by:

$$P_{dp} = V_{DD} \times f_c \times \int_0^{1/f_c} i_{dp}(t)dt \qquad 1.3$$

The current $i_{dp}(t)$ flows two times in one clock cycle so that the power dissipation is proportional to f_c. However, the energy consumed is wasted because it does not contribute to the propagation of signals.

P_{st} is the static power dissipation. It is due to leakage currents and current drawn from the supply due to the input voltage. The leakage current consists of MOS junction leakage currents caused by reverse biased pn-junction diodes. A typical value of this leakage current is 1 fA/device junction. This value is too small to have any effect on the static power. For example, if there are one million devices, the total contribution to the power would be ~0.01 µW. The second component of the static power, which is a function of the input voltage, is a result of the subthreshold current flowing through the MOS devices. In general, for one million devices integrated into a chip, this current can be quite significant. As P_{dp} and P_{st} are much smaller than P_{cd}, the total power dissipation is predominantly P_{cd}.

1.4.3 Power Dissipation of LSIs

It is difficult to express the power dissipation of LSIs with a single equation, since most LSIs consist of many functional blocks such as memories, input and output interface circuits, and combinatorial logic blocks. However, in general, the total power dissipation of an LSI chip of several logic functional blocks can be expressed as [38]:

$$P_t = P_{at} + P_{st} \qquad 1.4$$

where P_{at} is the power dissipation of the blocks in the active mode and P_{st} is the power dissipation of the blocks in the standby mode.

P_{at} is given in detail by:

$$P_{at} = \kappa_g \times n_g \times P_{cd,a} \qquad 1.5$$

where n_g is the total number of gates, κ_g is the ratio of operating gates to total gates, and $P_{cd,a}$ is the average power dissipation for an individual fundamental logic gate (such as NOT, NOR, NAND) and is given by:

$$P_{cd,a} = \alpha_i \times f_c \times C_{LA} \times V_{DD}^2 \qquad 1.6$$

where α_i is the probability that a switching transition occurs, and C_{LA} is the average load capacitance.

On the other hand, P_{st} can be expressed as:

$$P_{st} = (1 - \kappa_g) \times n_g \times P_{l,a} \qquad 1.7$$

$$P_{l,a} = I_{l,a} \times V_{DD} \qquad 1.8$$

where $I_{l,a}$ is the average leakage current for a single logic gate.

Because P_{at} is much larger than P_{st}, P_t can be approximated to P_{at}.

1.4.4 Techniques of Reducing the Power Dissipation

Note that in order to reduce the power dissipation, each factor involved in the power usage must be reduced. However, the signal handling capabilities *must* be maintained or increased, and the active areas *must* either be maintained or reduced if possible.

1.4.4.1 Power Supply Voltage

The most effective way to reduce P_{at} is to lower the supply voltage, as long as the throughput (and thus operating frequency) is not affected. The purpose of reducing the supply voltage can be looked upon as reducing the switching energy. A supply voltage reduction, however, may cause operating speed degradation, and the advantages of scaling are lost if the operating voltage is too low. Thus, the power dissipation can be minimized by operating the functional blocks at their lowest required voltages to maintain their high performance, while the input and the output interface circuits operate at the voltage required by the given systems. Another method is to introduce a pipelined scheme [46] so that the voltage requirements can be reduced without affecting the clock frequency and the circuit speed.

1.4.4.2 Power Dissipation in Gates

Power dissipation in the active mode also occurs when the gates are in operation. Thus, another way to reduce the power dissipation is to reduce the ratio of the operating gates to the total gates by stopping all idle circuits by disconnecting the power supplies. It would also be useful to reduce the total number of gates used in the system by either developing new architectures or improving the technology, since the presence of fewer gates means that less power is being consumed without degrading the throughput. A typical method of reducing the number of gates is to make use of a pass transistor logic circuit [47].

1.4.4.3 Gate Resizing

Another way to reduce power is through gate resizing. Gate resizing is done by replacing some gates in the circuit with devices in the gate library having smaller area, and therefore, smaller gate capacitance. Given that the power dissipated by a gate is proportional to its load, reducing that load leads to a reduction of the power dissipated by the circuit as well as a reduction of the chip area. Smaller gates are also slower; therefore, in order to preserve the timing behavior of the circuit, only gates that do not belong to a critical path can be slowed down. The main features of this technique are that it does not change the topology of the circuit under optimization, and that the resynthesized circuits are guaranteed to be as fast as the original implementations, yet smaller and substantially less power-consuming.

1.4.4.4   Amount of Computation/Algorithm

In systems, another key factor is the amount of computation required to implement a given algorithm such as motion estimation. Reducing the amount of computation by improving algorithms is quite an effective way to reduce throughput, in which no improved logic circuit performance is required. Reducing the amount of computation also reduces the frequency, resulting in power dissipation reduction. Furthermore, this technique is useful as it also allows the number of gates and chip area to be reduced.

1.4.4.5   Load Capacitance

Reducing the load capacitance of the logic gate is another important factor. There are many possible approaches to reduce the total load capacitance by shortening the wire lengths at different levels. At the chip level, a hierarchical bus structure should be used so that an individual bus line length is shortened. Circuitries that are frequently used should be connected to each other by short wires and disconnected from the main bus line. However, the drawback in adopting these techniques is the loss of flexibility of operations and applications. At the block level, word line lengths can be shortened, for example, by dividing the memory cell array into several blocks, and at the cell level, wire lengths can be shortened by optimizing the circuit structures, logic gate sizes, and floor layouts.

1.4.4.6   Parasitic Effects

The parasitic capacitance of CMOS LSI consists of the junction capacitance, the gate oxide capacitance, and the wiring capacitance. The junction capacitance is inversely proportional to the square root of the power-supply voltage, while the other two do not depend on it. As a result, the total parasitic capacitance naturally increases with decreasing supply voltage, and further, the junction capacitance becomes one of the dominant parasitics at low-voltage operation [48]. Therefore, reduction of the junction capacitance is a key issue for achieving high speed circuit operation at low voltage such as 1 V. In order to reduce the junction capacitance, punch-through stopper ion implantation can be introduced, with ion implantation adjusted to the junction depth and restricted severely under the channel area. Parasitic resistance for deep submicron MOSFET is dominated by the spreading and accumulation resistance [49,50]. They have to be significantly reduced, especially at low supply voltages, because they seriously degrade the device performance. Therefore, it is desirable that there is a high impurity concentration with very abrupt shallow junction profile, and that MOS devices are almost free from hot-carrier degradation under low voltage operation.

1.4.4.7 Power Dissipation in the Stand-by Mode

In order to reduce power dissipation in the stand-by mode, it is necessary to eliminate leakage currents in the CMOS circuitry. A low-power dissipation in the stand-by mode is strongly required for battery back-up portable multimedia devices. This can be achieved using the multiple threshold voltage technique. For example, switch drivers with high-threshold voltage can reduce the leakage currents so that the power dissipation in the stand-by mode can be kept as small as possible, while low-threshold voltage devices are used for the logic gates to maintain high performance [38].

1.4.4.8 Bus Architecture

Modification of the bus architecture is an important method of reducing the power usage. A new approach is a bus architecture that reduces the operating power by suppressing the bus signal swing to less than 1 V without increasing the stand-by current [51]. The total power dissipation in logic ULSIs becomes two thirds of that in the conventional architecture if the bus swing is one third of that in the logic stage. This is based on the assumption that the power consumed by the bus charging and discharging is one half of the total power consumed in the chip. The rise and fall times in this architecture are faster than those of a conventional CMOS bus driver circuit due to the reduced output swing. In the conventional architecture, the delay increases drastically when the supply voltage is below 2 V, while the delay in the new architecture is almost constant for signal swings ranging from 0.5 to 1 V. This architecture will relieve the constraint of the conventional supply voltage scaling while maintaining a high-speed transmission and a low-stand-by current.

1.4.4.9 Input/Output (I/O)

In recent years, the number of I/Os has increased very rapidly. For the state of the art, telecommunication switches with more than 300 I/Os are already in production and the need for a higher number of I/Os is persistent. However, the major limitation on the number of I/Os is the reliability problem that arises from the extreme heat generated by them. Again, lower power I/O designs will mean more I/Os and/or less packaging and cooling cost, hence, lower overall system cost. It has been reported [52] that a bipolar current-mode I/O circuit with a collector output and a common-base input results in a high-speed and low-power I/O.

In general, the power usage can be reduced by developing new architectures, new circuits, new device structures, new insulators with lower dielectric constants than those of silicon dioxide, and new computer-aided design (CAD) tools.

1.4.5 Future BiCMOS Directions

1.4.5.1 Decrease in the Power Supply Voltage

In addition to an increase in density, clear trends indicated in recent reports are for low-voltage and low-power memories, providing the key to meeting the increasing demand for battery-packed or battery-operated memories. The increase in demand in portable equipment will require lowering the power supply voltage levels further still. Figure 1–6 shows the memory

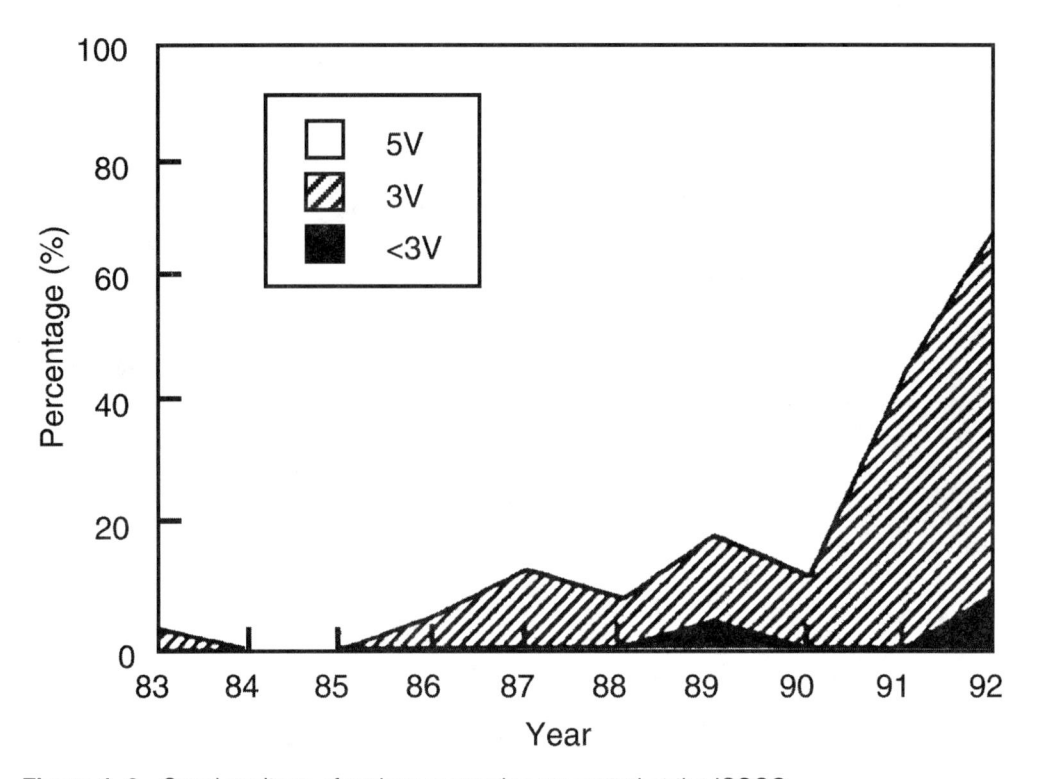

Figure 1–6 Supply voltage of various memories presented at the ISSCC.

LSIs supply voltage trends [38]. It can easily be seen that the mainstream supply voltage level is now 3 V, and in a few years' time, power supply voltages less than 3 V will be dominant. For standard DRAM, the internal operating voltage is expected to decrease to around 1.5 V for the 256-Mb or 1-Gb generation [53–55].

Figure 1–7 shows the switching energies of various electron devices compared to that of the brain cell [38]. The large disparity in the two-power dissipation suggests that in addition to the conventional scaling, new architectures such as massively parallel processing as in the brain are needed in order to increase the processing capability without increasing the switching energy per function. It is also worth noting that in order to have switching energy lower than 0.01 fJ, it may be necessary to operate at low temperatures.

1.4.5.2 Towards BiCMOS Technology

Figure 1–8 shows the development history for CMOS, bipolar ECL, and BiCMOS SRAMs in the last 10 years [30]. Low-power standard SRAMs with an access time slower than 10 ns to 20 ns and TTL I/O interface have been developed using CMOS technologies. Their memory capacities have been roughly quadrupled every three years. On the other hand, high-speed SRAMs with an access time faster than 10 ns were first developed using bipolar technolo-

Low-Voltage Low-Power Design

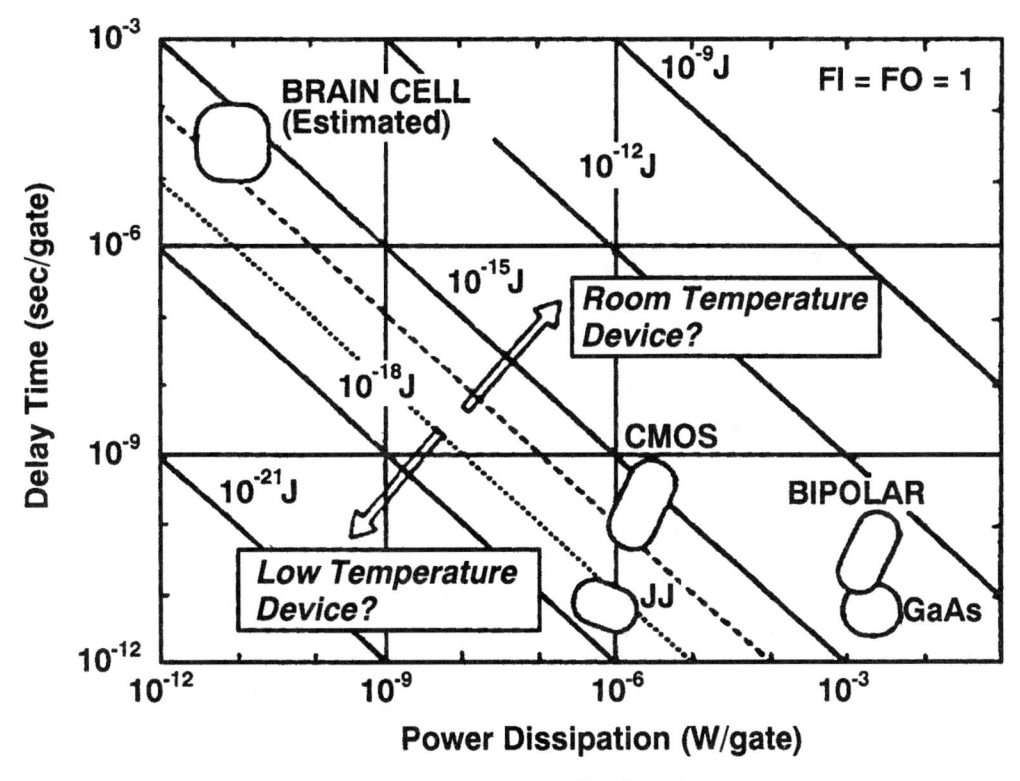

Figure 1-7 Switching energy of electron devices and brain cell.

gies and ECL I/O interface for up to 64-kb memory capacity. The submicron BiCMOS technology has been applied for ECL and TTL I/O high-speed SRAMs in order to prevent the increasing power dissipation crisis in bipolar ECL SRAMs.

Future BiCMOS SRAMs will still be applied to high-speed and large-capacity cache memories for super high-performance microprocessors with over 100 MHz clock frequency. However, for sub-half micron era, BiCMOS SRAMs must be able to perform under low voltage, while having a reduced process complexity.

1.4.5.3 Universal V_{DD} Concept

Although changing an external supply voltage is effective in reducing power dissipation, it is inconvenient to many users. DRAMs, which cover a wide operating voltage range from 1.5 to 3.3 or 5 V, have become increasingly important. This enables the memories to be operated with a variety of batteries having different supply voltage levels. The universal V_{DD} concept, which covers a wide operating voltage range from 1.5 to 3.6 V, has been proposed in a 64-Mb DRAM [56]. In general, scaled MOSFETs tailored to high speed at the minimum operating voltage suffer from higher oxide stress at the maximum operating voltage. Thus, it is usually difficult to

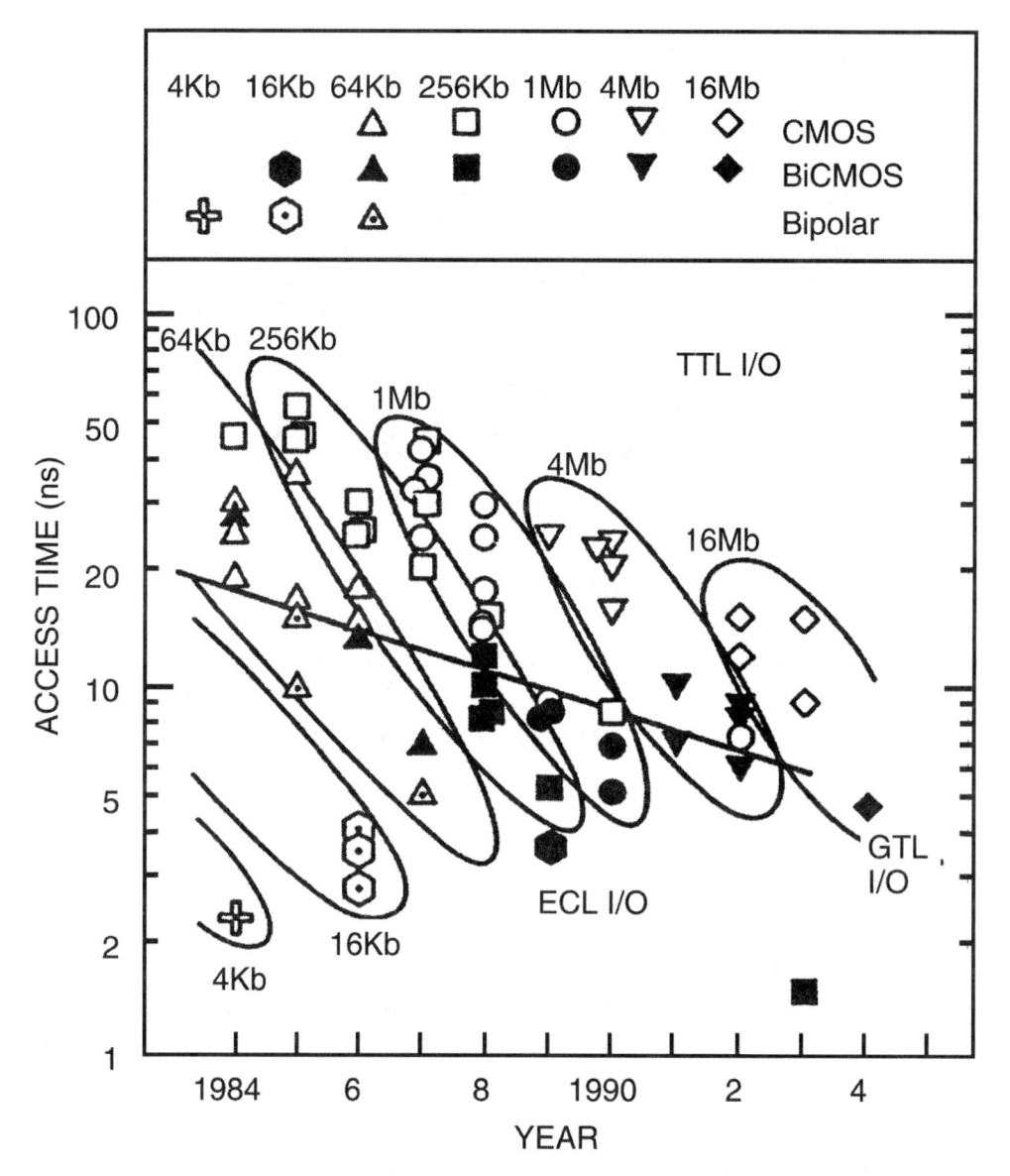

Figure 1-8 Development progresses of SRAMs' access time against the year.

achieve high speed and reliable operation, especially when the maximum operating voltage is more than twice the minimum one. In addition to solving the reliability problem, this universal V_{DD} concept will be a solution to the unacceptable large change in the operating speed arising from such a wide operating voltage. Thus, the universal V_{DD} concept will become an essential technique for future high-density memories.

1.5 Conclusions

The increased energy efficiency in contemporary electronic systems is a major requirement for portability, which will become the dominant feature of the electronic systems in the near future. Rapid development toward low power, mainly by lowering the power supply voltage because it offers substantial power savings and does not require the adoption of radically new circuit designs, has already been demonstrated in the real world.

Due to the large gain in speed and accuracy compared to CMOS and due to the great increase in possible circuit complexity compared to bipolar, BiCMOS has already captured the high-end market such as in the areas of SRAMs and gate arrays. The market share will further increase because of the advent of new challenging systems such as High Definition TeleVision (HDTV), high performance workstations, super-minicomputers and mobile communications. These products require the combination of logic with on-chip RAM and high-performance mixed analogue/digital circuitries which can only be achieved using BiCMOS technology. BiCMOS is and most certainly will be the mainstream technology for high-end applications with total systems being integrated into one silicon die for the best system performance and the lowest system costs.

There appears to be no revolutionary low power device or technology that is manufacturable or compatible with the mainstream circuit architectures today. Therefore evolutionary innovations and optimization for low power and low voltage plus continued scaling will fortunately be able to support the need of low-power applications for a long time into the future [57].

1.6 References

[1]   J. E. Lilienfeld, US Patent 1,745,175 (1930).

[2]   O. Heil, US Patent 439,457 (1935).

[3]   D. Kahng and M. M. Atalla, "Silicon-Silicon Dioxide Field Induced Surface Devices," *IRE Solid-State Devices Res. Conf.*, Carnegie Institute of Techn. Pittsburgh, Pa., 1960.

[4]   H. Momose, H. Shibata, Y. Mizutani, K. Kanzaki, and S. Kohyama, "High Performance 1.0 μm N-Well CMOS/Bipolar Technology," *Symp. VLSI Technology, Tech. Dig.*, pp. 40–41, 1983.

[5]   I. Walezyk and J. Rubinstein, "A Merged CMOS/Bipolar VLSI Process," *IEDM, Tech. Dig.*, pp. 59–62, 1983.

[6]   J. Mayamoto, S. Saitoh, H. Homose, H. Shibata, K. Kanzaki, and S. Kohyama, "A 1.0 μm N-Well CMOS/Bipolar Technology for VLSI Circuits," *IEDM, Tech. Dig.*, pp. 63–66, 1983.

[7]   H. C. Lin, J. C. Ho, R. R. Iyer, and K. Kwong, "CMOS-Bipolar Transistor Structure," *IEEE Trans. Electron Devices*, Vol. ED-6, No. 11, pp. 945–951, 1969.

[8]   H. Higuchi, G. Kitsukawa, T. Ikeda, Y. Nishio, N. Sasaki, and K. Ogiue, "Performance and Structures of Scale-Down Bipolar Devices Merged with CMOSFETs," *IEDM, Tech. Dig.*, pp. 694–697, 1984.

[9]   A. R. Alvarez, P. Meller, and B. Tien, "2 µm Merged Bipolar-CMOS Technology," *IEDM, Tech. Dig.*, pp. 761–764, 1984.

[10]   T. Ikeda, T. Nagano, N. Momma, K. Miyata, H. Higuchi, M. Odaka, and K. Ogiue, "Advanced BiCMOS Technology for High Speed VLSI," *IEDM, Tech. Dig.*, pp. 408–411, 1986.

[11]   H. Iwai, G. Sasaki, Y. Unno, Y. Niitsu, M. Norishima, Y. Sugimoto, and K. Kanzaki, "0.8 µm BiCMOS Technology with f_T Ion-Implanted Emitter Bipolar Transistor," *IEDM, Tech. Dig.*, pp. 28–31, 1987.

[12]   J. Miyamoto, S. Saitoh, H. Momose, H. Shibata, K. Kanzaki, and T. Iizuka, "A 28 ns CMOS RAM with Bipolar Sense Amplifiers," *IEEE ISSCC, Tech. Dig.*, pp. 245–248, 1984.

[13]   A. Bellaouar and M. I. Elmasry, *Low-Power VLSI Design: Circuits and Systems*, Kluwer Academic Publishers, 1995.

[14]   "Technology 1995: Solid State," *IEEE Spectrum*, pp. 35–39, January 1995.

[15]   D. Bearden, et al., "A 133 MHz 64-bit Four-Issue CMOS Microprocessor," *IEEE ISSCC, Tech. Dig.*, pp. 174–175, February 1995.

[16]   A. Charnas, et al., "A 64-bit Microprocessor with Multimedia Support," *IEEE ISSCC, Tech. Dig.*, pp. 178–179, February 1995.

[17]   MIPS Press release, 1994.

[18]   W. J. Bowhill et. al., "A 300 MHz 64-bit Quad-Issue CMOS RISC Microprocessor," *IEEE ISSCC, Tech. Dig.*, pp. 182–183, February 1995.

[19]   C. Small, "Shrinking Devices Put the Squeeze on System Packaging," *EDN*, Vol. 39, No. 4, pp. 41–46, February 1994.

[20]   T. Ikeda, T. Nagano, N. Momma, K. Miyata, H. Higuchi, M. Okada, K. Ogiue, "Advanced BiCMOS Technology For High Speed VLSI," *IEDM, Tech Dig.*, pp. 408–411, 1986.

[21]   R. H. Havemann, R. E. Eklund, R. A. Haken, D. B. Scott, H. V. Tran, P. K. Fung, T. E. Ham, D/P. Favreau, R. L. Virkus, "An 0.8 µm 256-kbit BiCMOS SRAM Technology," *IEDM, Tech Dig.*, pp. 748–751, 1987.

References

[22] T. Yuzuriha, T. Yamaguchi, J. Lee, "Submicron BiCMOS Technology Using 16 GHz f_T Double Poly-Si Bipolar Devices," *IEDM, Tech Dig.*, pp. 748–751, 1987.

[23] H. Klose, "BiCMOS – The Technology for Integrating Systems onto One Silicon IC," *Microelectronic Engineering*, Vol. 15, pp. 501–512, 1991.

[24] S. H. K. Embabi, A. Bellaouer, M. I. Elmasry, *BiCMOS Digital Integrated Circuit Design*, Kluwer Academic Publishers, M. A., 1993.

[25] S. M. Kang and Y. Leblebici, *CMOS Digital Integrated Circuits – Analysis and Design*, McGraw-Hill, 1996.

[26] A. Watanabe, T. Nagano, S. Shukuri, and T. Ikeda, "Future BiCMOS Technology for Scaled Supply Voltage," *IEDM, Tech. Dig.*, pp. 429–432, 1989.

[27] A. R. Alvarez, "Introduction to BiCMOS" in *BiCMOS Technology and Applications*, Ed. A. R. Alvarez, Kluwer Academic Publishers, 1989.

[28] A. R. Alvarez, "BiCMOS – Has The Promise Been Fulfilled?" in *BiCMOS Integrated Circuit Design: with Analog, Digital, and Smart Power Applications*, Ed. M. I. Elmasry, IEEE Press, pp. 15–18, 1994.

[29] A. G. Eldin, "An Overview of BiCMOS State-of-the-Art Static and Dynamic Memory Applications" in *BiCMOS Integrated Circuit Design: with Analog, Digital, and Smart Power Applications*, Ed. M. I. Elmasry, IEEE Press, pp. 257–267, 1994.

[30] M. Takada, K. Nakamura, T. Yamazaki, "High Speed Submicron BiCMOS Memory," *IEEE Trans. Electron Devices*, Vol. 42, No. 3, pp. 497–505, March 1995.

[31] K. Nakamura, S. Kuhara, T. Kimura, M. Takada, H. Suzuki, H. Yoshida and T. Yamazaki, "A 220 MHz pipelined 16-Mbit BiCMOS SRAM with PLL Proportional Self-Timing Generator," *IEEE ISSCC, Tech. Dig.*, pp. 258–259, 1994.

[32] G. Kitsukawa et al., "A 23 ns 1-Mbit BiCMOS DRAM," *IEEE J. Solid-State Circuits*, Vol. 25, No. 5, pp. 1102–1111, October 1990.

[33] H. Miwa et al., "A 17 ns 4-Mbit BiCMOS DRAM," *IEEE ISSCC, Tech. Dig.*, pp. 56–57, 1991.

[34] B. A. Ackland, "The Role of VLSI in Multimedia," *IEEE J. Solid-State Circuits*, Vol. SC-29, No. 4, pp. 381–388, April 1994.

[35] R. Kasai and T. Minami, "An Overview of Video Coding VLSIs," *IEICE Trans. Electron.*, Vol. SC-29, No. 4, pp. 1920–1929, December 1994.

[36] M. Yoshimoto, S. Nakagawa, T. Matsumura, K. Ishihara, and S. Uramoto, "ULSI Realization of MPEG2 Realtime Video Encoder and Decoder – An Overview," *IEICE Trans. Electron.*, Vol. E78-C, No. 12, pp. 1668–1681, December 1995.

[37] T. Inoue, J. Goto, M. Yamashina, K. Suzuki, M. Nomura, Y. Koseki, T. Kimura, T. Atsumo, M. Motomura, B. S. Shih, T. Horiuchi, N. Hamatake, K. Kumagai, T. Enomoto, H. Yamada, and M. Takada, "A 300 MHz 16-bit BiCMOS Video Signal Processor," *IEEE ISSCC, Tech. Dig.*, pp. 37–38, 258, February 1993.

[38] T. Enomoto, "Low Power Design Technology for Digital LSIs," *IEICE Trans. Electron.*, Vol. E79-C, No. 12, pp. 1639–1649, December 1996.

[39] R. Hadaway, T. Brown, K. Harris, and R. Foucault, "BiCMOS Technology for Telecommunications," *IEEE BCTM Proc.*, pp. 159–166, 1993.

[40] T. Kobayashi and T. Sakurai, " Self-Adjusting Threshold Voltage Scheme (SATS) for Low Voltage High Speed Operation," *IEEE CICC Proc.*, pp. 271–274, 1994.

[41] Y. Mii, S. Wind, Y. Taur, Y. Lii, D. Klaus, and J. Bucchigano, "An Ultra-Low Power 0.1 µm CMOS," *Symp. VLSI Technology, Tech. Dig.*, pp. 9–10, 1994.

[42] W. H. Chang, B. Davaari, M. R. Wordeman, Y. Taur, C. C. H. Hsu, and D. Rodriguez, "A High Performance 0.25 µm CMOS Technology (I) & (II)," *IEEE Trans. Electron Devices*, Vol. 39, No. 4, pp. 959–967, 1992.

[43] S. Shukuri, A. Watanabe, R. Izawa, T. Nagano, and E. Takeda, "The Guiding Principle for BiCMOS Scaling in ULSIs," *Symp. VLSI Technology, Tech. Dig.*, pp. 53–54, 1989.

[44] A. Bellaouar, S. H. K. Embabi and M. I. Elmasry, "Scaling of Digital BiCMOS Circuits," *IEEE J. Solid-State Circuits*, Vol. 25, No. 4, pp. 932–941, 1990.

[45] G. P. Rossel and R. W. Dutton, "Scaling Rules for Bipolar Transistors in BiCMOS Circuits," *IEDM, Tech. Dig.*, pp. 795–798, 1989.

[46] A. Chandrakasan, S. Sheng, and R. W. Brodersen, "Low-Power CMOS Design," *IEEE J. Solid-State Circuits*, Vol. 27, No. 4, pp. 472–484, April 1992.

[47] T. Kuroda, "Pass-Transistor Circuit Technology," *15th Workshop of Organization of Microelectronics, Proc.*, pp. 17–29, November 1995.

[48] P. Girard, C. Landrault, S. Pravossoudovitch, D. Severac, "A Technique to Reduce Power Consumption in CMOS Circuits," *ISIC-97*, pp. 526–529, 1997.

[49] H. Oyamatsu, M. Kinugawa, and M. Kakumu, "Design Methodology of Deep Submicron CMOS Devices for 1 V Operation," *Symp. VLSI Technology, Tech. Dig.*, pp. 89, 1993.

[50] K. K. Ng and W. T. Lynch, "Analysis of Gate-Voltage Dependent Series Resistance of MOSFETs," *IEEE Trans. Electron Device*, Vol. ED-34, pp. 503–506, 1987.

[51] Y. Nakagome et al., "A Sub-1-V Swing Bus Architecture for Future Low-Power ULSIs," *Symp. VLSI Circuits, Tech. Dig.*, pp. 82–83, June 1992.

[52] T. Kawamura, M. Suzuki, and H. Ichino, "An Extremely Low-Power Bipolar Current-Mode I/O Circuit for Multi-Gbit/s Interface," *Symp. VLSI Circuits, Tech. Dig.*, pp. 31–32, 1994.

[53] K. Itoh, "Trends in Mega-bit DRAM Circuit Design," *IEEE J. Solid-State Circuits*, Vol. 25, No. 3, pp. 778–789, June 1990.

[54] Y. Nakagome et al., "A 1.5 V Circuit Technology for 64-Mbit DRAMs," *Symp. VLSI Circuits, Tech. Dig.*, pp. 17–18, June 1990.

[55] Y. Nakagome et al. "An Experimental 1.5 V 64-Mbit DRAM," *IEEE J. Solid-State Circuits*, Vol. 26, No. 4, pp. 465–472, April 1991.

[56] Y. Nakagome et al., "Circuit Techniques for 1.5 - 3.6 V Battery-Operated 64-Mbit DRAMs," *ESSCIRC, Tech. Dig.*, pp. 157–160, September 1990.

[57] H. Oyamatsu, M. Kinugawa, and M. Kakumu, "Design Methodology of Deep Submicron CMOS Devies for 1 V Operation," *IEICE Trans. Electron.*, Vol. E79-C, No. 12, pp. 1720–1725, December 1996.

CHAPTER 2

BiCMOS Process Technology

2.1 Introduction

Complementary Metal Oxide Semiconductor (CMOS) has been the traditional fabrication technology for Very Large Scale Integration (VLSI) integrated circuits, memories in particular, because of its high-packing density and low-power features. However, in other applications, such as mainframe computers, high-speed analog-digital systems, and optical fiber network systems among others, better speed performance than that obtained by CMOS technology is required. Although bipolar-only devices can be used to achieve such a performance, they usually consume a lot of power. To circumvent this problem, BiCMOS devices are used because they are able to provide speed performance comparable to bipolar LSIs and integration density close to CMOS LSIs.

BiCMOS processes tend to be CMOS-intensive due to the need to reduce the power dissipation and achieve high packing density in VLSI and/or Ultra Large Scale Integration (ULSI) circuits [1]. Although the CMOS intensive nature of the BiCMOS processes will result in high-performance CMOS devices, the bipolar process integrated into the core CMOS process flow must be able to produce bipolar devices in an uncompromised form. Therefore, the major challenge associated with this integration approach is to obtain high-performance bipolar and CMOS transistors simultaneously at a minimal increase in process complexity.

In the following sections, the various process requirements and techniques for realizing a high-performance BiCMOS technology are presented. With these techniques, the merging and sharing of processes to satisfy the bipolar and CMOS requirements simultaneously while reducing the process complexity can be achieved. A typical submicron BiCMOS process flow and an analog/digital BiCMOS process design are described in the later sections. A new low-power, ultra-low capacitance BiCMOS process is also introduced and later in this chapter, we examine the future trends of the BiCMOS process technology.

2.2 Bipolar and CMOS Processes Convergence

In the submicrometer regime, the process flow of high-performance CMOS and bipolar devices has tended to converge. To comprehend the convergence, one must review the BiCMOS evolution from a CMOS perspective. Figure 2–1 illustrates various techniques to include the bipolar device into the conventional twin-well CMOS structure [2]. The most convenient way to merge an npn bipolar transistor is to use the n-well for the bipolar collector, and add one mask level (implant) for the bipolar base, as shown in Figure 2–1(a). In this technique, the bipolar emitter formation, the collector contact doping, and the nMOS n+ source/drain formation are accomplished together. Also, the extrinsic base doping region of the bipolar transistor and the pMOS p+ source/drain are formed concurrently. The major limitations of this structure are the high collector series resistance, because the lightly-doped pMOS n-well is used to form the bipolar collector, and the emitter junction depth dependence on the n+ source/drain depth, which is typically deeper than required for high-performance bipolar devices.

The second technique, as illustrated in Figure 2–1(b), overcomes the problem of high collector series resistance by forming a buried n+ layer under the n-well. Such a technique is normally employed in bipolar-only process to reduce the collector resistance. When applied to the BiCMOS processes, the presence of a heavily doped n+ layer under the n-well provides three advantages: (1) it aids in reducing the collector resistance; (2) it reduces the susceptibility of the CMOS to latch-up; and (3) it allows n-epitaxial layer to be used instead of p-epitaxial for better latch-up immunity. In this technique, the bipolar collector series resistance is further reduced by adding a deep n+ sinker in the collector contact region. Therefore, with the addition of just two additional mask levels (buried n+ and deep n+ sinker), the BiCMOS process can be optimized to realize a low collector resistance and reduce the CMOS latch-up susceptibility.

However, this approach still has a number of limitations. First, the packing density of the bipolar devices is limited by the doping level of the p-substrate. This limitation prevents punch-through between the collector of one bipolar transistor and another. However, increasing the doping level of the p-substrate, which allows closer packing of the bipolar devices, will cause an increase in the collector-to-substrate capacitance. Second, the n-epitaxial layer has to be counterdoped for isolation and the formation of p-wells for the nMOS devices. Counterdoping the n-type material may adversely affect the performance of the nMOS devices by degrading carriers mobility.

Figure 2–1(c) shows the third approach to merge the bipolar device into the conventional CMOS structure. Here twin n+/p buried layers are employed to allow tighter isolation spacing between adjacent wells. Using this technique, a reduction in the spacing between adjacent collectors can be achieved, but at the expense of a greater collector-to-substrate sidewall capacitance. Another advantage of this technique, as compared to the former, is that a twin-well CMOS structure can be implemented without heavily counterdoping the n-type epitaxial layer. This is possible because of the deposition of a near-intrinsic epitaxial layer [3] instead of the n-type epitaxial layer. Self-aligned p- and n-wells are then implanted into the thin intrinsic epitaxial layer and each is optimized separately. Another technology enhancement in this technique involves the usage of polycrystalline silicon (poly) for shallow bipolar emitters. A thin poly film deposited directly on the base serves as a diffusion source for the shallower emitter dopant. The

Bipolar and CMOS Processes Convergence

emitter active area is defined by a window in the base oxide (patterned just prior to poly deposition). Better bipolar performance can be obtained from polysilicon emitters because the emitters can be made narrower and shallower, thereby achieving both low transit time and emitter-to-base parasitic capacitance. Furthermore, this process can be easily integrated into a conventional CMOS process with only four additional mask levels (buried n+, deep n+ sinker, p-base, and polysilicon for the emitter).

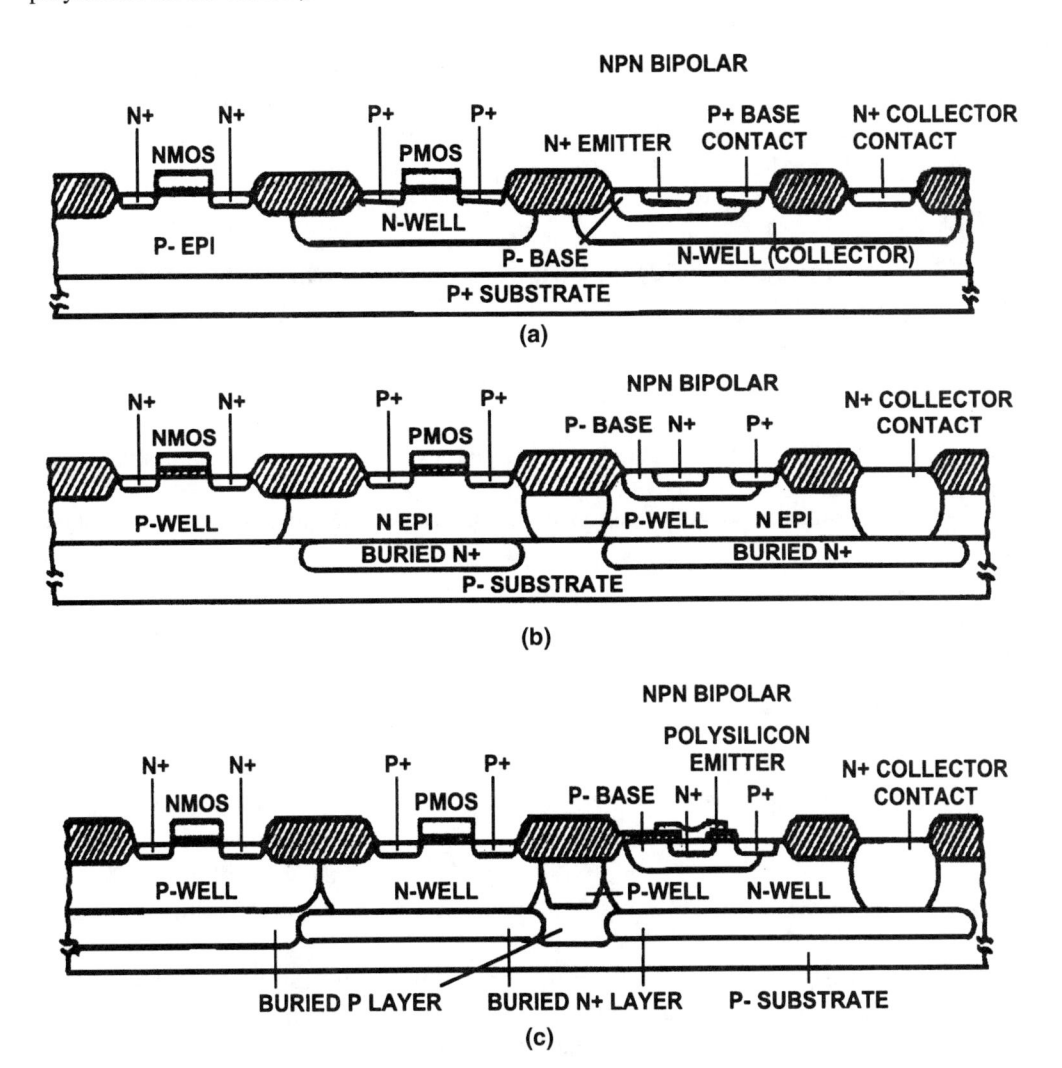

Figure 2-1 Different BiCMOS structures: (a) an npn bipolar transistor added directly to a basic n-well CMOS process; (b) a buried n+ layer and a deep n+ sinker added to reduce collector series resistance; (c) an optimized BiCMOS device structure, which includes self-aligned p and n+ buried layers for improved packing density.

Other performance enhancements include the sidewall oxide spacer technology. This method provides a self-aligned offset of the CMOS n+/p+ source/drain implants from the poly gate and the bipolar p+ extrinsic base implant from the poly emitter edges. The silicidation of the CMOS source/drain, poly emitter, bipolar extrinsic base and poly gate will reduce the series resistance. These two features will be elaborated on in sections 2.3.1 and 2.6.1.

2.3 CMOS Processing Issues of the Twin-Well BiCMOS Process

In high-performance submicron BiCMOS devices, the CMOS devices are usually fabricated using a CMOS twin-well process [4]. By using the twin-well approach, we can optimize separately the n-well and p-well doping concentrations to achieve uncompromised device characteristics for both the MOS and bipolar devices. The following subsections discuss several process issues that affect the CMOS transistor performance and reliability.

2.3.1 Various Source/Drain Structures and the Channel Profiles

As MOS devices are scaled down into the submicron regime, hot-electron generation due to impact ionization becomes a major problem, because the supply voltage does not reduce in proportion to the device technology. To circumvent this problem and to allow the circuits to operate at a conventional supply voltage of 5 V, specially engineered source/drain structures must be employed. These structures reduce hot electrons generation by reducing the electric field between the source and drain regions. Two structures that can relax the electric field are the lightly doped drain (LDD) and the double diffused graded drain structures [5–7]. The underlying principle of these two structures is to extend the drain depletion region further into the drain diffusion, thereby reducing the electric field. Figure 2–2(a) shows an nMOS LDD section, which utilizes the sidewall oxide spacer technology. The sidewall oxide spacer is formed by depositing a high quality oxide, typically 300 nm of Tetraethyl Orthosilicate (TEOS), and then performing an anisotropic etch. The etch results in high quality sidewall oxide being formed on the vertical edges of the polysilicon gate.

To form an n-type lightly doped structure, phosphorus with a typical concentration of 1 to 2×10^{13} atoms/cm^3 is implanted and self-aligned to the polysilicon gate edge prior to the formation of the sidewall oxide. After the sidewall oxide is formed, the more heavily doped source/drain impurity is implanted and self-aligned to the edge of the sidewall oxide. The sidewall oxide performs the function of a mask, preventing the heavily doped impurities from entering the more lightly doped region.

Figure 2–2(b) shows the double diffused graded drain structure. This technique also employs the oxide spacer technology. After the sidewall oxide formation, arsenic and phosphorus are implanted into the source/drain regions with typical doses of 5×10^{15} and 1×10^{15} atoms/cm^2, respectively. The phosphorus gets to diffuse further downwards, as well as inwards, beneath the sidewall oxide to "grade" the drain doping profile and thereby reducing the electric field. This technique can be easily integrated into a conventional CMOS process without any extra masking steps, whereas the LDD approach requires an additional mask for the LDD

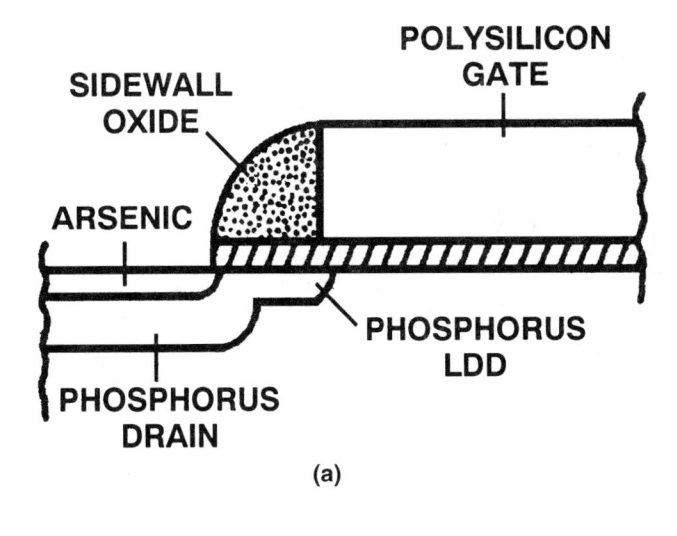

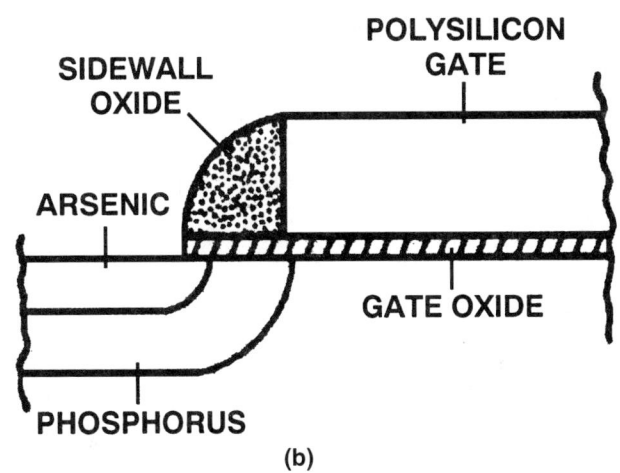

Figure 2–2 Two different drain structures to reduce hot-carrier generation: (a) lightly doped drain (LDD) structure; (b) double diffused graded drain structure.

source/drain implant. However, the LDD technique is more advantageous because it offers better electrical channel length control and device isolation. Better electrical channel length control is achieved as the channel length is predominantly determined by the gate etch.

For the graded drain approach, the variation in sidewall oxide thickness also contributes to the overall electrical gate length. The LDD technique offers better isolation because the heavily doped source and drain junction depths can be made shallower, since they do not have to be driven under the sidewall oxide to contact the active region. The implication is that devices can be packed closer since a tighter isolation pitch can be achieved. Another advantage offered by the LDD approach is that because the n+ source and drain junctions are only 0.22 µm deep, the threshold voltage implant functions as a punch-through stopper for the n+ source and drain junctions, thereby improving the short-channel performance and eliminating the need for a deep punch-through implant. Figure 2–3(a) shows the process cross-section of an LDD nMOS device, and Figure 2–3(b) shows the corresponding doping profile.

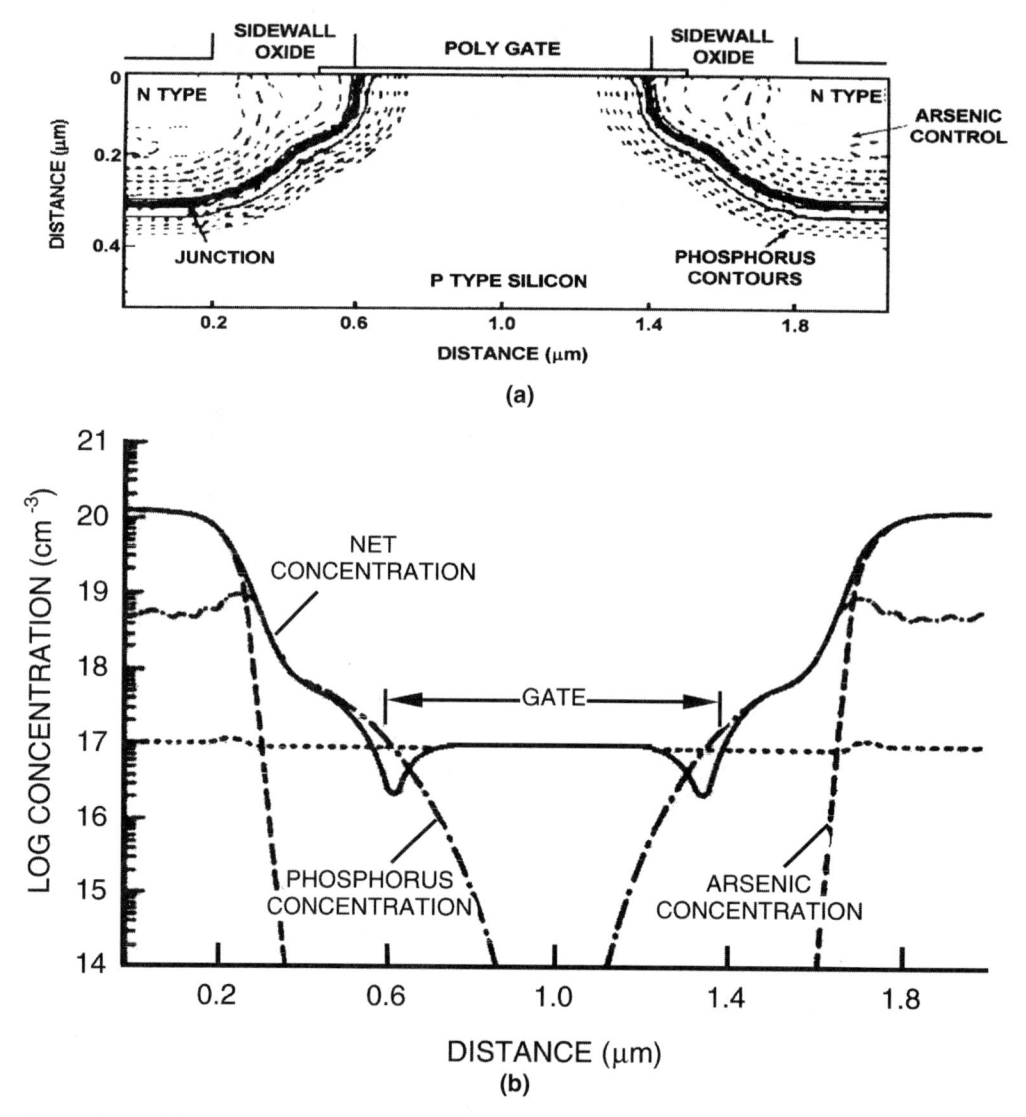

Figure 2–3 (a) Process cross-section of an 0.8 µm nMOS transistor with LDD sections; (b) doping profiles along the channel of the nMOS transistor in Figure 2–3(a).

For p-type devices, and because of the formation of buried channels, it is unnecessary either to employ a pMOS LDD or to drive the p+ source and drain junctions entirely beneath the sidewall oxide to provide a link with the active channel region. Figure 2–4(a) shows the structure of an 0.8 µm buried channel pMOS device, and its corresponding doping profile is shown in Figure 2–4(b).

Although LDD structures are not needed for pMOSFETs with channel length greater than 1 µm, they are necessary when devices are scaled to half-micrometer and below. This is because for deep submicron devices, the performance of the pMOS device becomes comparable to the nMOS device. As such, it is important to be able to control the leakage characteristics and the subthreshold properties of ultra-short buried channel pMOS device. However, when the device dimension is scaled to sub-half micrometer and below, the dual-implanted polysilicon gate approach is often used to improve the device subthreshold characteristics and to reduce its leakage currents. Dual-implanted gate approach gives rise to the formation of surface channel pMOS devices. Unlike the nMOS LDD devices, where the LDD sections are employed to reduce the hot-carrier generation, the LDD section in a pMOS device is used primarily to make the source and drain junctions shallower and to provide electrical conductivity between the p+ source and drain junctions and the active channel region.

2.3.2 Threshold Voltage Issues

The threshold voltage for CMOS devices is defined as the gate voltage that determines whether the device is on or off. The threshold voltage of the nMOS and pMOS devices can be adjusted to their appropriate values by implanting boron and phosphorus into the nMOS and pMOS channel regions, respectively, for the dual gate CMOS devices, and boron into the nMOS and pMOS channel regions for the n+ gate CMOS devices. In the case of n+ gate CMOS devices, the boron dopants increase the nMOS threshold voltage above the value predetermined by the p-well doping. On the other hand, the pMOS threshold voltage becomes less negative than the value determined by the n-well. The boron counterdopes the n-well just beneath the surface and, being light, forms a layer of finite thickness instead of an infinitesimal layer near the surface. This gives rise to the formation of a buried p-channel.

Very often for low-voltage BiCMOS applications, low-threshold voltage CMOS devices are required [8–10], as such low-threshold voltage MOS devices are often included as one of the process option modules [11]. No additional masking steps, and hence no additional cost, are involved; only threshold voltage implant dosage changes are required.

2.4 Bipolar Process Techniques

This section examines the different base, emitter and collector designs used to realize high-performance BiCMOS technologies. As stated earlier, the major challenge associated with the integration of bipolar processing steps into a baseline CMOS process is to obtain high-

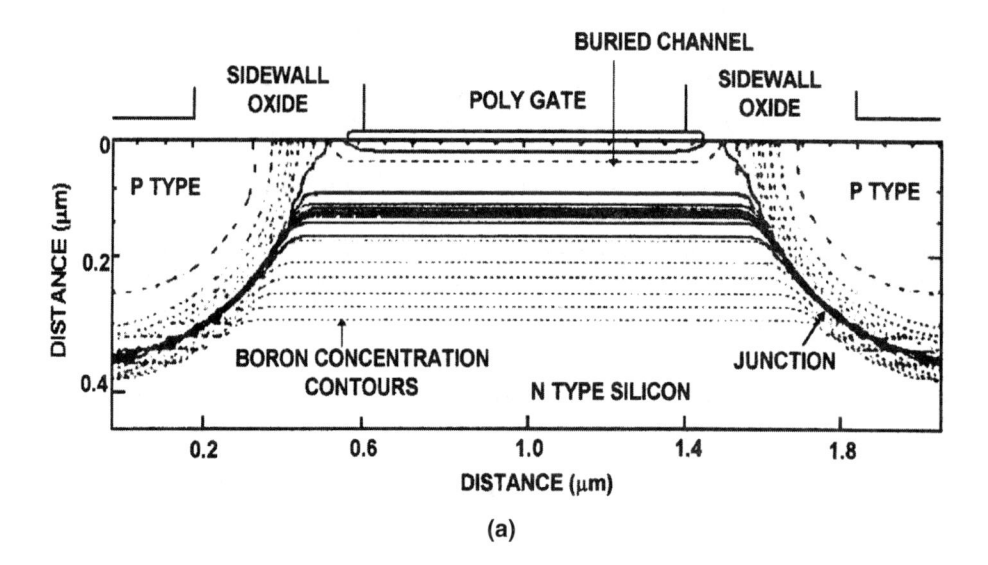

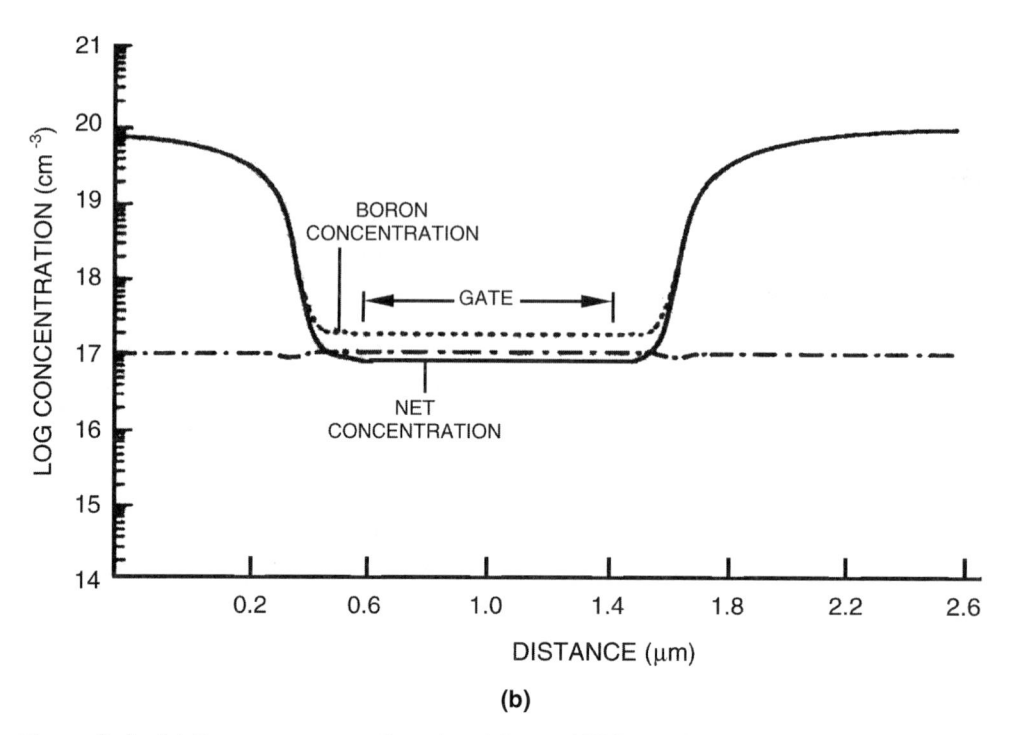

Figure 2–4 (a) Process cross-section of an 0.8 μm pMOS transistor illustrating the buried channel region that serves as a source drain extension under the sidewall oxide; (b) doping profile along the channel of the pMOS transistor in Figure 2–4(a).

Bipolar Process Techniques

performance bipolar and CMOS devices simultaneously for a minimal increase in the process complexity. In the subsequent subsections, we will look into the pros and cons associated with the base, emitter and collector designs of each structure, as illustrated in Figures 2–5(a–d).

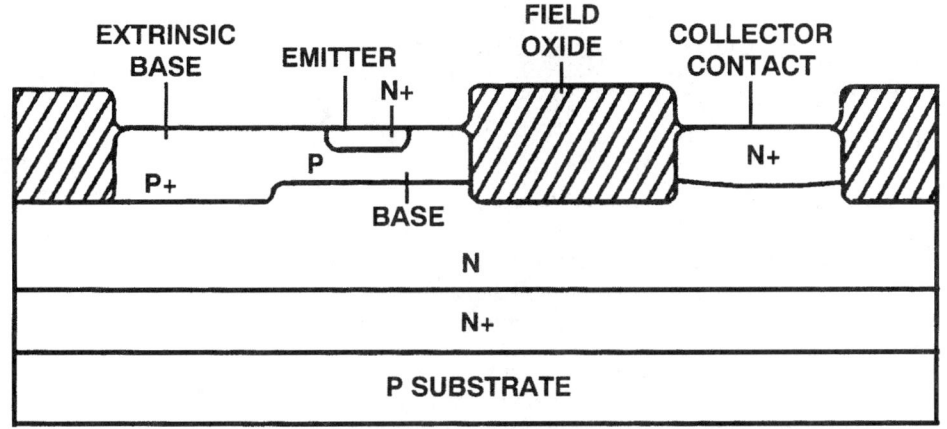

Figure 2–5(a) Cross-section of a bipolar transistor with a directly implanted emitter. During the fabrication process, both the n+ collector contact and the p+ extrinsic base are formed during the respective n+/p+ source/drain implants.

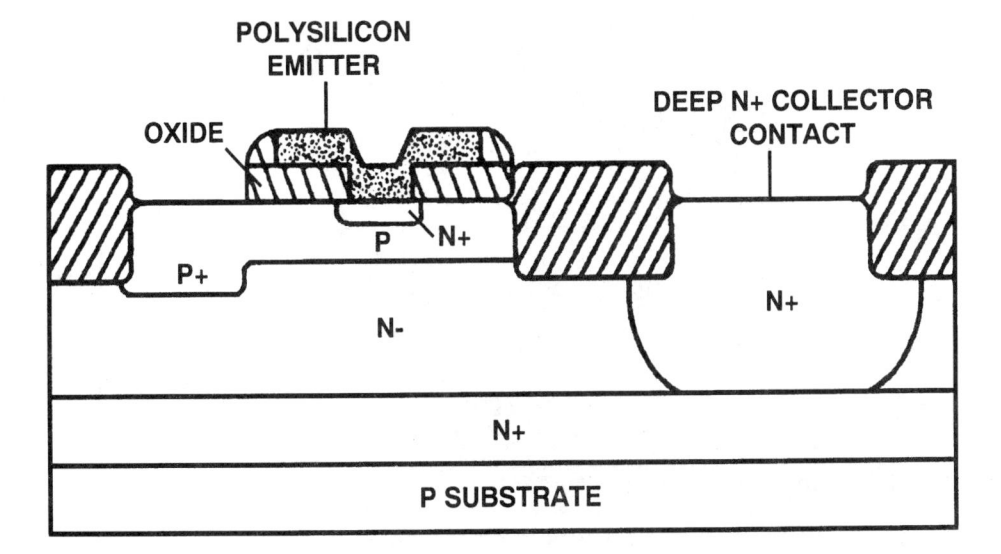

Figure 2–5(b) Cross-section of a bipolar transistor having a single level polysilicon emitter with an underlying protective oxide. A deep n+ collector sinker is used to reduce the collector series resistance.

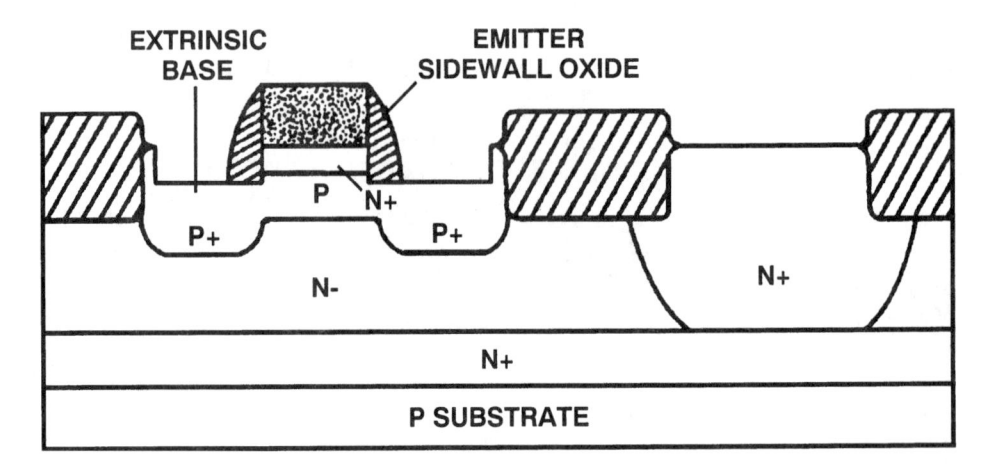

Figure 2–5(c) Cross-section of a bipolar transistor with a self-aligned single level polysilicon emitter. The extrinsic base diffusion is offset by a sidewall oxide to maintain adequate emitter-base breakdown voltage.

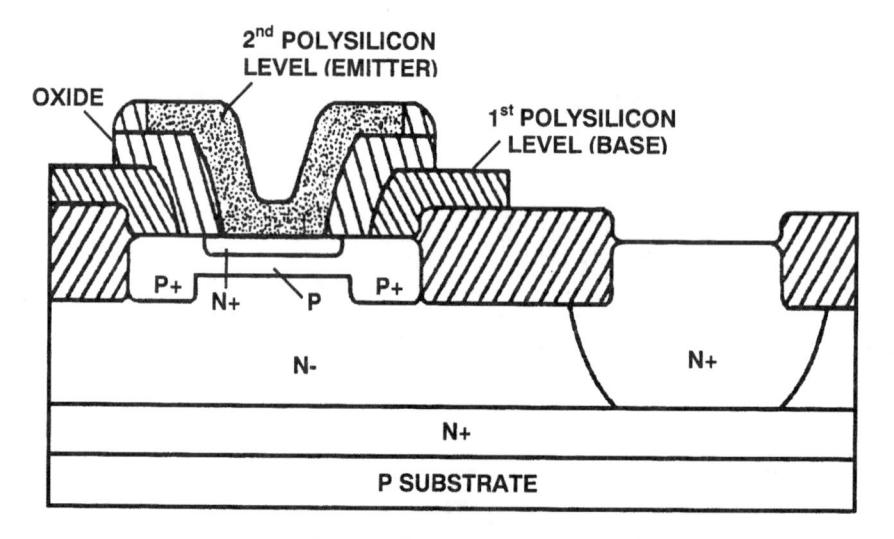

Figure 2–5(d) Cross-section of a bipolar transistor utilizing the double polysilicon emitter process. The first layer of polysilicon is used to contact the base region and the second layer of polysilicon is the source for the self-aligned polysilicon emitter.

2.4.1 Base Design Techniques

For high-performance BiCMOS applications, the bipolar device is expected to have a high cut-off frequency f_T, a low-base resistance, and a small parasitic capacitance which includes C_{BE}, C_{BC} and C_{CS}. The base profile must be optimized to prevent collector-emitter punch-through and increase the emitter-base breakdown voltage. For BiCMOS processes utilizing the bipolar options shown in Figure 2–5(a–c), the extrinsic base region is formed during the p+ source/drain implant for the CMOS.

In order to achieve a high cut-off frequency, the base-width of the bipolar must be decreased. However, this decrease leads to an increase in the base resistance which, in turn, degrades the device characteristics. By using self-alignment schemes, it is possible to reduce the spacing between the extrinsic base and the emitter, [12], thereby lowering the base resistance. Furthermore, self-aligned transistors tend to have smaller parasitic areas, resulting in reduced parasitic capacitance and enhanced speed performance [13]. Another approach to minimizing the extrinsic base capacitance is to use local interconnects to form contacts to the base region. This method reduces the base junction area because it no longer needs to accommodate the metal contacts.

2.4.2 Collector Design Techniques

When the BiCMOS circuit is driving high capacitance loads, the circuit delay is strongly dependent on the collector resistance of the bipolar transistor [14]. Hence minimizing the bipolar collector resistance is one of the crucial techniques to obtaining high-performance BiCMOS circuits. The more practical way to form the collector contact is by using the n+ source/drain implant as shown in Figure 2–5(a). However, due to the presence of the lightly doped n-well between the collector contact and the buried layer, the resultant collector resistance is very high. To reduce the collector resistance, usually a deep n+ contact diffusion (also known as a plug or sinker) is employed. The diffusion penetrates the epitaxial layer to contact the underlying buried layers (refer to Figure 2–5(b)). The drawback of this approach is that a relatively large collector-to-base spacing must be maintained to prevent degradation of the collector-base breakdown voltage, because substantial lateral diffusion of the deep n+ contact will occur.

In the submicron regime, the scalability of the bipolar transistor is limited by the real estate consumed by the lateral diffusion of the deep n+ collector. An alternative way to obtain a scalable, low-resistance path to the buried n+ layer is to employ a deep n+ polysilicon plug contact as shown in Figure 2–6 [15]. The sidewall oxide lining of the polysilicon plug eliminates the lateral diffusion which, in turn, permits scaling of the transistor.

Another parameter that needs to be optimized for minimum collector resistance is the epitaxial layer thickness. Reducing the depth of the epitaxial layer can increase f_T and allows better current handling capability before base pushout, but at the expense of decreasing the breakdown voltage of the collector-emitter junction, BV_{CEO} [16]. One solution to increase the collector current density before base pushout occurs is to increase the n-well concentration [16]. However, degradation of the overall circuit performance may occur because of higher collector-base capacitance for the bipolar transistor, and higher junction capacitance for the CMOS devices. This can be solved by adding a phosphorus implant through the emitter opening to raise the well concentration of the emitter region [17–18]. By doing so, base pushout can be prevented while having a minimal increase in the collector-base capacitance.

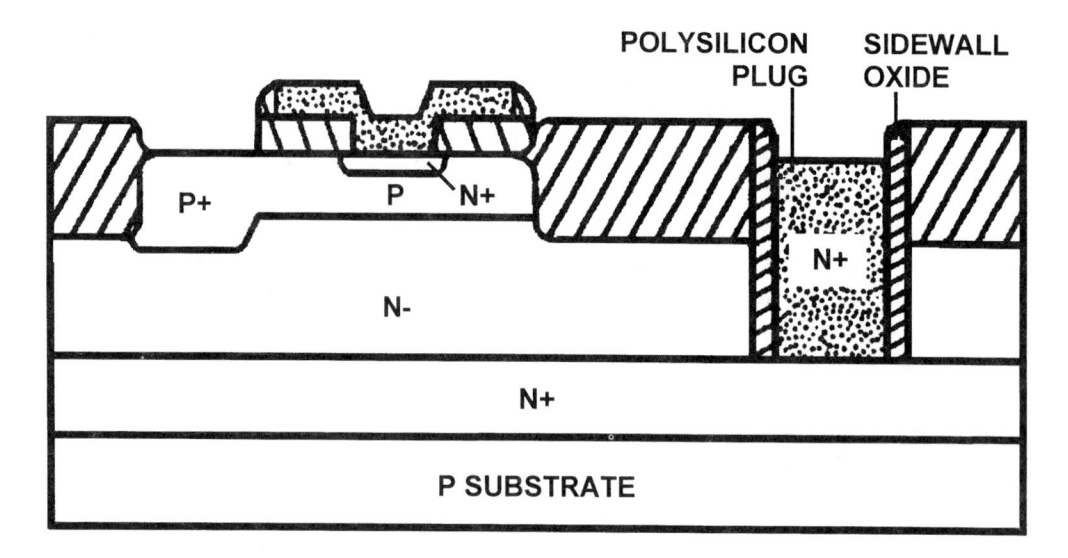

Figure 2–6 Cross-section of a bipolar transistor with an n+ polysilicon plug used to minimize the collector series resistance.

2.4.3 Emitter Design Techniques

The different techniques of forming the emitter of the bipolar transistor range from a directly implanted emitter, as illustrated in Figure 2–5(a), to a single or double level polysilicon emitter as shown in Figure 2–5(b–d). While the directly ion-implanted emitter approach is not scalable for high-performance applications, it is generally used in applications where the f_T requirement is less than 8 GHz. An example of such applications is in 1-Mb BiCMOS DRAMs [19–20]. Figure 2–5(b) illustrates a single level polysilicon emitter with oxide spacers. This structure is also known as a triple-diffused bipolar transistor [21]. The bipolar collector-substrate capacitance is reduced with the use of a shallow buried layer and deep collector sinker. The poly emitter process produces a well-controlled emitter profile. A poly emitter facilitates the formation of a shallow base and provides high emitter efficiency. But in this approach, the emitter is not self-aligned to the base region. Two potential problems arising due to non self-alignment are: (1) encroachment of the extrinsic base on the emitter region, and (2) substrate exposure to the emitter poly etch. The first problem occurs due to the use of p+ implant in the extrinsic base region. The implant is self-aligned to the emitter poly edge. The highly doped p+ base region in close proximity to the n+ emitter forms a poor quality junction due to high leakage and low breakdown. This degrades the bipolar performance. The second problem can cause etching of the silicon substrate in an exposed emitter opening, thus shorting the device during salicidation. To prevent such a situation from happening, the design rules must provide adequate overlap of the emitter poly in the base region.

Self-alignment schemes employing single level polysilicon and double levels polysilicon are shown in Figure 2–5(c–d). In the double polysilicon process, the first layer of polysilicon forms the base electrode, and it can be employed to form low sheet resistors. The first layer of the polysilicon will simultaneously form the polysilicon gates of the CMOS devices. The second polysilicon layer forms the emitter of the bipolar transistor, which is self-aligned to the extrinsic base. It contacts the silicon (Si) substrate without contacting the first poly level. The npn transistors with a f_T of 16 GHz have been fabricated employing the double polysilicon emitter process.

Recently, Motorola introduced an advanced SRAM technology using a new triple-level polysilicon [13]. The first two levels serve the same functions as those described in a double polysilicon emitter process. The third poly level essentially provides high-value load resistors in the four-transistor CMOS memory cell. Figure 2–7 shows a simplified illustration of the triple poly SRAM, and Figure 2–8 illustrates a detailed cross-section of the Motorola triple-poly BiCMOS device.

2.5 BiCMOS Isolation Issues

Isolation considerations in BiCMOS circuits are crucial factors in determining the overall speed performance and packing density of the circuits. The major issue in BiCMOS isolation revolves around the well formation strategy to lower the collector series resistance in order to increase the operating speed of the device. Most high-performance submicron BiCMOS processes use the CMOS twin-well, where the n-well and p-well doping profiles are optimized by separate diffusions. And the background doping of the epitaxial layer must be low enough ($< 1 \times 10^{15}$ atoms/cm^3) to eliminate the need for excessive counterdoping during n- or p-wells formation, which leads to degradation of carrier mobility.

2.5.1 Latch-up Phenomenon

Parasitic bipolar transistors exist in any CMOS circuits. Figure 2–9 shows the different parasitic transistor configurations and their interconnections. The pnp and npn bipolar transistors are combined to form a pnpn thyristor in a twin-well CMOS inverter. Positive feedback can be started from spurious noise, voltage overshoots, electro-static discharge, or the application of signal levels to the inputs before power-up [22]. The pnpn thyristor switches to a low impedance state resulting in a high-current flow between V_{SS} and V_{DD}.

The techniques to reduce the latch-up include fabricating the source/drain regions as close as possible to the V_{DD}/V_{SS} contacts, using a heavily doped substrate, and employing retrograde well structures. Some methods to eliminate the pnpn path are a combination of deep trench isolation structures and epitaxial layers on heavily doped substrate and silicon-on-insulator (SOI) substrates [22].

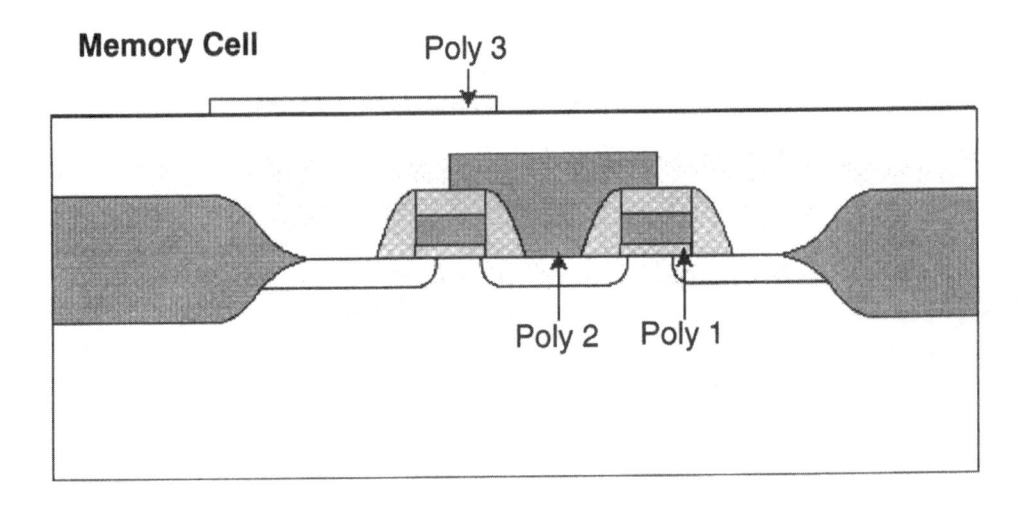

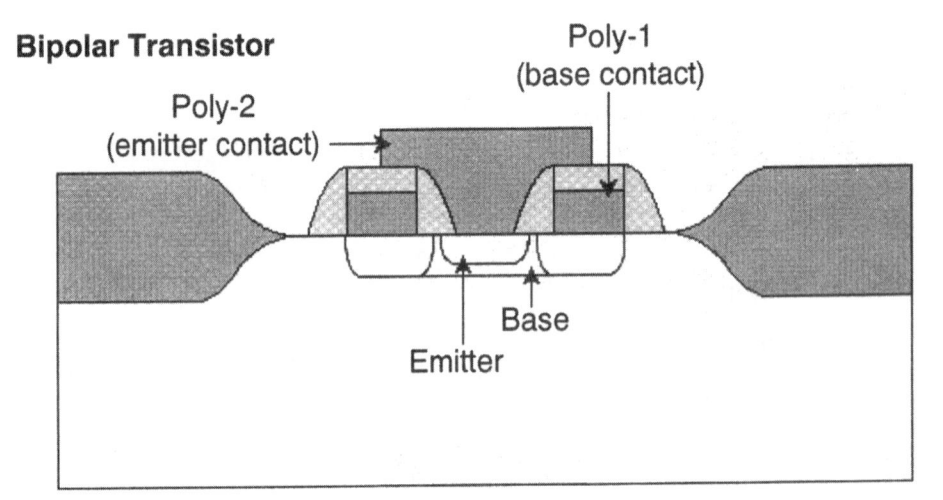

Figure 2–7 Triple-poly SRAM with the first level polysilicon as the bipolar base, the second level polysilicon as the bipolar emitter, and the third level polysilicon as the high-value load resistor.

2.5.2 Trench Isolation

The implementation of bipolar trench isolation helps to increase the bipolar packing density and reduce the collector-to-substrate capacitance for improved performance. Figure 2–10 illustrates the trench-isolation submicron BiCMOS structure. The trenches are lined with oxide, which acts as an insulation barrier, and filled with polysilicon. They prevent latch-up and since the trenches are extended beyond the n+ buried layer, the sidewall capacitance is reduced. How-

BiCMOS Isolation Issues

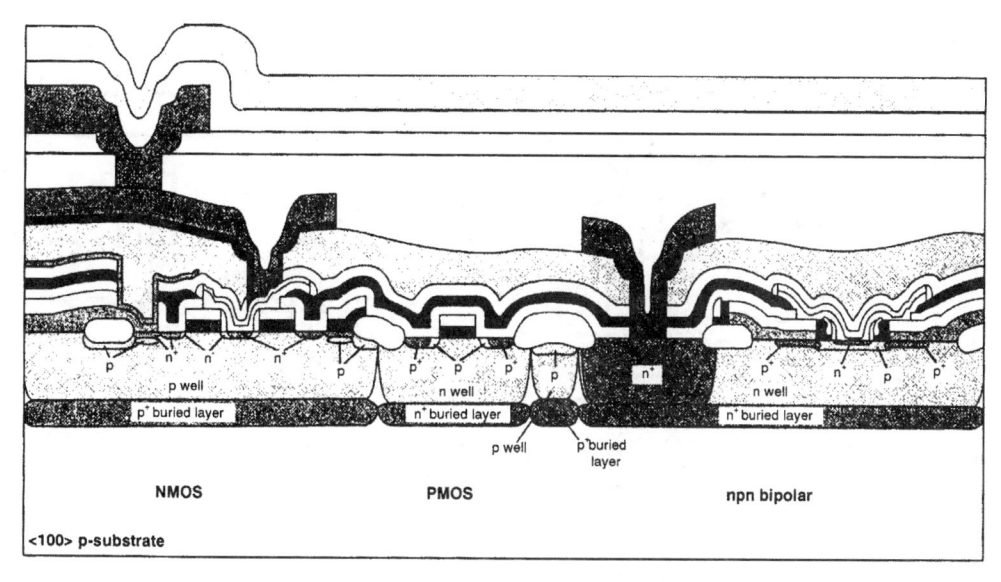

Figure 2–8 Detailed cross-section of the Motorola triple-poly BiCMOS device.

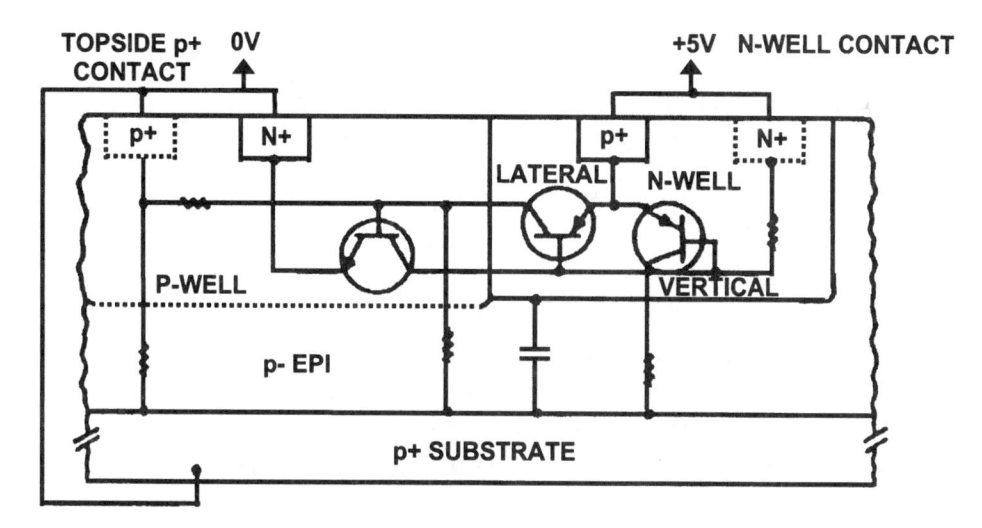

Figure 2–9 Cross-section of a twin-well CMOS showing the pnpn latch-up circuit.

ever, deep-trench technology has been affected by fabrication difficulties. Deep trenches cannot be manufactured reliably from run-to-run, and they make it difficult to meet the throughput demands of full production.

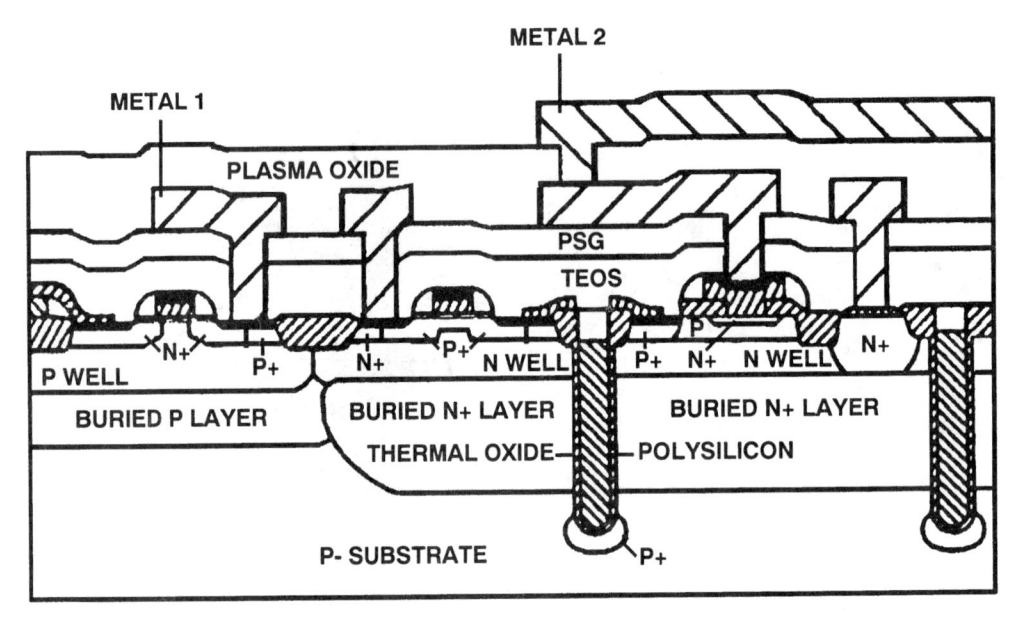

Figure 2–10 Cross-section of a high-performance submicron BiCMOS process illustrating the trench-isolated bipolar devices, polysilicon emitter, and silicided contacts for low sheet-resistance.

Moderate-depth trench isolation for CMOS was reported recently [23]. It employs a shallow epilayer and shallow trenches (< 3 μm in depth). This effectively helps to reduce the complexity of trench etch-and-refill. CMOS circuits having this trench-isolation technique exhibit a holding voltage greater than 10 V, ensuring latch-up free operation in circuits operated with a 5 V supply.

2.5.3 Epitaxial Layer

The growth of a lightly-doped, thin epitaxial layer over the twin-buried layers employed in BiCMOS devices presents a difficult challenge because both the vertical and lateral autodoping occur concurrently. As a result, the epi/substrate widens, and the epitaxial layer may not reach the targeted resistivity. Autodoping must, in general, be minimized to prevent the need to counterdope both the n- and p-wells excessively later [24].

For the intrinsic-doped epitaxial layer, which is of a lighter background doping concentration, autodoping becomes a major concern especially if the epitaxial layer is thin. The lateral autodoping can be reduced by increasing the time and temperature of the stabilization period between in-situ etching of the surface and actual deposition of the epitaxial layer [25].

2.5.4 Buried Layers

Buried layers help to reduce the collector series resistance and improve latch-up immunity of the device. The device is usually heavily doped with a concentration of about 1×10^{19} atoms/cm^3. The p+ buried layers in BiCMOS devices serve to maintain adequate punch-through breakdown-voltage values between adjacent n+ buried layers. Hence the p+ buried layer concentration must be high enough to provide isolation between adjacent n+ buried layers, but not so great as to produce high sidewall capacitance for the n+ buried layer [15].

A major drawback with the use of buried layers, beside adding to the process complexity, is that the buried n+ regions are recessed by approximately 100 to 200 nm, because a silicon step is required for alignment of the subsequent mask level. Additional topography is also introduced by the self-aligned twin-well process, and if this step occurs over the n-well, the focal plane of the n-well will be 200 to 400 nm lower than the p-well. However, buried layers will still be indispensable in scaled submicron BiCMOS device structures.

2.5.5 Active Device Isolation

The standard LOCal Oxidation of Silicon (LOCOS) technology used for the isolation of active device regions is limited by oxide and boron encroachment, as well as by nonplanar surface topographics and the field-thinning effect. Hence, alternative isolation techniques must be employed for submicron devices. The polysilicon buffer LOCOS process is one of several approaches [26–28] to solve the problem of active area encroachment of LOCOS. Figure 2–11 shows the key processing steps in the poly-buffered LOCOS process. Essentially, a polysilicon buffer layer is inserted between the pad oxide and the nitride oxidation mask of the conventional LOCOS process. This polysilicon layer, in addition to the pad oxide, aids in relieving stress, thus permitting thinner oxide and thicker nitride to be used during field oxidation. In this way, active area encroachment is reduced without causing defects.

Another advanced active area isolation technique is the SideWAll-Masked Isolation (SWAMI) technique, introduced by Chiu et al. [26] in 1982. It is a bird's-beak free-isolation scheme and has received much attention since. SWAMI pad-oxide and Chemical Vapour Deposition (CVD) nitride layers are formed and etched similarly to those in the conventional LOCOS. Then groves are etched in the silicon to an approximate depth of half the required field-oxide thickness. This results in recesses having sidewalls inclined at about 60°. During field oxidation, these sloping sidewalls act as stress relief regions. Next a second stress-relief oxide layer is grown, followed by the deposition of a second CVD nitride and a CVD oxide. The composite CVD SiO_2-CVD Si_2N_4-thermal SiO_2 layer is anisotropically etched in the field region, leaving only portions of it on the sidewalls of the recessed silicon. The CVD oxide on the sidewalls forms spacers that protect the second nitride layer. After etching of the oxide spacers, the final structure has sidewalls inclined at an angle of 60° and the sidewalls are surrounded by the second nitride and oxide. Subsequently, channel-stop implant is carried out followed by field oxide growth.

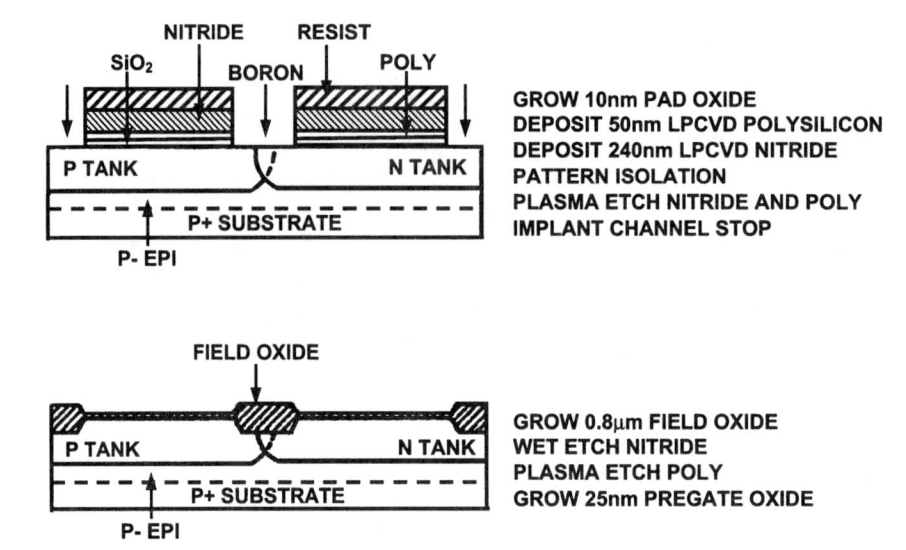

Figure 2–11 Key steps in the polysilicon buffer LOCOS process.

The thin sidewall nitride is bent upwards because of the expansion of the silicon oxide. By proper selection of the distance of the foot of the nitride on the floor of the recessed silicon, the encroachment of the bird's beak into the active area can be minimized. Figure 2–12 illustrates the SWAMI process in both two- and three-dimensional views.

2.6 BiCMOS Interconnect Issues

2.6.1 Silicidation

As BiCMOS devices are scaled down into the submicron regimes, the intrinsic series-resistance effects will negatively impact the speed performance of the devices. Hence it becomes essential to minimize the series drain, source, gate, emitter, base, and collector resistance in order to realize the full performance advantages of the scaled BiCMOS devices [30]. The channel conductance of MOS devices increases with decreasing channel length. Unless the source/drain series resistance is minimized, the saturation transconductance will be degraded. One such technique to reduce the intrinsic series resistance is the use of self-aligned silicidation, or salicide [31]. This technology will simultaneously aid in the reduction of the polysilicon and the diffusion sheet resistance. The salicide processing steps are shown in Figure 2–13.

The process sequence is as follows:

1. After the formation of the source and drain regions, polysilicon sidewall spacers are formed as shown in Figure 2–13(b).
2. Metal is deposited for the formation of silicides.

BiCMOS Interconnect Issues 45

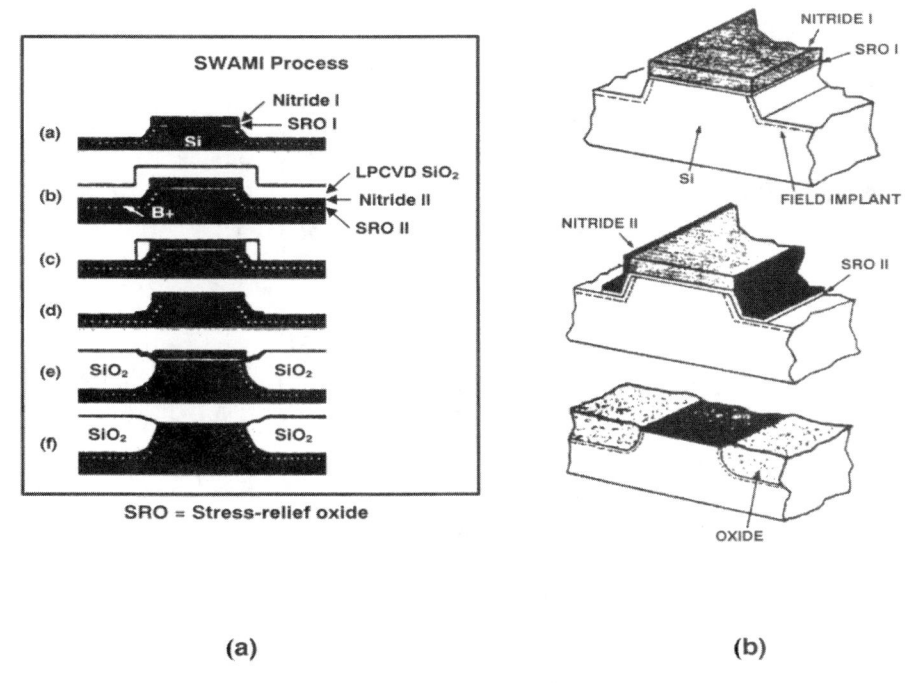

Figure 2–12 (a) The SWAMI process developed by Hewlett-Packard [26]; (b) perspective views of the SWAMI process. [29]

3. The wafer is subjected to high temperature, which causes the silicide reaction to occur wherever the metal is in contact with silicon. Elsewhere, the metal remains unreacted as shown in Figure 2–13(c).
4. Unreacted metal is removed through the use of silicide resistant etchant, leaving behind silicide films on the source, gate, and drain contact.
5. A dielectric layer is deposited onto the silicide, and contact openings for the silicide layer are made, as illustrated in Figure 2–13(d).
6. Metal is deposited into the contact openings, as shown in Figure 2–13(e).

The titanium silicide ($TiSi_2$) and group-VIII metal silicides are usually used for the self-aligned ohmic contacts, as well as local interconnects to silicon. $TiSi_2$ is attractive for salicide application because it exhibits low resistivity. Furthermore, transistors incorporating titanium silicide gate electrodes are more resistant to high field-induced hot-electron degradation than are conventional polysilicon gate devices [33]. Figure 2–14 shows the fabrication of MOS devices using self-aligned $TiSi_2$ process, which is essentially the same as that described in Figure 2–13, except that the metal used to form the silicide is defined.

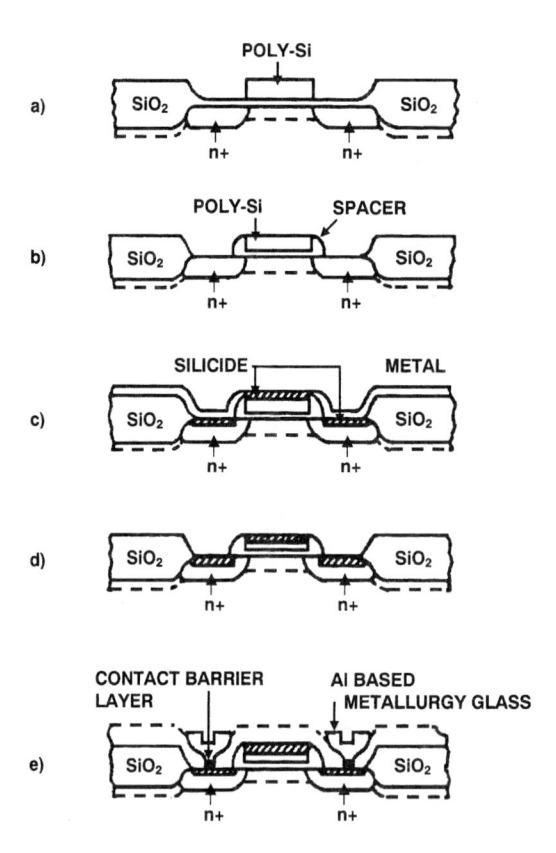

Figure 2–13 Salicide processing steps and the final structure. [32]

When implementing the self-aligned titanium silicide into a process flow, it is important to consider the subsequent processing temperatures after the silicidation. If subsequent heat treatments are not optimized, the active devices can exhibit characteristics that are in fact inferior to those achievable without salicidation [34–35]. Another problem associated with $TiSi_2$ contacts is the generation of defects at the edge of the $TiSi_2$ film due to stresses in the film [36]. It has been reported that once the thickness of the $TiSi_2$ film exceeds 100 nm, defects will start to form. Other viable alternatives to self-aligned titanium silicide contacts include self-aligned cobalt silicide contacts [37] and buried-oxide MOS contact structure [38].

2.6.2 Local Interconnect

The use of traditional buried contacts (see Figure 2–15) in submicron BiCMOS technology is undesirable because dopant out diffusion from the polysilicon connecting the gates and emitters to the diffusion areas will affect the isolation integrity and the active device characteris-

Low-Voltage Low-Power Design

BiCMOS Interconnect Issues

- **POLY GATE PATTERN AND ETCH, REACHTHROUGH IMPLANT TO FORM LIGHTLY DOPED DRAIN EXTENSION**
- **SIDEWALL OXIDE DEPOSITION AND ETCH, S/D IMPLANT AND ANNEAL**

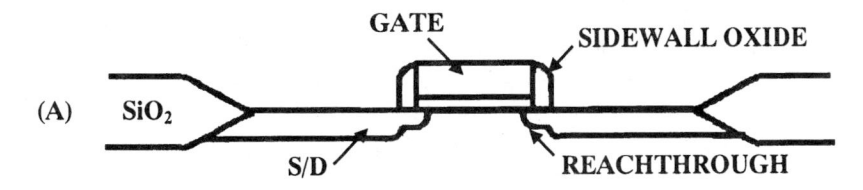

(A) SiO$_2$, GATE, SIDEWALL OXIDE, S/D, REACHTHROUGH

- **HF DEGLAZE**
- **SPUTTER DEPOSITION OF TITANIUM**

(B) Ti

- **TITANIUM/SILICON REACTION**
- **TITANIUM NITRIDE STRIP**
- **ANNEAL**

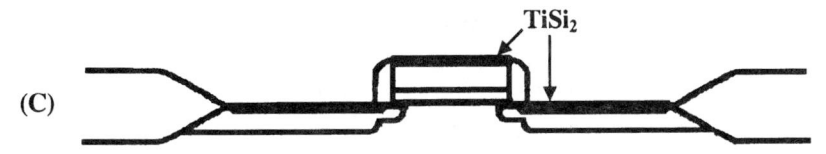

(C) TiSi$_2$

Figure 2–14 Fabrication of MOS transistor using the self-aligned TiSi$_2$ process.

tics. Furthermore, since the polysilicon layer is used as both a gate material and an interconnect material, the poly interconnect will not be able to cross over regions where a polysilicon gate exists without making contact to the gate.

A new technology is therefore required to achieve higher packing density through the direct connection of n-type polysilicon gates and emitters to both the n+ and p+ diffusion regions. The Local Interconnect (LI) technology [40] has been developed for that purpose. In the LI technology, refractory metal-silicides (for example, WSi$_2$, TaSi$_2$, MoSi$_2$, or TiSi$_2$) are deposited on top of a polysilicon layer for the purpose of electrical connections, as well as to reduce the relatively high resistivity of heavily doped polysilicon. The metal-silicide layer can pass over polysilicon gates without any restriction. In other words, there are no areas of the chip surface over which the metal line is restricted from being routed.

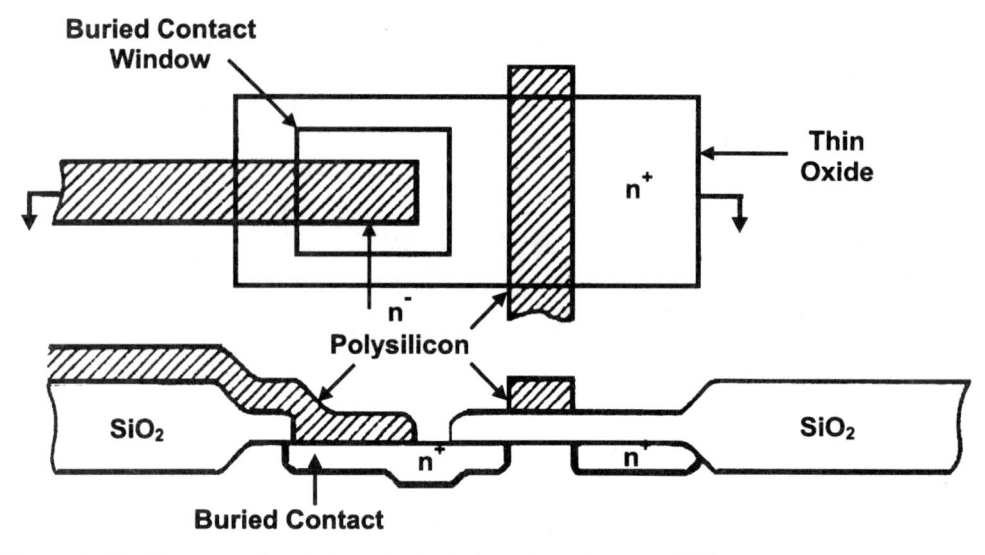

Figure 2–15 Cross-sectional view of a buried-contact structure. [39]

Figure 2–16 shows a self-aligned titanium silicide (TiSi$_2$) process being modified to utilize the TiN layer for local interconnection between the gate and the source/drain junction. This technique does not require any extra mask level since it utilizes the sixth by-product, that is, TiN, of the self-aligned TiSi$_2$ process, which is usually discarded. After the titanium/silicon reaction has taken place, the conductive TiN layer is patterned and etched to form the local interconnect. After stripping the photoresist, a high temperature (about 800 °C) anneal in argon lowers the resistivities of TiSi$_2$ and TiN to their final values of 1.0 ohm/square and 15 ohm/square, respectively. Figure 2–17 demonstrates how the TiN layer provides local interconnection between the drain and source regions of the first stage to the common gate of the second stage of the CMOS inverter without having to utilize extra area for the contacts and metal straps.

Local interconnects have allowed junctions to be "extended" over the isolation regions, thus achieving minimum geometry junctions to reduce capacitance and thereby increase speed performance [40]. Other forms of the LI technology include the utilization of CoSi$_2$ layer as an LI [41] and the selective CVD tungsten local interconnect technology [42].

2.6.3 Metallization and Planarization

With the scaling of devices down to submicron regime, planarization of the oxide layers used between metal layers must be improved in order to compensate for the reduced depth of focus associated with higher resolution lithographic tools. Furthermore, the techniques used must be compatible with low temperature below 800 °C, which is crucial to maintain shallow junction depths, critical doping profiles, and low silicide-to-diffusion contact resistance. The metallization process must be able to produce metal lines that can withstand high current densities without failing. To accommodate the above requirements, the typical high temperature

BiCMOS Interconnect Issues 49

- POLY GATE FORMATION, LDD IMPLANT
- SIDEWALL OXIDE FORMATION, S/D IMPLANT AND ANNEAL

(a)

- TITANIUM DEPOSITION
- TITANIUM/SILICON REACTION
(b) - LI PATTERN

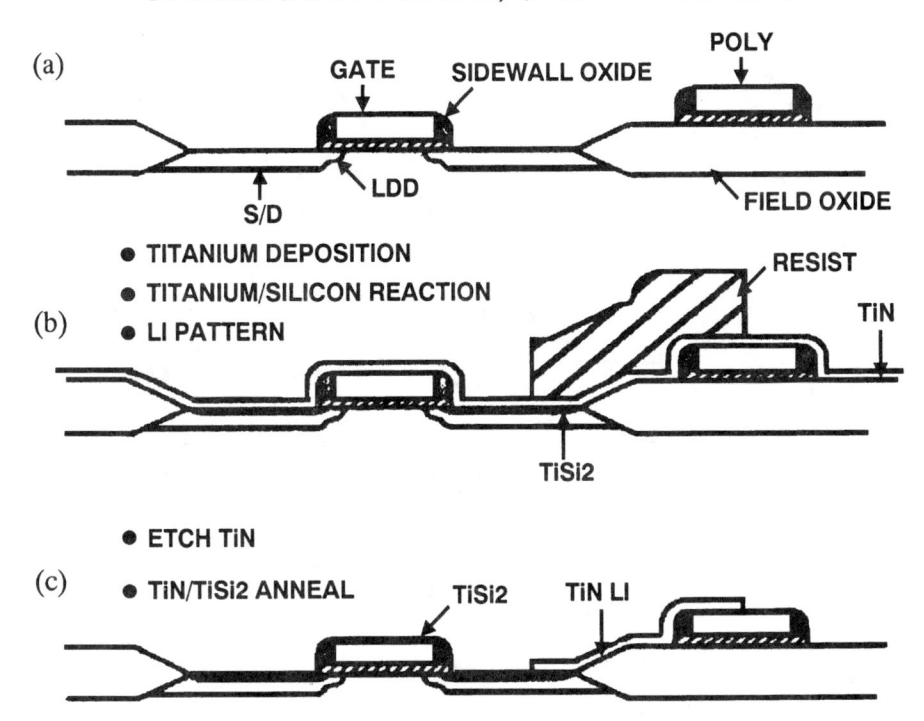

- ETCH TiN
(c) - TiN/TiSi2 ANNEAL

Figure 2–16 The self-aligned titanium silicide process being modified to utilize the TiN layer for local interconnect.

(greater than 950 °C) Borophosphosilicate glass (BPSG) or Phosphosilicate glass (PSG) reflow processes have given way to the low temperature planarization techniques. An example is the resist etchback planarization [43] technique illustrated in Figure 2–18.

The first step in this technique involves depositing a conformal oxide over the underlying interconnect level and then coating the entire wafer surface with photoresist, after which the wafer is etched anisotropically into the conformal oxide layer using a plasma etch with approximately 1:1 etch ratio between oxide and photoresist. In the process of plasma etch, the photoresist and oxide in those regions with higher topography are etched first, while oxide in those regions with lower topography remains protected by the photoresist. Etching stops once the oxide over the highest topography is removed. Thereafter, any remaining photoresist is stripped, resulting in the structure shown in Figure 2–18(b). Next, a second oxide layer is deposited, and then followed by the via formation and definition of the upper level of interconnect as shown in Figure 2–18(c). The end result is a well-planarized surface that allows the vias and interconnects to be patterned accurately.

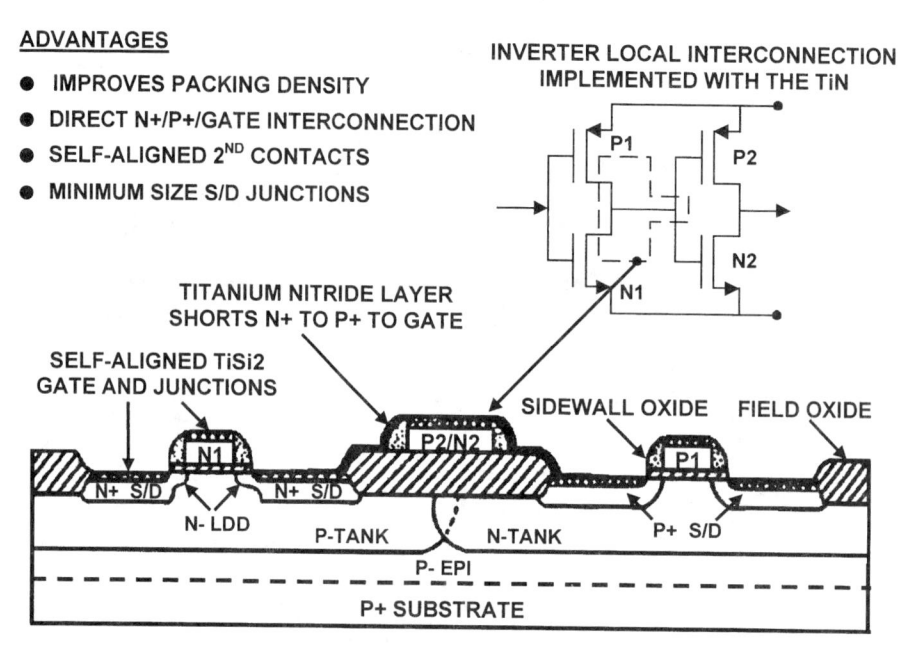

Figure 2–17 Cross-sectional view of a CMOS inverter illustrating the local interconnection of the drain and source junctions to the n-type polysilicon gate.

2.7 Classification of BiCMOS Technologies

At present, there are three main classes of BiCMOS technologies, namely, (1) low-cost, medium-speed 5 V digital, (2) high-cost, high-performance 5 V digital and (3) analog/digital technologies. While the low-cost BiCMOS Integrated Circuits (ICs) are fabricated using a slightly modified single-well CMOS process, the high-performance BiCMOS ICs are based on a modified twin-well CMOS process. The analog/digital BiCMOS ICs are fabricated with processes modified to accommodate a larger voltage of around 10 to 30 V for analog applications. Table 2–1 gives a comparison of the three categories of BiCMOS technologies with respect to the CMOS and bipolar technologies, which have the same minimum feature size. It has been assumed that all processes come with LDD source/drain regions and a double-level metallization.

In the following subsections, we will focus on a submicron BiCMOS process flows for both high-performance digital and analog/digital applications.

2.7.1 Process Flow of a High-Performance 5 V, 0.8 µm Digital BiCMOS

This subsection discusses a typical 0.8 µm high-performance BiCMOS process flow [45]. Table 2–2 illustrates how a bipolar process is merged into a baseline submicron CMOS process [4,46]. The process commonality between the BiCMOS and CMOS flows has been adopted wherever possible, and the additional process steps required for the bipolar device fabrication are introduced with minimal changes to the CMOS transistor fabrication sequence.

Classification of BiCMOS Technologies

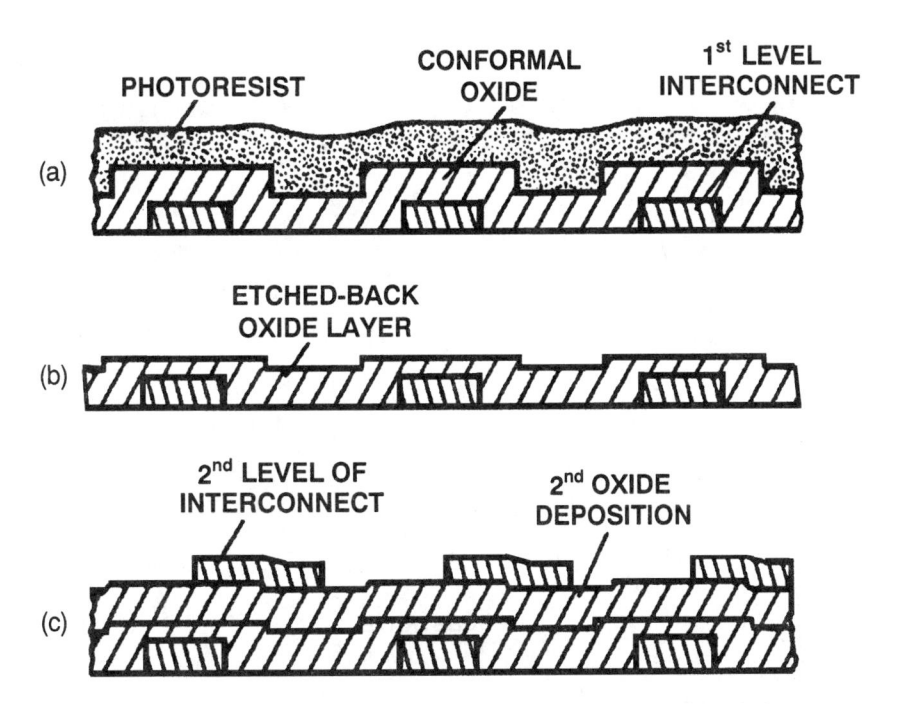

Figure 2–18 Planarization of inter-level oxides using the resist etchback technique.

Table 2–1 Comparison of different BiCMOS, CMOS and bipolar technologies in terms of process complexity. [44]

Step	CMOS	BiCMOS High Performance	BiCMOS Low Cost	BiCMOS Analog	Bipolar
Masks	12	15	13	16	13
Reactive Ion Etching	11	12	11	12	11
Epitaxy	Optional	Required	Optional	Required	Required
Furnace	16	19	16	19	16
Implant	8	12	9	13	7
Metal	2	2	2	2	2
Total	49	61	51	63	50

Table 2–2 Comparison between CMOS and BiCMOS process flows for high-performance BiCMOS technology.

CMOS Baseline Flow	Changes for Bipolar
p$^+$ substrate	p$^-$ substrate
	*Buried n$^+$/p layer
p$^-$ thick epitaxy	Intrinsic thin epitaxy
n and p well formation	Additional deep n well implant
n and p well drive	Reduced drive (thinner epitaxy)
Scaled LOCOS process	
Anti-Kooi/pregate oxidation	
	*p base formation
	*p resistor (pattern/implant)
	*Deep n$^+$ sinker (pattern/implant)
Blanket CMOS threshold implant	
Gate oxidation	
Poly deposition	Thinner poly layer
	*Emitter pattern/etch
	Poly deposition
Doping from diffusion source	Doping from ion implant
Pattern/etch poly	
LDD pattern/implant	
Sidewall oxide deposition/etch	
Pattern/implant n$^+$/p$^+$ source/drain	
Source/drain anneal	Anneal optimized for emitter
Silicide formation	
Low temperature back-end processing	

* Denotes extra mask required in comparison to CMOS baseline.

The starting material for the IC fabrication is a lightly-doped of about 10 ohm-cm p-type substrate. Then buried n+ layers are formed by implanting antimony into the etched oxide openings. This is followed by a high temperature anneal of around 1250 °C. During annealing, re-oxidation of the surface is performed to allow for further processing. Figure 2–19(a) shows the buried n+ implant step. Next, as illustrated in Figure 2–19(b), a selective boron punch-through implant of about 1 to 2×10^{13}/cm^2 is performed to maintain adequate breakdown between the buried n+ regions.

The self-aligned boron implant results in p-buried layers to be formed between buried n+ layers. Following that, all the surface oxide is removed prior to epitaxy, and a short hydrochloric acid (HCL) etch is done to remove surface defects before growing a thin (1 to 1.5 μm) near intrinsic epilayer, with a concentration of about 1×10^{15} dopant atoms/cm^3. The use of such a low background doping concentration in the epilayer makes it unnecessary to perform excessive

Classification of BiCMOS Technologies

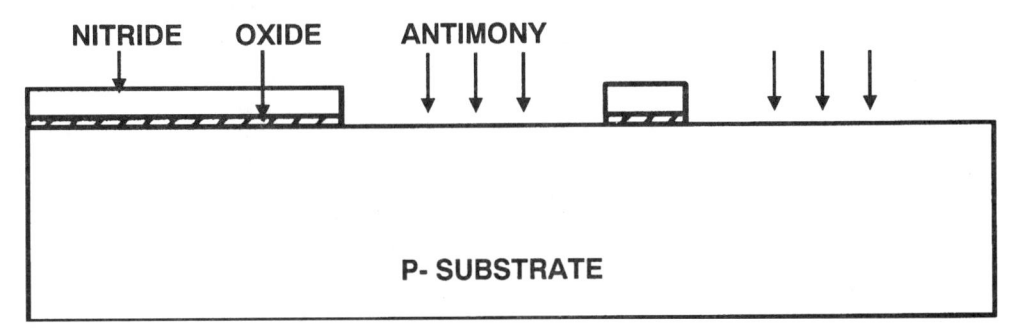

Figure 2–19(a) Device cross-section of the BiCMOS process flow showing buried n+ implant.

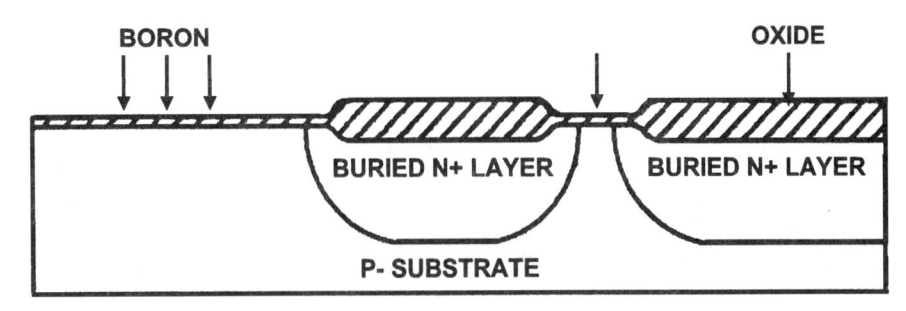

Figure 2–19(b) Device cross-section of the BiCMOS process flow showing buried p implant self-aligned to buried n+ layers.

counterdoping during n- or p-wells formation which would degrade the carriers mobility. The twin-well formation is performed next, as shown in Figure 2–19(c). The thick oxide prevents the p-well dopants from counterdoping the n-well. The p-wells formed are self-aligned to the n-well edges.

After the dopant's implantation, a well drive-in step is performed. Then all the oxide from the surface is stripped and a thin pad oxide (about 10 nm), capped by a polysilicon buffer pad (about 50 nm) and thick nitride (about 240 nm) are grown. The active regions are defined by using a semi-recessed LOCOS isolation process to form the field oxide between them. A blanket boron channel stop is implanted prior to the field-oxide growth, as shown in Figure 2–19(d).

Deep collector regions are patterned and implanted with phosphorus, as shown in Figure 2–19(e). Next, deep collector regions are covered with photoresist and windows defining the p-type base and resistor regions are opened and implanted, as illustrated in Figure 2–19(f).

Then polysilicon emitter formation takes place and the split polysilicon process [47] is used to form the active emitter. First, the pad oxide is removed, and then a thin gate oxide is grown, followed by depositing a thin polysilicon layer. The CMOS gate oxide is protected from

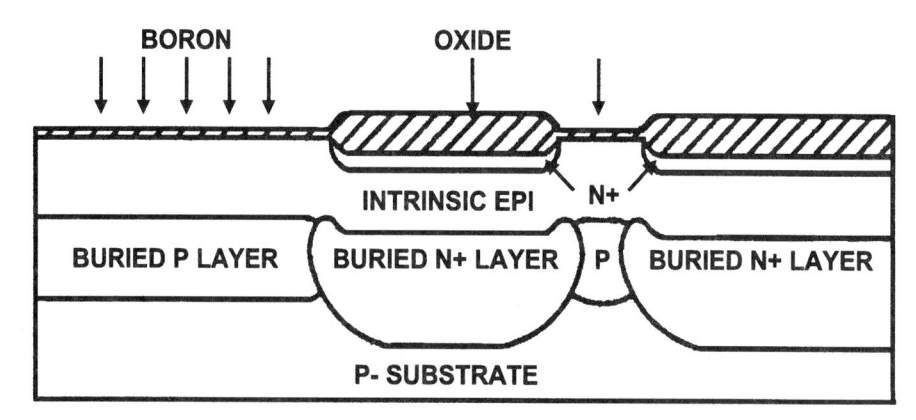

Figure 2–19(c) Device cross-section of the BiCMOS process flow showing p-well implant self-aligned to the oxide-masked n-well.

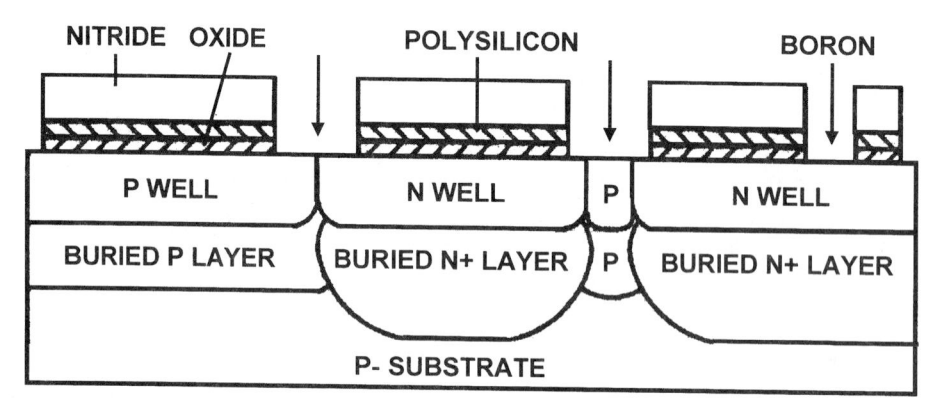

Figure 2–19(d) Device cross-section of the BiCMOS process flow showing channel stop implant after etching of field isolation regions.

deleterious emitter processing effects by the first polysilicon layer. The active emitter window is then patterned and anisotropically etched through the polysilicon and underlying oxide. A second polysilicon layer is deposited and it contacts the base through the emitter window, as shown in Figure 2–19(g). Subsequently, arsenic or phosphorus dopants are implanted into the second polysilicon layer, which will later serve as a diffusion source for the emitter junction. Figure 2–19(h) shows the device after emitter formation.

The second polysilicon layer is then patterned and etched anisotropically to define the CMOS polysilicon gates and the emitter polysilicon contacts. Following that, selective phosphorus implant is performed to form n-type, shallow LDD regions, which are self-aligned to the polysilicon gate edges, as illustrated in Figure 2–19(i). A CVD-oxide layer is deposited and

Classification of BiCMOS Technologies

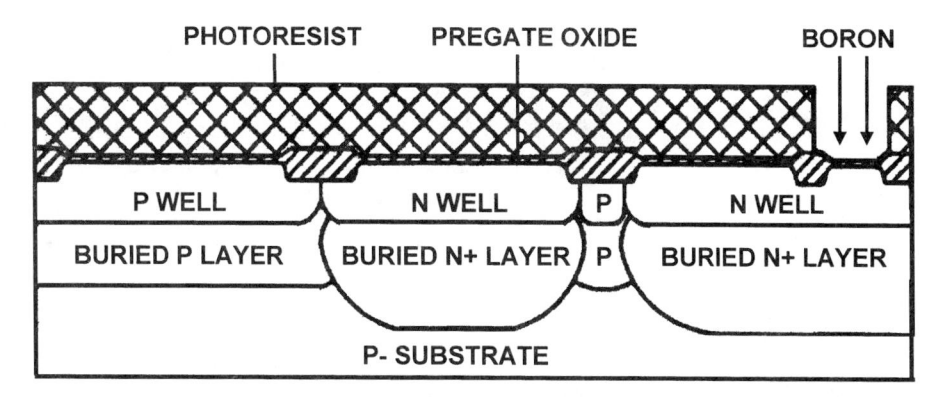

Figure 2-19(e) Device cross-section of the BiCMOS process flow showing patterned deep collector implant.

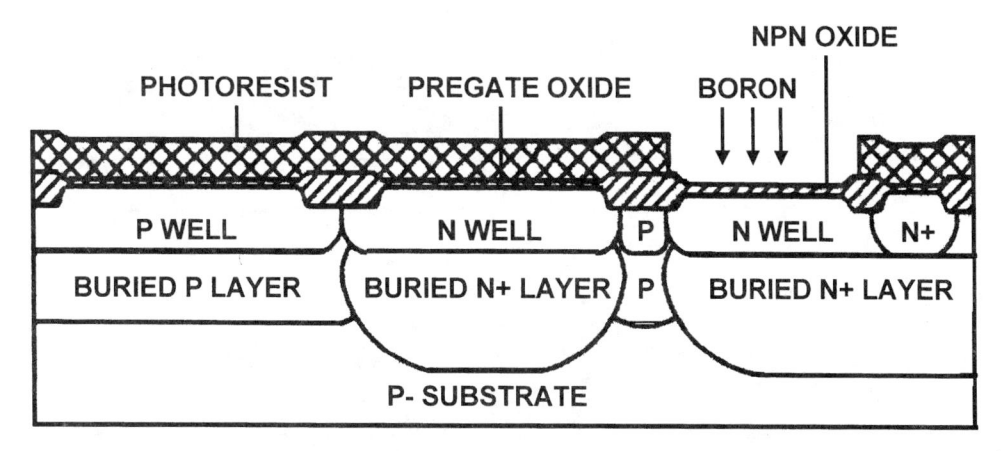

Figure 2-19(f) Device cross-section of the BiCMOS process flow showing patterned base and p-type resistor implant.

anisotropically etched back to form the sidewall oxide spacers. Subsequently, n+ and p+ source/drains regions and the respective well contacts are patterned and implanted, as shown in Figure 2-19(j and k).

Next, silicidation of the polysilicon gates, source/drain regions, extrinsic base contact, collector contact, polysilicon emitter, and well contacts with $TiSi_2$ is performed simultaneously. TiN local interconnects are formed, as shown in Figure 2-19(l), as a result of the silicidation process. Finally, a double-level metal interconnect is implemented and followed by the deposition of the passivation layers. The completed 0.8 μm BiCMOS transistor is shown in Figure 2-19(m).

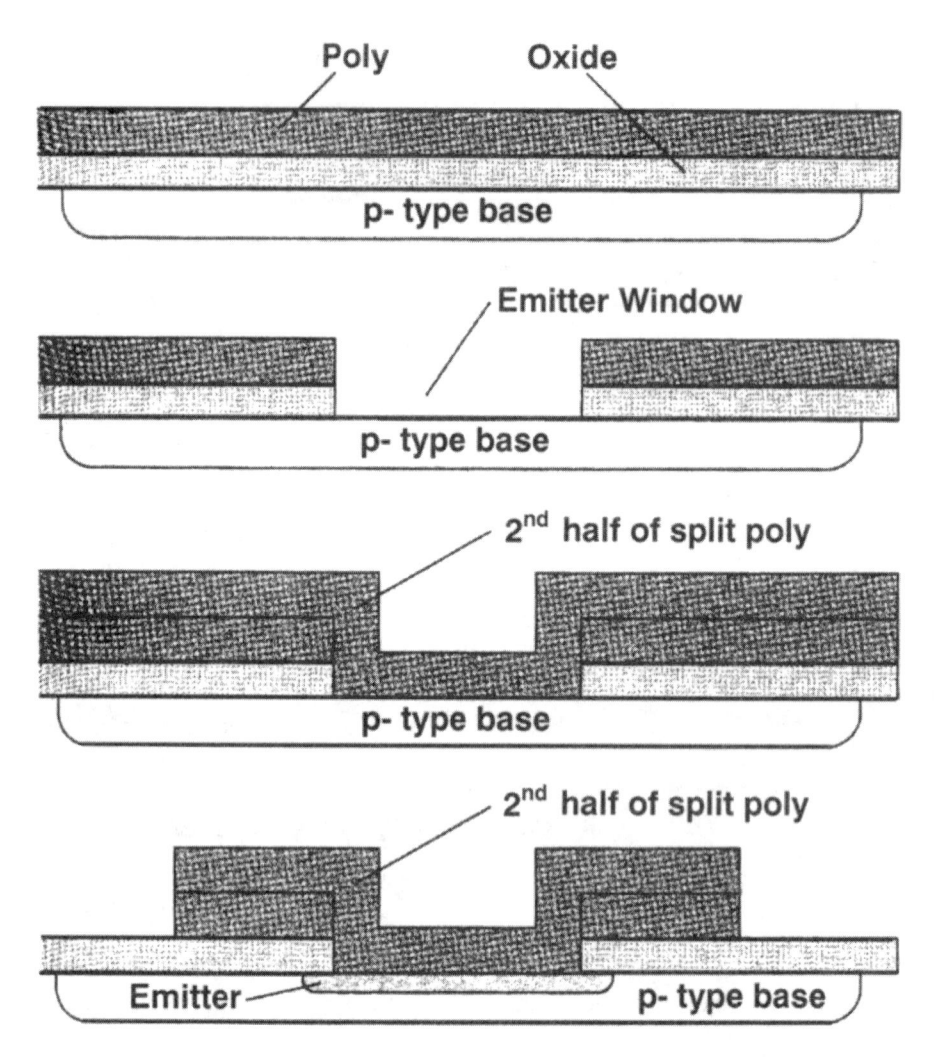

Figure 2–19(g) Bipolar poly-emitter formation using a split polysilicon process.

2.7.2 Process Changes for the 0.5 µm Digital BiCMOS Technology

In order to adapt the 0.8 µm high-performance BiCMOS process to shorter BiCMOS technologies such as 0.5 µm and below, process changes and special circuit techniques must be employed to avoid degradation in the BiCMOS performance.

2.7.2.1 Complementary Bipolar (CB) Technology

For both low-voltage operations and analog applications, ICs incorporating both npn and pnp transistors are able to overcome the limitations of a high-performance npn-only digital

Classification of BiCMOS Technologies

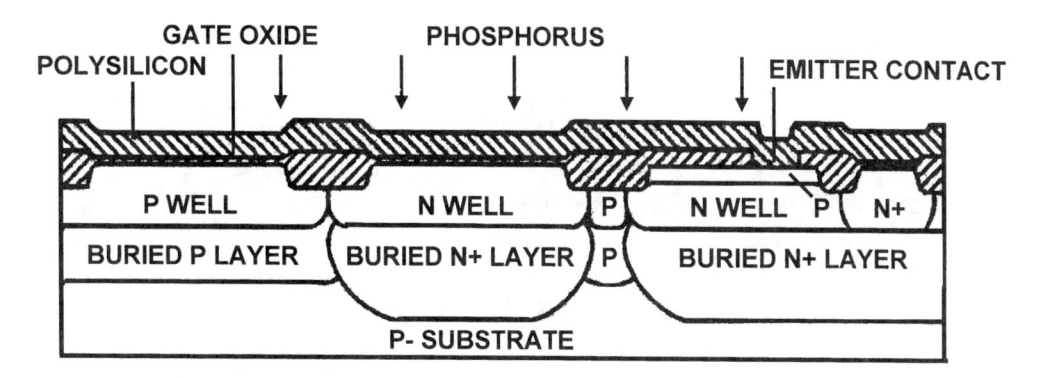

Figure 2-19(h) Device cross-section of the BiCMOS process flow after emitter formation.

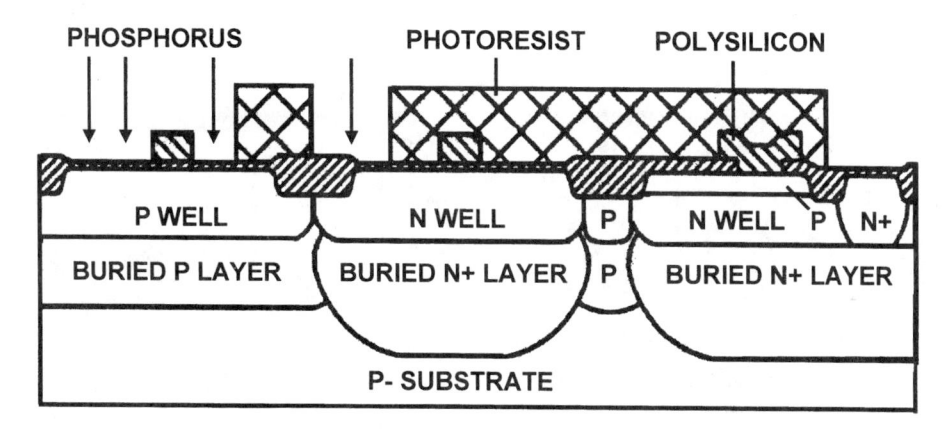

Figure 2-19(i) Device cross-section of the BiCMOS process flow illustrating n-type LDD implants.

BiCMOS IC. In general, these shortcomings are (1) the lack of high-speed, substrate-isolated pnp transistors, (2) the lack of capacitors, (3) the lack of low and high diffused sheet resistance, and (4) the lack of circuit symmetry. The substrate-isolated pnp device can be formed in a p-well that is isolated from the substrate by means of the buried layer. Recently, AT&T has developed several CB processes for the 5 V analog and mixed analog-digital applications, which include the Complementary Bipolar IC V (CBIC-V) and Complementary Biploar IC U (CBIC-U) processes. In these processes, the pnp devices are isolated from the substrate by means of the junction isolation. The more advanced CBIC-V process features vertical pnp with f_T of 5 GHz and npn with f_T of 13 GHz [48].

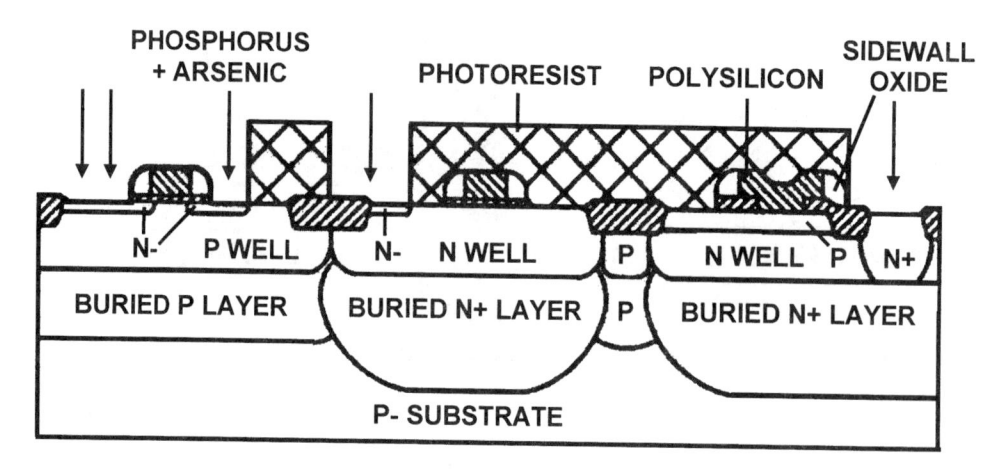

Figure 2–19(j) Device cross-section of the BiCMOS process flow showing n+ source/drain and well contact pattern and implant.

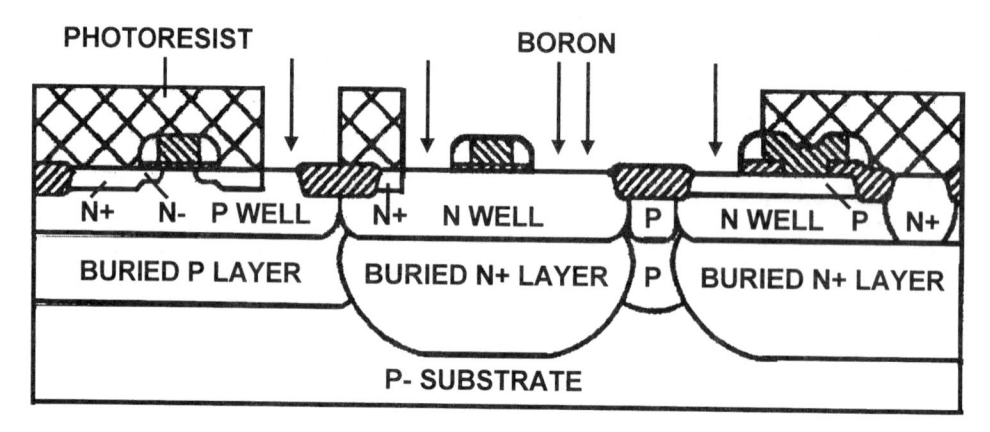

Figure 2–19(k) Device cross-section of the BiCMOS process flow showing p+ source/drain, well contact and extrinsic base contact pattern and implant.

2.7.2.2 CMOS Transistor Process Changes

When devices are scaled down to deep submicron dimensions, the gate oxide thickness must be reduced to below 100 Å, the source/drain junction depths must be made shallower to reduce problems with punch-through, and the well concentration must be increased to counter short-channel effects. Furthermore, the implementation of a punch-through stopper implant is necessary to improve the short-channel characteristics for the CMOS devices. Surface channel pMOS transistors [49–51] are favoured since better device characteristics such as low off-state leakage currents can be obtained as compared to buried channel pMOS transistors.

Classification of BiCMOS Technologies

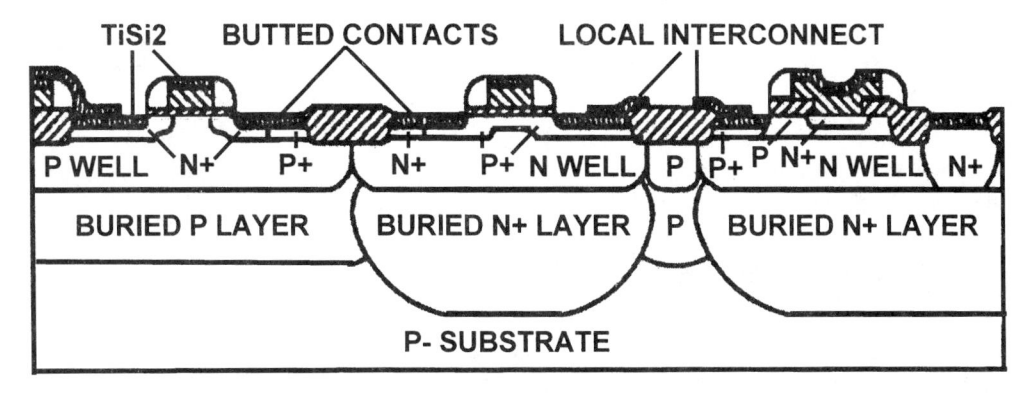

Figure 2–19(l) Device cross-section of the BiCMOS process flow showing silicide and local interconnections.

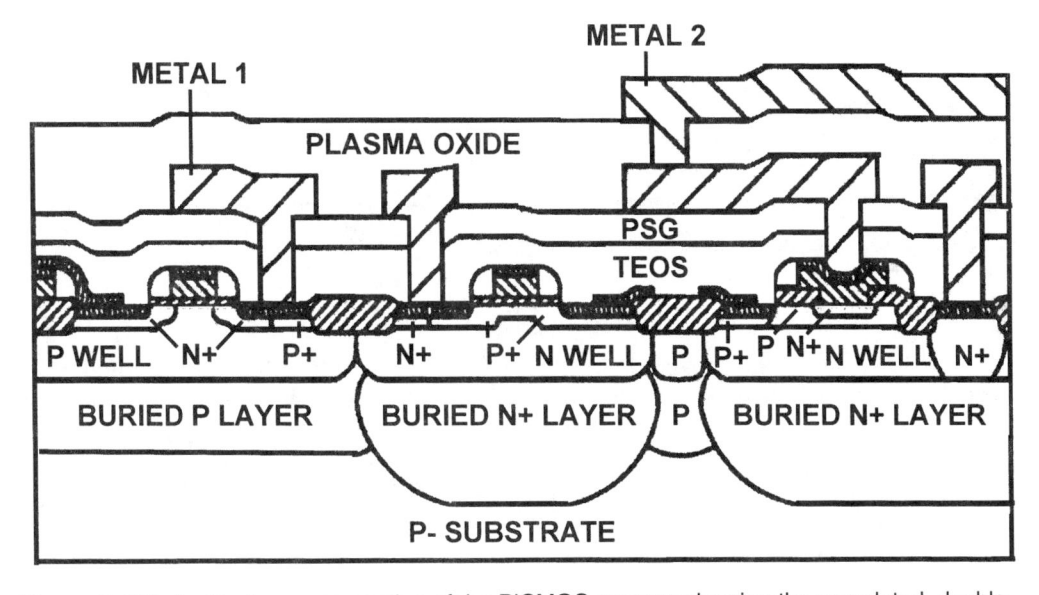

Figure 2–19(m) Device cross-section of the BiCMOS process showing the completed double-level metal 0.8 μm device.

2.7.2.3   NPN Profile Optimization

With the scaling of the device technology, the base-width of the npn transistor must also be scaled in order to improve the cut-off frequency, f_T, of the transistor. This can be achieved with the use of low implant energy for the base implant. For narrow base-width transistors, the emitter junction must be kept shallow to ensure substantial bipolar action. To achieve this, arsenic is often used as the emitter dopant. Rapid Thermal Anneals (RTAs) are employed to obtain a high degree of dopant activation while maintaining a shallow junction depth.

The performance of the npn transistor can also be improved by optimizing the epitaxial layer thickness and the n-well doping level while at the same time maintaining adequate margin for the collector-base breakdown voltage (BV_{CBO}). Note that reducing the epitaxial layer thickness improves f_T by reducing the collector transit time [52], and increasing the n-well doping causes a reduction in the collector resistance.

2.7.3 An Analog/Digital BiCMOS Process Flow

This subsection describes a process sequence, developed by Texas Instruments [53,54], for fabricating medium voltage (20 V) analog/digital BiCMOS ICs. The starting material is a p-type <100> orientation substrate. The wafer surface is first oxidized, followed by the definition of n+ buried layers. These buried layers are formed by antimony implant and diffusion. The surface oxide is then stripped, as shown in Figure 2–20(a).

Figure 2–20(a) Formation of the n+ buried layers.

The wafer surface is treated to a short HCL etch to remove any defects and a p-epi layer is deposited next. Following the epitaxial deposition, the wafer is re-oxidized and definition of windows for the n-well regions is performed. The n-wells are formed by phosphorus ions implantation and diffusion, as shown in Figure 2–20(b). The n+ collector regions are then formed by the n+ collector heat cycle using phosphorus dopants, which help to complete the n-well drive-in.

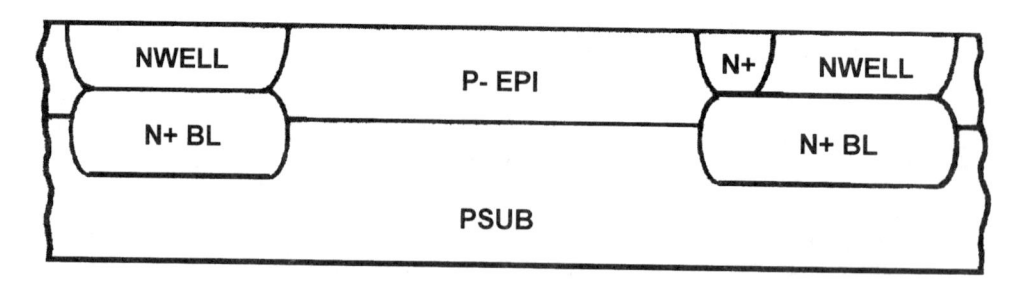

Figure 2–20(b) Formation of the n-well and n+ collector regions following the p-epi deposition.

Classification of BiCMOS Technologies

After the formation of the n-well and n+ collector regions, the surface oxide is stripped and a thin pad oxide is grown. The base regions of the npn bipolar transistors are patterned and then formed by ion implantation using boron, as illustrated in Figure 2–20(c). The implant is diffused in an inert ambient to anneal the silicon for improved bipolar performance. This step aids to reduce the base resistance.

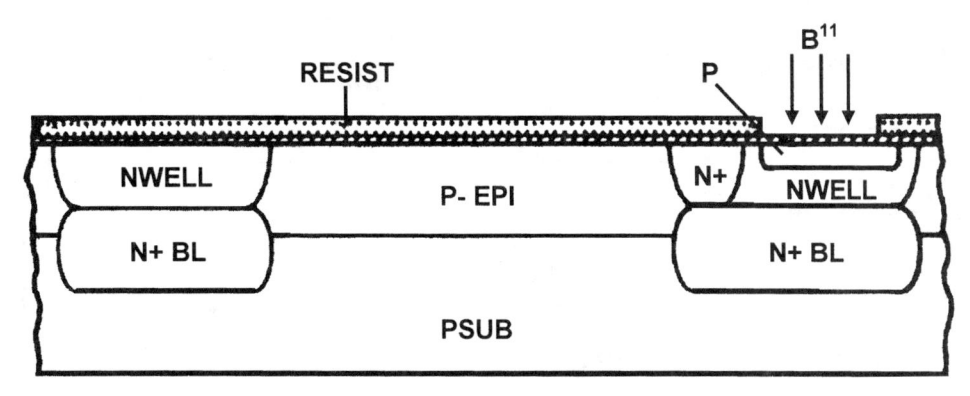

Figure 2–20(c) Formation of the npn bipolar base region.

Next, a nitride film is deposited over the pad oxide and an active mask is used to define these regions. After the nitride has been etched, a blanket phosphorus channel-stop (c/s) implant is carried out to increase the threshold voltage of the parasitic thick field-oxide of the pMOS devices. Following that, another masking step is used to pattern the regions where a boron channel stop implant is performed in the p-type regions to raise the threshold voltage of the parasitic nMOS field-oxide devices. The photoresist is stripped, and a 1 µm thick field oxide is grown in the nonactive regions, as shown in Figure 2–20(d).

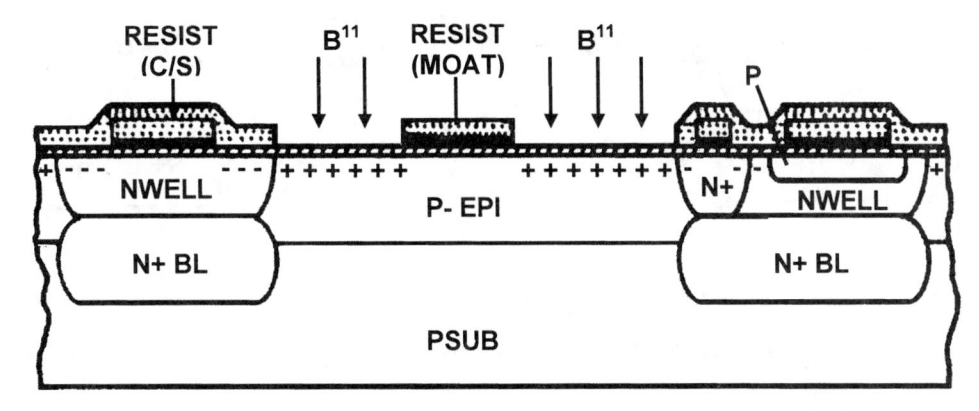

Figure 2–20(d) Formation of the moat regions and subsequent channel stop implants.

The nitride and the underlying pad oxide are stripped and a thin gate oxide of approximately 350 to 500 Å is grown. Subsequently a threshold-adjust implant is performed. The first level polysilicon (poly-1) is next deposited, doped patterned and etched to form the MOS gates electrodes and the bottom plates of the poly-to-poly capacitors (Figure 2–20(e)).

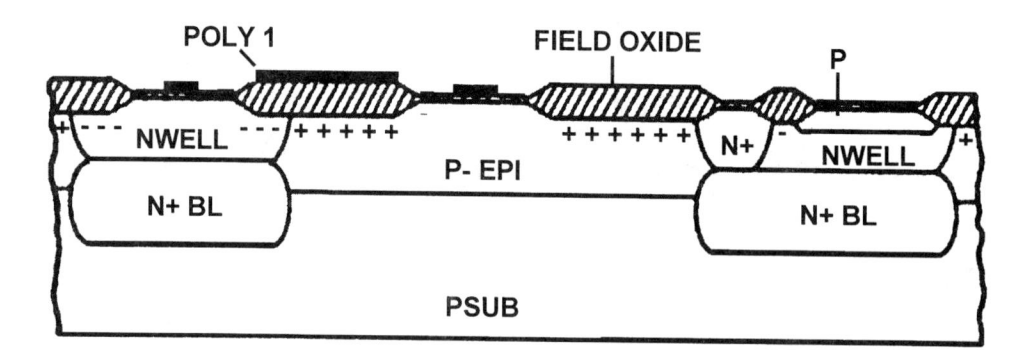

Figure 2–20(e) Formation of the first polysilicon layer for CMOS gates and MOS capacitors.

Next, a 30 to 100 nm thick capacitor Inter-Level Dielectric (ILD) is formed over poly-1 by utilizing either an Oxide-Nitride-Oxide (ONO) film or an oxide alone. The second polysilicon layer (poly-2) is deposited and doped to form a high sheet resistance film that can be used to fabricate high valued resistors. A mask is used to shield those high resistivity regions while the remainder of the polysilicon is n+ doped to form the capacitor top plates as illustrated in Figure 2–20(f and g).

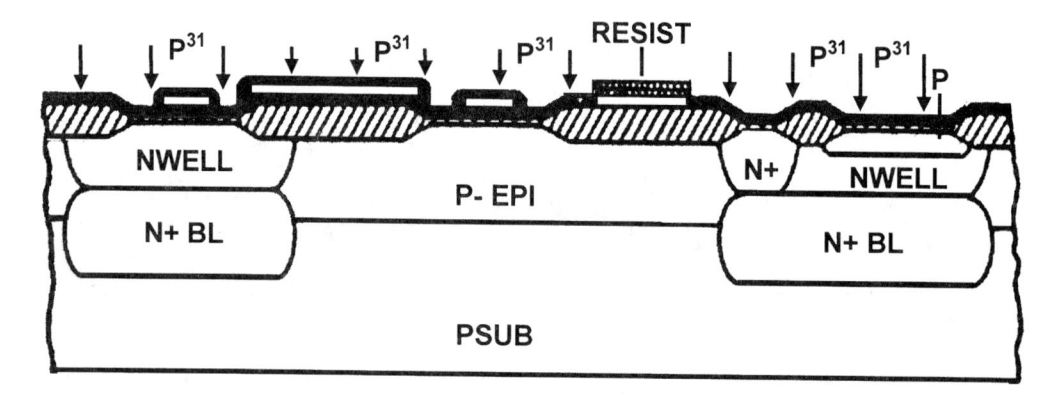

Figure 2–20(f) Deposition and definition of the second polysilicon layer for resistors and top capacitor plates.

Classification of BiCMOS Technologies

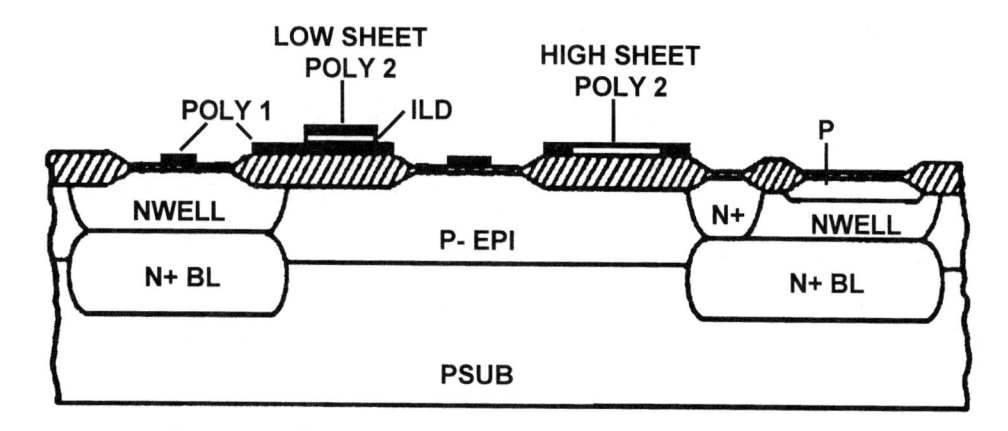

[Figure: Cross-section showing POLY 1, LOW SHEET POLY 2 with ILD, HIGH SHEET POLY 2, P region, NWELL, N+ BL, P-EPI, N+, NWELL, N+ BL, PSUB]

Figure 2–20(g) Formation of the second polysilicon layer for resistors and top capacitor plates.

The remaining portion of the process flow follows that of the conventional CMOS and bipolar device formation. First, we have the formation of the nMOS transistor LDD regions using phosphorus implant as shown in Figure 2–20(h). Next the n+ and p+ source/drain regions are independently formed with arsenic and boron implants, respectively, as illustrated Figure 2–20(i). These implants are offset from the gate by the sidewall spacers, and this helps to reduce the overlap capacitance.

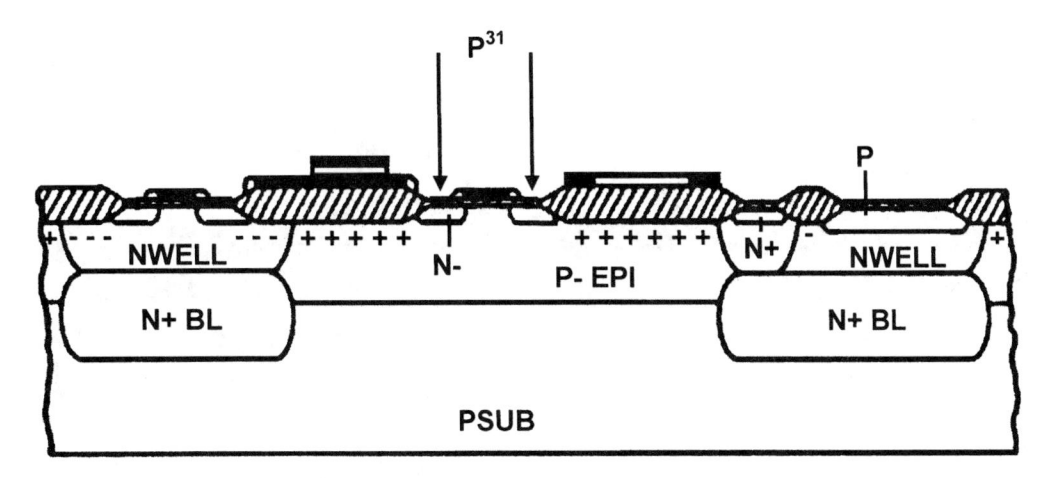

[Figure: Cross-section showing P^{31} implant, NWELL, N−, P-EPI, N+, NWELL, N+ BL regions, PSUB]

Figure 2–20(h) Formation of the LDD nMOS source/drain regions and the oxide sidewall spacers.

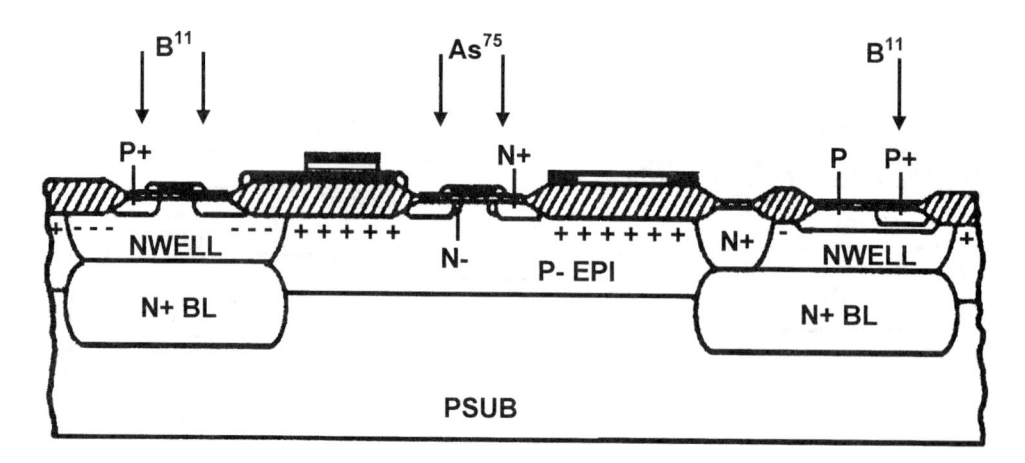

Figure 2–20(i) Formation of the CMOS source/drain regions and the bipolar base contact region.

Then with a photoresist mask, the bipolar emitter and collector contact regions are formed with the use of a phosphorus implant, as shown in Figure 2–20(j). It should be noted that polysilicon emitters are usually avoided in analog/digital processes since they yield higher emitter-contact resistance values than directly implanted emitters. The last part of the process is to form contact holes and the metal interconnections.

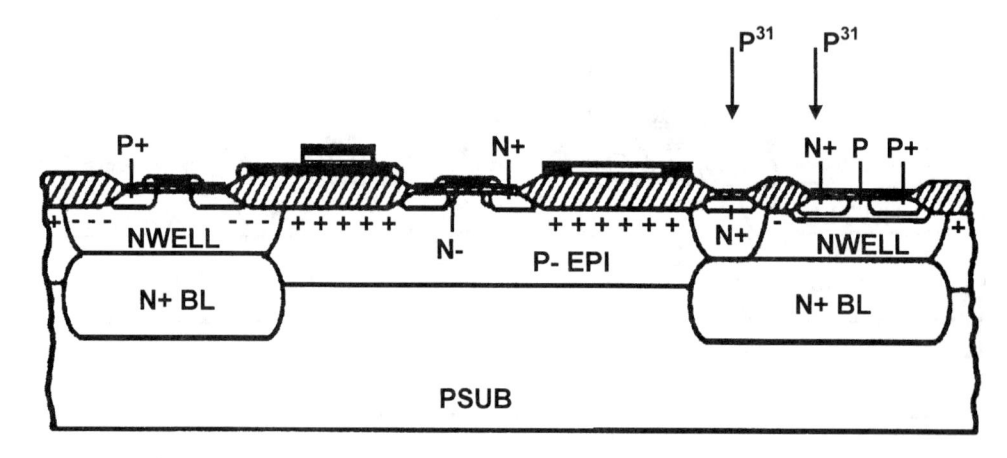

Figure 2–20(j) Formation of the directly implanted n+ emitter and collector contact regions.

2.8 A New Low-Power Ultra Low-Capacitance BiCMOS Process

A new BiCMOS process to fabricate ultra low-capacitance npn bipolar transistor, together with the conventional 10 GHz, single-poly, npn, and MOS devices, was reported in [55]. This process was based on the conventional, single-poly, BiCMOS process [56], with the inclusion of the following three additional modules: (1) shallow and deep trenches, (2) the poly-ridge emitter transistor (PRET) [57,58], and (3) strap connections to the active areas. Table 2–3 compares the new process flow with the standard BiCMOS process and lists out the additional steps required.

Table 2–3 Comparison between the new process flow and the conventional BiCMOS process.

Standard BiCMOS	Additional for PRET
Substrate: Cz, p type, 20 Ωcm Buried n-layer Implant: As, 3×10^{15} cm^{-2}, 100 keV, anneal Buried p-layer implant (not shown) Epitaxial growth Si: 1 μm, n-type, 10^{16} cm^{-3} As	
	20 nm padoxide growth + 200 nm nitride deposition 550 nm shallow trench etch Channel stop implant: B, 4×10^{12} cm^{-2}, 50 keV 3.3 μm deep trench etch Channel stop implant: B, 10^{13} cm^{-2}, 25 keV 30 nm sacrificial oxide growth and wet etch Trench fill (30 nm thermal oxide and 1 μm TEOS) CMP Nitride and padoxide removal (1% HF)
20nm gate oxide + 20 nm α-Si deposition collector plug implant (P, 2×10^{15} cm^{-2}) and anneal N-ch threshold voltage adjust and anti-punch-through P-ch threshold voltage adjust and anti-punch-through Base implant (2.4×10^{13} cm^{-2}) Collector enhancement implant (10^{12} cm^{-2}, 180 keV)	
	Oxide removal (single-poly npn)
Base anneal (900 °C, 20 mins.) 350 nm poly-1 deposition 1000 Ω resistor implant: B, 3.3×10^{14} cm^{-2}, 35 keV Emitter poly implant (As, 10^{16} cm^{-2}) in single poly bipolar devices	
	Emitter poly implant (As, 1.5×10^{16} cm^{-2}) in PRET 40 nm nitride deposition Poly-1 mask-1, etch poly-1
See Figure 2–21(a)	
	Etch gate oxide wet chemically Poly-2 deposition (200 nm) Poly-2 anisotropic etchback
Poly-1 mask-2	
See Figure 2–21(b)	
Poly-1 etch and reoxidation (900 °C, 30 mins.) LDD nMOS LDD pMOS (B, 2.4×10^{13} cm^{-2}) - also linkup bipolar oxide spacer formation	
	60 nm poly-3 strap deposition and patterning
Source/drain (base and collector contacts) Implants and anneal (900 °C, 25 mins.) Salicidation TiSi$_2$ (excluding resistors) Oxide deposition and planarization W-plugged contacts AlCu deposition and patterning	

(continued)

Table 2–3 Comparison between the new process flow and the conventional BiCMOS process.

Standard BiCMOS	Additional for PRET
	See Figure 2–21(c)
Intermetal planarization	
W-plugged vias	
AlCu deposition and patterning	
Passivation	

2.8.1 Process Description

In this subsection, the additional processing steps, besides the conventional BiCMOS processing sequence, will be discussed. In this process, isolation is achieved using shallow and deep trenches. The advantage of employing such trenches is significant capacitance reduction. Figure 2–21(a) shows the cross-sectional view of the PRET, single-poly, and pMOS devices before patterning. Once both the shallow and deep trenches have been defined, they are filled with 30 nm of thermal oxide, followed by 1 μm of TEOS and then planarized by Chemical-Mechanical Polishing (CMP). The deep trenches, which cut through the buried layer, cause the collector-substrate capacitance to scale with their dimensions.

Figure 2–21(a and b) indicates that two poly masks are used in the fabrication of the PRET. The first poly mask defines the poly rim against which the poly emitter ridge will lean. Next, 200 nm of polysilicon (poly-2) is deposited and etched back. The poly-2 layer is removed entirely from the wafer surface except for the ridges at the PRET active area. The second poly-1 mask defines the MOS gates and emitters of the PRET and the conventional single-poly devices. By having two poly masks, widening of gates by poly ridges is avoided.

As shown in Figure 2–21(c), the strap connections to the active areas are made by the deposition of polysilicon. After the implantation and annealing, the silicidation converts the polysilicon into silicide. Hence the construction of the strap allows it to perform a dual role of an interconnect and a silicide which, at the same time, reduces the series resistance of the collector.

The basic BiCMOS process constitutes 19 mask steps. The end products are CMOS devices, single-poly npn bipolar devices, two levels of metallization, passivation and 1000 ohm/square resistors. The deep trench, PRET emitter and strap modules require three additional mask steps, bringing the total number of mask steps to 22. However, the increase in the number of mask steps is easily offset by the area reduction achieved by strap contacted active areas and the utilization of the poly-ridge emitter transistor [57,58]. This area reduction, in turn, results in capacitance reduction.

2.8.2 Shallow Trench Isolation (STI)

The advantages of employing Shallow Trench Isolation (STI) include (1) reduction of the peripheral contribution to the base-collector capacitance C_{CB}, and (2) better definition of the active region with LOCOS, allowing smaller dimensions to be achieved without the occurrence of an enhanced bird's beak [59]. An experiment was conducted to measure C_{CB} for four differ-

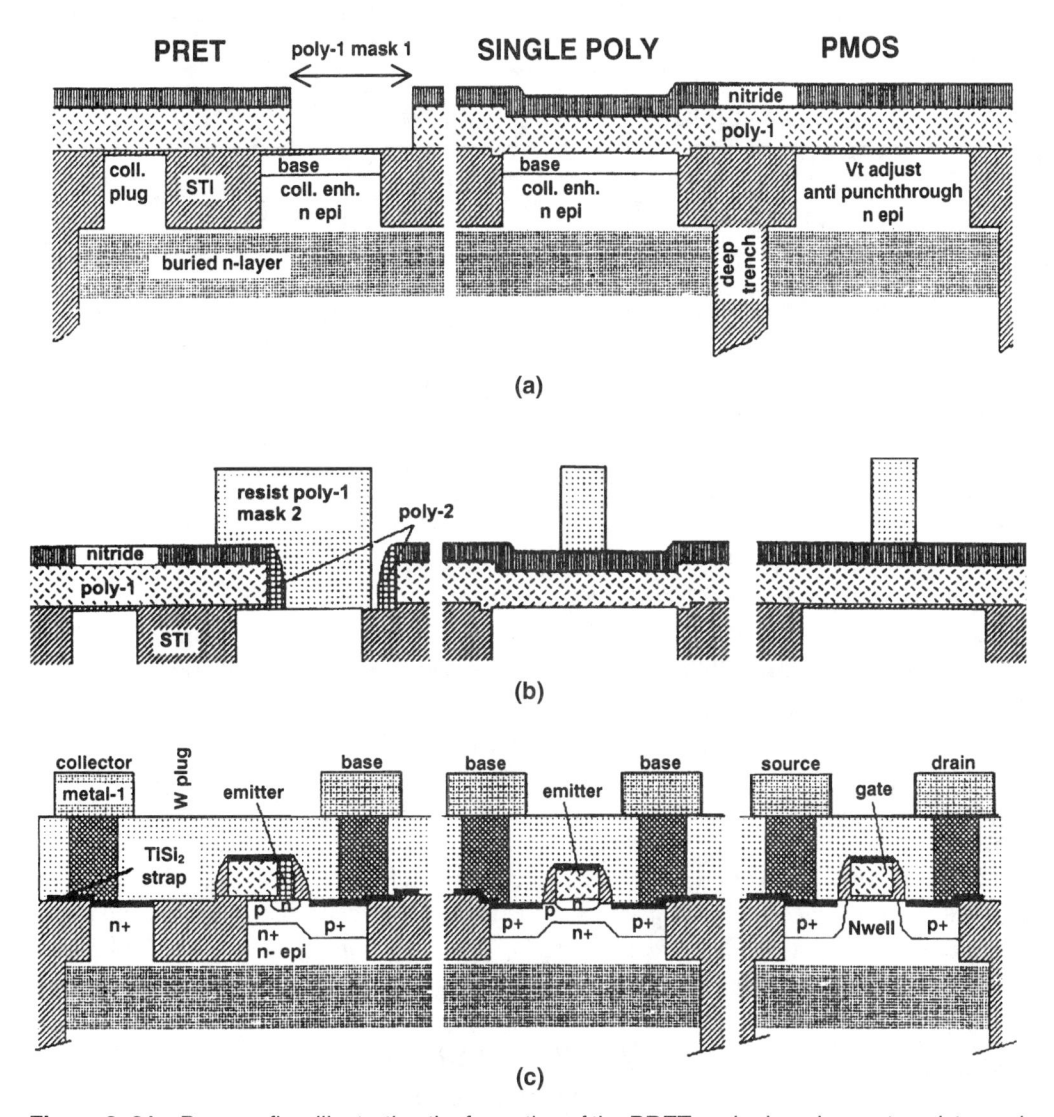

A New Low-Power Ultra Low-Capacitance BiCMOS Process

Figure 2–21 Process flow illustrating the formation of the PRET, a single-poly npn transistor and a pMOS transistor.

ent LOCOS and two STI variants. The results are shown in Table 2–4. Czochralski wafers with p-type <100> orientation and selectively Sb-implanted and covered with 1.5 µm As-doped epi with a resistivity of 0.5 ohm-cm were used. The results prove that shallow trench is indeed effective in reducing the base-collector capacitance.

Table 2-4 Base-collector capacitance (in fF/μm) for different isolation schemes.

Isolation style	Perimeter C_{BC} (fF/μm) Measured	Perimeter C_{BC} (fF/μm) Simulated
LOCOS: 850 nm, 1000 °C	0.25	—
LOCOS: 630 nm, 1000 °C	0.25	0.24
LOCOS: 550 nm, 1000 °C	0.22	—
LOCOS: 600 nm trench, 1.23 μm oxide thickness, bird's head planarization	0.22	0.22
Isotropic trench 400 nm deep, 250 nm lateral undercut	0.15	0.13
Anisotropic trench 550 nm deep	0.08	0.08

2.9 Manufacturing Considerations

The main disadvantage of the BiCMOS process is cost. Most memory BiCMOS processes add performance enhancement features (buried layer and epitaxy, polysilicon emitter, deep collector and so on) to a core CMOS process. Because of the greater manufacturing complexity, extra cost is incurred. The mask count is a good yardstick of the relative process complexity. For high-performance BiCMOS, the mask count and the total process steps typically increase by 25% as compared to those of CMOS. Most of the additional processing steps is in the front end. Due to the complexity of the technology, BiCMOS imposes more stringent requirements on cycle time, defect density, and process control. Hence it is inevitable that BiCMOS chip manufacturers must charge a high price for their products. In return, the BiCMOS circuits must provide superior performance at a considerable reduction in power dissipation, as compared to that of the conventional bipolar devices. In the analog market, the ability to integrate large mixed systems provides the basis for utilizing BiCMOS circuits [11].

2.10 Future Trends in BiCMOS Technology

Recent years have seen the scaling of devices into the submicron regimes. For BiCMOS gates, several issues have to be addressed with regard to scaling the device dimensions. First, the scaled MOS devices in such BiCMOS gates will require lower supply voltages. However, at low supply voltages (< 3.3 V) the performance of the BiCMOS gates are severely degraded [60]. This can be solved by using special circuit techniques and/or configurations, such as the CBiCMOS [61], BiNMOS [62], FS-BS-CMOS [63], to name a few (see Chapter 4 for more details of these circuits and/or configurations). Second, the scaling of the bipolar devices relative to the MOS devices in the BiCMOS gates must be optimized. Finally, improvements to the existing BiCMOS device structures or new BiCMOS device structures must be developed. The latter will be dealt with in greater details in the following two sub-sections.

2.10.1 Bipolar Device Structure Improvements

In order to improve the performance of high-speed bipolar transistors for high-performance BiCMOS LSIs, single or double-polysilicon emitter bipolar devices (as discussed in section 2.4.3) were employed. Recently, Si/SiGe heterojunction bipolar transistors have been used [64–66] to achieve further speed improvement. The double-polysilicon self-aligned bipolar transistor structure was first introduced over 10 years ago [67,68] and soon became a common feature of most high-performance bipolar devices used in BiCMOS technologies [69–72]. The schematic cross-section of a typical double-poly self-aligned bipolar transistor is reproduced in Figure 2–22 for discussion.

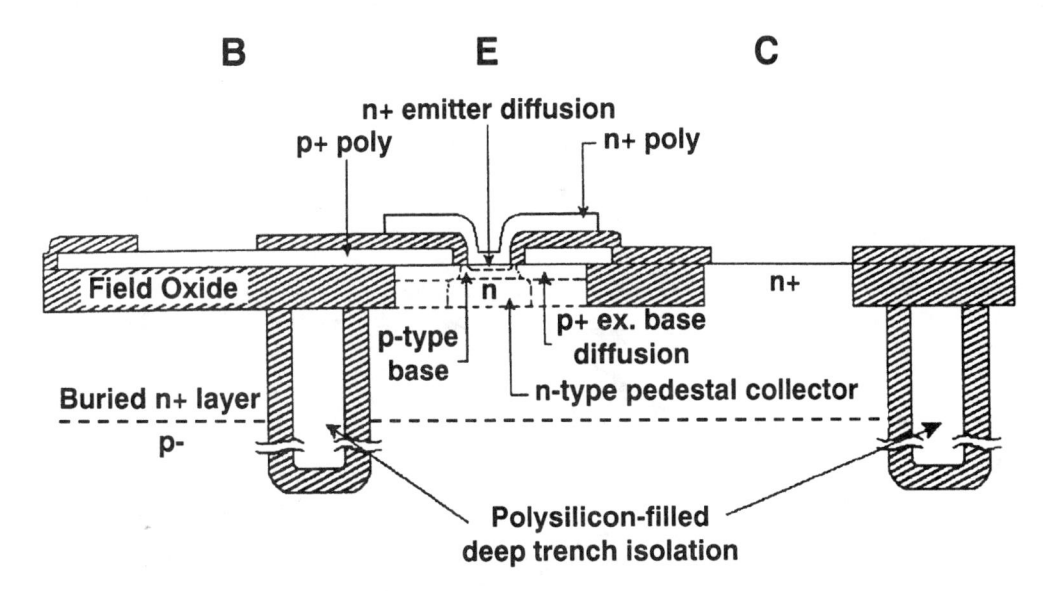

Figure 2–22 Cross-sectional view of a double-polysilicon self-aligned bipolar device. [73]

The polysilicon emitter has allowed the formation of an exceptionally shallow emitter-base junction while simultaneously improving the device current gain and performance. Furthermore, the poly emitter enables the emitter-base junction depth to be scaled, which has been necessary in order to allow reduction of the device base width. Figure 2–23 shows that reducing the base width results in lowering of the base transit time, hence resulting in better speed performance. However, the emitter resistance will become a limiting factor for small device dimensions, as a result of the increasing current density. Thus future emitter processing improvements will need to focus on the reduction of the emitter resistance while maintaining adequate current gain. In addition, the reliability of the emitter contact may become questionable with increasing emitter resistance after being subjected to prolonged stressing at high forward current densities [74].

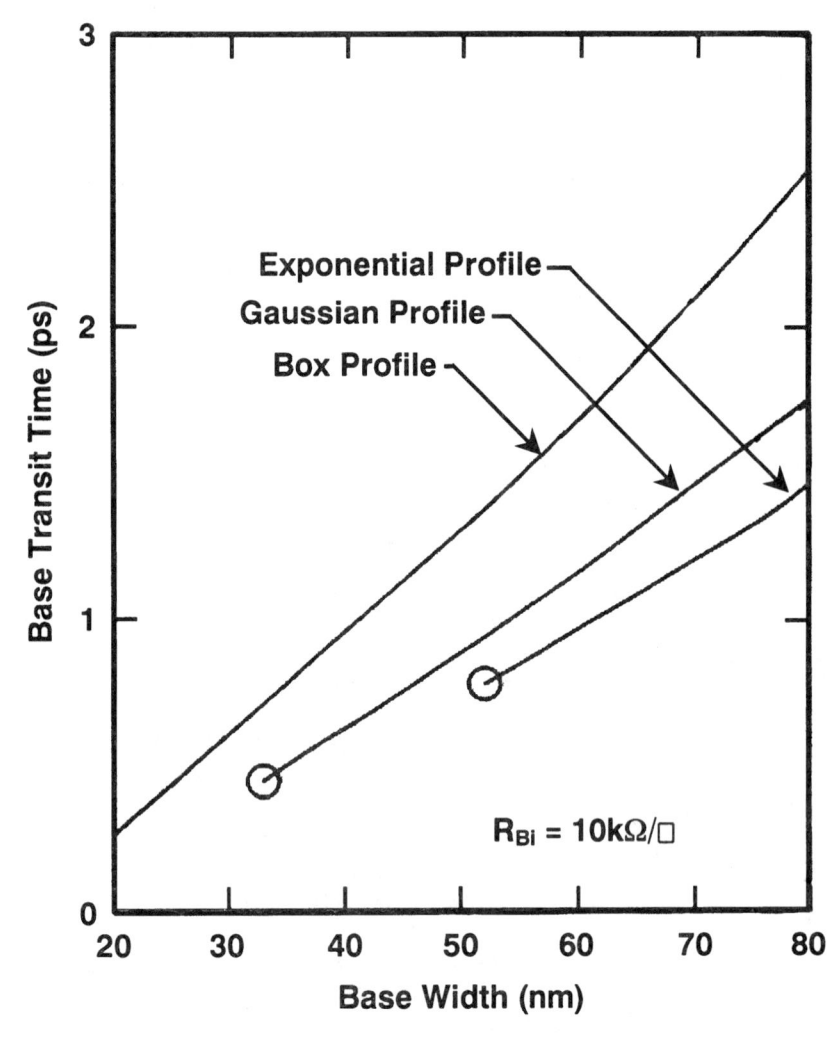

Figure 2–23 Base transit time against the base-width for three different types of base dopant profiles [75]. Open circles denote the point where the emitter-base (E-B) junction concentration has reached $5 \times 10^{18}/cm^3$, which is the maximum practical E-B junction concentration.

In the self-aligned bipolar structures, the extrinsic base contacts also impact the performance of the devices. The ideal extrinsic base contact should contribute minimally to the collector-base capacitance (C_{CB}), as well as the overall base resistance. Figure 2–24 illustrates the extrinsic and link base regions, showing the different resistive and capacitive parasitic components.

Manufacturing Considerations 71

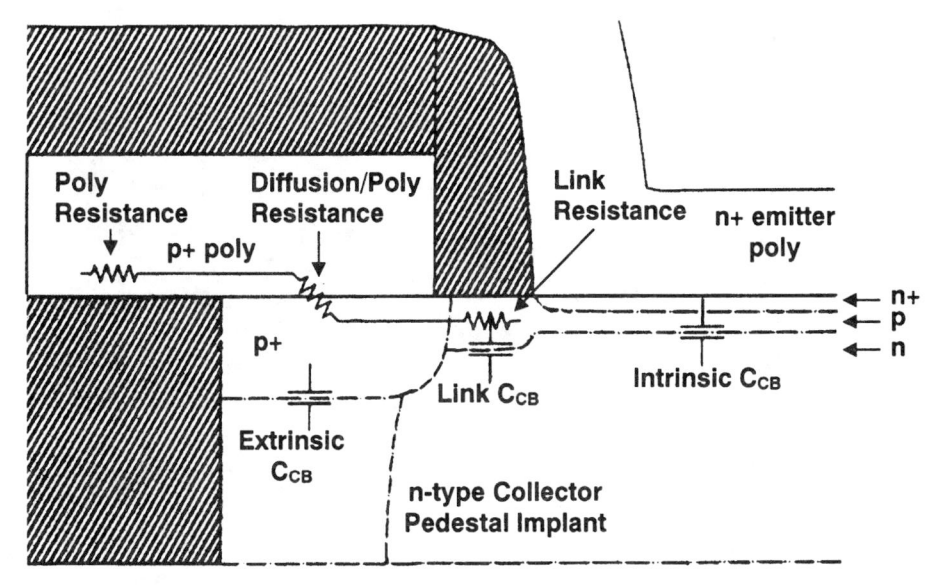

Figure 2-24 Schematic drawing of the extrinsic and link base regions, illustrating the connections of the various resistive and capacitive parasitic components. [73]

In practical fabrication process, the extrinsic base capacitance depends strongly on the background dopant concentration, the device epilayer thickness and the sharpness of the transition to the n+ subcollector layer. The major techniques to reduce the collector-base capacitance include (1) minimizing the area of the extrinsic base junction, (2) reducing the n-type doping concentration under the extrinsic base junction, and (3) decreasing the junction depth. The extrinsic base junction area can be minimized by employing the double-poly device structure, whereby the contact to the base is made via the p+ extrinsic base poly and this contact area can be made very small. Furthermore, it is possible to further minimize the extrinsic C_{CB} component in a double-poly device structure by self-aligning the emitter opening to the edge of the field oxide [71]. As such it is possible to reduce not only C_{CB}, but also the overall device size. Another way to reduce C_{CB} is with the use of special device layouts, as shown in Figure 2-25.

The emitter-base junction is butted directly against the field oxide on two sides [68,76], with the base contact made only on the non-butted sides. By butting the emitter right against a deep trench filled with dielectric, C_{CB}, C_{CS} and the overall device size can be reduced even further. However, reduced BV_{CEO} and punch-through issues are the major considerations for these types of devices. These issues could be resolved with the use of shallow trench isolation (STI) technique.

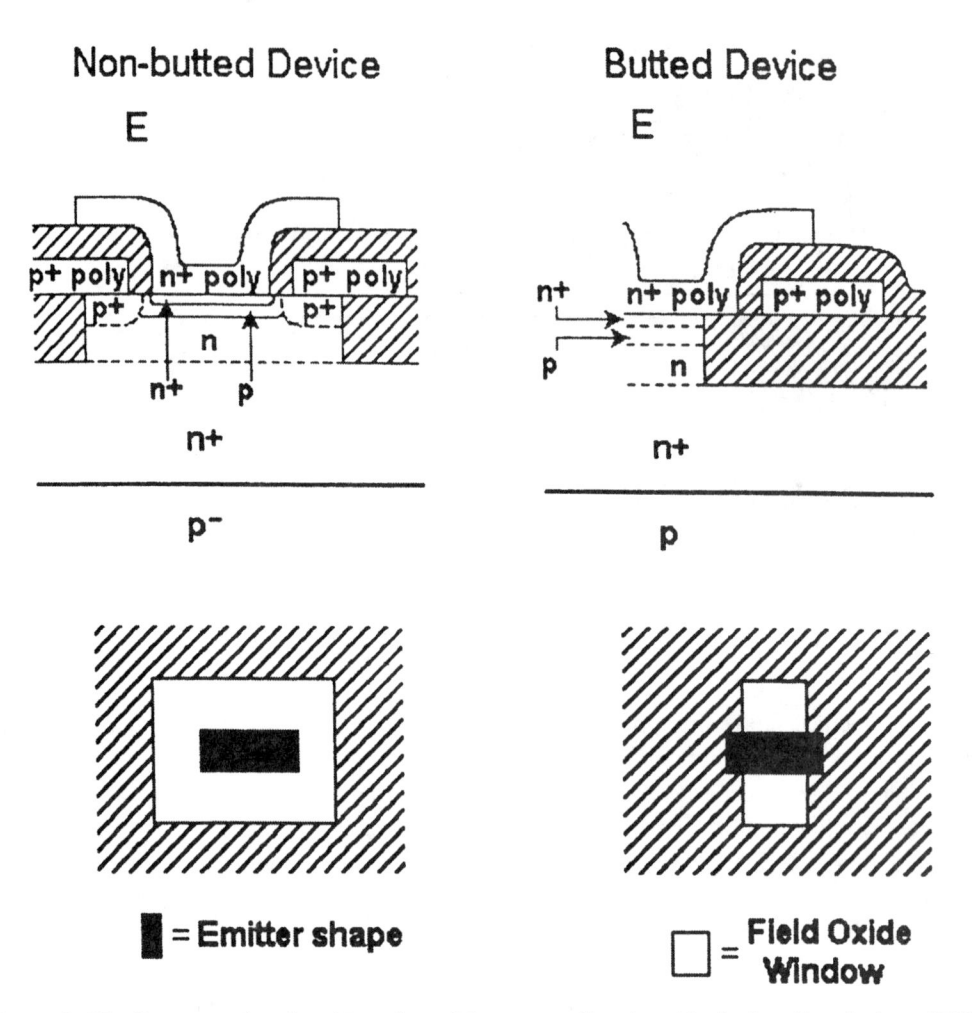

Figure 2–25 Cross-sectional and top view of the conventional and butted emitter devices. [73]

2.10.2 Silicon-On-Insulator (SOI) Technology

SOI technology has been developed to increase the operating speed of bipolar devices by reducing the parasitic area and the depth of deep trenches used for isolation [77,78]. The SOI offers a way of scaling the device substructure, which previously was difficult to achieve using bulk silicon. Very small emitter lateral bipolar devices can be fabricated with minimal parasitic [79–82]. A lateral bipolar structure for BiCMOS circuits [82] is shown in Figure 2–26.

An n+ p n- n+ layer for emitter, base, and collector regions was fabricated. The p-type base region is connected to the p++ polysilicon layer through the polysilicon sidewall. The oxide layer between p++ polysilicon and the collector region acts to minimize the base-collector

Conclusions

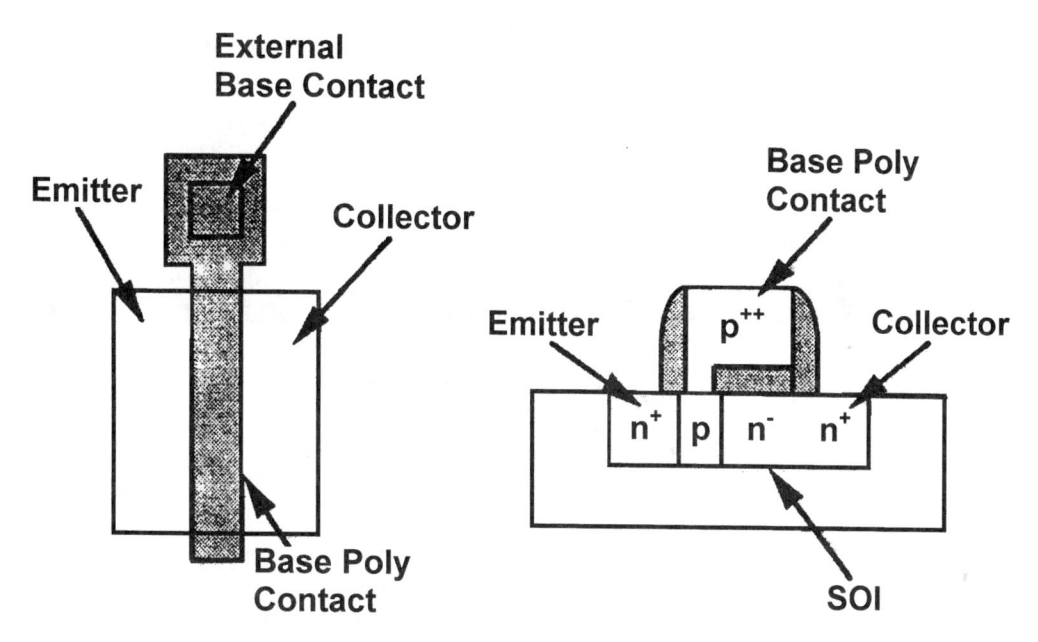

Figure 2–26 A lateral bipolar transistor using the SOI technology.

capacitance. Figure 2–27 shows the schematic cross-section of an ECL-CMOS SRAM [77] using the SOI technology. Bipolar transistors with a cut-off frequency of 27 GHz and CMOS memory cells with an area of 58 μm^2 have been integrated on a SOI substrate. Figure 2–28 compares the conventional BiCMOS device structure on Si substrate with a SOI-BiCMOS device structure [83] for future high-speed logic and high-speed RAM LSIs.

Deep trench isolation regions filled with CVD oxide have been employed to reduce defect density generated in a silicon film during the thermal process. The above-mentioned SOI-BiCMOS devices offer promising possibilities in future technologies. However, it has been proven difficult to make a good base contact in SOI-CMOS process, and incorporating a polysilicon emitter is likely to create problems.

2.11 Conclusions

BiCMOS technology has emerged as the best method of merging the most distinctive features of bipolar and CMOS devices into a design methodology far superior to either of the two separate components. BiCMOS circuits have been employed extensively in high-speed SRAMs [84–87]. Also, they have been used in advanced microprocessors [88–92]. Early BiCMOS technologies used relatively simple bipolar devices, usually single-poly structures [93–98]. Enhanced features such as deep trench isolation and buried n+ sub-collectors were often not included in order to reduce the process complexity. The single-poly devices were more CMOS-

74　　　　　　　　　　　　　　　　　　　　　　　Chapter 2　•　BiCMOS Process Technology

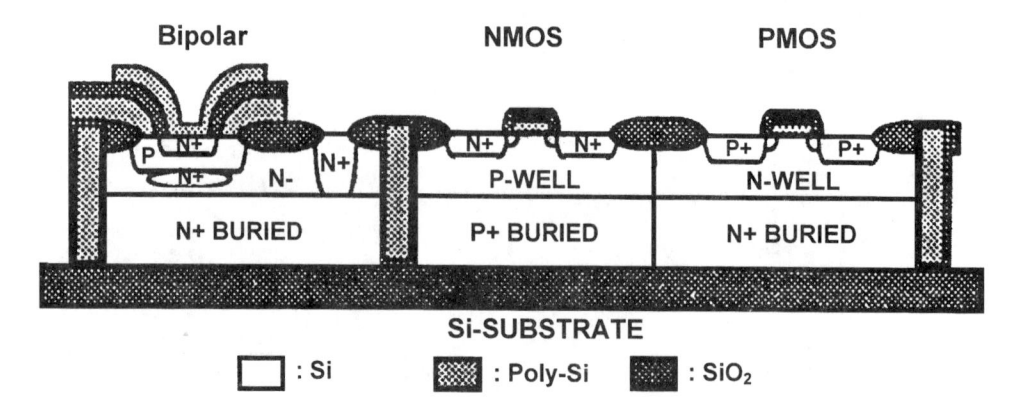

Figure 2–27 Cross-section of an ECL-CMOS SRAM utilizing the SOI BiCMOS device structure. [77]

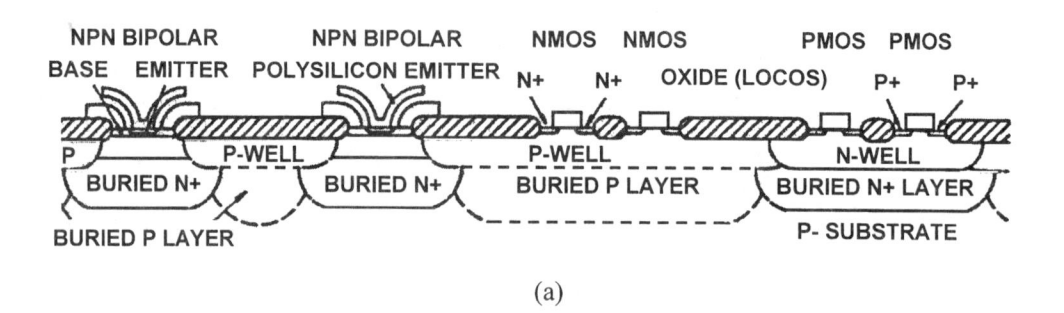

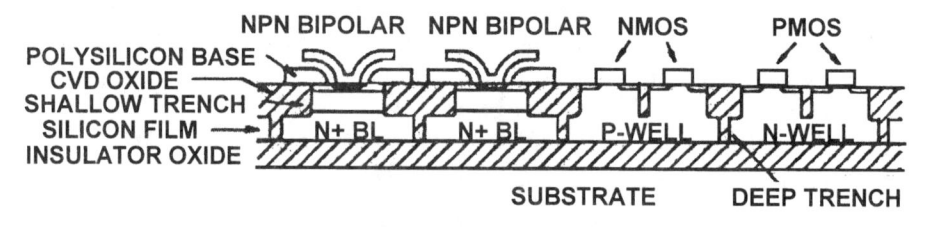

Figure 2–28 Present and future BiCMOS device structure (a) conventional BiCMOS device structure on Si substrate (b) future SOI-BiCMOS device structure.

like in their structure and hence could be incorporated easily into the CMOS process. However, as CMOS technologies continue to improve with the scaling of lithographic dimensions, it becomes necessary to substantially improve the bipolar performance in order to maintain the high-speed performance of BiCMOS devices. Scaling of device dimensions must be accompanied by lowering of supply voltages for reliability purposes. However, there will come a point when, at a low supply voltage, it may no longer be possible, due to process complexity and/or manufacturing cost, to improve on the CMOS performance by adding in high-performance bipolar devices. At that point of time, innovative circuit techniques [61–63] will be required to bring the supply voltages further down into the sub-2 V regimes.

2.12 References

[1] R. H. Havemann and R. H. Eklund, "Process Integration Issues for Submicron BiCMOS Technology," *Solid-State Tech.*, Vol. 35, No. 6, pp. 71–76, June 1992.

[2] A. R. Alvarez, *BiCMOS Technology and Applications*, 2nd ed., Kluwer Academic Publishers, United States of America, 1993.

[3] H. Tran, D. Scott, P. Fung, R. Havemann, R. Eklund, T. Ham, R. Haken, and A. Shah, "An 8 ns 256-kbit ECL SRAM with CMOS Memory Array and Battery Backup Capability," *IEEE J. Solid-State Circuits*, Vol. SC-23, No. 10, pp. 1041–1047, October 1988.

[4] R. Chapman et al., "An 0.8 µm CMOS Technology for High-Performance Logic Applications," *IEDM, Tech. Dig.*, pp. 362–365, December 1986.

[5] S. Ogura et al., "Design and Characteristics of the Lightly Doped Drain Source (LDD) Insulated Gate Field Effect Transistor," *IEEE Trans. Electron Devices*, Vol. ED-27, pp. 1359–1367, 1980.

[6] H. Katto et al., "Hot-Carrier Degradation Modes and Optimization of LDD MOSFETs," *IEDM, Tech. Dig.*, pp. 774–777, December 1984.

[7] K. Balasubramanyam et al., "Characterization of As-P Double Diffused Drain Structure," *IEDM, Tech. Dig.*, pp. 782–785, December 1984.

[8] R. B. Ritts, P. A. Raje, J. D. Plummer, K. C. Sarawat, and K. M. Cham, "Merged BiCMOS Logic to Extend the CMOS/BiCMOS Performance Crossover Below 2.5 V supply," *IEEE J. Solid-State Circuits*, Vol. 26, No. 11, pp. 1606–1613, November 1991.

[9] H. Momose, Y. Unno, and T. Maeda, "Supply Voltage Design Tradeoffs Between Speed and nMOSFET Reliability of Half-Micrometer BiCMOS Gates," *IEEE Trans. Electron Devices*, Vol. 38, No. 3, p. 566, March 1991.

[10] H. Momose, Y. Unno, and T. Maeda, "A Supply Voltage Design for Half-Micrometer BiCMOS Gates," *Symp. VLSI Technology, Tech. Dig.*, pp. 53–54, 1989.

[11] D. Stolfa et al., "A BiCMOS 0.8 µm Process with a Tool-Kit for Mixed-Mode Design," *IEEE CICC, Proc.*, pp. 24.2.1–24.2.4, 1993.

[12] H. Nakashiba et al., "An Advanced PSA Technology for High-Speed Bipolar LSI," *IEEE Trans. Electron Devices*, Vol. ED-27, No. 8, pp. 1390–1394, August 1980.

[13] C. Lage, "BiCMOS Memories: Increasing Speed While Minimising Process Complexity," *Solid-State Technology*, Vol. 35, No. 8, pp. 31–34, August 1992.

[14] E. Greeneich and K. Mclaughlin, "Analysis and Characterization of BiCMOS for High-Speed Digital Logic," *IEEE J. Solid-State Circuits*, Vol. 23, No. 2, pp. 558–565, April 1988.

[15] R. Chapman et al., "Submicron BiCMOS Well Design for Optimum Circuit Performance," *IEDM, Tech. Dig.*, pp. 756–759, December 1988.

[16] T. Ikeda et al., "High-Speed BiCMOS Technology with a Buried Twin-Well Structure," *IEEE Trans. Electron Devices*, Vol. ED-34, No. 6, pp. 1304–1309, June 1987.

[17] P. I. Van Wijnen et al., "A Localised Collector Implant to Improve Uniformity of Polysilicon Bipolar Transistors," *BCTM, Proc.*, pp. 160–163, September 1989.

[18] R. C. Taft, J. D. Hayden, and C. D. Gunderson, "Optimization of Two-Dimensional Collector Doping Profiles for Submicron BiCMOS Technologies," *IEDM, Tech. Dig.*, pp. 869–872, December 1991.

[19] Y. Kobayashi et al., "Bipolar CMOS Merged Structure for High Speed 1-Mbit DRAM," *IEDM, Tech. Dig.*, pp. 802–804, December 1986.

[20] G. Kitsukawa et al., "An Experimental 1-Mbit BiCMOS DRAM," *IEEE J. Solid-State Circuits*, Vol. SC-22, No. 5, pp. 657–662, October 1987.

[21] S. A. Shahriar et al., "A Triple Diffused Approach for High-Performance 0.8 µm BiCMOS Technology," *Solid-State Technology*, Vol. 35, No. 10, pp. 33–40, October 1992.

[22] S. Wolf, *Silicon Processing for the VLSI Era Volume 2: Process Integration*, Chp. 6, p. 414, Lattice Press, California, 1991.

[23] Y. Niitsu et al., "Latch-Up-Free CMOS Structure Using Shallow Trench Isolation," *IEDM, Tech. Dig.*, p. 509, 1985.

[24]   J. Borland et al., "Silicon Epitaxial Growth for Advanced Device Structures," *Solid-State Technology*, Vol. 31, No. 1, pp. 111–119, January 1988.

[25]   G. Srinivasan, "Kinetics of Lateral Auto-Doping in Silicon Epitaxy," *Journal of Electrochemical Society*, Vol. 125, No. 1, pp. 146–151, January 1978.

[26]   K. Chiu et al., "A Bird's Beak Free Local Oxidation Technology Feasible for VLSI Circuits Fabrication," *IEEE Trans. Electron Devices*, Vol. ED-29, No. 4, pp. 536–540, April 1982.

[27]   K. Wang et al., "Direct Moat Isolation for VLSI," *IEEE Trans. Electron Devices*, Vol. ED-29, No. 4, pp. 541–547, April 1982.

[28]   N. Matsukawa et al., "Selective Polysilicon Oxidation Technology for VLSI Isolation," *IEEE Trans. Electron Devices*, Vol. ED-29, No. 4, pp. 561–573, April 1982.

[29]   T. Stultz et al., *On Laser-Solid Interactions and Transient Thermal Processing of Materials*, Ed. I. Narayan, p. 463, North-Holland, New York, 1983.

[30]   D. Scott, W. Hunter and H. Shichijo, "A New Transmission Line Model for Silicided Source/Drain Diffusion: Impact on VLSI Circuits," *IEEE Trans. Electron Devices*, Vol. ED-29, No. 4, pp. 651–661, April 1982.

[31]   M. Alperin et al., "Development of the Self-Aligned Titanium Silicided Process for VLSI Applications," *IEEE Trans. Electron Devices*, Vol. ED-32, No. 2, pp. 141–149, February 1985.

[32]   C. Y. Ting, "Silicide for Contacts and Interconnects," *IEDM, Tech. Dig.*, pp.110–114, 1984.

[33]   S. T. Chang and K. Y. Chiu, "Reduced Oxide Charge Trapping and Improved Hot-Electron Reliability in Submicrometer MOS Devices Fabricated by Titanium Salicide Process," *IEEE Electron Device Letters*, Vol. 9, No. 5, pp. 244–246, 1988.

[34]   R. Haken, "Application of the Self-Aligned Titanium Silicide Process to VLSI nMOS and CMOS Technologies," *Journal of Vacuum Science Technology*, Vol. B3, No. 6, pp. 1657–1663, December 1985.

[35]   D. Scott et al., "Titanium Disilicide Contact Resistivity and Its Impact on 1 µm CMOS Circuit Performance," *IEEE Trans. Electron Devices*, Vol. ED-34, No. 3, pp. 562–574, March 1987.

[36]   L. V. D Hove et al., *Electrochemical Society Meeting, Ext. Abstract*, p. 312, Spring, Atlanta, Georgia/Chicago Illinois, 1988.

[37] S. Wolf, *Silicon Processing for the VLSI Era Volume 2: Process Integration*, Chp. 3, p. 150, Lattice Press, California, 1991.

[38] W. Lynch, "Self-Aligned Contact Schemes for Source-Drains in Submicron Devices," *IEDM, Tech. Dig.*, p. 354, 1987.

[39] R. C. Jaeger, *Introduction to Microelectronic Fabrication*, Addison-Wesley, Reading, MA, 1987.

[40] T. E. Tang et al., "Titanium Nitride Local Interconnect Technology for VLSI," *IEEE Trans. Electron Devices*, Vol. ED-34, No. 3, pp. 682–688, March 1987.

[41] R. Wolters and L. V. D. Hove, "Feasibility of $CoSi_2$, TiW and TiW(N) as Local Interconnection in a Self-Aligned $CoSi_2$ Technology," *IEEE VMIC, Proc.*, p. 149, Santa Clara, CA, 1988.

[42] V. V. Lee, S. Verdonckt-Vandebroek, and S. S. Wong, "Selective CVD Tungsten Local Interconnection Technology," *IEDM, Tech. Dig.*, p. 450, 1988.

[43] T. Bonifield et al., "A 1 Micron Design Rule, Double Level Metallization Process," *IEEE VMIC, Proc.*, pp. 71–77, June 1986.

[44] S. Wolf, *Silicon Processing for the VLSI Era Volume 2: Process Integration*, Chp. 7, p. 530, Lattice Press, California, 1991.

[45] R. H. Havemann et al., "An 0.8 μm 256-kbit BiCMOS SRAM Technology," *IEDM, Tech. Dig.*, pp. 748–751, December 1987.

[46] R. Chapman et al., "0.5 μm CMOS for High-Performance at 3.3 V," *IEDM, Tech. Dig.*, p. 52, December 1988.

[47] A. R. Alvarez et al., "2 Micron Merged Bipolar - CMOS Technology," *IEDM, Tech. Dig.*, pp. 761–764, December 1984.

[48] A. Feygonson et al., "CBIC-V, A New Very High Speed Complementary Silicon Bipolar IC Process," *BCTM, Proc.*, p. 173, 1989.

[49] R. Eklund et al., "A 0.5 μm BiCMOS Technology for Logic and 4-Mbit Class SRAM's," *IEDM, Tech. Dig.*, pp. 425–428, December 1989.

[50] J. Kirchgessner et al., "An Advanced 0.4 μm BiCMOS Technology for High-Performance ASIC Applications," *IEDM, Tech. Dig.*, pp. 97–100, December 1991.

[51] S. W. Sun et al., "A 0.4 µm Fully Complementary BiCMOS Technology for Advanced Logic and Microprocessor Applications," *IEDM, Tech. Dig.*, pp. 85–88, December 1991.

[52] M. Nanba, T. Shiba, T. Nakamura, and T. Toyabe, "An Analytical and Experimental Investigation of the Cut-Off Frequency f_T of High-Speed Bipolar Transistors," *IEEE Trans. Electron Devices*, Vol. 35, No. 7, pp. 1021–1028, July 1988.

[53] S. Weber, "TI Soups Up LinCMOS Process with 20 V Bipolar Transistors," *Electronics*, pp. 59–60, February 4th, 1988.

[54] R. A. Haken et al., "BiCMOS Process Technology," in *BiCMOS Technology and Applications*, Ed. A. R. Alvarez, Chp. 3, p. 63, Norwell, MA, Kluwer Academic, 1989.

[55] W. V. D. Wel et al., "A Low-Power, Ultra-Low Capacitance BiCMOS Process Applied to a 2 GHz Low-Noise Amplifier," *IEEE Trans. Electron Devices*, Vol. 43, No. 9, pp. 1539–1546, September 1996.

[56] J. L. D. Jong et al., "Single Polysilicon Layer Advanced Super High-Speed BiCMOS Technology," *IEEE BCTM, Proc.*, p. 182, Minneapolis, MN, September 1989.

[57] W. V. D. Wel et al., "Low-Power, Ultra-Low Capacitance Bipolar Transistor Compatible with Mainstream CMOS," *ESSDERC, Proc.*, p. 425, Den Hadg, The Netherlands, September 1995.

[58] W. V. D. Wel et al., "Poly-Ridge Emitter Transistor (PRET): Simple Low-Power Option to a Bipolar Process," *IEDM, Tech. Dig.*, p. 453, Washington, DC, December 1993.

[59] P. A. V. D. Plas et al., "Geometry Dependent Bird's Beak Formation for Submicron LOCOS Isolation," *ESSDERC, Proc.*, p. 131, Berlin, Germany, September 1989.

[60] H. Momose et al., "Lower Submicron FCBiCMOS (Fully Complementary BiCMOS) Process with RTP and MeV Implanted 5 GHz Vertical PNP Transistor," *Symp. VLSI Technology, Tech. Dig.*, pp. 55–57, 1989.

[61] S. S. Rofail and K. S. Yeo, "New Complementary BiCMOS Digital Gates for Low-Voltage Environments," *Solid-State Electronics*, Vol. 39, No. 5, pp. 681–687, 1996.

[62] R. F. Krick, L. T. Clark, D. J. Deleganes, K. L. Wong, R. Fernando, G. Debnath, and J. Banik, "A 150 MHz 0.6 µm BiCMOS Superscalar Microprocessor," *IEEE J. Solid-State Circuits*, Vol. 29, No. 12, pp. 1455–1462, December 1994.

[63] A. Bellaouar, M. I. Elmasry, and S. H. K. Embabi, "Bootstrapped Full-Swing BiCMOS/BiNMOS Logic Crcuits for 1.2 - 3.3 V Supply Voltage Regime," *IEEE J. Solid-State Circuits*, Vol. 30, No. 6, pp. 629–636, June 1995.

[64] D. L. Harame et al., "A High-Performance Epitaxial SiGe-Base ECL BiCMOS Technology," *IEDM, Tech. Dig.*, pp. 19–22, 1992.

[65] G. L. Patton et al., "63 - 75 GHz f_T SiGe-Base Heterojunction Bipolar Technology," *Symp. VLSI Technology, Tech. Dig.*, pp. 49–50, 1990.

[66] K. Imas et al., "A Cryo-BiCMOS Technology with SiGe Heterojunction Bipolar Transistors," *BCTM, Tech. Dig.*, pp. 90–93, 1990.

[67] H. Nakashiba et al., "An Advanced PSA Technology for High-Speed Bipolar LSI," *IEEE Trans. Electron Devices*, Vol. ED-27, pp. 1390–1394, 1980.

[68] D. D. Tang et al., "1.25 µm Deep-Groove-Isolated Self-Aligned Bipolar Circuits," *IEEE J. Solid-State Circuits*, Vol. SC-17, pp. 925–931, 1982.

[69] T. Chen et al., "A Submicron High-Performance Bipolar Technology," *Symposium on VLSI Technology, Tech. Dig.*, pp. 87–88, 1989.

[70] J. Cressler et al., "A Scaled 0.25 µm Bipolar Technology using Full E-Beam Lithography," *IEEE Electron Device Letters*, Vol. 13, pp. 262–264, 1992.

[71] T. Shiba et al., "A 0.5 µm Very High-Speed Silicon Bipolar Devices Technology U-Groove Isolated SICOS," *IEEE Trans. Electron Devices*, Vol. 38, No. 11, pp. 2505–2511, 1991.

[72] F. Sato et al., "Sub-20 ps ECL Circuits with 50 GHz f_{max} Self-Aligned SiGe HBTs," *IEDM, Tech. Dig.*, pp. 397–399, 1992.

[73] J. D. Warnock, "Silicon Bipolar Device Structures for Digital Applications: Technology Trends and Future Directions," *IEEE Trans. Electron Devices*, Vol. 42, No. 3, pp. 377–389, March 1995.

[74] D. D. Tang et al., "On the Very High Current Degradation Si NPN Transistors," *IEEE Trans. Electron Devices*, Vol. ED-37, No. 7, pp. 1698–1706, 1990.

[75] K. Suzuki, "Optimum Base Doping Profile for Minimum Base Transit Time," *IEEE Trans. Electron Devices*, Vol. 38, No. 9, pp. 2128–2133, 1991.

[76] J. Warnock et al., "High-Performance Bipolar Technology for Improved ECL Power Delay," *IEEE Electron Device Letters*, Vol. 12, pp. 315–317, 1991.

[77] T. Hiramoto et al., "A 27 GHz Double Polysilicon Bipolar Technology on Bonded SOI with Embedded 58 µm^2 CMOS Memory Cells for ECL-CMOS SRAM Application," *IEDM, Tech. Dig.*, pp. 39–42, 1992.

[78] H. Nishizawa et al., "A Fully SiO$_2$-Isolated Self-Aligned SOI-Bipolar Transistor for VLSI's," *IEEE BCTM, Proc.*, pp. 53–58, 1991.

[79] M. Rodder and D. Antoniadis, "Silicon-On-Insulator Bipolar Transistors," *IEEE Electron Device Letters*, Vol. 4, pp. 193–195, 1983.

[80] B. Y. Tsaur et al., "Fully Isolated Lateral Bipolar-MOS Transistors Fabricated in Zone-Melting-Recrystallized Si Films on SiO$_2$," *IEEE Electron Device Letters*, Vol. 4, pp. 269–271, 1983.

[81] N. Higaki et al., "A Thin-Base Lateral Bipolar Transistor Fabricated on Bonded SOI," *Symp. VLSI Technology, Tech. Dig.*, pp. 53–54, 1992.

[82] G. G. Shahidi et al., "A Novel High-Performance Lateral Bipolar on SOI," *IEDM, Tech. Dig.*, pp. 663–666, 1991.

[83] H. Nishizawa et al., "An Advanced Dielectric Isolation Structure for SOI-CMOS/BiCMOS VLSIs," *Int. Symp. Semiconductor Wafer Bonding: Science, Technology Applications, PV 93-29, The Electron Chemical Society*, pp. 176–188, 1993.

[84] H. Kato et al., "A 9 ns 4-Mbit BiCMOS SRAM with 3.3 V Operation," *IEEE ISSCC, Tech. Dig.*, pp. 210–211, 1992.

[85] K. Nakamura et al., "A 6 ns 4-Mbit ECL I/O BiCMOS SRAM with LV/TTL Mask Option," *IEEE ISSCC, Tech. Dig.*, pp. 212–213, 1992.

[86] N. Tamba et al., "A 1.5 ns 256-kbit BiCMOS SRAM with 11 k 60 ps Logic Gates," *IEEE ISSCC, Tech. Dig.*, pp. 246–247, 1993.

[87] K. Nakamura et al., "A 220 MHz Pipelined 16-Mbit BiCMOS SRAM with PLL Proportional Self-Timing Generator," *IEEE ISSCC, Tech. Dig.*, pp. 258–259, 1994.

[88] F. A. Nofel et al., "A Three Million Transistor Microprocessor," *IEEE ISSCC, Tech. Dig.*, pp. 108–109, 1992.

[89] O. Nishii et al., "A 1000 MIPS BiCMOS Microprocessor with Superscalar Architecture," *IEEE ISSCC, Tech. Dig.*, pp. 114–115, 1992.

[90] S. Tanaka et al., "A 120 MHz BiCMOS Superscalar RISC Processor," *Symp. VLSI Circuits, Tech. Dig.*, pp. 9–10, 1993.

[91] A. Saini, "Design of the Intel Pentium Processor," *IEEE Int. Conf. Computer Design, Tech. Dig.*, pp. 258–261, 1993.

[92] J. Schutz, "A 3.3 V 0.6 µm BiCMOS Superscalar Microprocessor," *IEEE ISSCC, Tech. Dig.*, pp. 202–203, 1994.

[93] T. Y. Chiu et al., "Non-Overlapping Super Self-Aligned BiCMOS with 87 ps Low-Power ECL," *IEDM, Tech. Dig.*, pp. 752–755, 1988.

[94] M. P. Brassington et al., "An Advanced Single-Level Polysilicon Submicrometer BiCMOS Technology," *IEEE Trans. Electron Devices*, Vol. 36, No. 4, pp. 712–719, 1989.

[95] N. Rovedo et al., "Process Design for Merged Complementary BiCMOS," *IEDM, Tech. Dig.*, pp. 485–488, 1990.

[96] E. D. Johnson et al., "A High-Performance 0.5 µm BiCMOS Technology with 3.3 V CMOS Devices," *Symp. VLSI Technology, Tech. Dig.*, pp. 89–90, 1990.

[97] W. R. Burger, C. L. Lage, B. Landau, M. Delong, and J. Small, "An Advanced 0.8 µm Complementary BiCMOS Technology for Ultra-High Speed Circuit Performance," *BCTM, Proc.*, pp. 78–81, 1990.

[98] A. Chantre et al., "An Investigation of Non-Ideal Base Currents in Advanced Self-Aligned "Etched-Polysilicon" Emitter Bipolar Transistors," *IEEE Trans. Electron Devices*, Vol. 38, No. 6, pp. 1354–1361, 1991.

CHAPTER 3

MOS Device Modeling

The main objective of this chapter is to present and evaluate the main models of the MOS device, highlight the assumptions made in developing each model, and accurately describe its performance in a scaled technology environment. The continuous device miniaturization process justifies the need to understand the performance of ultra-small devices. Such an understanding is necessary to fully exploit these devices in the optimum circuit design for future VLSI/ULSI. For this reason, much coverage is given in this chapter to the analytical and experimental characterization of sub-half micron MOS devices. A methodology has been presented to show how to construct a device model for a given wafer, with the aid of device characterization tools and successfully retrieved experimental data.

Recently, hybrid-mode devices, employing lateral pnp bipolar transistors in a pMOSFET structure, have been able to capitalize on the advantages of both types of devices, that is, the lateral bipolar device associated with the MOS structure. This device structure has been used frequently in CMOS/BiCMOS circuit design for low-voltage, low-power applications [1,2]. In order to understand the characteristics of this structure, it is necessary for us to devote a good part of this chapter to covering the trends and general features of the various device/process parameters of scaled pMOSFETs operating in a hybrid-mode environment.

3.1 The Threshold Voltage Models

The threshold voltage of an MOS device is defined as the gate-source voltage at which the surface potential ϕ_s is equal to $2|\phi_f|$. This condition defines the channel strong inversion and separates the two states of the MOSFET operation. The threshold voltage of a MOSFET can be given by:

$$V_{TO} = V_{FB} + \phi_s + \frac{\theta}{C_{OX}}\sqrt{2q\varepsilon_{si}N_{sub}\phi_s} \qquad 3.1$$

where V_{TO} is the threshold voltage at zero source-drain voltage and zero source-bulk voltage. θ is modeled by extending the classical threshold voltage expression using the charge-sharing model [3,4] to account for the reduction of its value when the channel length is reduced.

If the bulk-source is biased, the threshold voltage can be rewritten as:

$$V_T = V_{TO} + f(V_{BS}) \qquad 3.2$$

The function $f(V_{BS})$ models the effects of nonzero V_{BS} in general. These effects vary according to the kind of substrate bias, its doping concentration profile, and the scaled technology. If the bulk-source junction is reverse-biased, $f(V_{BS})$ becomes:

$$f(V_{BS}) = \gamma_B \left(\sqrt{|V_{BS}|+2|\phi_f|} - \sqrt{2|\phi_f|}\right) \qquad 3.3a$$

where the body effect coefficient γ_B is given by:

$$\gamma_B = \frac{\sqrt{2q\varepsilon_{si}N_{sub}}}{C_{OX}} \qquad 3.3b$$

Scaling the technology adds, in general, modulating effects linking the device potentials to its threshold voltage. The semi-empirical short-channel model (SPICE LEVEL 3) accounts for the short- and narrow-channel effects (L ≤ 2 µm, W ≤ 2 µm) including the Drain-Induced Barrier Lowering (DIBL) effect. The threshold voltage takes the form [5]:

$$V_T = V_{FB} + 2|\phi_f| - \eta V_{DS} + \gamma_B F_S \sqrt{|V_{BS}|+2|\phi_f|} + F_N(|V_{BS}|+2|\phi_f|) \qquad 3.4$$

γ_B in this expression is given by Eq. 3.3b. Equation 3.4 includes the static feedback effect coefficient η (due to the DIBL effect) [6]:

$$\eta = \omega \frac{8.15 \times 10^{-22}}{C_{OX} L_{eff}^3} \qquad 3.4a$$

where ω is an empirical coefficient. The correction factor for short-channel effects (derived based on a trapezoidal-shaped charge under the channel) can be obtained from [7]:

$$F_S = 1 - \frac{X_j}{L_{eff}}\left[\frac{L_d+W_c}{X_j}\left\{1-\left(\frac{W_D}{X_j+W_D}\right)^{\frac{1}{2}}\right\} - \frac{L_d}{X_j}\right] \qquad 3.4b$$

The Threshold Voltage Models

where W_c, the depletion layer width of a cylindrical junction, is given by:

$$W_c = 0.0831353 + 0.8013929 \frac{W_D}{X_j} - 0.0111077 \left(\frac{W_D}{X_j}\right)^2 \qquad 3.4c$$

The correction factor for a narrow-channel MOS is given by:

$$F_N = 2\pi\varepsilon_{si} \frac{\delta}{4C_{OX} W_{eff}} \qquad 3.4d$$

where δ is the narrow width factor for adjusting the threshold voltage.

BSIM (Berkeley Short-Channel IGFET Model) is another short-channel model [8]. It accounts for, among other short-channel effects, the nonuniform doping in the channel surface and subsurface regions effect. According to the model, the threshold voltage is given by:

$$V_T = V_{FB} + \phi_s + K_1\sqrt{|V_{BS}| + \phi_s} - K_2(|V_{BS}| + \phi_s) - \eta' V_{DS} \qquad 3.5$$

The parameters K_1 and K_2, model the effect of nonuniform doping of the substrate on the threshold voltage. The factor η' models both the DIBL and the channel length modulation effects.

Although they accurately predict the threshold voltage for channel lengths down to 1 μm, the above models fail to account for the reduction in the threshold voltage for deep-submicrometer MOSFETs. The experimental data show that in the very short-channel length range, the previously reported exponential dependence of the threshold voltage shift, ΔV_T, on the channel length [9,10] and the linear dependence on the drain voltage no longer hold true.

A short-channel threshold voltage model for both Lightly Doped Drain (LDD) and non-LDD MOSFETs has been reported [11]. The model is derived using a quasi-two-dimensional approach and can predict the accelerated V_T reduction observed in the very short-channel range. V_T is given by:

$$V_T = V_{TO} - \frac{[2(V_{bi} - \phi_s) + V_{DS}]}{2\cosh\left(\frac{L}{2l_c}\right) - 2}$$

$$= V_{TO} - \Delta V_T \qquad 3.6$$

where l_c is the characteristic length defined, in terms of the physical parameters and a fitting parameter m, as:

$$l_c = \sqrt{\frac{\varepsilon_{si} T_{OX} X_{dep}}{\varepsilon_{OX} m}} \qquad 3.6a$$

For $l_c \ll L$, and relatively small values of V_{DS}, the threshold voltage shift ΔV_T is given by [11]:

$$\Delta V_T = [2(V_{bi} - \phi_s) + V_{DS}]\left(e^{-\frac{L}{2l_c}} + 2e^{-\frac{L}{l_c}}\right) \qquad 3.6b$$

The latest SPICE model that caters for short-channel devices is the BSIM3v3 model [12]. This model developed in 1995 takes into account the effects of short- and narrow-channel effects, as well as lateral and vertical nonuniform doping effects on the threshold voltage. The threshold voltage of this model is given by [12]:

$$V_T = V_{FB} + \phi_s + K_1\sqrt{\phi_s} + K_1\left(\sqrt{\phi_s - |V_{BS}|} - \sqrt{\phi_s}\right)$$
$$- K_2 |V_{BS}| + K_1\left(\sqrt{1 + \frac{N_{LX}}{L_{eff}}} - 1\right)\sqrt{\phi_s} - \Delta V_T \qquad 3.7$$

The first three terms represent the ideal threshold voltage of a long channel length and width device with uniform substrate doping. The fourth and fifth terms account for the vertical nonuniform doping effect, whereas the sixth term models the lateral nonuniform doping effect on the threshold voltage. N_{LX} is the lateral nonuniform doping parameter. ΔV_T contains numerous complex terms to account for DIBL, charge-sharing and narrow width effects on the threshold voltage.

Hybrid-mode devices employing lateral pnp BJT in a pMOS structure have received much attention recently [13–19] due to their high current gain and simple technology. A threshold voltage model for deep-submicron LDD MOSFET in a Bi-MOS hybrid-mode environment has been reported [19]. The proposed model made it possible to evaluate the effects of independently biasing the source, drain, gate, and body on the device performance. The threshold voltage takes the form [19]:

$$V_T = V_{TO} - \eta V_{SDeff} - \gamma V_{SBeff} \qquad 3.8$$

where V_{SDeff} and V_{SBeff} are defined to take into account the voltage drop in the LDD region [11]. The introduction of η and γ enables the effects of both the drain-induced barrier lowering (DIBL) and the forward source-body bias on the threshold voltage to be studied separately. Both η and γ can be obtained experimentally and modeled analytically to describe the threshold voltage reduction for a specific technology and process parameters. It has been shown that η, fairly constant with V_{SD}, is sensitive to the channel length. Forward biasing the source-body has been shown to affect the threshold voltage through: the channel length modulation, the reduction of the depletion layer thickness under the gate, and the lowering of the source-body barrier height, which triggers hole injection into the channel region.

3.2 The MOSFET Current Models

The MOS device possesses two primary regions of operation, which are (1) the subthreshold or weak inversion region and (2) the strong inversion region. The former occurs when the gate voltage is less than the device threshold voltage, and the latter occurs when the gate voltage is greater than the threshold voltage. The current flow in the first region is due to diffusion of carriers whereas in the second region, carriers drift across from the source to the drain. The zone where the subthreshold region and the strong inversion region meet is termed the transition region. Figure 3–1 shows pictorially the two primary regions of operation and the transition zone.

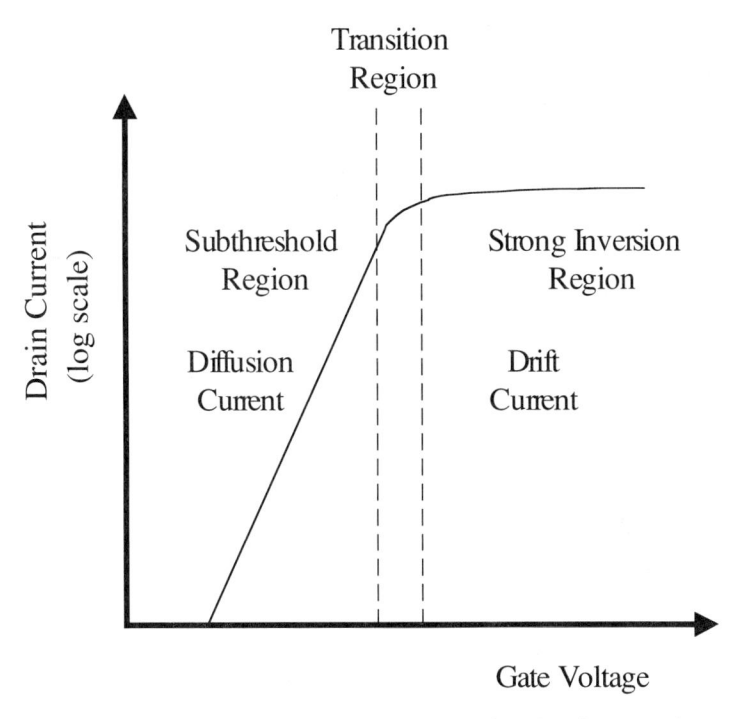

Figure 3–1 Plot of drain current against the gate voltage showing the two primary regions of MOSFET operation and the transition region where both diffusion and drift currents co-exist.

Long-channel MOSFETs behavior is usually modeled by one-dimensional analysis using *Gradual-Channel Approximation* (GCA). This approximation assumes that (1) the vertical gate field induces the formation of the channel and (2) the drain-to-source voltage acts in the horizontal direction to drive the mobile charges, thus constituting the source-to-drain current. This assumption of two independent one-dimensional effects permits the results of the two-terminal MOS capacitor (MOS-C) analysis [20] to be employed when deriving the I-V equations of long-channel MOSFETs.

SPICE provides a few levels of MOSFET models for circuit simulation. The Level-1 MOSFET model is often used when accuracy is less important than computational time. This GCA-based quadratic model, also known as the Shichman and Hodges model [21], takes into account channel length modulation. The equations for Level-1 nMOSFET model are as follows:

Cutoff region, $V_{GS} < V_T$

$$I_{DS} = 0 \qquad 3.9a$$

Linear region, $V_{GS} \geq V_T$ and $(V_{DS} < V_{GS} - V_T)$

$$I_{DS} = \frac{k'}{2} \frac{W}{L_{eff}} \left[2(V_{GS} - V_T)V_{DS} - V_{DS}^2 \right](1 + \sigma V_{DS}) \qquad 3.9b$$

Saturation region, $V_{GS} \geq V_T$ and $V_{DS} \geq (V_{GS} - V_T)$

$$I_{DS} = \frac{k'}{2} \frac{W}{L_{eff}} (V_{GS} - V_T)^2 (1 + \sigma V_{DS}) \qquad 3.9c$$

where the expression for V_T is as given in Eqs. 3.2, 3.3a, and 3.3b, σ represents the channel length modulation effect factor, and $k' = \mu \frac{\varepsilon_{ox}}{T_{ox}} = \mu C_{ox}$.

The weaknesses of the Level-1 MOSFET model include: (1) it loses on accuracy as the GCA adopted in the derivation of the model equations breaks down for L < 2 µm; (2) the number of fitting parameters is too small; and (3) there is no equation for subthreshold conduction.

Level-2 MOSFET model, which is more accurate, eliminates some of the simplifying assumptions of GCA and takes into account the bulk depletion charge in modeling the device characteristic. The model calculates the drain current of MOS devices based on the two primary regions of operation, that is,

Cutoff region, $V_{GS} < V_T$

$$I_{DS} = I_{DS(ON)} e^{(V_{GS} - V_{ON})\left(\frac{q}{nkT}\right)} \qquad 3.10a$$

where $I_{DS(ON)}$ is the current in strong inversion for $V_{GS} = V_{ON}$ and

$$V_{ON} = V_T + \frac{nkT}{q} \qquad 3.10b$$

$$n = 1 + \frac{qN_{FS}}{C_{ox}} + \frac{C_d}{C_{ox}} \qquad 3.10c$$

The MOSFET Current Models

Turn-on region, $V_{GS} \geq V_T$

$$I_{DS} = \frac{k'}{(1-\sigma V_{DS})} \frac{W}{L_{eff}} \left\{ \left(V_{GS} - V_{FB} - |2\phi_f| - \frac{V_{DS}}{2} \right) V_{DS} \right. \\ \left. - \frac{2}{3}\chi \left[(V_{DS} - V_{BS} + |2\phi_f|)^{1.5} - (-V_{BS} + |2\phi_f|)^{1.5} \right] \right\} \quad \text{3.10d}$$

where the factor χ models the short channel effect.

Equation 3.10a describes the subthreshold conduction (the region where carriers diffuse from the source to the channel region). The on-voltage, V_{ON}, is the threshold potential above which the device enters the strong inversion region and below which the device operates in the weak inversion region. The parameter N_{FS} indicates the number of fast superficial states and acts as a fitting parameter, which determines the slope of the subthreshold I-V curve. C_d represents the depletion capacitance.

Level-2 MOSFET model takes into account three short-channel effects. First, unlike in the Level-1 model, where the surface carrier mobility has been taken to be a constant, Level-2 model takes into consideration the degradation of the mobility of carriers with increasing gate voltage. The effective mobility of the carriers is given by:

$$\mu_{eff} = \mu_o \left(\frac{\varepsilon_{si}}{C_{OX}} \times \frac{E_{crit}}{V_{GS} - V_T - U_1 V_{DS}} \right)^{U_2} \quad \text{3.11}$$

where E_{crit} represents the gate-to-channel critical field, U_1 represents the contribution of the drain voltage to the gate-to-channel field and U_2 is an exponential fitting parameter. Second, Level-2 MOSFET model accounts for the channel length modulation effect by using the empirical parameter, σ (similar to that of the Level-1 model). However if σ is not specified in the .MODEL statement, the model will calculate σ as follows:

$$\sigma = \frac{\Delta L}{L_{eff} V_{DS}} \quad \text{3.12a}$$

where:

$$\Delta L = \sqrt{\frac{2\varepsilon_{si}}{qN_{sub}}} \left\{ \frac{V_{DS} - V_{Dsat}}{4} + \sqrt{1 + \left(\frac{V_{DS} - V_{Dsat}}{4} \right)^2} \right\} \quad \text{3.12b}$$

V_{Dsat} is the saturation voltage required for the pinch-off of the channel at the drain side and is given by:

$$V_{Dsat} = V_{GS} - V_{FB} - |2\phi_f| + \chi^2 \left\{ 1 - \sqrt{1 + \frac{2}{\chi^2}(V_{GS} - V_{FB})} \right\} \quad \text{3.12c}$$

The calculation of V_{Dsat} in Eq. 3.12c is done with the assumption that the channel charge near the drain drops to zero when the device enters saturation. However, in actual fact, the charge concentration there must be greater than zero in order to sustain the saturation current. Furthermore, the carriers in the channel will reach their saturation velocity, v_{sat}, before the channel charge approaches zero. If v_{sat} is specified in the MODEL statement, then the amount of channel length shortening, ΔL will be computed as follows, rather than that stated in Eq. 3.12b:

$$\Delta L = X_d \left\{ \sqrt{\left(\frac{X_d v_{sat}}{2\mu_{eff}}\right)^2 + V_{DS} - V_{Dsat}} - \frac{X_d v_{sat}}{2\mu_{eff}} \right\} \quad \text{3.13a}$$

where:

$$X_d = \sqrt{\frac{2\varepsilon_{si}}{qN_{sub}N_{eff}}} \quad \text{3.13b}$$

N_{eff} is a fitting parameter.

Level-2 MOSFET model has incorporated modifications to cater for several short-channel effects and is capable of computing the subthreshold current. However, the model requires a large amount of CPU computational time due to the 1.5 power terms in the strong inversion current equation (see Eq. 3.10d). Furthermore, there will occasionally be convergence problems in the Newton-Raphson algorithm used in SPICE.

Level-3 MOSFET model was developed using a semi-empirical approach to deal with short- and narrow-channel effects. It was also formulated to overcome the computational disadvantage that Level-2 MOSFET model faces by simplifying the current equations to a more manageable level. The majority of the Level-3 model equations are empirical.

The subthreshold current model in the Level-3 model is similar to the one used in Level-2. The strong inversion current equation is given by:

$$I_{DS} = \mu_{eff} C_{OX} \frac{W}{L_{eff}} \left(V_{GS} - V_T - \frac{1+F_B}{2} V_{DS} \right) V_{DS} \quad \text{3.14a}$$

where the parameter F_B has the form:

$$F_B = \frac{\gamma_B F_S}{4} \left(|2\phi_f| + V_{BS} \right)^{-\frac{1}{2}} + F_N \quad \text{3.14b}$$

The empirical parameter F_B describes the dependence of the bulk depletion charge on the geometrical considerations of the MOSFET. F_S is related to alteration of the charge distribution in the bulk region for short-channel MOSFETs, while F_N is the correction factor for narrow-channel MOSFETs.

The MOSFET Current Models

The carrier mobility reduction due to the normal field is modeled in Level-3 by the following:

$$\mu_s = \frac{\mu_o}{1+\xi(V_{GS}-V_T)} \qquad 3.15a$$

where ξ is the mobility degradation factor.

Furthermore, Level-3 MOSFET model also accounts for the effect of the lateral electric field, when the user specifies the maximum drift velocity of carriers, v_{sat}, in the .MODEL statement, with the following equation:

$$\mu_{eff} = \frac{\mu_s}{1+\mu_s \dfrac{V_{DS}}{v_{sat} L_{eff}}} \qquad 3.15b$$

The saturation voltage takes into consideration the velocity saturation effect and is given by:

$$V_{Dsat} = V_e + V_c - \sqrt{V_e^2 + V_c^2} \qquad 3.16a$$

$$V_e = \frac{V_{GS}-V_T}{1+F_B} \qquad 3.16b$$

$$V_c = \frac{v_{sat} L_{eff}}{\mu_s} \qquad 3.16c$$

The channel length modulation effect model is slightly different from that in Level-2 MOSFET model. Level-3 model determines the channel length reduction by using two different equations, which in turn depend on whether v_{sat} is specified in the MODEL statements. The equations are:

$v_{sat} = 0$

$$\Delta L = \sqrt{\frac{2\varepsilon_{si}}{qN_{sub}}} \sqrt{K_p(V_{DS}-V_{Dsat})} \qquad 3.17a$$

where K_p is the saturation field factor.

$v_{sat} > 0$

$$\Delta L = \frac{E_p \varepsilon_{si}}{qN_{sub}} + \sqrt{\left(\frac{E_p \varepsilon_{si}}{qN_{sub}}\right)^2 + K_p \frac{2\varepsilon_{si}}{qN_{sub}}(V_{DS}-V_{Dsat})} \qquad 3.17b$$

E_p represents the lateral electric field at the pinch-off point and is approximated by:

$$E_p = \frac{V_c(V_c + V_{Dsat})}{L_{eff} V_{Dsat}} \qquad 3.17c$$

Level-3 MOSFET model achieves approximately the same level of accuracy as that of the Level-2 model but the computational time required is much less and the number of iterations required are far fewer. However, as Level-3 model approaches the modeling process from a semi-empirical manner, whereby the model employs approximate device physics and depends on the proper choice of the empirical parameters to match the experimental data, a more physically-based short- and narrow-channel MOSFET model is still required. This requirement has been met with the development of the BSIM model in 1987.

The BSIM model [8] was formulated based on the device physics of small-geometry MOS transistors, and it takes care of both the weak and strong inversion regimes. Its features include:

1. Vertical field dependence of carrier mobility
2. Carrier velocity saturation
3. Drain-induced barrier lowering
4. Charge sharing by source and drain
5. Channel length modulation
6. Channel narrowing
7. Subthreshold conduction
8. Nonuniform doping profile for ion-implanted devices
9. Geometric dependencies

BSIM has approximately 60 parameters. BSIM2 [22,23], introduced in 1990, is an extension of the BSIM model. It has a parameter count of 120. In retrospect, it is generally felt that while Levels 1–3 MOSFET models have too few parameters, BSIM and BSIM2 possess too many. Hence a model with a moderate number of parameters is required. BSIM3 with approximately 38 parameters succeeded in doing just that.

The latest BSIM3v3 [12] model has been developed based upon finding solutions to Poisson's equation using GCA and Quasi-Two Dimensional Approximation. This model has the following features:

1. Threshold voltage roll-off
2. Nonuniform doping effect (in both lateral and vertical directions)
3. Mobility reduction due to vertical field
4. Bulk charge effect
5. Carrier velocity saturation
6. Drain-Induced Barrier Lowering (DIBL)
7. Channel length modulation
8. Substrate Current induced Body Effect (SCBE)

The MOSFET Current Models

9. Subthreshold conduction
10. Source/drain parasitic resistances
11. Narrow width effect

One of its enhancements, as compared to previous BSIM version, is a unified I-V model to describe the current and output conductance characteristics from subthreshold to strong inversion regions, as well as from linear to saturation regimes. This ensures that the current, conductance, and their derivatives are continuous in all transition regions. As a result, the convergence and computation efficiencies are greatly improved.

To enable the formulation of a unified I-V expression, an effective (V_{GS}-V_T) function, V_{GSteff} has been introduced to describe the channel charge characteristics from subthreshold to strong inversion:

$$V_{GSteff} = \frac{2n_s V_t \ln\left[1 + \exp\left(\frac{V_{GS} - V_T}{2n_s V_t}\right)\right]}{1 + 2n_s C_{OX}\sqrt{\frac{2\phi_s}{q\varepsilon_{si} N_{sub}}} \exp\left(-\frac{V_{GS} - V_T - 2V_{off}}{2n_s V_t}\right)} \qquad 3.18$$

where V_t is the thermal voltage, V_{off} is the offset voltage [10], and n_s is the subthreshold swing parameter.

BSIM3v3 uses a unified mobility model based on the expression for V_{GSteff} given by:

$$\mu_{eff} = \frac{\mu_o}{1 + (U_a + U_c V_{bseff})\left(\frac{V_{GSteff} + 2V_T}{T_{OX}}\right) + U_b\left(\frac{V_{GSteff} + 2V_T}{T_{OX}}\right)^2} \qquad 3.19$$

V_{bseff} has been used to set an upper bound for the body bias value during simulations. U_a, U_b, and U_c represent the first-, second-, and third-order mobility degradation coefficients respectively. Unified expressions for the linear and saturation drain current from the subthreshold to the strong inversion region have been derived separately and then linked together. The overall current equation for both the linear and saturation current valid from the subthreshold to the strong inversion region is given as:

$$I_{DS} = \frac{I_{dso}(V_{DSeff})}{1 + \frac{R_{ds} I_{dso}(V_{DSeff})}{V_{DSeff}}} \left(1 + \frac{V_{DS} - V_{DSeff}}{V_A}\right)\left(1 + \frac{V_{DS} - V_{DSeff}}{V_{ASCBE}}\right) \qquad 3.20$$

where

$$V_{DSeff} = V_{Dsat} - \frac{1}{2}\left(V_{Dsat} - V_{DS} - \varsigma + \sqrt{(V_{Dsat} - V_{DS} - \varsigma)^2 + 4\varsigma V_{Dsat}}\right) \qquad 3.20a$$

$$V_{Dsat} = \frac{E_{crit} L (V_{GSteff} + 2V_t)}{A_{bulk} E_{crit} L + V_{GSteff} + 2V_t} \qquad 3.20b$$

$$I_{dso}(V_{DSeff}) = \frac{W_{eff}\mu_{eff}C_{OX}V_{GSteff}\left(1 - A_{bulk}\dfrac{V_{DSeff}}{2(V_{GSteff} + 2V_t)}\right)V_{DSeff}}{L_{eff}\left[1 + \dfrac{V_{DSeff}}{E_{crit}L_{eff}}\right]} \qquad 3.20c$$

E_{crit} corresponds to the critical electrical field at which the carrier velocity becomes saturated, A_{bulk} is the parameter modeling the bulk charge effect, ς is an extracted constant, V_A represents the Early Voltage at $V_{DS} = V_{Dsat}$, and V_{ASCBE} is the Early Voltage due to the substrate current-induced body effect.

An alternative engineering model for short-channel MOSFETs has been developed by K.Y. Toh et al. [24]. This model provides a simple closed-form I-V relationship between the terminal voltages and the drain current, and it was formulated based on the piecewise carrier drift velocity model [25]. This model provides drain current equations for $V_{GS} > V_T$; however, it does not account for subthreshold conduction. The drain current in the ohmic and saturation regions is given by

$V_{GS} \geq V_T$ and $0 < V_{DS} < V_{Dsat}$

$$I_{DS} = \frac{\mu_{eff}C_{OX}W}{L_{eff}} \frac{1}{1 + \dfrac{V_{DS}}{E_{crit}L_{eff}}} \left(V_{GS} - V_T - \frac{1}{2}V_{DS}\right)V_{DS} \qquad 3.21a$$

$V_{GS} \geq V_T$ and $V_{DS} \geq V_{Dsat}$

$$I_{DS} = \frac{1}{1 + \dfrac{E_{crit}L_e}{V_{GS} - V_T}} v_{sat}C_{OX}W(V_{GS} - V_T) \qquad 3.21b$$

where

$$L_e = L_{eff} - X_d \qquad 3.21c$$

$$X_d = \frac{1}{A}\ln\left[\frac{A(V_{DS} - V_{Dsat}) + E_d}{E_{crit}}\right] \qquad 3.21d$$

$$\frac{E_d}{E_{crit}} = \sqrt{1 + \left[\frac{A(V_{DS} - V_{Dsat})}{E_{crit}}\right]^2} \qquad 3.21e$$

$$A^2 = \frac{\varepsilon_{ox}}{\varepsilon_{si}}\frac{1.5}{X_j T_{OX}} \qquad 3.21f$$

The MOSFET Current Models

$$V_{Dsat} = \left(1 - \frac{1}{1 + \frac{E_{crit}L_e}{V_{GS} - V_T}}\right)(V_{GS} - V_T) \qquad 3.21g$$

This model is relatively simple for computer simulation, and it has not compromised on the device physics. However, its weaknesses are, first, it fails to cover the entire range from weak to strong inversion, and second, there will be convergence problems due to the piecewise current equations.

There has not been a complete SPICE current model for hybrid-mode devices until recently. The dc model [19] considers the different current components that flow with varying terminal voltage biases and addresses each of them separately. Basically, when an MOS device is operating in the hybrid mode, its drain current consists of three components, namely, (1) the space-charge current, (2) the lateral bipolar current, and (3) the principal MOS current.

The space-charge current of a pMOSFET that flows under high drain potential and with no gate bias has been modeled as [19,26]:

$$I_{space} = \frac{\alpha A_{space} \varepsilon_{si} \mu_p V_{SDeff}^2}{L_{effav}^3}, \qquad \frac{\mu_p V_{SDeff}}{L_{effav}} < \nu_{sat} \qquad 3.22a$$

$$I_{space} = \frac{\alpha A_{space} \varepsilon_{si} \nu_{sat} V_{SDeff}}{L_{effav}^2}, \qquad \frac{\mu_p V_{SDeff}}{L_{effav}} \geq \nu_{sat} \qquad 3.22b$$

α is a fitting parameter that is process dependent. L_{effav} is the average sum of the effective channel lengths taken near the Si/SiO$_2$ interface and at the extreme of the gate depletion region.

The lateral bipolar current that flows in very short-channel MOS devices under high-forward source-body bias condition cannot be neglected. This has been modeled as [19]:

$$I_{bipolar} = \frac{q\mu_p V_t P_e A_{bipolar}}{L_e(1-\zeta)}(1 - e^{-\frac{\zeta L_e \nu_{sat}}{D_p}}) \qquad 3.23$$

where P_e is the hole concentration at the edge of the source/emitter of the Bi-MOS structure and ζ is a fitting parameter.

The principal MOS current equations of a p-channel device operating in a hybrid-mode environment (adapted from K.Y. Toh et al. [24] described previously) is given by:

$V_{SG} \geq V_T$ and $V_{SD} < (V_{SG} - V_T)$

$$I_{SD} = \frac{\mu_{peff} C_{OX} W}{L_{eff}} \cdot \frac{1 + \lambda V_{SBeff}}{1 + \frac{V_{SDeff}}{E_{crit}L_{eff}}}\left(V_{SG} - V_T - \frac{1}{2}V_{SDeff}\right)V_{SDeff} \qquad 3.24a$$

$V_{SG} \geq V_T$ and $V_{SD} \geq (V_{SG} - V_T)$

$$I_{SD} = \frac{1 + \lambda V_{SBeff}}{1 + \frac{E_{crit}L_d}{V_{SG} - V_T}} v_{sat} C_{OX} W(V_{SG} - V_T)$$ 3.24b

where L_d is the electrical channel length for the MOS current.

The effective carrier mobility, the carrier velocity saturation, the channel length modulation, and the parasitic source and drain series resistances have been accounted for in the MOS current model. The factor $(1 + \lambda V_{SBeff})$ has been introduced to account for the increase in hole injection into the channel region due to the lowering of the barrier height. More in-depth discussions of the space-charge, lateral bipolar, and MOS current models will be given in subsequent sections.

3.3 The MOSFET in a Hybrid Mode Environment

It is well known that bipolar transistors exhibit lower noise and higher speed performance as compared to MOSFETs. However, MOS devices have lower power dissipation, higher packing density and can operate at very low currents. Hybrid-mode devices, employing lateral pnp BJT in a pMOS structure, have been able to capitalize on the advantages of both types of devices.

Scaling the MOS device to deep submicrometer is the main trend for present and future ULSI systems. Hence improvement in circuit performance employing hybrid-mode devices will be very significant, because the lateral bipolar transistor, generic to the CMOS process, exhibits higher current gain for smaller channel lengths.

An analytical dc model [19] encompassing the threshold voltage, the device currents, the transconductance, and the output conductance has already been derived for submicrometer LDD pMOSFETs (device drawn lengths of 0.4 μm to 0.7 μm) operating in the hybrid-mode environment. In this chapter, we will discuss the modifications made to extend the analytical dc model down to quarter-micron devices, and highlight the effects of temperature variations on the hybrid-mode device characteristics. The main motivation for this work is that, according to the Semiconductor Industry Association (SIA) roadmap of 1994, the 0.25 μm generation will be the mainstream technology from the year 1998 onwards. Moreover, our present findings have shown that with the rapid scaling of channel length at elevated temperatures under high-level injection and at large source-drain bias, the secondary bipolar current component becomes comparable to the principal MOS current component.

The hybrid-mode pMOS device used in our discussion is schematically shown in Figure 3–2.

Essentially, this device has four electrodes, whereby the source, body, and drain terminals of the pMOSFET correspond to the emitter, base, and collector of the lateral BJT device respectively. When the source-body of the pMOS device is forward biased, the total drain current com-

The MOSFET in a Hybrid Mode Environment 97

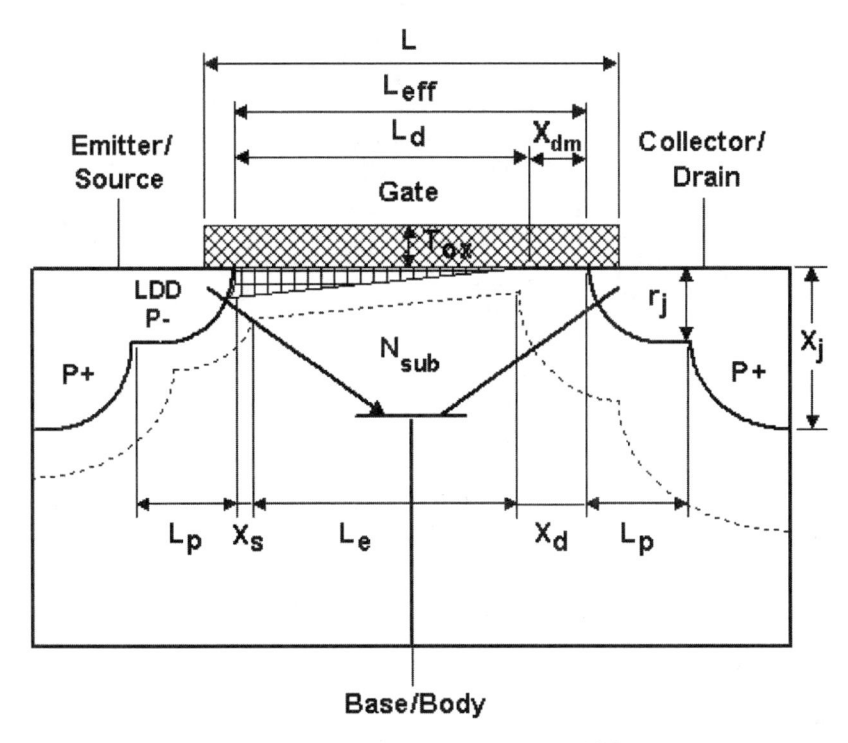

Figure 3-2 Cross-sectional view of the hybrid-mode LDD pMOS device.

prises a surface MOS current component, and two bulk components, which are the lateral bipolar current and the space-charge current. Figure 3–3 illustrates these current components.

3.3.1 Surface P-Channel for Sub-Half Micron Devices

Historically, heavily-doped n-type polysilicon has been used as the gate electrode for CMOS devices. However, with the scaling of devices down to sub-half micrometer and below, the fabrication of p-channel devices with n-doped polysilicon gate gives rise to severe short-channel effects.

First, the submicrometer p-channel devices are highly susceptible to punch-through effects. This vulnerability is caused in part by the boron impurity that is used to form the source and drain regions. Boron, known to diffuse much more rapidly in silicon than phosphorus or arsenic (dopants used to form the n+ source and drain junctions), creates deeper p+ source and drain junctions. The deeper junctions accentuate the devices' punch-through susceptibility. Second, the threshold voltage roll-off is much more severe in n+ polysilicon gate devices than in p+ polysilicon gate devices. Third, n+ polysilicon gate devices exhibit larger subthreshold swing, which implies that they are more susceptible to leakage currents in the off state.

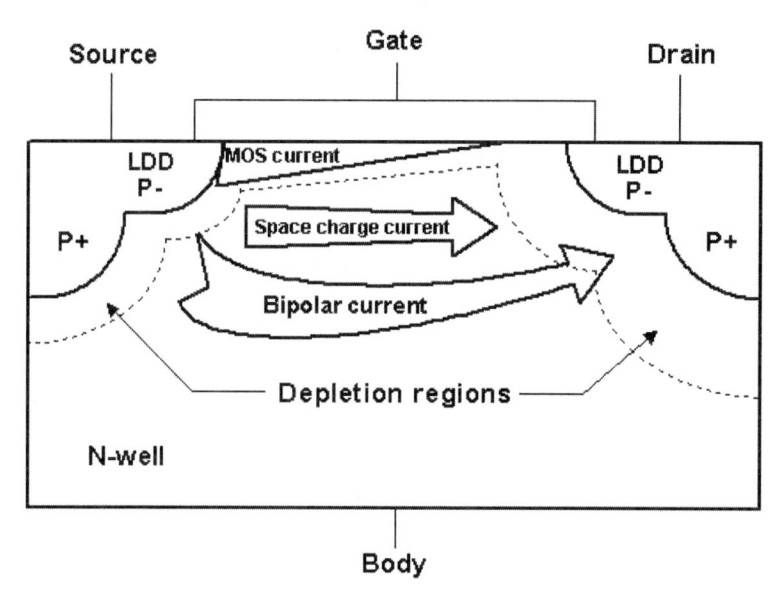

Figure 3–3 The current components flowing in the pMOS device during the hybrid-mode operation.

The latter two short-channel effects are related to the threshold adjust implant dopant used for n+ polysilicon gate devices. To comprehend this, we first compare the band diagrams of n+ and p+ polysilicon gate devices, as shown in Figure 3–4.

From Figure 3–4, it is evident that due to the negative flat-band voltage, the overall threshold voltage for the n+ polysilicon p-channel device becomes more negative, whereas that for the p+ polysilicon p-channel device becomes less negative. Therefore, the threshold voltage of the n+ polysilicon p-channel device becomes higher than desired, whereas that of the p+ polysilicon p-channel device drops lower than desired. Thus in order to adjust the threshold potential to the appropriate level, a p-type dopant (boron) is used as the p-channel threshold adjust implant for n+ polysilicon gate devices and n-type dopant (phosphorus) is used for p+ polysilicon gate devices. Implanting boron atoms near the surface of a n+ polysilicon gate device results in a severe shortcoming, that is, boron (being light) forms not an infinitely thin sheet of charge at the surface as one would expect ideally, but a layer of finite thickness. The lightly-doped implanted p-layer extends across from the p+ source to the drain near the surface.

It has been found [28] that with n+ polysilicon gate and boron as the p-channel threshold adjust implant, the potential minimum will be shifted away from the silicon surface to some distance beneath the surface. Mobile carriers, which are always found concentrated near the proximity of the potential minimum, will now flow in the channel formed beneath the surface of the device. Hence n+ polysilicon gate devices are also known as buried-channel devices. For p+ polysilicon gate devices, since the threshold adjust implant and the channel dopants are of the same species, the potential minimum remains near the surface, therefore they are often called surface-channel devices.

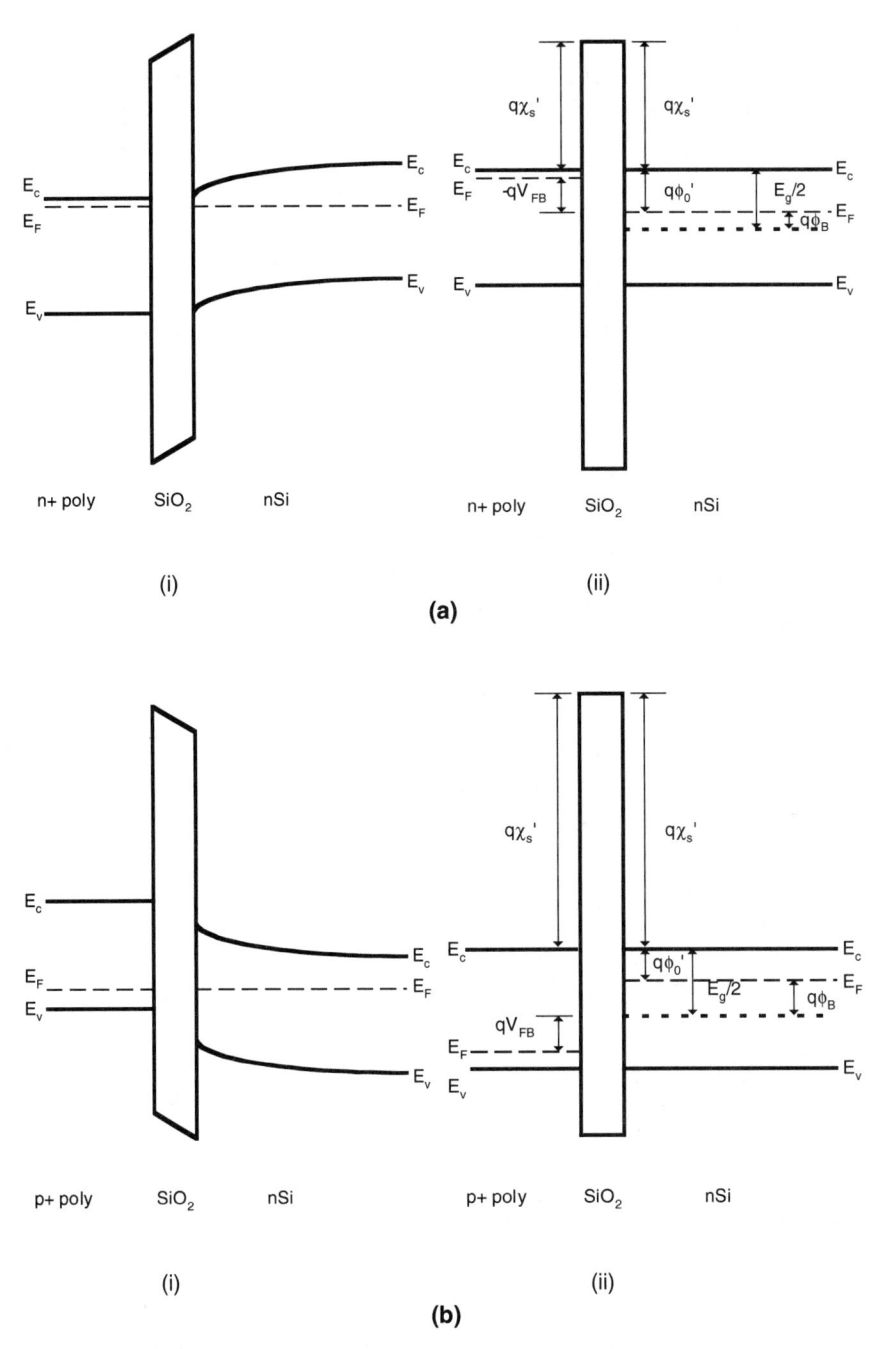

Figure 3–4 (adopted from [27]) (a) Energy-band diagrams of n+ polysilicon gate on an n-type substrate for: (i) $V_{GB} = 0$ V; (ii) flat-band condition. (b) Energy-band diagrams of p+ polysilicon gate on an n-type substrate for: (i) $V_{GB} = 0$ V; (ii) flat-band condition.

In retrospect, with the rapid scaling of transistors, the use of buried p-channel devices becomes less feasible due to severe short-channel effects. The employment of surface p-channel transistors would be the alternative trend. This leads to what is known as the dual polysilicon gate approach. However, circuit designers may still be willing to sacrifice leakage problems for speed by using buried p-channel devices, where the mobility of the carriers is higher as compared to their surface counterparts [28].

3.3.2 Device Fabrication

The transistors used in our study have been fabricated using the 0.25 μm process flow. To counter short-channel effects, the dual-doped polysilicon gate approach utilizing surface p-channel transistors has been adopted. The retrograde twin well process has been used to improve the isolation and increase the packing density. Transistor sizes ranging from 20 μm down to 0.25 μm are available for characterization.

The transistor formation begins with the deposition of a thick resist to cater for high energy well implants. The retrograde n-well for the p-transistors is formed by the implantation of phosphorus (2×10^{13} cm^{-2}, 490 keV) tilted at 7° to minimize channeling effects [29]. The n-well junction depth is about 1 μm with a sheet rho of approximately 675 Ω/□. Following that is the p-channel punch-through implant (5×10^{12} cm^{-2}, 140 keV) and the p-channel threshold adjust implant (3×10^{12} cm^{-2}, 70 keV). A 50 Å nitrided (N$_2$O) gate oxide is next grown. The N$_2$O gate oxide prevents the diffusion of dopants from the p+ polysilicon gate into the silicon. After the polysilicon gate definition, boron difluoride (BF$_2$) pLDD implant (2×10^{14} cm^{-2}, 20 keV) is subsequently carried out. Next LDD spacers are formed by the deposition of a 1500 Å tetraethylorthosiliane (TEOS) liner, followed by anisotropic etchback. The p+ source and drain regions are formed using BF$_2$ (3×10^{15} cm^{-2}, 30 keV) for shallower junction formation and anneal using a short thermal cycle. The p+ sheet rho is approximately 100 Ω/□. Note that for surface p-channel transistors, the polysilicon gate is initally undoped but will be heavily p-doped during the pLDD, source and drain implants. After the transistors formation, the wafers undergo back-end processing to the first metal layer and were then sent for electrical characterization.

3.3.3 Model Parameters Extraction

There are three important model parameters that one should determine before proceeding to develop the analytical model. These are:

1. The pLDD junction depth, r_j
2. The p+ implant junction depth, X_j
3. The surface doping concentration, N_{sub}

An attempt has been made to obtain the the p+ implant junction depth using Scanning Electron Microscopy (SEM) by breaking a die through the transistor tile and then etching the sample for a few seconds using the 6 nitric acid : 6 acetic acid : 1 HF (661) etchant. Figure 3–5 illustrates the microphotograph of one of the devices in the transistor tile. Due to the fact that the pLDD implant dose is of the same order of magnitude as the n-well implant, the pLDD regions cannot be picked up by the etchant. Hence only the p+ implants are visible after etching. Figure 3–5 however, provides only an approximation of the depth of the p+ regions (approximately 0.2 µm).

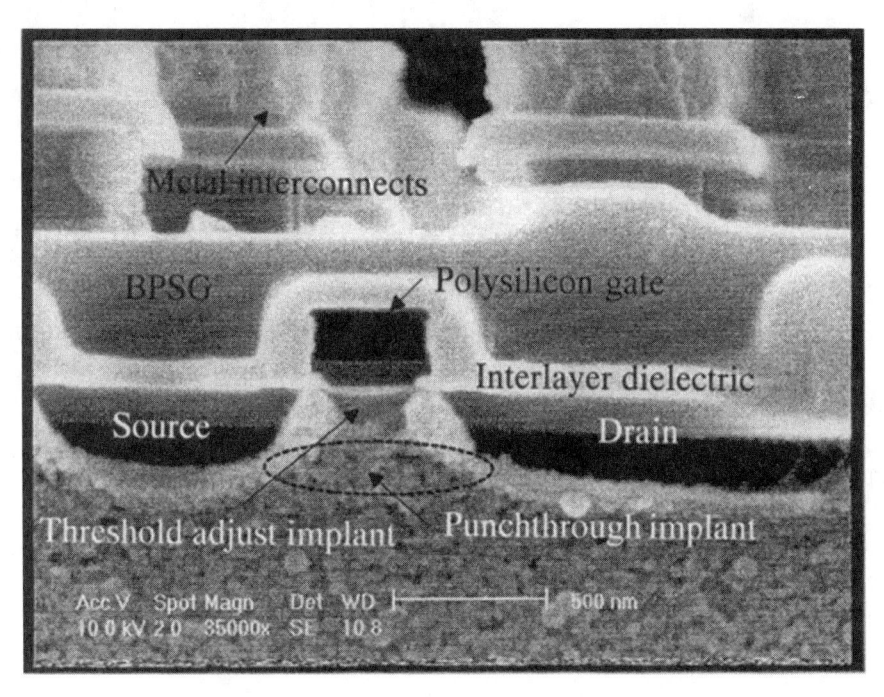

Figure 3–5 SEM microphotograph showing the cross-sectional view of a p-transistor structure.

In order to obtain fairly accurate values for the above mentioned three dimensions, methods like TRansport of Ions in Matter (TRIM) [30] simulation and TMA TSUPREM-4 [31] simulation have been used.

TRIM calculates the transport of ions into solids for a wide range of implant energy using a full quantum mechanical treatment of ion-atom collisions. TRIM employs the range algorithms by J.P. Biersack [32] and the stopping theory by J.F. Ziegler [33]. A TRIM simulation was run using the actual implant energy employed in the fabrication of the wafers. The result is shown in Figure 3–6.

The simulation result from TRIM shows that with an implant energy of 30 keV and a tilt of 7°, the implanted boron ions' stopping distance from the surface is approximately 0.1457 µm. This is in agreement with the junction depth range set by the Semiconductor Industry Association (SIA) for the 0.25 µm generation in 1994 [34]. Simultaneously, TSUPREM-4 simulation has been carried out, and the result indicates that the pLDD junction depth is approximately 0.075 µm and the p+ region junction depth, approximately 0.15 µm.

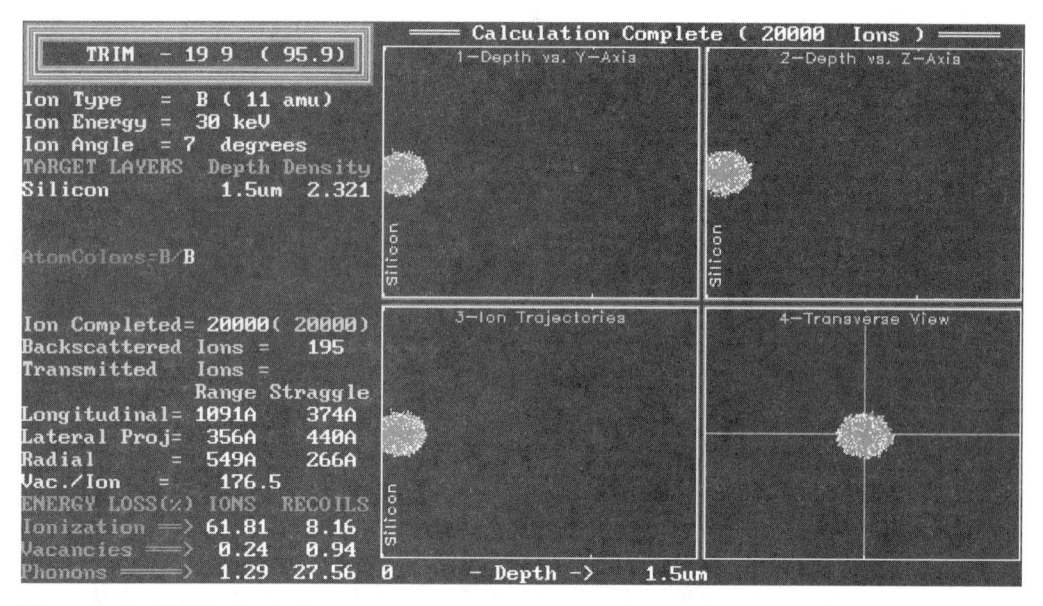

Figure 3-6 TRIM simulation result showing the extent of boron ions penetration into silicon for an implant energy of 30 keV and a tilt of 7°.

The final model parameter that needs to be extracted is N_{sub}. It is obtained similarly via TSUPREM-4 simulation by performing a 1-D cutline across the final device structure and then zooming into the silicon oxide/silicon interface to estimate the surface doping concentration. This is shown pictorially in Figure 3–7 and N_{sub} is found to be in the order of 10^{17} atoms/cm³.

3.3.4 Sub-Half Micron DC Model Formulation

The dc model for the 0.25 μm technology transistors consists of two main components: (1) the threshold voltage and (2) the device currents and their respective derivatives. The following sections give a detailed description of the formulation of the temperature-dependent 0.25 μm technology model.

3.3.4.1 Threshold Voltage Model

3.3.4.1.1 Extraction of the Experimental Threshold Voltage The Transconductance peak method [35] is used in the extraction of the device threshold voltage. This method has been adopted among others [36–39] because the values of the threshold voltage extracted are relatively insensitive to interface states, normal field mobility degradation, and series resistances [35]. Furthermore, this method is less sensitive to noise in the measurements because the second derivative acts as a high-pass filter.

In this method, a peak in the second derivative of the drain current with respect to the gate voltage occurs at a value that has been shown [35] to be related to the surface threshold poten-

The MOSFET in a Hybrid Mode Environment 103

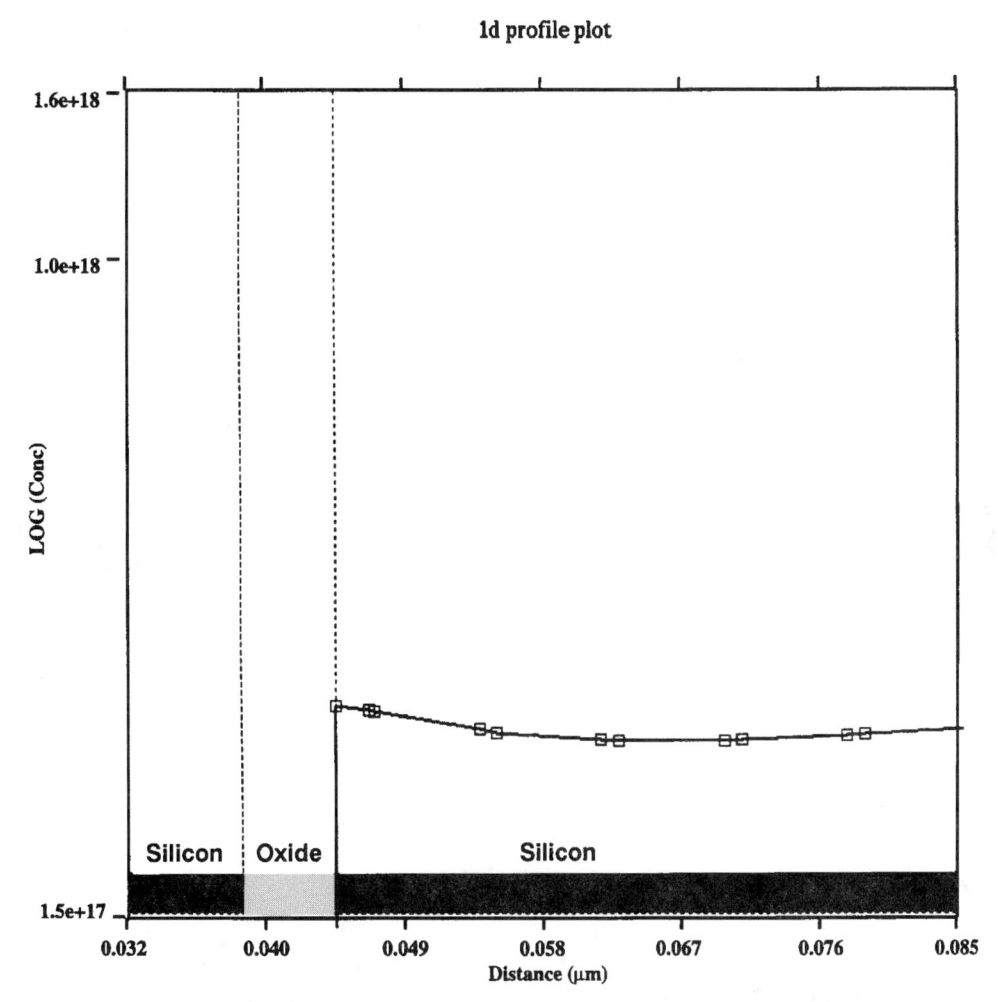

Figure 3-7 TSUPREM-4 simulation illustrating the doping concentration around the silicon oxide/silicon interface region.

tial. The drain current (I_{SD}) is differentiated once with respect to the gate voltage to give the transconductance (g_m), and once again to give the curve labeled g_m'. The position on the V_{SG} axis where the peak in the g_m' curve occurs defines the threshold voltage. A detailed explanation on how the threshold voltage is obtained using the transconductance peak method is given in the following paragraph. Figure 3–8 shows sample plots of I_{SD}, g_m, g_m', g_m'', and how the threshold voltage is obtained using the peak of g_m'.

The peak of the g_m' curve occurs at the point where the g_m'' curve crosses the zero x-axis. This point lies between V_{SG1} and V_{SG2} (Figure 3–8) when the value of g_m'' changes from a pos-

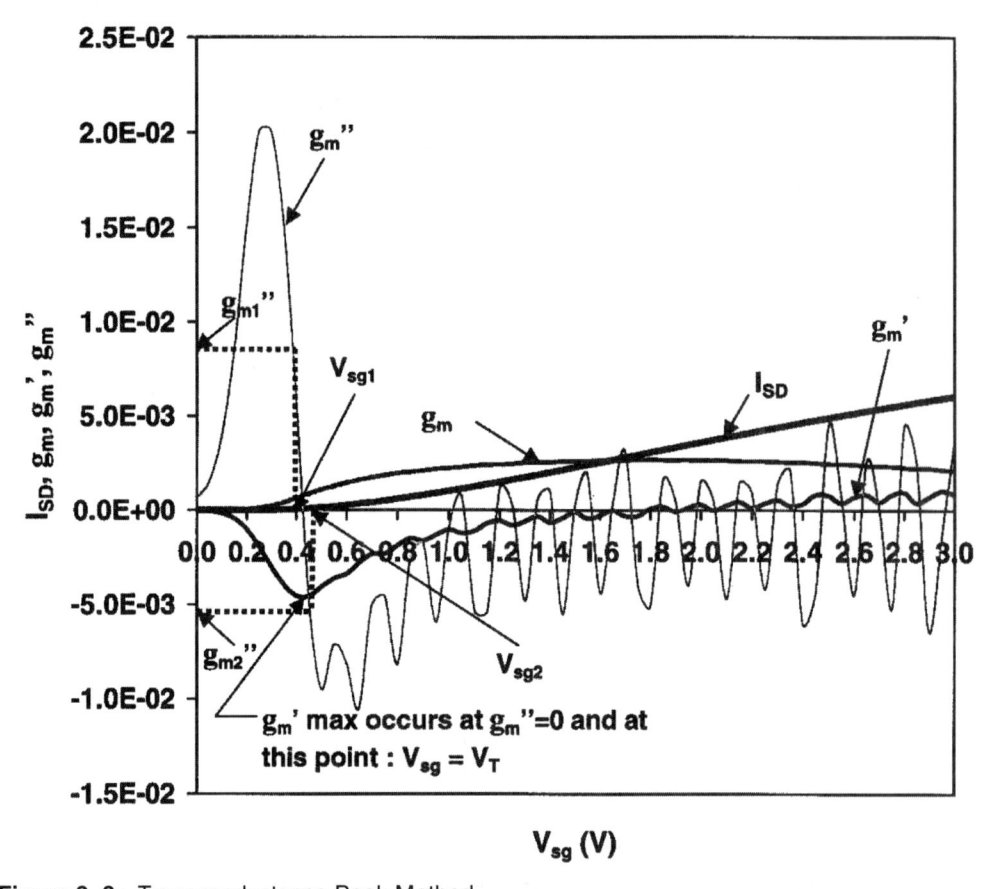

Figure 3–8 Transconductance Peak Method.

itive value, $g_{m1}"$, to a negative value, $g_{m2}"$. The threshold voltage that corresponds to this zero crossing point can then be obtained accurately by a simple computation as shown below.

The values of V_{SG1} and V_{SG2} are very close to each other, so we can approximate the $g_m"$ curve between these two points to be linear. Thus by similar Δ rules,

$$\frac{g_{m1}"}{V_T - V_{SG1}} = \frac{|g_{m2}"|}{V_{SG2} - V_T}$$

$$g_{m1}"(V_{SG2} - V_T) = |g_{m2}"|(V_T - V_{SG1})$$

$$V_T (|g_{m2}"| + g_{m1}") = g_{m1}" V_{SG2} + |g_{m2}"| V_{SG1}$$

The MOSFET in a Hybrid Mode Environment

$$\therefore V_T = \frac{g_{m1}'' V_{SG2} + |g_{m2}''| V_{SG1}}{g_{m1}'' + |g_{m2}''|}$$

3.3.4.1.2 Impact of short-channel effects on the classical threshold voltage model
The following assumptions are made in deriving the threshold voltage model:

1. The interface-trapped charges and oxide-trapped charges are ignored.
2. The oxide thickness T_{OX}, junction depth X_j, and the source, drain and channel (or well) doping concentrations remain unchanged when the device is scaled.
3. The characteristic length l_c [11] remains relatively constant for different channel lengths.

The experimental measurements of the device threshold voltage have shown that as the device is scaled, the threshold voltage decreases. However, this trend is not predicted by the classical threshold voltage expression given by:

$$V_T = V_{FB} - \phi_s - \frac{qN_{sub} W_D}{C_{OX}} \qquad 3.25$$

where:
V_{FB} is the voltage required for the flat-band condition,
ϕ_s is the surface potential,
N_{sub} represents the substrate doping,
W_D represents the width of the depletion region under the gate, and
C_{ox} is the gate oxide capacitance per unit area.

Deviation from the prediction of Eq. 3.25 is due to several short-channel effects. Three of the more crucial effects are described below.

For relatively long-channel devices, one can assume the depletion region under the gate to be a rectangular-shaped volume. This approximation, however, does not take into account the charges in the overlapping depletion regions of the source and drain junctions that help to terminate the built-in fields. But for deep submicrometer devices, and due to the greater amount of overlap, the electric field lines emanating from the gate charges will not all be terminated by the charges in the depletion region under the gate. Some lines will be terminated by the charges in the overlapping depletion regions, as illustrated in Figure 3–9.

Yau [41] first investigated this effect and assumed the depletion region to be a trapezoid. To account for this change in geometry of the gate-induced depletion region, the charge-sharing factor θ has been introduced in the classical threshold voltage equation.

In short-channel MOSFETs, the depletion regions of the drain and the source diffusions are so close to each other that significant field penetration from the drain to the source results, causing the lowering of the potential barrier at the source end (Figure 3–10).

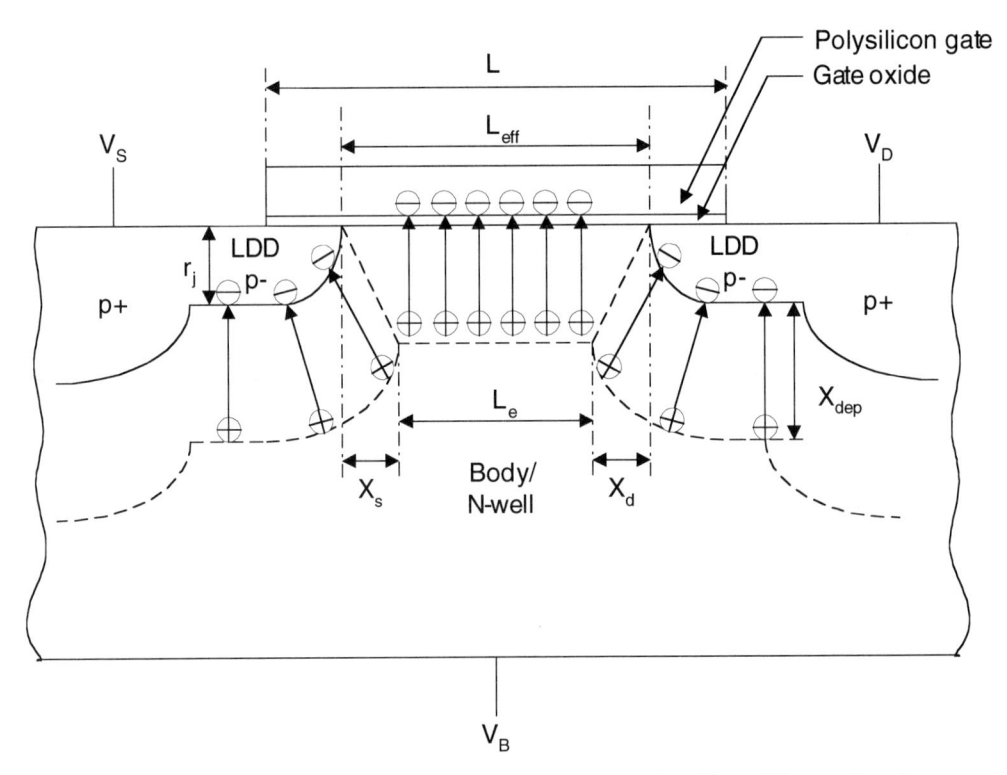

Figure 3-9 Short-channel LDD p-channel transistor showing the effect of the overlapping source and drain depletion regions on the gate field. (adopted from [40])

Hence more carriers from the source end can surmount the reduced barrier height and be injected into the channel region, consequently giving rise to an increase in the source to drain current.

The forward biasing of the source-body junction triggers the lateral bipolar action for short-channel devices. By doing so, the source-body potential barrier is further lowered (in addition to the same effect caused by DIBL), leading to greater carrier injection into the channel region. This in turn brings about a decrease in the threshold voltage.

3.3.4.1.3 Temperature dependent hybrid-mode device threshold voltage model

The threshold voltage model has been formulated by extending the classical threshold voltage equation to account for the above-mentioned short-channel effects that occur with the scaling of technology.

The proposed threshold voltage model including short-channel effects and temperature variations is given by:

$$V_T = V_T(T = T_{nom}) + (TC_1 + TC_2 V_{SBeff})\left(\frac{T}{T_{nom}} - 1\right) \qquad 3.26$$

The MOSFET in a Hybrid Mode Environment 107

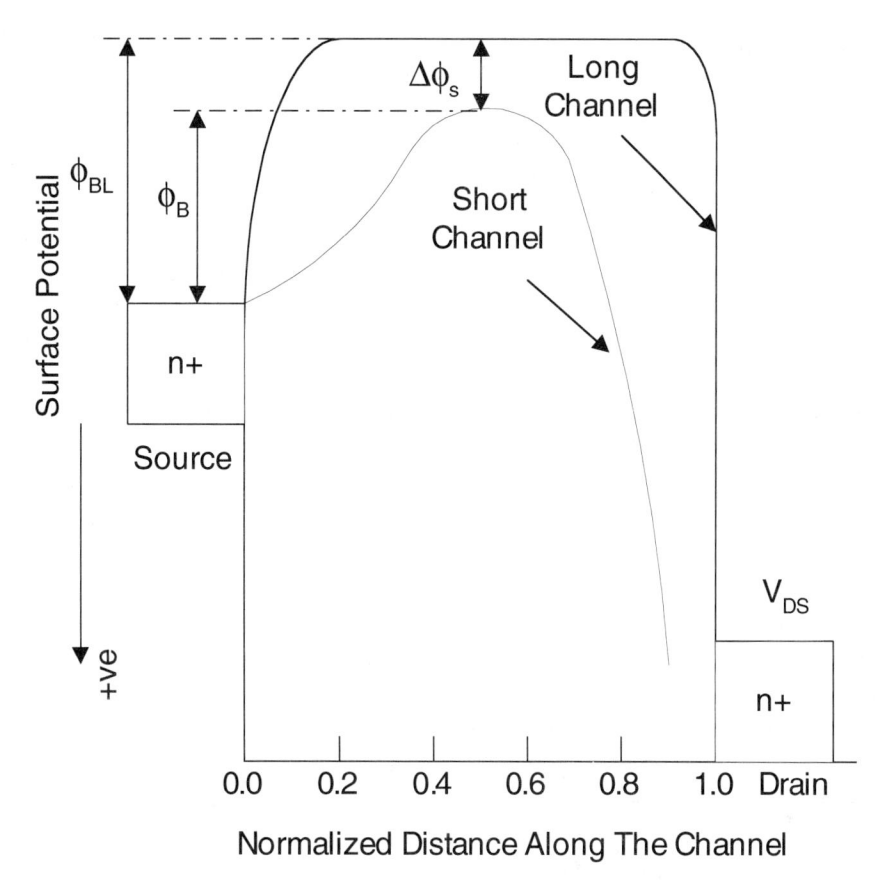

Figure 3–10 Potential distribution along the channel surface for a long- and short-channel nMOSFET, showing the lowering of the source potential due to DIBL. (adopted from [42])

where:

$$V_T(T = T_{nom}) = V_{TO} - \eta V_{SDeff} - \gamma V_{SBeff} \qquad 3.26a$$

$$V_{TO} = -V_{FB} + \frac{qD_i}{C_{OX}} + \phi_S + \frac{\theta}{C_{OX}}\sqrt{2q\varepsilon_{si}N_{sub}\phi_S} \qquad 3.26b$$

$$V_{FB} = \frac{E_g}{2} + \frac{\phi_S}{2} \qquad 3.26c$$

$$\theta = 1 - \varkappa r_j \qquad 3.26d$$

$$\varkappa = 23e^{-22L^2} + 5.3e^{-0.1L} \qquad 3.26e$$

T and T_{nom} are in kelvins, r_j and L are in μm. V_{TO} is the threshold voltage at zero drain and body biases, and it contains the charge-sharing factor, θ [3,4], to account for the lowering of the threshold voltage due to termination of the built-in field by the overlapping depletion regions of the source and drain junctions. For LDD devices, to account for the source and drain series resistance, the effective source-drain and source-body voltages are modified as follows [11]:

$$V_{SDeff} = \frac{V_{SD}}{1 + \frac{\Delta L_p}{l_c}} \qquad 3.27a$$

$$V_{SBeff} = \frac{V_{SB}}{1 + \frac{\Delta L_p}{l_c}} \qquad 3.27b$$

The DIBL factor and the forward source-body effect factor have been modeled using the methodology outlined in [19], and they are given by:

$$\eta = 0.8e^{-70L^2} + 0.25e^{-10L} \qquad 3.28$$

$$\gamma = 0.5(1 - e^{-10L^{1.6}}) - V_{SB}e^{-50L} \qquad 3.29$$

L is in μm. The experimentally extracted DIBL factor showed that it is insensitivity to temperature. A similar finding has been reported in Huang et al. [43]. The forward source-body effect factor has also been found to be a weak function of temperature (experimental data to justify this will be shown later). Hence the device threshold voltage is modeled using two components: the first encompassing the various short-channel effects at room temperature, where T_{nom} is the ambient room temperature, and the second is the temperature compensation term.

The effect of scaling coupled with DIBL and forward source-body effects at two different temperatures is shown in Figure 3–11.

The threshold roll-off for L < 0.4 μm is due to the severe charge-sharing effect of very short-channel devices. The graph reaffirms that the DIBL effect and the forward source-body effect are rather insensitive to temperature. The plot of the threshold voltage against temperature depicted in Figure 3–12 also supports the latter claim.

It can also be seen that the threshold voltage changes linearly with temperature for the two different channel lengths. It decreases with temperature due to Fermi level shifts [26]. Figures 3–13 and 3–14 illustrate the match of the analytical threshold voltage model with the experimental data for the range of source-drain and source-body biases considered in the experiment.

The threshold voltage shift at room temperature is shown in Figure 3–15. This shift is defined as:

$$\Delta V_T(T = T_{nom}) = \eta V_{SD} + \gamma V_{SB} \qquad 3.30$$

Figure 3–15 records the percentage contribution of ηV_{SD} and γV_{SB} to the room temperature threshold voltage shift respectively for each technology. For L = 0.25 μm, at large source-drain bias and forward source-body bias, the DIBL effect and the forward source-body effect have approximately equal contributions in the reduction of the threshold voltage. But for L = 1.0

The MOSFET in a Hybrid Mode Environment

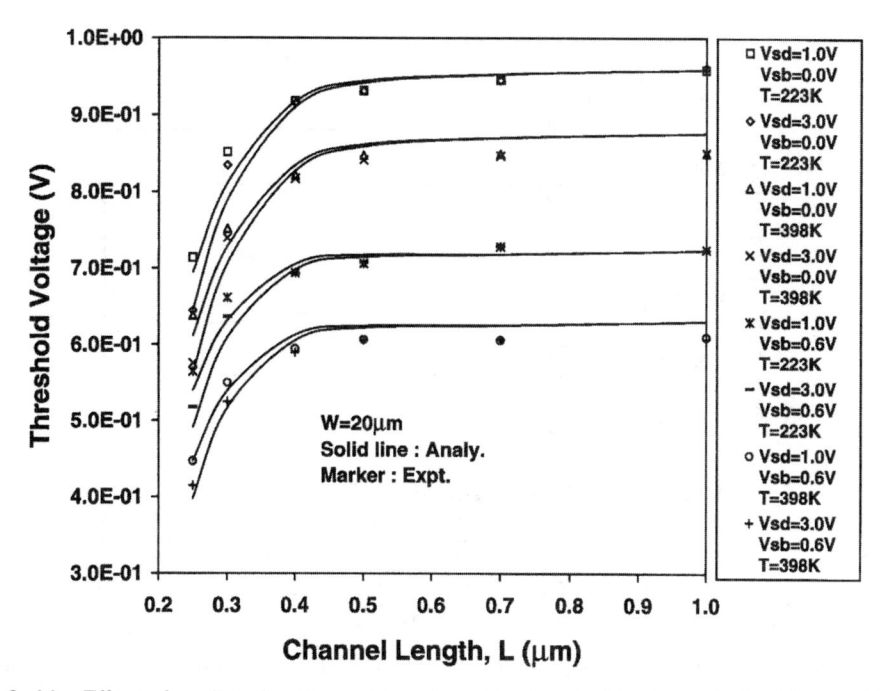

Figure 3–11 Effect of scaling the channel length on the threshold voltage for two different V_{SD}, V_{SB}, and temperature values.

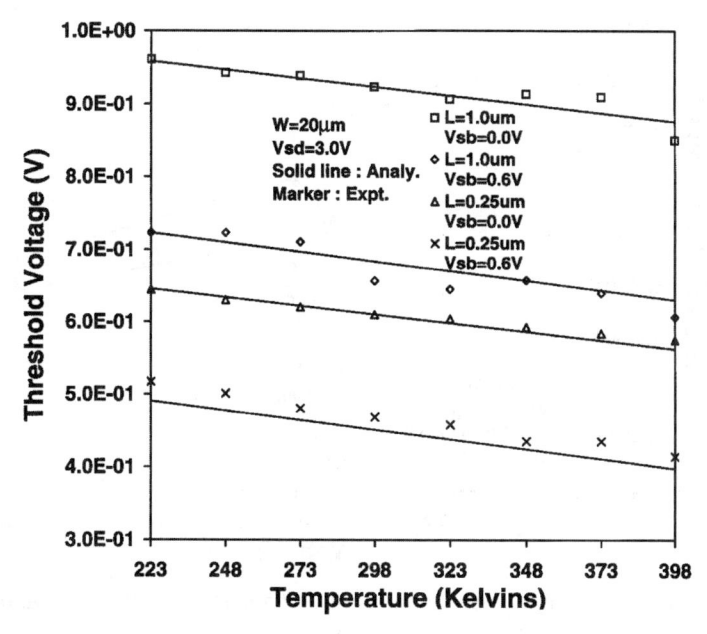

Figure 3–12 Effect of temperature variation on the threshold voltage for two different V_{SB} and channel lengths.

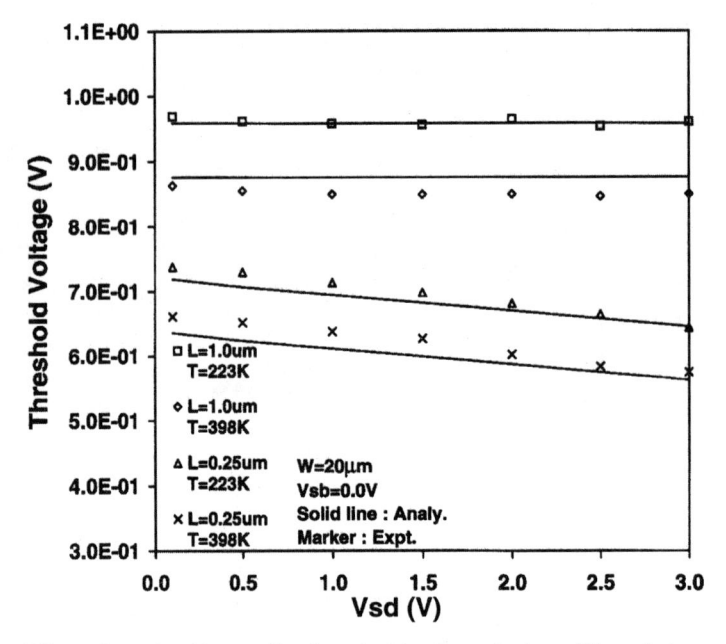

Figure 3–13 Effect of varying V_{SD} on the threshold voltage for two different channel lengths and temperature values.

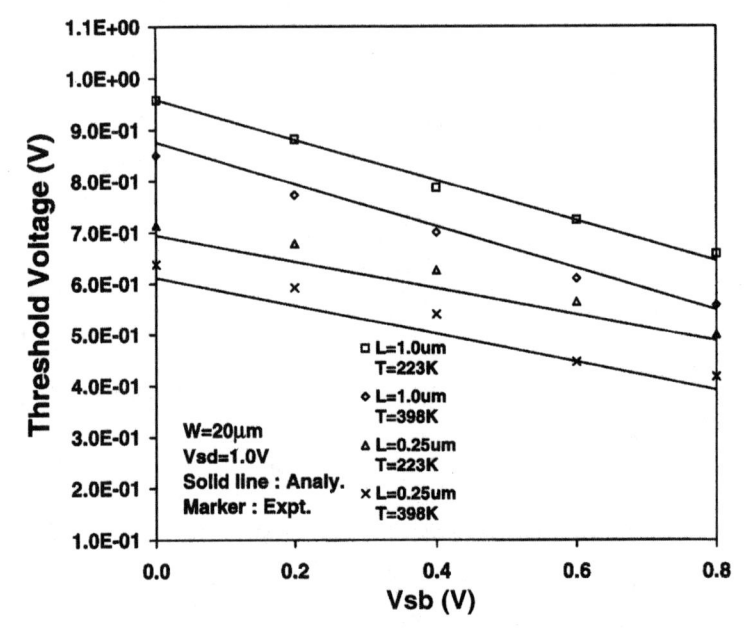

Figure 3–14 Effect of varying V_{SB} on the threshold voltage for two different channel lengths and temperature values.

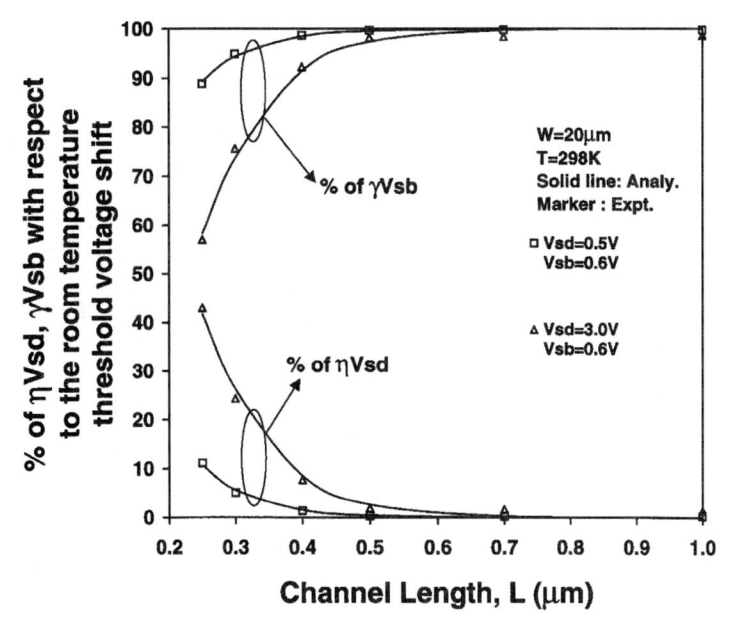

Figure 3-15 Effect of channel scaling on the percentages of η V_{SD} and γV_{SB} with respect to the room temperature threshold voltage shift for two different V_{SD}.

µm, the DIBL effect is nonexistent and the forward source-body effect is the sole cause for the lowering of the threshold voltage.

The temperature-dependent threshold voltage shift is depicted in Figure 3–16. ΔV_{T1} is obtained by finding the difference between the threshold voltages at $V_{SB} = 0.6$ V and $V_{SB} = 0.0$ V for each temperature, and ΔV_{T2} is found by subtracting the threshold voltage for each temperature from that at T_{nom} for a particular V_{SB}. The data for ΔV_{T1} confirms that the forward source-body effect is a weak function of temperature. The data for ΔV_{T2} reveals that the threshold voltage shift due solely to temperature is linear over the range of temperatures concerned.

3.3.4.2 Device Currents Model

3.3.4.2.1 Extraction of the device currents

When the source-body of a pMOS device is forward biased (with certain applied gate and drain biases), its total drain current comprises three main components: (1) the principal MOS current, (2) the lateral bipolar current, and (3) the space-charge current. The space-charge current is obtained by setting both the source-gate and source-body voltages to zero, and varying the source-drain potential. The bipolar current is measured, with the source-gate potential at zero bias, by sweeping the source-body voltage for various values of source-drain bias, and subtracting the corresponding space-charge current. The MOS current is found by subtracting both the space-charge and the bipolar currents from the measured total drain current at different source-gate, source-body and source-drain voltages.

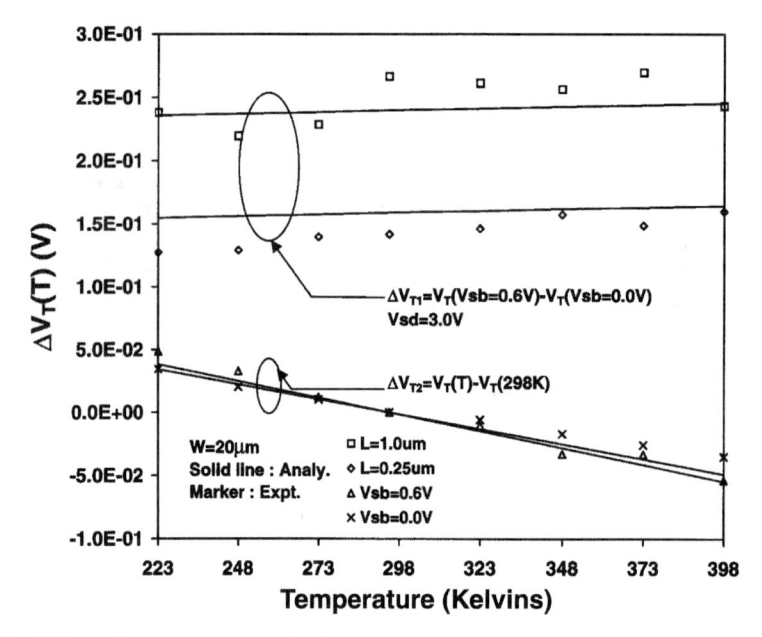

Figure 3–16 Effect of temperature variations on the temperature-dependent threshold voltage shifts for two different channel lengths and V_{SB}.

3.3.4.2.2 Space-charge current For very small spacing between the drain and source regions of a MOSFET and high substrate resistivity, the depletion regions of the two reverse-biased p-n junctions will spread and eventually touch each other giving rise to the punch-through condition [44] shown in Figure 3–17.

Under the punch-through condition, the depletion region will span the entire effective channel length, L_{eff}. This region will consist mainly of ionized donors with a small number of free carriers. As such, very little current is expected to transverse from the source to the drain with a negative drain bias. However, due to extremely short channel lengths and high substrate resistivity, carriers from the source can surmount the source-body barrier height by thermionic emission and then be swept across from the source to the drain by the strong longitudinal electric field. This phenomenon constitutes the space-charge current flow.

The space-charge current has been extracted using the methodology outlined in Rofail and Yeo [19] and modeled as a function of temperature by Sze et al. [26–27,44–45]:

$$I_{space} = \frac{\alpha A_{space} \varepsilon_{si} \mu_p V_{SDeff}^2}{L_{effav}^3}, \quad \frac{\mu_p V_{SDeff}}{L_{effav}} < v_{sat} \quad \quad 3.31a$$

$$I_{space} = \frac{\alpha A_{space} \varepsilon_{si} v_{sat} V_{SDeff}}{L_{effav}^2}, \quad \frac{\mu_p V_{SDeff}}{L_{effav}} \geq v_{sat} \quad \quad 3.31b$$

The MOSFET in a Hybrid Mode Environment

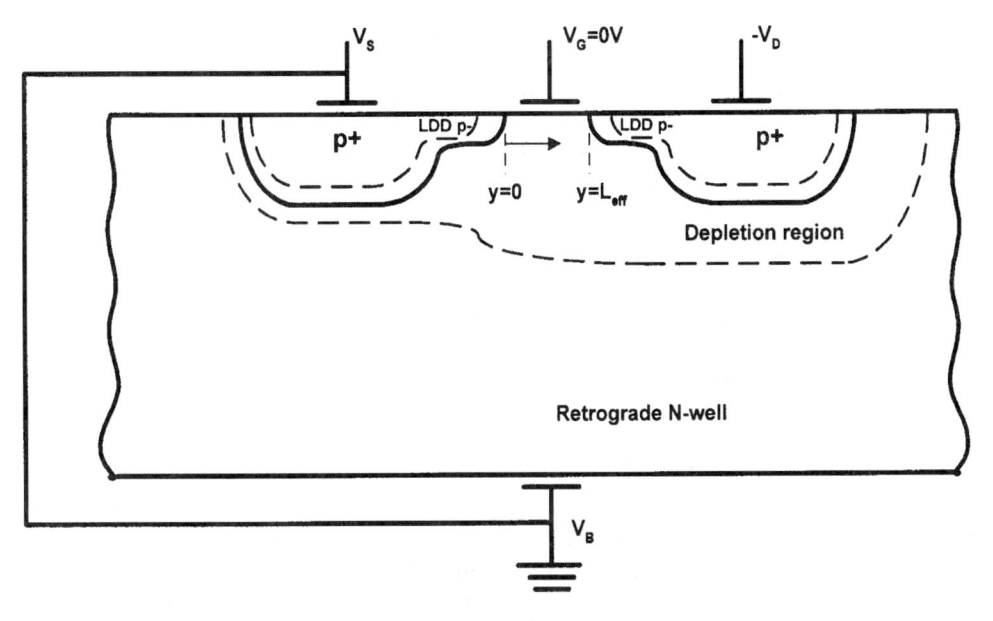

Figure 3–17 Schematic diagram of a pMOSFET showing the punch-through condition at zero gate, source, body biases and varying drain bias. (adopted from [44])

where:

$$\mu_p = 54.3\left(\frac{T}{T_{nom}}\right)^{-0.57} + \frac{1.36\times10^8 T^{-2.23}}{1+\left[\frac{N_{sub}}{2.35\times10^{17}\left\{\frac{T}{T_{nom}}\right\}^{2.4}}\right]\times 0.88\times\left(\frac{T}{T_{nom}}\right)^{-0.146}} \qquad 3.31c$$

$$\alpha = 5.0\times10^{-14}\, e^{\frac{T}{19}} \qquad 3.31d$$

$$N_{sub} = 0.9121\times n_i\, e^{\frac{E_F-E_{Fi}}{kT}} \qquad 3.31e$$

$$n_i = \sqrt{(N_c N_v)}\, e^{-\frac{Eg}{2kT}} \qquad 3.31f$$

$$N_c = 2\left(\frac{2\pi m_e^* kT}{h^2}\right)^{3/2}\times 10^{-6} \qquad 3.31g$$

$$N_v = 2\left(\frac{2\pi m_h^* kT}{h^2}\right)^{3/2} \times 10^{-6} \qquad 3.31h$$

$$m_e^* = (1.036 + 4.263 \times 10^{-4} T) \times m_o \qquad 3.31i$$

$$m_h^* = (0.58 + 7.72 \times 10^{-4} T) \times m_o \qquad 3.31j$$

$$E_g = 1.170 - \frac{4.73 \times 10^{-4} T^2}{T + 288} \qquad 3.31k$$

$$\nu_{sat} = \frac{6 \times 10^6}{1 + 0.8 e^{\frac{T}{600}}} \qquad 3.31l$$

T and T_{nom} are in kelvins. The empirical parameter α is found to increase exponentially with temperature. L_{effav} is the average sum of the effective channel lengths taken near the Si/SiO$_2$ interface and at the extreme of the gate depletion region.

It has been found that the space-charge component is not negligible for the quarter-micron devices because it will contribute to static power dissipation. Figure 3–18 illustrates the effect of varying the source-drain bias and temperature on the space-charge current for quarter-micron devices.

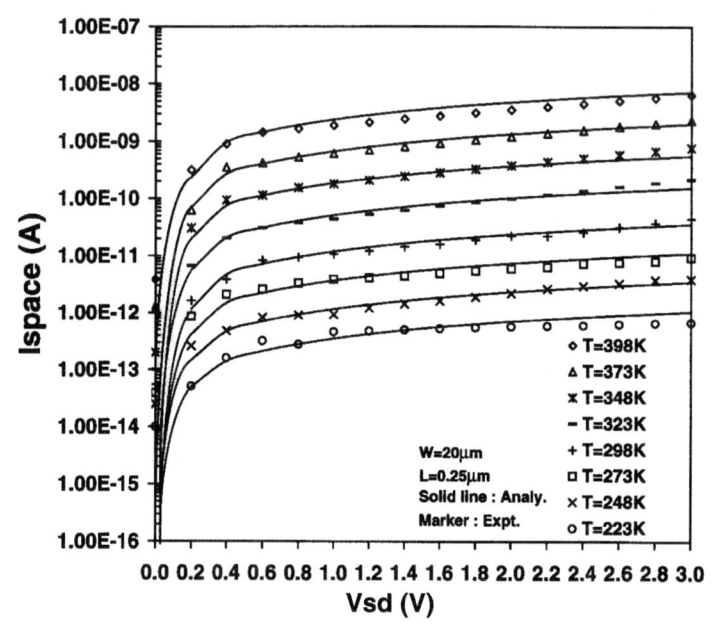

Figure 3–18 Effect of varying V_{SD} and temperature on the space-charge current of the quarter-micron device.

The MOSFET in a Hybrid Mode Environment

The source-drain bias has a modulating effect on the space-charge current flow. The temperature increment accelerates thermionic emission, which in turn results in a higher space-charge current flow.

Figure 3–19 shows that the space-charge current increases more than three decades of current when the temperature increases from 223 K to 398 K. However, a slight decrease in the analytical value of the space-charge current can be observed at room temperature. This is due to the two competing exponential terms in the N_{sub} equation (see Eqs. 3.31e and 3.31f). The intrinsic carrier concentration, n_i, increases with temperature, whereas the term $e^{\frac{E_F - E_{Fi}}{kT}}$ decreases with temperature. As a result, there is an unexpected peak in N_{sub} at room temperature. This phenomenon will affect all other current components, as well as the output conductance.

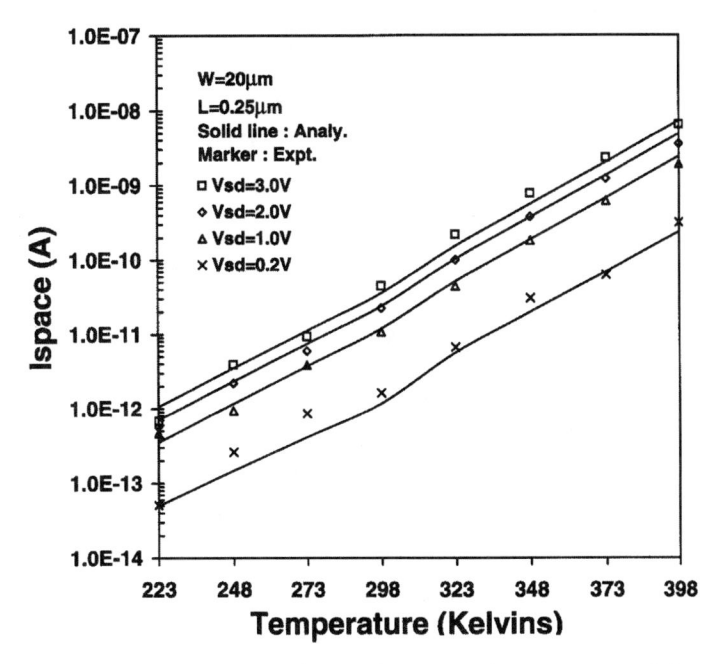

Figure 3–19 Effect of temperature variation on the space-charge current of the quarter-micron device for different V_{SD}.

3.3.4.2.3 Bipolar current

Under forward source-body bias, high source-drain bias, and for very short-channel lengths, lateral bipolar action occurs beneath the channel region. The bipolar current equation has been derived [19] based on the modified transport equation [46]. The key equations involved are:

$$I_{bipolar} = \frac{q\mu_p V_t P_e A_{bipolar}}{L_e(1-\zeta)}(1 - e^{-\frac{\zeta L_e v_{sat}}{D_p}}) \qquad 3.32$$

where:

$$P_e = \frac{n_i^2}{N_{sub}} e^{\frac{V_{j1}}{n_f V_t}} \qquad 3.32a$$

$$V_{j1} = V_{SBeff} - I_s R_{well}(e^{\frac{V_{j1}}{n_f V_t}} - 1) \qquad 3.32b$$

$$V_{j2} = V_{SBeff} - V_{SDeff} - I_s R_{well}(e^{\frac{V_{j2}}{n_f V_t}} - 1) \qquad 3.32c$$

$$X_s = \sqrt{\frac{2\varepsilon_{si}}{qN_{sub}}(V_{bi} - V_{j1})} \qquad 3.32d$$

$$X_d = \sqrt{\frac{2\varepsilon_{si}}{qN_{sub}}(V_{bi} - V_{j2})} \qquad 3.32e$$

$$L_e = L_{eff} - X_s - X_d \qquad 3.32f$$

$$R_{well} = \varrho_s \frac{L}{W} \qquad 3.32g$$

$$\zeta = \frac{V_{SDeff}}{V_{SDeff} + 0.00016 e^{7L}} \qquad 3.32h$$

ζ models the dependence of the bipolar current on the drain potential and the device drawn length. ζ approaches 1 for increasing drain potential and decreasing channel length.

The ζ expression given above has been modified to extend the model to quarter-micron devices and to include the effect of temperature variations. It is given as:

$$\zeta = \zeta_1(V_{SB})\zeta_2(T) \frac{V_{SDeff}}{V_{SDeff} + 1.6 \times 10^{-9} e^{700L}} \qquad 3.33$$

where:

$$\zeta_1(V_{SB}) = 1 - e^{-12.1\left(1 + \frac{\Delta L_p}{l_c}\right) V_{SBeff}} \qquad 3.33a$$

$$\zeta_2(T) = 0.755 + 0.245\left\{1 - e^{-0.418\left(6.21 - \frac{T}{100}\right)^{2.25}}\right\} \qquad 3.33b$$

T is in kelvins and L is in µm. ζ has been modified to include the effects of the source-drain and the source-body biases, the device technology and temperature. At room temperature, it approaches unity at high source-drain and source-body biases and for shorter channel lengths.

The MOSFET in a Hybrid Mode Environment	117

When the temperature increases from 223 K to 398 K, ζ decreases gradually from the unity asymptote.

Increasing the source-drain bias affects the lateral bipolar current, as illustrated in Figure 3–20. Figure 3–21 shows that the bipolar current increases exponentially with temperature such that at higher temperatures and larger source-body biases, the lateral bipolar current is comparable to the principal MOS current. For $V_{SB} = 0.8$ V and T = 398 K, the ratio of the experimental lateral bipolar current to the experimental MOS current is approximately 1:16. This phenomenon can be explained by the fact that an increase in the operating temperature will cause (1) the intrinsic carrier densities to increase, (2) the bandgap energies to decrease, and (3) the carrier mobility to decrease. The first two effects dominate the last, thus resulting in an overall increase in the lateral bipolar current with temperature. Furthermore, it has been reported [13] that the injection barrier for holes in a hybrid-mode lateral BJT is given by:

$$\psi_{Bh} = \left(V_{bi} - \frac{\Delta E_g}{q}\right) - V_{SB} \qquad 3.34$$

Eq. 3.34 suggests that the hole injection barrier in hybrid-mode devices decreases with temperature.

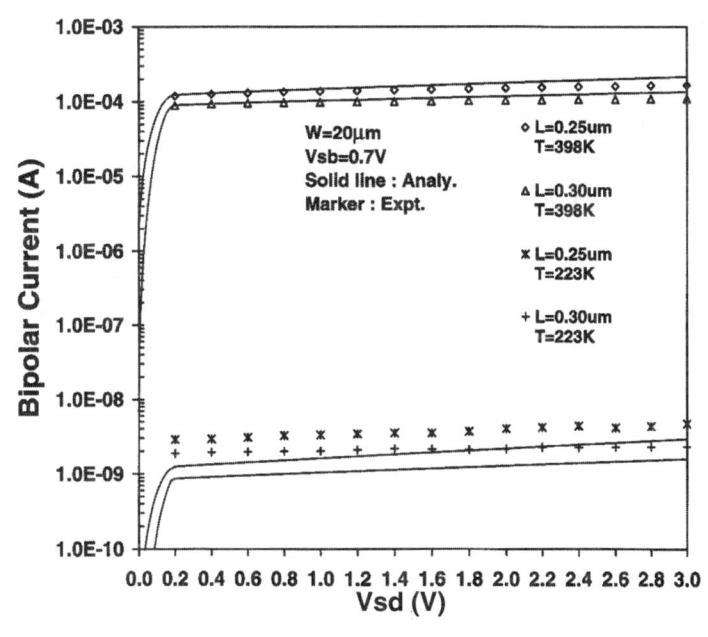

Figure 3–20 Effect of variation of V_{SD} on the lateral bipolar current for two different channel lengths and temperature values.

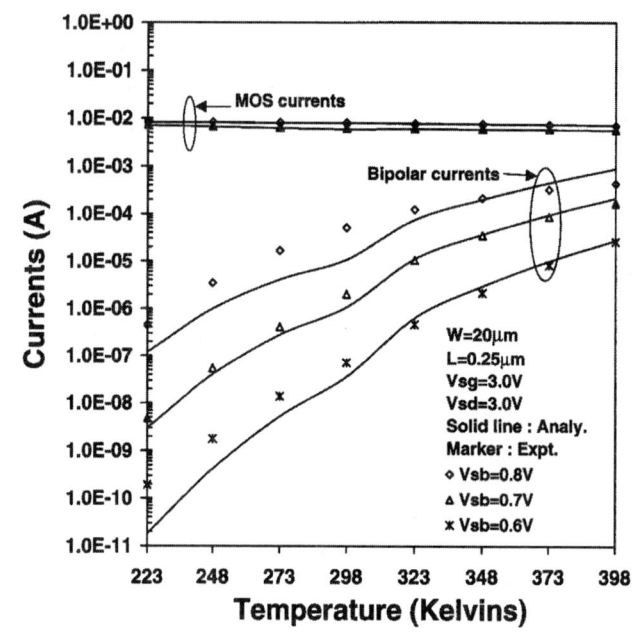

Figure 3–21 Comparison of the effects of temperature variation on the MOS and lateral bipolar currents of the quarter-micron device for different V_{SB}.

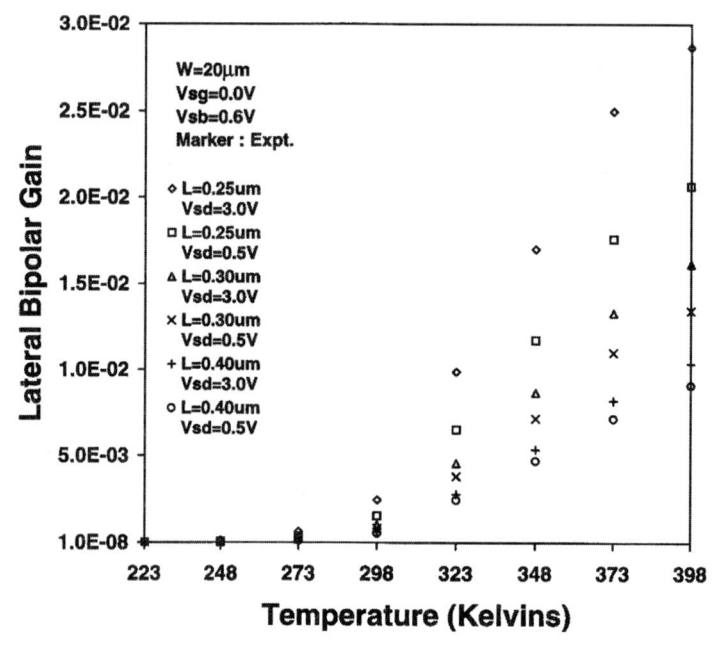

Figure 3–22 Effect of temperature variation on the lateral bipolar gain.

Figure 3–22 depicts the effect of temperature coupled with scaling on the lateral bipolar gain, which is defined by:

$$\text{Lateral Bipolar Gain} = \frac{I_{bipolar}}{I_{body}} \qquad 3.35$$

$I_{bipolar}$ is the collector current of the lateral pnp transistor, whereas I_{body} is the base current. The gain is fractional because of the addition of the p-channel punch-through implant during the transistor formation. This implant will impede the flow of carriers in the vicinity of the active region by increasing the rate of recombination. However, the implant is necessary to suppress short-channel effects in the 0.25 μm CMOS process.

3.3.4.2.4 MOS Current

The MOS current model, based on the engineering model for short-channel MOS devices [24] takes into consideration high-field effects [25,47–48], which is crucial for deep submicrometer devices. These effects are described by the piecewise carrier drift velocity model [25], which states that:

$$v = \begin{cases} \dfrac{\mu_{peff} E}{1 + \dfrac{E}{E_{crit}}} & \text{for } E < E_{crit} \\ v_{sat} & \text{for } E \geq E_{crit} \end{cases} \qquad 3.36a,b$$

where

$$E_{crit} = \frac{2 v_{sat}}{\mu_{peff}} \qquad 3.36c$$

The MOS current in the ohmic and saturation regions can be written as [24,26]:

$$I_{SD} = \frac{\mu_{peff} C_{OX} W}{L_{eff}} \cdot \frac{1 + \lambda V_{SBeff}}{1 + \dfrac{V_{SDeff}}{E_{crit} L_{eff}}} \left(V_{SG} - V_T - \frac{1}{2} V_{SDeff} \right) V_{SDeff} \qquad 3.37a$$

$$I_{SD} = \frac{1 + \lambda V_{SBeff}}{1 + \dfrac{E_{crit} L_d}{V_{SG} - V_T}} \cdot v_{sat} C_{OX} W (V_{SG} - V_T) \qquad 3.37b$$

where:

$$L_d = L_{eff} - X_{dm} \qquad 3.37c$$

$$X_{dm} = \sqrt{\frac{2 \varepsilon_{si}}{q N_{sub}} (V_{bi} - \phi_s + V_{SDeff} - V_{SBeff})} \qquad 3.37d$$

The factor $(1+\lambda V_{SBeff})$ accounts for the increase in the source-drain current due to the lowering of the source-body barrier height. The value of λ that gives a good fit to the experimental data is 0.3.

The piecewise MOS current model [19] has been extended to quarter-micron devices with the temperature effects taken into consideration.

Much work has already been done to model the effect of temperature on the carrier mobility of MOSFETs [49–51]. The recent work on a temperature-dependent MOSFET inversion layer carrier mobility model for device and circuit simulation [52] has been adopted to account for the major scattering mechanisms in the MOSFET inversion layer. This is given by:

$$\frac{1}{\mu_{peff}} = \frac{1}{\mu_1}\left(\frac{10^6}{E_{eff}}\right)^{\alpha_1} + \frac{1}{\mu_2}\left(\frac{10^6}{E_{eff}}\right)^{\alpha_2} + \frac{1}{\mu_3}\left(\frac{10^{18}}{N_{sub}}\right)^{-1}\left(\frac{10^{-12}}{Q_{inv}}\right)^{\alpha_3} \qquad 3.38$$

where:

$$E_{eff} = \frac{1}{\varepsilon_{si}}(Q_{dep} + \varphi Q_{inv}) \qquad 3.38a$$

$$Q_{dep} = \sqrt{2q\varepsilon_{si} N_{sub}(\phi_S - V_{SBeff})} \qquad 3.38b$$

$$Q_{inv} = C_{OX}(V_{SG} - V_T) \qquad 3.38c$$

$$\varphi = 0.42 - 2.9\times 10^{-4} T \qquad 3.38d$$

T is in kelvins. Figure 3–23 demonstrates that a reasonably good agreement between the experimental data and the analytical model of the MOS current has been achieved for a wide range of technology in the triode and saturation regions for temperatures between 223 K and 398 K.

Figure 3–24 illustrates that the MOS current decreases with temperature. In the triode region, devices with channel lengths of 0.25 µm and 1.0 µm show approximately equal enhancement (2 to 4 µA/K) of the drain current when the temperature reduces from 398 K to 223 K. This is due to the common sensitivity of the carrier mobility to temperature. In the saturation region, the saturation velocity is fairly sensitive to temperature and could increase as much as 30% [26]. The effect of a greater reduction in the electrical channel length of the MOS current, L_d, for L = 0.25 µm results in approximately 6 µA/K increase in the MOS current with lower temperature. In retrospect, the gradual decline in the MOS current with temperature can be compensated for by the exponential increase in the lateral bipolar current shown earlier in Figure 3–21. At higher temperatures and larger source-body bias, the bipolar current will have approximately the same order of magnitude as the MOS current, thereby complementing the MOS current in enhancing the overall device performance.

3.3.4.3 Device Parameters

There are two important device parameters correlated to the device I-V characteristics that should be studied. These are the device transconductance and the output conductance.

The MOSFET in a Hybrid Mode Environment

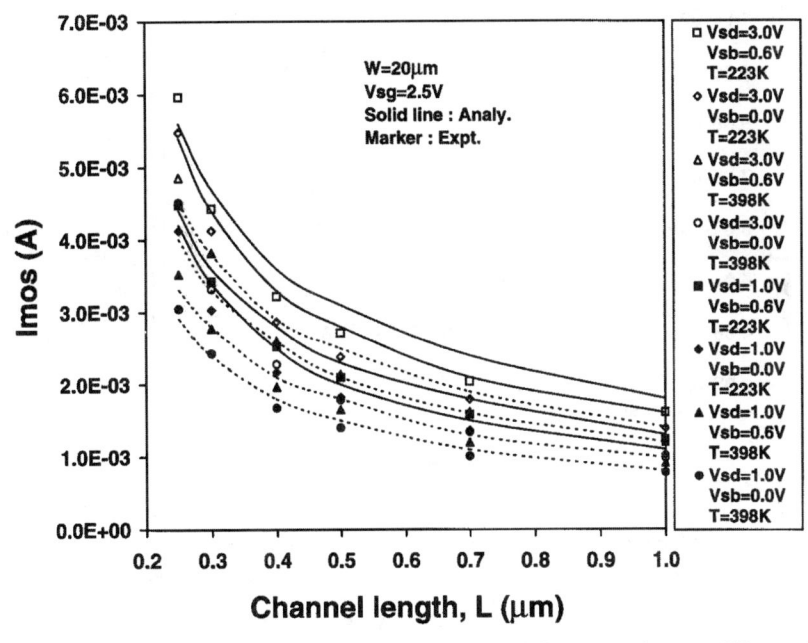

Figure 3-23 Effect of scaling the channel length on the MOS current for two different V_{SD}, V_{SB} and temperatures.

3.3.4.3.1 Device transconductance
For a MOSFET operating in a hybrid-mode environment, its device transconductance can be written as:

$$\Delta I_{SD} = g_{m1} \Delta V_{SG} + g_{m2} \Delta V_{SB} \qquad 3.39$$

where:

$$g_{m1} = \frac{\delta I_{SD}}{\delta V_{SG}} \quad \text{at constant } V_{SB} \text{ and } V_{SD} \qquad 3.39a$$

$$g_{m2} = \frac{\delta I_{SD}}{\delta V_{SB}} \quad \text{at constant } V_{SG} \text{ and } V_{SD} \qquad 3.39b$$

$$\frac{\delta I_{SD}}{\delta V_{SG}} = \frac{\delta I_{mos}}{\delta V_{SG}} + \frac{\delta I_{bipolar}}{\delta V_{SG}} + \frac{\delta I_{space}}{\delta V_{SG}} \qquad 3.39c$$

$$\frac{\delta I_{SD}}{\delta V_{SB}} = \frac{\delta I_{mos}}{\delta V_{SB}} + \frac{\delta I_{bipolar}}{\delta V_{SB}} + \frac{\delta I_{space}}{\delta V_{SB}} \qquad 3.39d$$

Figure 3–25 shows the effect of temperature on g_{m1} in the triode and saturation regions for L = 0.25 μm and L = 1.0 μm. The plot shows that g_{m1} of the quarter-micron device increases by approximately 2.2 μA/K and 1.5 μA/K in the saturation and triode regions respectively, when

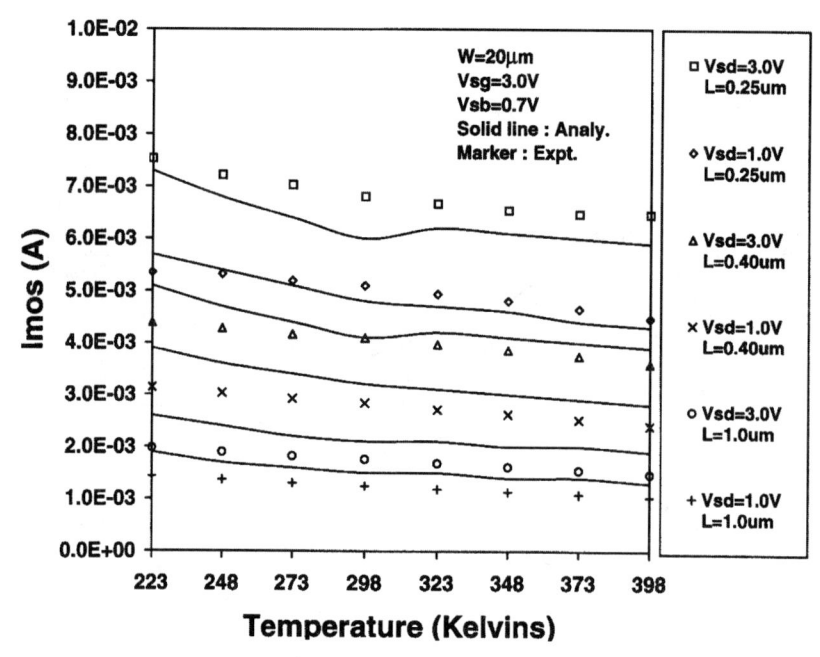

Figure 3–24 Effect of temperature variation on the MOS current for two different V_{SD} and various technologies.

the temperature reduces from 398 K to 223 K. For L = 1.0 µm, the respective rates are 1.5 µA/K and 0.67 µA/K. These figures are in agreement with the explanation stated in the previous section for the variation of the MOS current with temperature.

Figure 3–26 illustrates the variation of g_{m2} with temperature. As evident from the graph, g_{m2} in general increases with temperature. Furthermore g_{m2} in the triode region is larger than g_{m2} in the saturation region. This is because in the triode region, the output resistance is relatively smaller; hence the drain current has a strong dependence on both the drain and the body potentials. In the saturation region, when the velocity of the carriers starts to saturate, this dependence begins to diminish.

3.3.4.3.2 Output Conductance

The output conductance of a hybrid-mode device is defined as:

$$g_{sd} = \frac{\delta I_{SD}}{\delta V_{SD}} \quad \text{at constant } V_{SG} \text{ and } V_{SB} \qquad 3.40a$$

$$\frac{\delta I_{SD}}{\delta V_{SD}} = \frac{\delta I_{mos}}{\delta V_{SD}} + \frac{\delta I_{bipolar}}{\delta V_{SD}} + \frac{\delta I_{space}}{\delta V_{SD}} \qquad 3.40b$$

The effect of temperature on g_{sd} in the triode and saturation regions for L = 0.25 µm and L = 1.0 µm is shown in Figure 3–27. The results reveal that scaling the channel length to 0.25 µm

The MOSFET in a Hybrid Mode Environment

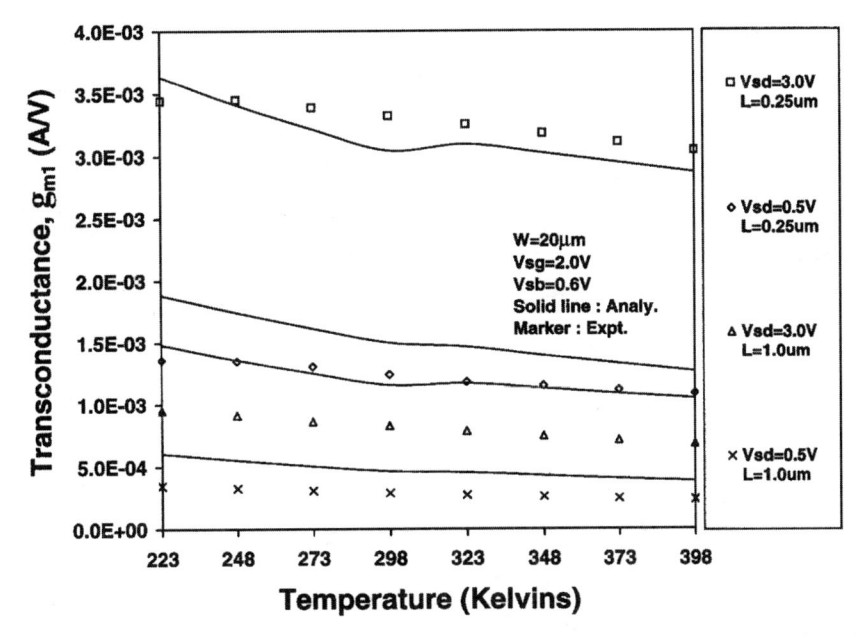

Figure 3–25 Effect of temperature variation on g_{m1} in the triode and saturation regions for L = 0.25 μm and L = 1.0 μm.

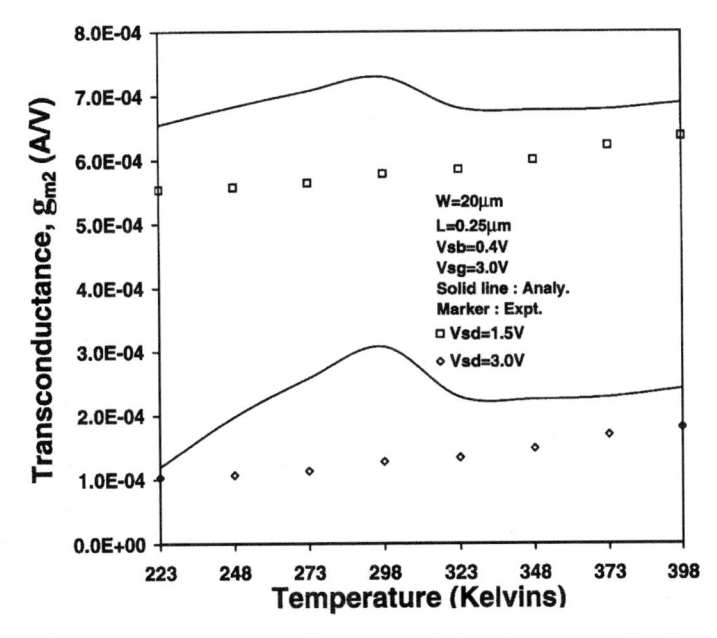

Figure 3–26 Effect of temperature variation on g_{m2} in the triode and saturation regions for L = 0.25 μm.

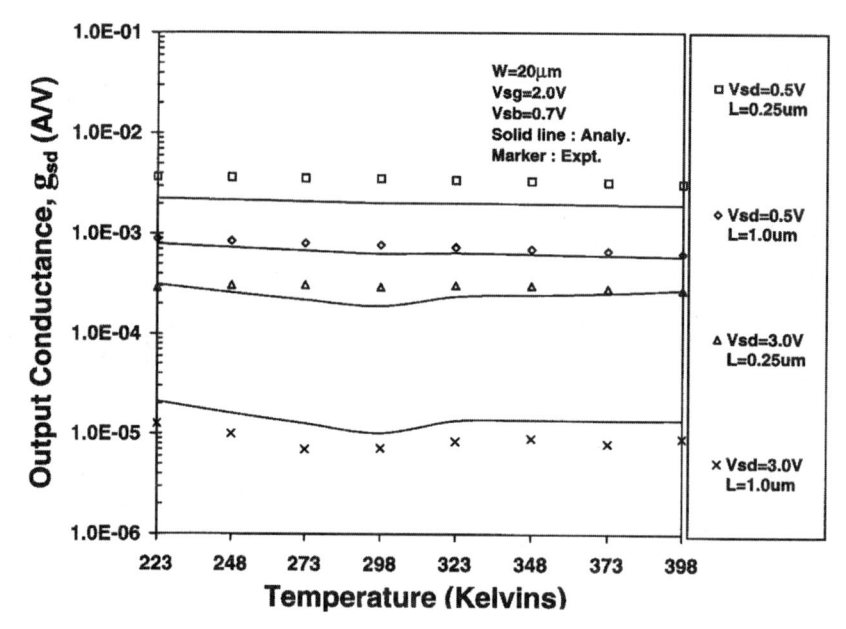

Figure 3-27 Effect of temperature variation on g_{sd} in the triode and saturation regions for L = 0.25 μm and L = 1.0 μm.

increases the saturation output conductance by approximately one order of magnitude. This is due to the nonsaturation of the drain current for the quarter-micron devices and is desirable in digital applications for faster switching. On the other hand, the lowering of temperature brings about a very minimal increase in the saturation output conductance for L = 0.25 μm.

3.4 Concluding Remarks

The earlier dc model has been successfully extended down to quarter-micron devices and the effect of temperature variation taken into account. Comparison of the two models reveals that the components of the dc model remain relatively intact. Modifications made to the earlier model reflect the adaptation to a new set of devices fabricated using a different technology and process. Trends predicted in earlier work have been confirmed in the present work.

With the extracted experimental results as the baseline, and using various simulation tools, a temperature-dependent dc model for LDD pMOSFETs operating in the hybrid-mode environment has been derived. This model caters to a wide range of channel lengths (1.0 μm to 0.25 μm) and a temperature range from 223 K to 398 K.

The threshold voltage has been modeled as two components: One includes short-channel effects and another, the temperature compensation term. The threshold voltage decreases with temperature due to Fermi level shifts. DIBL and forward source-body effects have been found to be weak functions of temperature. The forward source-body effect dominates for L = 1.0 μm, whereas both the DIBL effect and the forward source-body effect are comparable for L = 0.25

µm. The threshold voltage shifts linearly with temperature. Both the space-charge and the lateral bipolar currents have been found to be direct exponential functions of temperature.

On the contrary, the MOS current decreases linearly with temperature. Hence, operating the transistor at high temperatures would mean degrading the MOS current and increasing leakage at zero gate bias. On the other hand, elevated temperatures are able to increase the lateral bipolar current to approximately the same order of magnitude as the principal MOS current. For $V_{SB} = 0.8$ V and T = 398 K, the ratio of the experimental lateral bipolar current to the experimental MOS current of the quarter-micron device is approximately 1:16. This secondary current component can be utilized to enhance the switching speed of the transistor. In general, the transconductance associated with the variation of the source-gate voltage, and the output conductance, g_{m1} and g_{sd} respectively, decrease with temperature. On the other hand, the transconductance associated with the variation of the source-body voltage, g_{m2}, exhibits the opposite trend due to the exponential increase of the lateral bipolar and space-charge currents.

3.5 Summary

Numerous device models have been developed over the years; the latest is the University of California Berkeley (UCB) BSIM3v3 SPICE model. This model contains numerous enhancements compared to the earlier BSIM model, such as the polysilicon gate depletion effect and the lateral and vertical nonuniform doping effect. However, there has not been a complete dc SPICE model for devices operating in the hybrid-mode environment, that is, the body of the device is forward-biased with respect to the source, until recently.

The model reported by Rofail and Yeo [19] allows the effects of independently biasing the source, drain, gate, and body on the device performance to be investigated, and it models the different current components that flow when the device is operating in the hybrid-mode environment.

Three important model parameters have been extracted using SEM and process simulations. The reported model has been extended down to quarter-micron devices fabricated using the dual-gate CMOS process. As dual-gate CMOS process results in surface p-channel devices, the earlier model has been modified to adapt it to this new process.

It has been observed that for the quarter-micron devices, both the DIBL effect and the forward source-body effect have approximately equal contributions to the lowering of the threshold voltage. But for the longer channel devices, the DIBL effect is greatly diminished and the forward source-body effect is the major contributor to the lowering of the threshold voltage.

With the source-body of the device forward-biased, the device is said to be operating in the hybrid-mode environment and its drain current consists of the space-charge current, the lateral bipolar current, and the principal MOS current. These currents are extracted from the total drain current by varying the source, drain, gate, and body biases and then performing the necessary decomposition.

The results revealed that only the quarter-micron devices exhibit a space-charge current of appreciable magnitudes under large source-drain bias and at elevated temperatures. Also, the space-charge current increases exponentially with temperature. The bipolar current of the quarter-micron device has been similarly found to increase exponentially with temperature under large source-drain bias and high-level injection. Under very high-level injection and temperatures beyond 390 K, the magnitude of the bipolar current is comparable to the principal MOS current. However, the added advantage offered by the lateral bipolar current is somewhat negated by the degradation of the principal MOS current and the higher off-state leakage at higher temperature.

3.6 References

[1] K. S. Yeo and S. S. Rofail, "A 1.1 V Full-Swing Double Bootstrapped BiCMOS Logic Gates," *IEE Proceedings - Circuits, Devices, and Systems*, Vol. 143, No. 1, pp. 41–45, 1996.

[2] S. S. Rofail and K. S. Yeo, "Novel Low-Voltage BiCMOS Digital Circuits Employing a Lateral PNP BJT in a pMOS Structure," *IEE Proceedings – Circuits, Devices, and Systems*, Vol. 143, No. 2, pp. 83–90, Apr. 1996.

[3] C. R. Viswanathan, B. C. Burkey, G. Lubberts and T. J. Tredwell, "Threshold Voltage in Short-Channel MOS Devices," *IEEE Trans. Electron Devices*, Vol. ED-32, p. 932, 1985.

[4] D. K. Ferry, L. A. Akers, and E. W. Greeneich, *Ultra Large Scale Integrated Microelectronics*, Englewoods Cliffs, NJ: Prentice Hall, 1988.

[5] A. Vladimirescu, and S. Liu, "The Simulation of MOS Integrated Circuits using SPICE2," Memo. No. UCB/ERL M80/7, Univ. California, Berkeley, Oct. 1980.

[6] H. Masuda, M. Nakai and M. Kubo, "Characterization and Limitations of Scaled Down MOSFET's Due to Two Dimensional Field Effect," *IEEE Trans. Electron Devices*, Vol. ED-26, pp. 980–986, 1979.

[7] R. L. M. Dang, "A Simple Current Model for Short-Channel IGFET and Its Application to Circuit Simulation," *IEEE J. Solid-State Circuits*, Vol. SC-14, pp. 358–367, 1979.

[8] B. J. Sheu, D. L. Schafetter, P. K. Ko and M. C. Jeng, "BSIM: Berkeley Short-Channel IGFET Model for MOS Transistors," *IEEE J. Solid-State Circuits*, Vol. SC-22, pp. 558–566, 1987.

[9] T. Toyabe and S. Asai, "Analytical Models of Threshold Voltage and Breakdown Voltage of Short-Channel MOSFET's Derived from Two-Dimensional Analysis," *IEEE J. Solid-State Circuits*, Vol. SC-14, p. 375, 1979.

[10] D. R. Poole and D. L. Kwong, "Two-Dimensional Analysis Modeling of Threshold Voltage of Short-Channel MOSFETs," *IEEE Electron Device Letter*, Vol. EDL-5, p. 443, 1984.

[11] Z. H. Liu, C. H. Hu, J. H. Huang, T. Y. Chan, M. C. Jeng, P. K. Ko, and Y. C. Cheng, "Threshold Voltage Model for Deep-Submicrometer MOSFETs," *IEEE Trans. Electron Devices*, Vol. 40, No. 1, pp. 86–94, Jan. 1993.

[12] Y. Cheng, M. Chan, K. Hui, M. C. Jeng, Z. Liu, J. Huang, K. Chen, J. Chen, R. Tu, P. K. Ko and C. Hu, "BSIM3v3 Manual (Final Version)," University of California, Berkeley, Dec. 1996.

[13] S. Verdonckt-Vandebroek, S. S. Wong, C. S. Woo and P. K. Ko, "High-Gain Bipolar Action in a MOSFET Structure," *IEEE Trans. Electron Devices*, Vol. 38, No. 11, pp. 2487–2495, Nov. 1991.

[14] S. Verdonckt-Vandebroek, J. You, C. S. Woo and S. S. Wong, "High-Gain Lateral PNP Bipolar Action in a pMOSFET Structure," *IEEE Electron Device Letter*, Vol. 13, No. 6, pp. 312–313, June 1992.

[15] S. A. Parke, C. H. Hu, and P. K. Ko, "Bipolar-FET Hybrid-Mode Operation of Quarter-Micrometer SOI MOSFETs," *IEEE Electron Device Letter*, Vol. 14, No. 5, pp. 234–236, May 1993.

[16] K. Joardar, "An Improved Analytical Model for Collector Currents in Lateral Bipolar Transistors," *IEEE Trans. Electron Devices*, Vol. 41, No. 3, pp. 373–381, Mar. 1994.

[17] H. Jih-Shin, H. Tzuen-Hsi and C. Ming-Jer, "High Gain PNP Gated Lateral Bipolar Action in a Fully Depleted Counter-Type Channel pMOSFET Structure," *Solid-State Electronics*, Vol. 39, No. 2, pp. 261–267, 1996.

[18] Y. Zhixin, M. J. Deen and D. S. Malhi, "Gate-Controlled Lateral PNP BJT: Characteristics, Modeling and Circuit Applications," *IEEE Trans. Electron Devices*, Vol. 44, No. 1, pp. 118–128, Jan. 1997.

[19] S. S. Rofail and K. S. Yeo, "Experimentally Based Analytical Model of Deep-Submicron LDD pMOSFETs in a Bi-MOS Hybrid-Mode Environment," *IEEE Trans. Electron Devices*, Vol. 44, No. 9, p. 1417, Sept 1997.

[20] S. Wolf, *Silicon Processing for The VLSI Era Volume 3 – The Submicron MOSFET*, Chp. 4, Lattice Press, 1995.

[21] G. Merkel, J. Borel and N.Z. Cupcea, "An Accurate Large Signal MOS Transistor Model for Use in Computer-Aided Design," *IEEE Trans. Electron Devices*, Vol ED-19, 1972.

[22] M. C. Jeng, "Design and Modeling of Deep Submicrometer MOSFETs," PhD. Dissertation, University of California, Berkeley, 1989.

[23] J. S. Duster, M. C. Jeng, P. K. Ko and C. Hu, "User's Guide for the BSIM2 Parameter Extraction Program and the SPICE3 with BSIM Implementation," Industrial Liaison Program, Software Distribution Office, University of California, Berkeley, May 1990.

[24] K. Y. Toh, P. K. Ko, and R. G. Meyer, "An Engineering Model for Short-Channel MOS Devices," *IEEE J. Solid-State Circuits*, Vol. 23, No. 4, pp. 950–958, Aug.1988.

[25] B. Hoefflinger, H. Sibbert and G. Zimmer, "Model and Performance of Hot-Electron MOS Transistor for VLSI," *IEEE Trans. Electron Devices*, Vol. ED-26, p. 513, Apr. 1979.

[26] S. M. Sze, *Physics of Semiconductor Devices*, John Wiley & Sons, 1981.

[27] D. L. Pulfrey and N. G. Tarr, *Introduction to Microelectronic Devices*, Englewood Cliffs, NJ: Prentice Hall, 1989.

[28] G. J. Hu and H. B. Richard, "Design Tradeoffs Between Surface and Buried-Channel FETs," *IEEE Trans. Electron Devices*, Vol. ED-32, No. 3, pp. 584–588, March 1985.

[29] S. Wolf and R. N. Tauber, *Silicon Processing for the VLSI Era Volume 1 – Process Technology*, Lattice Press, June 1987.

[30] TRIM Instruction Manual, Version 95.4, March 7, 1995.

[31] TMA TSUPREM-4 User's Manual, Version 6.5, May 1997.

[32] J. P. Biersack and L. Haggmark, *Nucl. Instr. and Meth.*, Vol. 174, p. 257, 1980.

[33] J. F. Ziegler, "The Stopping and Range of Ions in Matter," Vol. 2, No. 6, Pergamon Press, p. 1977, 1985.

[34] The National Technology Roadmap for Semiconductors, Semiconductor Industry Association (SIA), 1994.

[35] R. V. Booth, M. H. White, H. S. Wong, and T. J. Krutsick, "The Effect of Channel Implants on MOS Transistor Characterization," *IEEE Trans. Electron Devices*, Vol. ED-34, No. 12, pp. 2501–2508, Dec. 1987.

[36] T. J. Krutsick, M. H. White, H. S. Wong and R. V. Booth, "An Improved Method of MOSFET Modeling and Parameter Extraction," *IEEE Trans. Electron Devices*, Vol. ED-34, No. 8, pp. 1676–1680, Aug. 1987.

[37] A. B. Fowler and A. M. Hartstein, "Techniques for Determining Threshold," *Surface Science*, Vol. 98, p. 169, 1980.

[38] J. Y. C. Sun, M. R. Wordeman and S. Laux, "On the Accuracy of Channel Length Characterization of LDD MOSFETs," *IEEE Trans. Electron Devices*, Vol. ED-33, No. 10, p. 1556, 1986.

[39] C. G. Sodini, T. W. Ekstedt and J. L. Moll, "Charge Accumulation and Mobility in Thin Dielectric MOS Transistors," *Solid-State Electronics*, Vol. 25, No. 8, p. 831, 1982.

[40] T. A. DeMassa and H. S. Chien, "Threshold Voltage of Small-Geometry Si MOSFETs," *Solid-State Electronics*, Vol. 29, No. 4, pp. 409–419, 1986.

[41] L. D. Yau, "A Simple Theory to Predict the Threshold Voltage of Short-Channel IGFETs," *Solid-State Electronics*, Vol. 17, pp. 1059–1063, 1974.

[42] S. G. Chamberlain and S. Ramanan, "Drain-Induced Barrier-Lowering Analysis in VLSI MOSFET Devices Using Two-Dimensional Numerical Simulations," *IEEE Trans Electron Devices*, Vol. ED-33, No. 11, pp. 1745–1752, Nov. 1986.

[43] J. H. Huang, Z. H. Liu, M. C. Jeng, K. Hui, M. Chan, P. K. Ko and C. Hu, "BSIM3 Manual (Version 2.0)," University of California, Berkeley, March 1994.

[44] Paul Richman, *MOS Field-Effect Transistors and Integrated Circuits*, John Wiley and Sons, 1973.

[45] K. Lee, M. Shur, T. A. Fjeldly and T. Ytterdal, *Semiconductor Device Modeling for VLSI*, Englewood Cliffs, NJ: Prentice Hall, 1993.

[46] G. Persky, "Thermionic Saturation of Diffusion Currents in Transistors," *Solid-State Electronics*, Vol. 15, pp. 1345–1351, 1972.

[47] C. G. Sodini, P. K. Ko and J. L. Moll, "The Effect of High Fields on MOS Device and Circuit Performance," *IEEE Trans. Electron Devices*, Vol. ED-31, No. 10, pp. 1386–1393, Oct. 1984.

[48] Y. A. El-Mansy, "MOS Device and Technology Constraints in VLSI," *IEEE Trans. Electron Devices*, Vol. ED-29, p. 567, Apr. 1982.

[49] D. S. Jeon and D. E. Burk, "MOSFET Electron Inversion Layer Mobilities – A Physical Based Semi-Empirical Model for a Wide Temperature Range," *IEEE Trans. Electron Devices*, Vol. ED-36, pp. 1456–1463, Aug. 1989.

[50] C. L. Huang and G. S. Gildeblat, "Measurement and Modeling of the N-Channel MOSFET Inversion Layer Mobility and Device Characteristics in the Temperature Range 60 - 300 K," *IEEE Trans. Electron Devices*, Vol. ED-37, p. 1289, 1990.

[51] H-J Wildau, H. Bernt, D. Friedrich, W. Seifert, P. Staudt-Fischbach, H. G. Wagemann and W. Windbracke, "The Inversion Layer of Subhalf-Micrometer N- and P-Channel MOSFETs in the Temperature Range 208 - 403 K," *IEEE Trans. Electron Devices*, Vol. 40, No. 12, pp. 2318–2324, Dec. 1993.

[52] B. Cheng and J. Woo, "A Temperature-Dependent MOSFET Inversion Layer Carrier Mobility Model for Device and Circuit Simulation," *IEEE Trans, Electron Devices*, Vol. 44, No. 2, pp. 343–345, Feb. 1997.

CHAPTER 4

Low-Voltage BiCMOS Digital Circuits

4.1 Introduction

This chapter presents an in-depth analysis and the development of a new generation of BiCMOS circuits for present and future VLSI requirements. The work is motivated by the increasing demand in both the speed and low-power consumption at low-power supply voltages. First, two techniques are investigated to improve the performance of the conventional BiCMOS logic gates for low-voltage applications. The first technique uses the source-well tie pMOS/npn combination. The second technique employs the quasi-reduction of the bipolar turn-on voltage. These two techniques are demonstrated in section 4.2.

Next, in section 4.3, full-swing BiCMOS logic circuits using the common-emitter (CE), gated-diode (GD), and emitter-follower driver configurations are presented. These configurations have been used to design full-swing logic circuits that give high performance at low-voltage supply. In fact, the full-swing complementary MOS/bipolar logic (FS-CMBL) circuits can operate down to a power supply voltage of 1.6 V.

In section 4.4, we introduce the Merged BiCMOS (MBiCMOS) logic circuits. These circuits achieve performance leverage over the CMOS in the sub-2 V regime without utilizing any additional devices and/or processing steps. They have the advantage of low-power dissipation, consume the same power as the conventional BiCMOS gate, and with only one additional process step, they outperform the CMOS circuit down to a power supply voltage of 1.9 V.

Another class of full-swing BiCMOS circuits that uses the multi-drain and multi-collector structures is discussed in section 4.5. Although these circuits require a complementary BiCMOS technology, the comparative evaluation has shown that they are still efficient even for supply voltages less than 2 V. One major advantage of using these circuits is that they have symmetrical rise and fall times.

Section 4.6 covers the quasi-complementary BiCMOS circuits (QC-BiCMOS). This type of circuits requires a high performance complementary BiCMOS process. Two key techniques are used for low-voltage applications: (1) having a separation between the base of the pull-up bipolar and the base of the quasi-pnp transistor, and (2) carefully designing the base discharging circuit to enable the QC-BiCMOS to outperform the CMOS even below a power supply voltage of 3 V.

The Schottky full-swing BiCMOS/BiNMOS logic circuits are discussed in section 4.7. The idea is based on the concept of confining the excess minority charge in the base of the pull-up npn BJT (Q1) so that the charge will keep Q1 conducting until it saturates. This technique has successfully extended the operating voltage range of the Schottky BiCMOS/BiNMOS circuits down to a supply of 2 V.

Section 4.8 covers different kinds of feedback BiCMOS circuits. In general, the R + N type and the feedback (FB) type BiCMOS logic gates degrade for power supply voltage below 3.3 V. While circuits using the capacitively coupled feedback technique extend the power supply operating range down to 1.5 V, the complementary feedback BiCMOS (FB-BiCMOS) circuit has been used for operating below 1.5 V.

Section 4.9 presents the high-β BiCMOS (Hβ-BiCMOS) logic circuits. These circuits have low-input capacitance and achieve high-speed operation at sub-2 V power supply voltage. It has been shown that if the Hβ-BiCMOS circuit is combined with a BiCMOS charge pump, the operating power supply voltage can be further reduced.

Section 4.10 describes the transient saturation full-swing technique. As the name implies, the bipolar devices are saturated during the transient period. This method is by far the most promising for BiCMOS VLSI in the sub-2 V regime.

Finally, in section 4.11, a number of circuits employing the bootstrapping technique is demonstrated. This method is very effective in extending the operating voltage range of the BiCMOS circuits down to the sub-1.5 V regime.

4.2 Source-Well Tie and Quasi-Reduction of Bipolar Turn-On Voltage Techniques

This section discusses two techniques to improve the performance of the conventional BiCMOS logic gates for reduced supply operations. First, a BiCMOS configuration employing the source-well tie pMOS/npn pull-down combination [1] is analyzed to show the improved performance at low supply voltages. The same configuration, without the source-well tie, is also analyzed and compared with the former. Second, a design technique to quasi-reduce the output bipolar transistor turn-on voltage is discussed [2].

4.2.1 pMOS/NPN Pull-Down Technique With and Without Source-Well Tie

Figure 4–1(a) shows the conventional BiCMOS circuit. Its performance degrades as the supply voltage is scaled down due to the reduction of the available gate-to-source voltage to turn

Source-Well Tie and Quasi-Reduction of Bipolar Turn-On Voltage Techniques 133

on the nMOS device. Thus the ability of the nMOS to supply the base current of the npn BJT during the pull-down phase is significantly reduced by lowering the supply voltage. This situation is worsened by the back-gate bias (body) effect on the pull-down MOSFET.

Figure 4–1(b) shows a pMOS/npn pull-down configuration [3] used to improve the low-voltage logic performance of the BiCMOS. The n-well of the pMOS is tied to the V_{DD} rail as it would be in a CMOS logic environment. The CMOS inverter (shown in Figure 4–1(b)) supplies the gate voltage of the pull-down pMOS device. Such a design eliminates the forward diode drop at the gate of the pull-down MOSFET, as compared to that of the conventional BiCMOS circuit. The pMOS device is thus able to supply more current to the output npn transistor due to a higher gate-to-source voltage available at its input. A major setback of the pMOS/npn pull-down configuration is that the source terminal of the pull-down pMOS device is connected to the output node. As the output voltage in dynamic conditions swings from a high logic of (V_{DD} − V_{BE}) to a low logic of V_{BE}, the threshold voltage of this transistor will be affected by the back-gate bias effect during the pull-down transition. This back-gate bias effect will, in turn, affect the ability of the pMOS device to drive the base of the pull-down npn transistor.

A technique to solve the above mentioned problem is presented in Figure 4–1(c). In this design, the source and the n-well of the pMOS device are tied together to the output node. This configuration helps to eliminate the increase in the threshold voltage due to the back-gate bias effect.

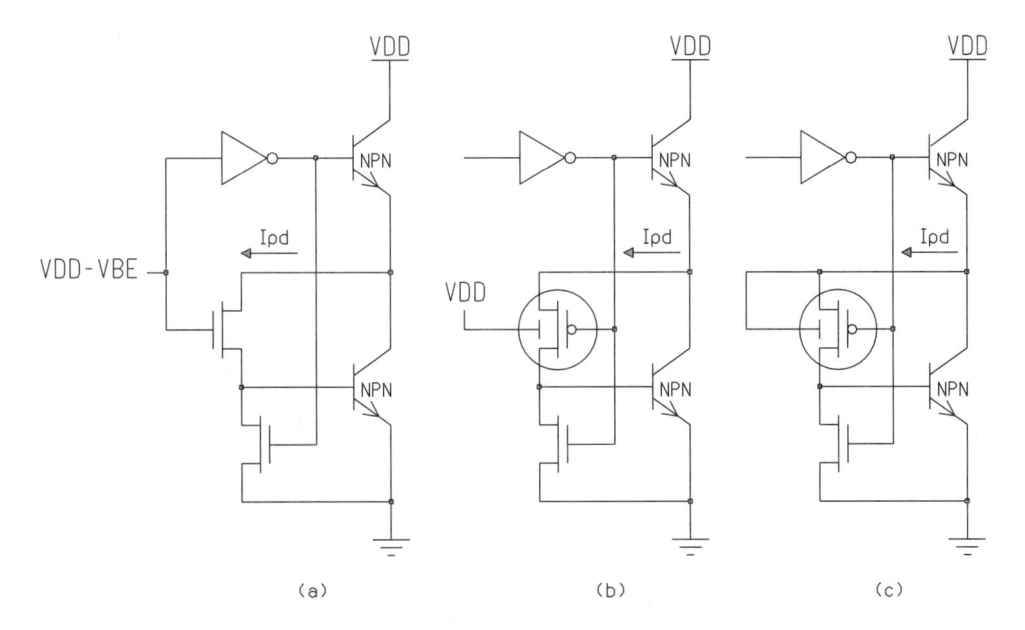

Figure 4–1 BiCMOS inverter gate configuration (a) conventional (b) pMOS/npn pull-down (c) pMOS/npn pull-down with source-well tied together.

The simulation results shown in Figure 4–2 confirm that with the n-well being electrically tied to the source, a greater amount of the base current can be achieved as compared to that in the pMOS/npn pull-down without the source-well tie-up. Figure 4–3 shows the period of the ring oscillator measured as a function of the supply voltage with an output load of 4 pF. The pMOS/npn pull-down with the source-well tied together shows superior low-voltage performance as compared to the other two configurations.

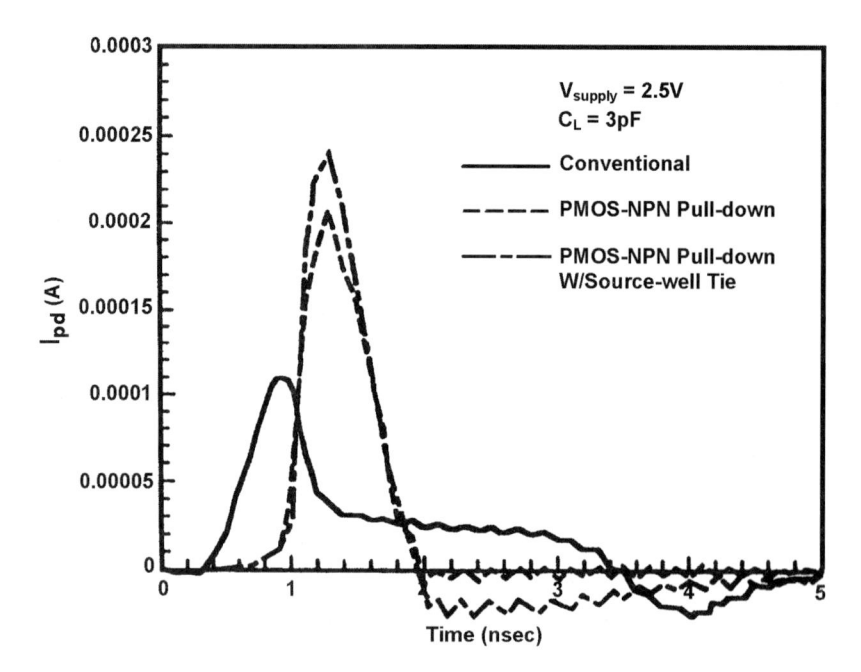

Figure 4–2 Simulated pull-down base drive current waveforms for the BiCMOS configurations shown in Figure 4–1.

Next, the temperature effects on the BiCMOS gate performance are investigated. The two main parameters responsible for the poor performance at a reduced voltage in a conventional BiCMOS logic gate are the threshold voltage and the forward emitter-base junction V_{BE} of the pull-down MOSFET. Both parameters are known to be strongly influenced by temperature fluctuations. Figure 4–4 shows the average pair delay time of a BiCMOS ring oscillator as a function of the supply voltage at two different temperatures of 130 °C and 0 °C.

As shown in Figure 4–4, the delay decreases at high temperatures for supply voltages below 2.5 V. This decrease can be explained by the fact that both the V_{BE} of the bipolar npn and the threshold voltage of the MOSFET decrease with temperature [4,5]. Furthermore, the bipolar forward gain has a positive temperature coefficient [6]. All these factors contribute to a better low-voltage performance at higher temperatures.

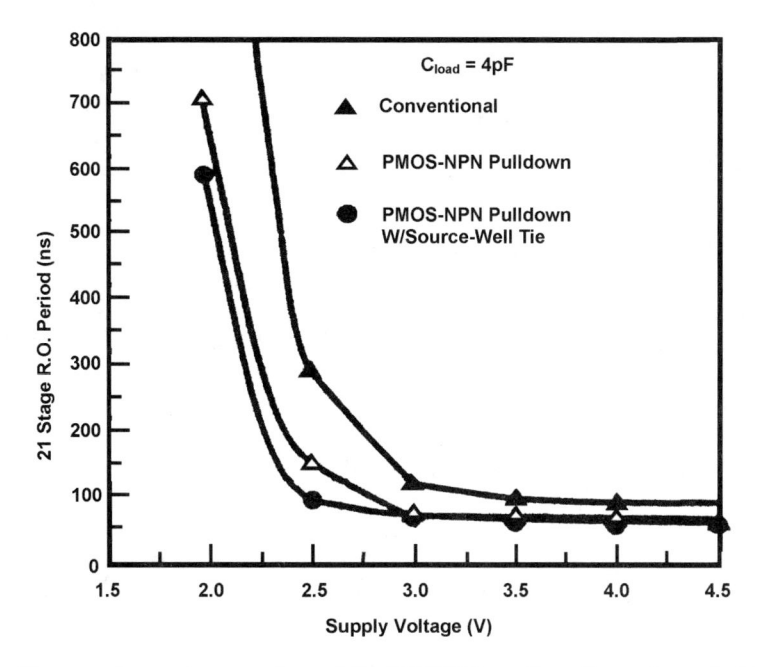

Source-Well Tie and Quasi-Reduction of Bipolar Turn-On Voltage Techniques 135

Figure 4–3 Measured speed comparison of the BiCMOS configurations shown in Figure 4–1 at room temperature.

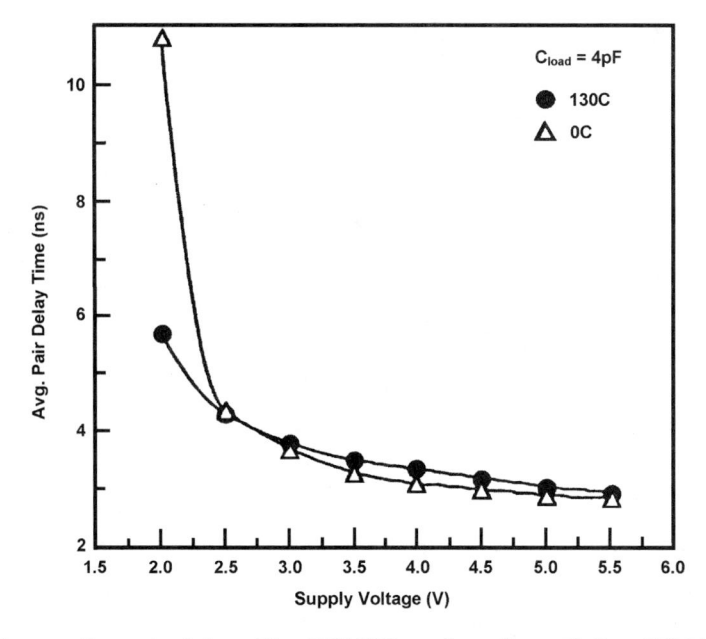

Figure 4–4 Propagation gate delay of the BiCMOS configuration with the pMOS/npn pull-down (source-well tie together) at 130 °C and 0 °C.

In Figure 4–5, we show the pair delay times of various BiCMOS gate configurations at −10 °C. It is expected that the low-voltage performance of the various BiCMOS gates will be poor at this temperature due to the negative temperature coefficients of the emitter-base voltage of the bipolar device. Also, low temperature degrades the gain of the bipolar transistor. However, Figure 4–5 shows that despite the above mentioned limitations, the pMOS/npn pull-down with the source-well tied together performs better at 2.5 V supply as compared to the pMOS/npn pull-down without the source-well tie-up.

In retrospect, a novel BiCMOS configuration employing the source-well tie pMOS/npn pull-down combination has been shown to improve the BiCMOS gate performance at reduced supply voltages. The source-well tie pMOS/npn outperforms its counterpart (without source-well tie) and the conventional BiCMOS in the low-voltage supply regime, at both high and low temperatures.

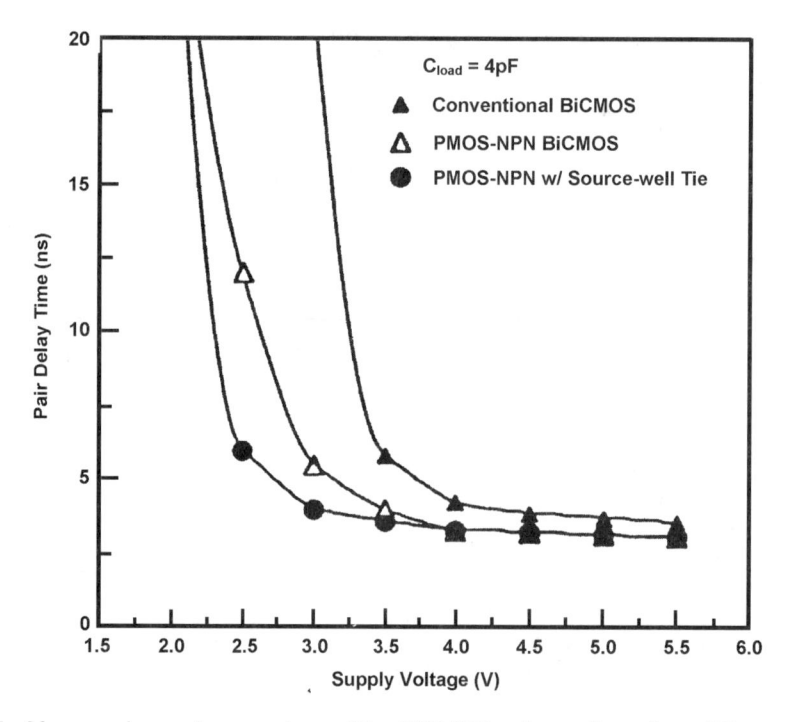

Figure 4–5 Measured speed comparison of the BiCMOS gate configuration of Figure 4–1 at −10 °C.

4.2.2 Quasi-Reduction of Bipolar Turn-On Voltage Technique

One important characteristic that determines the performance of the conventional BiCMOS circuit is the charge-up time (bipolar turn-on time), which is determined by the current

Source-Well Tie and Quasi-Reduction of Bipolar Turn-On Voltage Techniques 137

drivability of the MOSFETs. When the power supply voltage is reduced, the charge-up time degrades because the gate-to-source voltage of the MOS device driving the bipolar transistor is decreased. A design technique has been employed [2] to quasi-reduce the bipolar transistor turn-on voltage (V_{BEon}) from V_{bi}, where V_{bi} is the emitter-base built-in voltage, to about 0.2 V (denoted as V_a).

The basic low-voltage operational bipolar FET (LV-BiFET) employing the quasi bipolar turn-on voltage technique is shown in Figure 4–6. It consists of one p-type MOSFET, one depletion mode n-type MOSFET and two npn bipolar transistors.

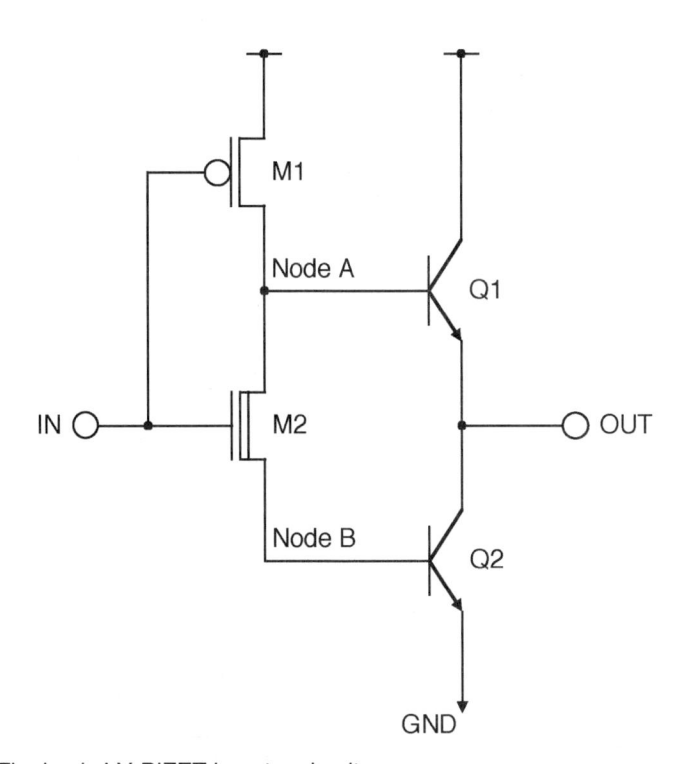

Figure 4–6 The basic LV-BiFET inverter circuit.

4.2.2.1 Circuit Operation

During the pull-up phase, the input voltage goes from high to low and causes the transistor M1 to turn on and the transistor M2 to turn off. Consequently, Q1 conducts and Q2 turns off. Hence, the output node voltage rises to a high level given by ($V_{DD} - V_{BE}$) (Figure 4–7(a)). For the pull-down phase, the input voltage transits from low to high. As such, M1 turns off and M2 turns on. The base node voltage of Q2 will be brought close to that of Q1. The output node will sink the current through Q2 to ground, thereby resulting in a low-output logic of approximately 0 V.

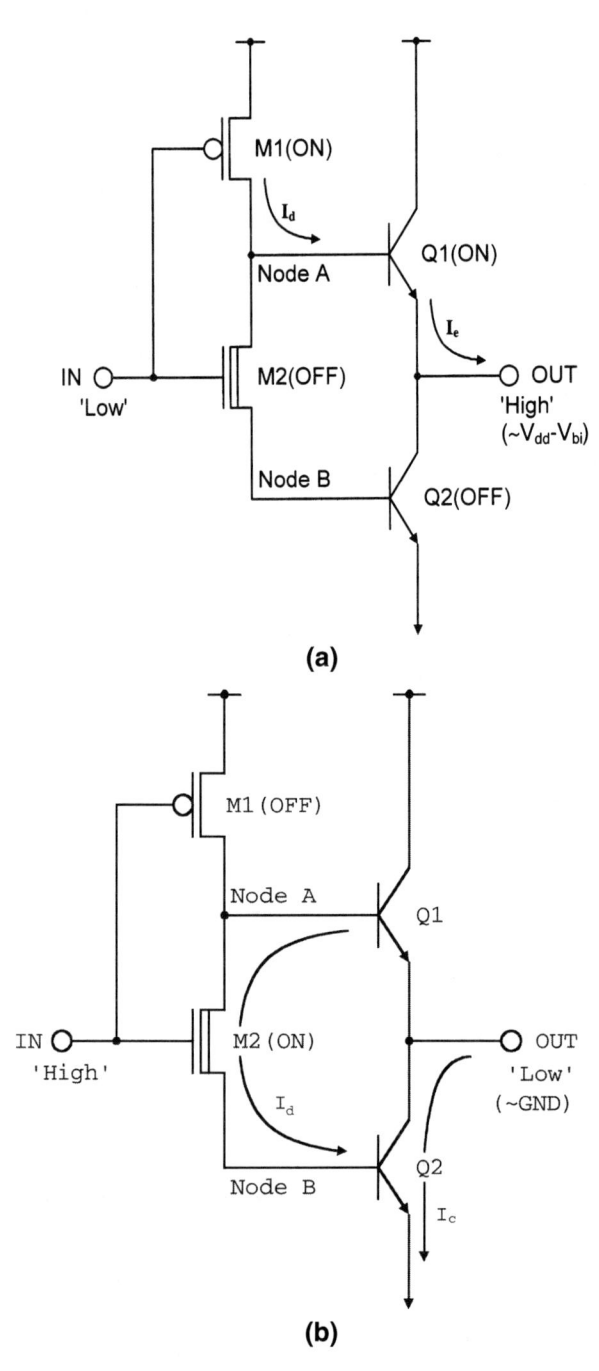

Figure 4–7 (a) Pull-up operation (b) pull-down operation of the LV-BiFET.

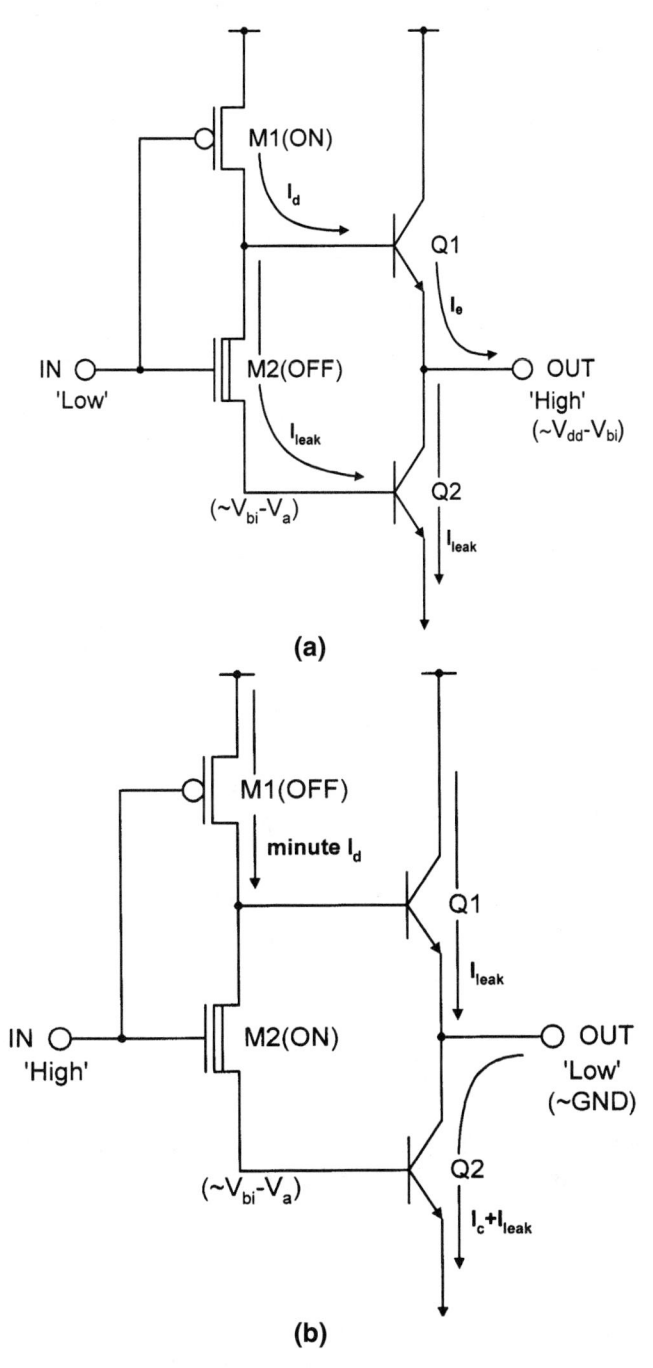

Source-Well Tie and Quasi-Reduction of Bipolar Turn-On Voltage Techniques

Figure 4–8 Retaining operation (a) pull-up transition (b) pull-down transition.

Figure 4–8 shows the retaining operation of the LV-BiFET during the pull-up and pull-down phases. During the pull-up operation, M1 turns on and its drain current charges up the base voltage of the transistor Q1 to V_{DD}. Meanwhile, the base voltage of Q2 is kept at $(V_{bi} - V_a)$ due to the subthreshold current of M2 and the base leakage current of Q2. Therefore, the output voltage is maintained at its high level. For the pull-down phase, M2 turns on and equalizes the base node voltages of Q1 and Q2. Consequently, Q2 conducts and Q1 turns off. Although M1 is off, its subthreshold current serves to retain the base nodes of Q1 and Q2 at $(V_{bi} - V_a)$, in spite of the very small leakage currents in Q1 and Q2. Figure 4–9 shows the voltage and current characteristics of the LV-BiFET inverter at a power supply voltage of 1.6 V.

In Figure 4–9(a), we can see that the low level voltage of node A is about 0.7 V and that of node B is about 0.6 V. Hence, the turn-on voltage of Q1 and Q2 has been quasi-reduced to about 0.2 V. In order for the LV-BiFET to perform at high-speed with low-power consumption, optimization of the threshold voltage of M1 and the capacitance of nodes A and B must be done.

4.2.2.2 Simulation Results and Analysis

SPICE simulations, using the 0.55 μm BiCMOS technology, were performed on the LV-BiFET, BiNMOS, BiCMOS, CMOS, and NTL circuits [7]. Figure 4–10 gives a detailed description of the various circuits. Note that all the circuits have an input capacitance of 0.1 pF. Figure 4–11 shows the voltage characteristics of various BiCMOS/CMOS circuits for a fan-out of 4. The LV-BiFET, BiNMOS, and CMOS circuits are simulated at a power supply voltage of 1.6 V while the NTL circuit is simulated at 1.9 V. Figures 4–12 and 4–14 show the dependence of the delay time and the power consumption on the power supply voltage. The LV-BiFET achieves a 40% reduction in the delay time and a 45% improvement in the power consumption over the CMOS circuit, as well as a 40% reduction in delay time and a 35% improvement in power consumption over the BiNMOS circuit at a power supply voltage of 1.6 V. Figures 4–13 and 4–15 show the dependence of the inverter delay time and power consumption on the fan-out. The LV-BiFET shows the least power consumption with respect to the fan-out and its inverter delay time is second only to that of the NTL circuit. Figure 4–16 shows the simulated power-delay characteristics for the various circuits.

The LV-BiFET shows the lowest power-delay product, and it achieves an improvement of 70% over CMOS, 60% over BiNMOS and 55% over NTL. Table 4–1 summarizes the delay, power, area, and delay-power product comparison of the various circuits.

4.3 Full-Swing BiCMOS Logic Circuits with Complementary Emitter-Follower Driver Configuration

4.3.1 Introduction

In BiCMOS logic circuits, there are three generic types of drivers: the common-emitter (CE), the gated-diode (GD), and the emitter-follower (EF) driver configurations. They all use switching MOSFETs to supply the base current and the BJTs to drive the output nodes. The comparative evaluation of these driver configurations has shown that the EF with full-swing

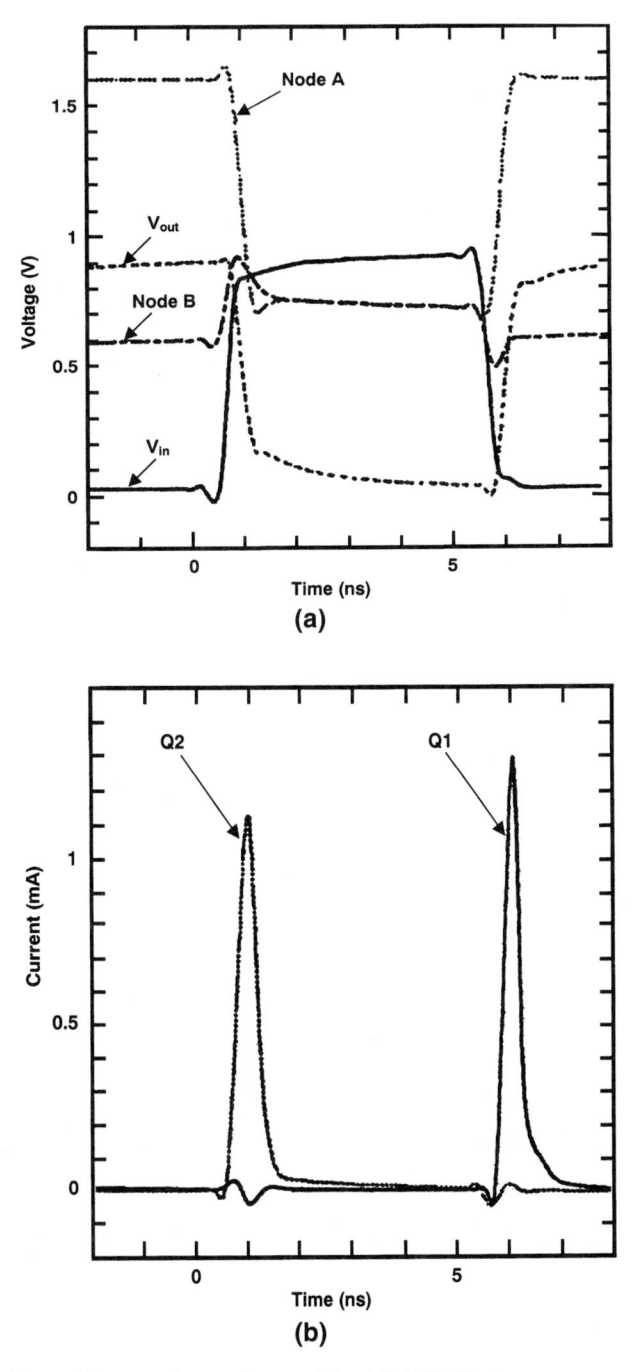

Figure 4–9 (a) Voltage (b) current waveforms of the LV-BiFET at a power supply voltage of 1.6 V and a fan-out of 4.

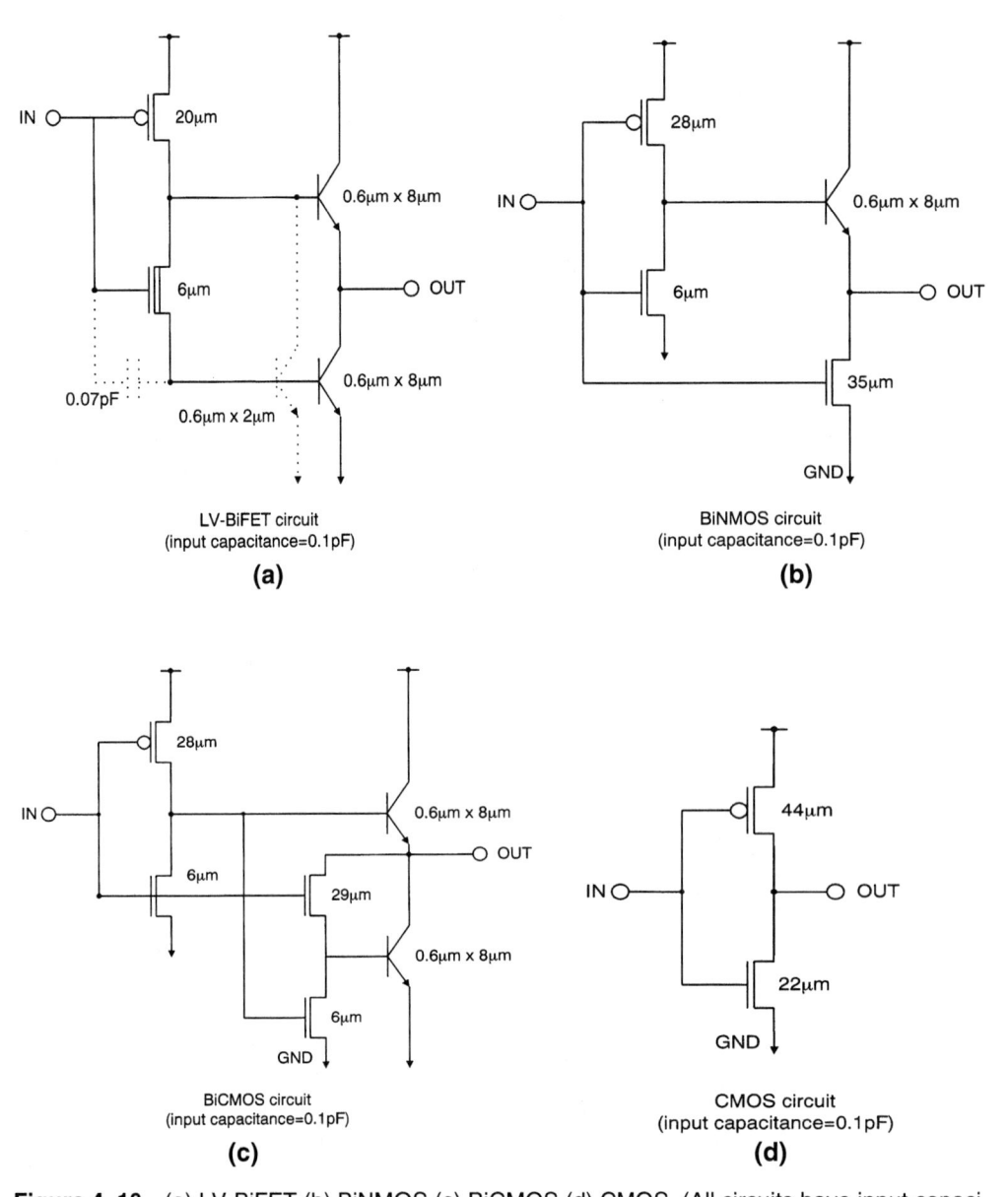

142　　　　　　　　　　　　　　　　　　Chapter 4 • Low-Voltage BiCMOS Digital Circuits

Figure 4–10 (a) LV-BiFET (b) BiNMOS (c) BiCMOS (d) CMOS. (All circuits have input capacitance of 0.1 pF.)

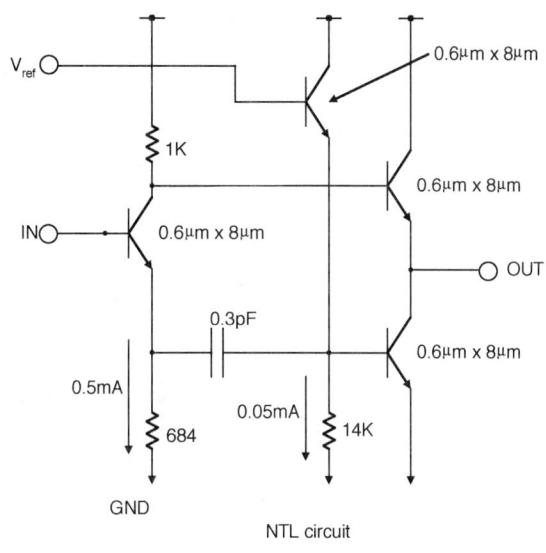

Figure 4–10 (cont.) (e) NTL circuits

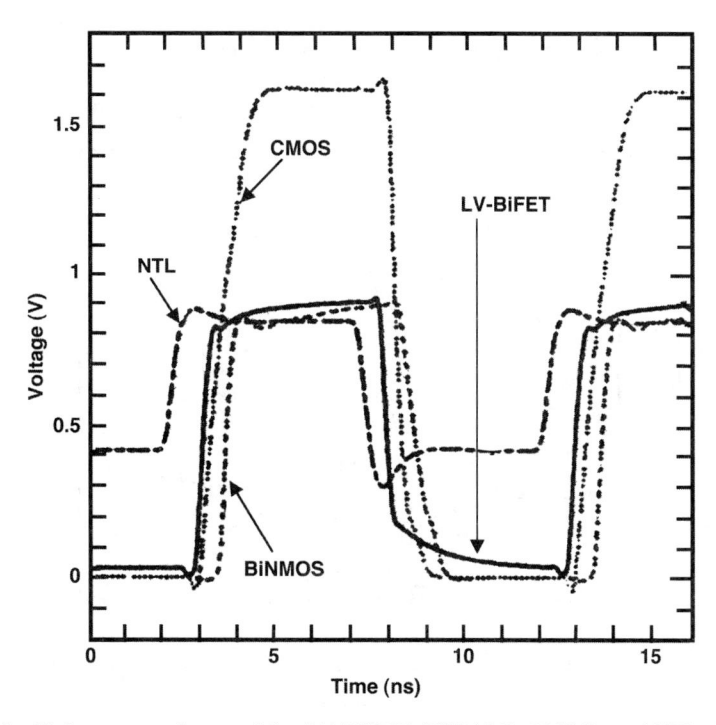

Figure 4–11 Voltage waveforms of the LV-BiFET, BiNMOS, CMOS, and NTL circuits.

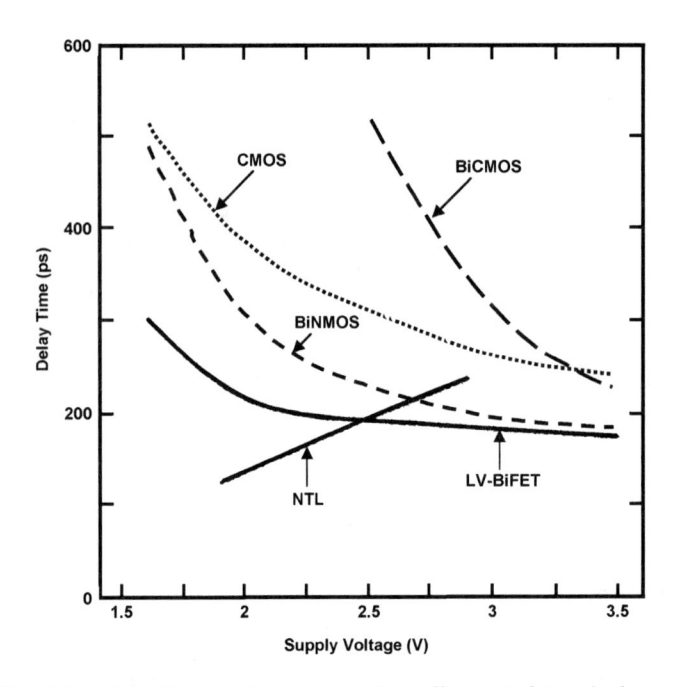

Figure 4–12 Simulated delay time vs. the supply voltage (fan-out of 4 and a frequency of 100 MHz).

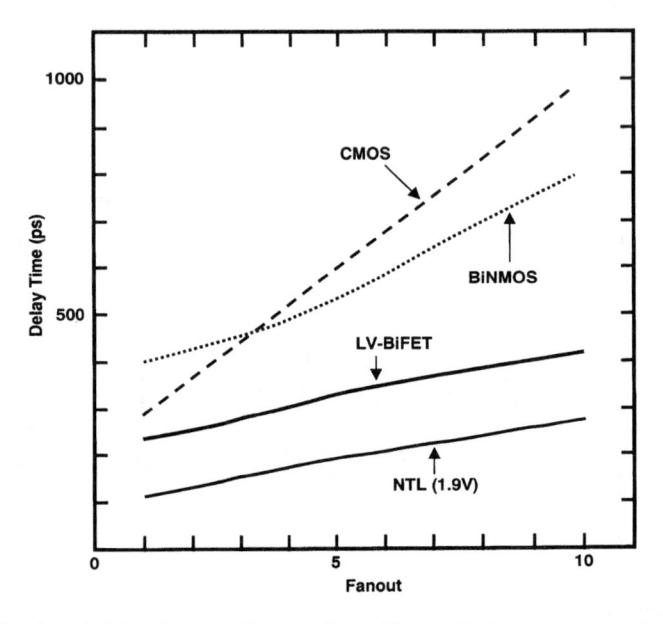

Figure 4–13 Simulated delay time vs. the number of fan-out at a power supply voltage of 1.6 V and a frequency of 100 MHz. Open circles represent the delay time at a load of 1 pF.

Full-Swing BiCMOS Logic Circuits with Complementary Emitter-Follower Driver Configuration 145

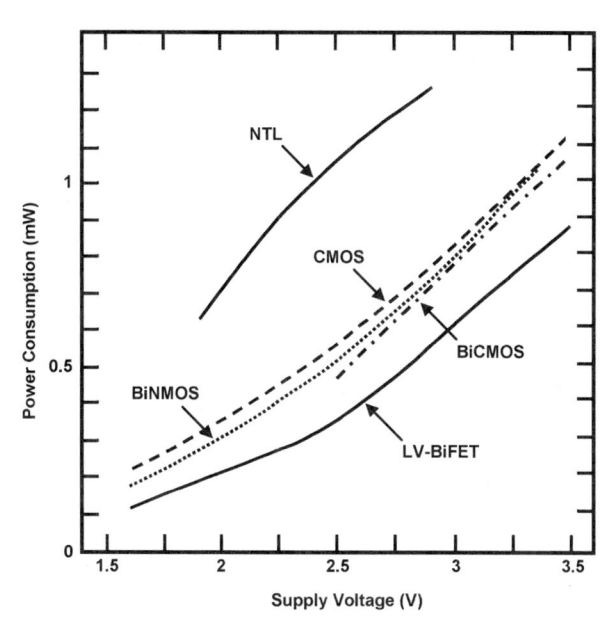

Figure 4–14 Simulated power consumption vs. the supply voltage (fan-out of 4 and a frequency of 100 MHz).

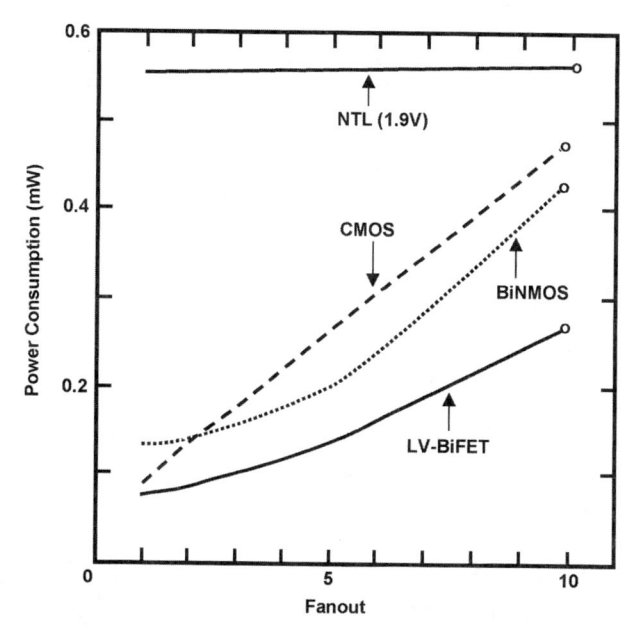

Figure 4–15 Simulated power consumption vs. the number of fan-out at a power supply voltage of 1.6 V and a frequency of 100 MHz. Open circles represent the power consumption at a load of 1 pF.

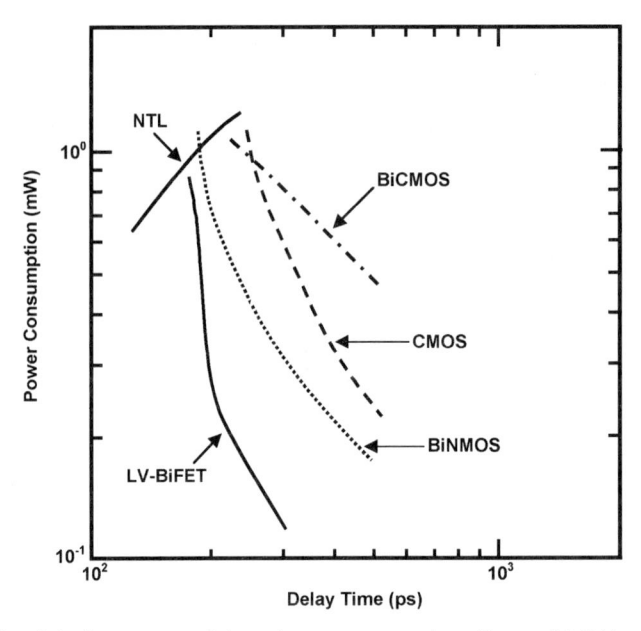

Figure 4–16 Simulated power vs. delay at a power supply voltage of 1.6 V ~ 3.5 V (fan-out of 4 and a frequency of 100 MHz).

Table 4–1 Performance comparison of the LV-BiFET, BiNMOS, BiCMOS, NTL, and CMOS circuits.

	LV-BiFET	BiNMOS	CMOS	BiCMOS	NTL
Delay (ps)	304	492	515	519	126
Power (mW)	0.119	0.176	0.221	0.466	0.632
Area (normalized)	3.5	2.5	1	4	7
Delay-Power Product (pJ)	0.036	0.087	0.114	0.242	0.080

techniques has improved the BiCMOS performance for reduced supply voltages and scaled technologies [8,9]. Furthermore, it was demonstrated experimentally that circuits utilizing a complementary EF for efficient driving, and a switched base-emitter shunting to achieve a full swing have improved performance over the conventional BiCMOS and the CMOS circuits [10]. Various full-swing complementary MOS/bipolar logic (FS-CMBL) circuits with complementary EF have been implemented and compared to illustrate the performance leverage over the conventional BiCMOS circuits for low voltage operations [11]. The next few sections will discuss the various types of driver configurations.

4.3.2 BiCMOS Driver Configurations

4.3.2.1 Common-Emitter Driver Configuration

Figure 4–17(a) shows the common-emitter (CE) configuration where the pull-down BJT operates in the CE mode. During the pull-down operation, the input goes from high to low. The pMOSFET in the pull-down circuitry turns on and soon charges the base node of the output BJT transistor. The BJT becomes active and sinks a large current from the output. The base (or drain) current at that point in time is large because the MOSFET is operating in the saturation region with the drain-source bias of ($V_{DD} - V_{BE}$) shown in Figure 4–17(b). When the output is fully discharged, the BJT will be saturated since the base potential will be higher than the collector potential. This circuit has an approximately full logic swing of V_{DD} to V_{CEsat}. Furthermore, this driver configuration achieves quite a high speed due to the large near constant base current during switching. Its disadvantages are, first, its logic function is noninverting, so extra circuitry is required in order to achieve inversion. Second, because a substantial drain current flows during the nonswitching period, dc power is dissipated. Third, because the BJTs are saturated, the speed of the BiCMOS is degraded unless the base-collector junction is clamped using devices such as a Schottky diode.

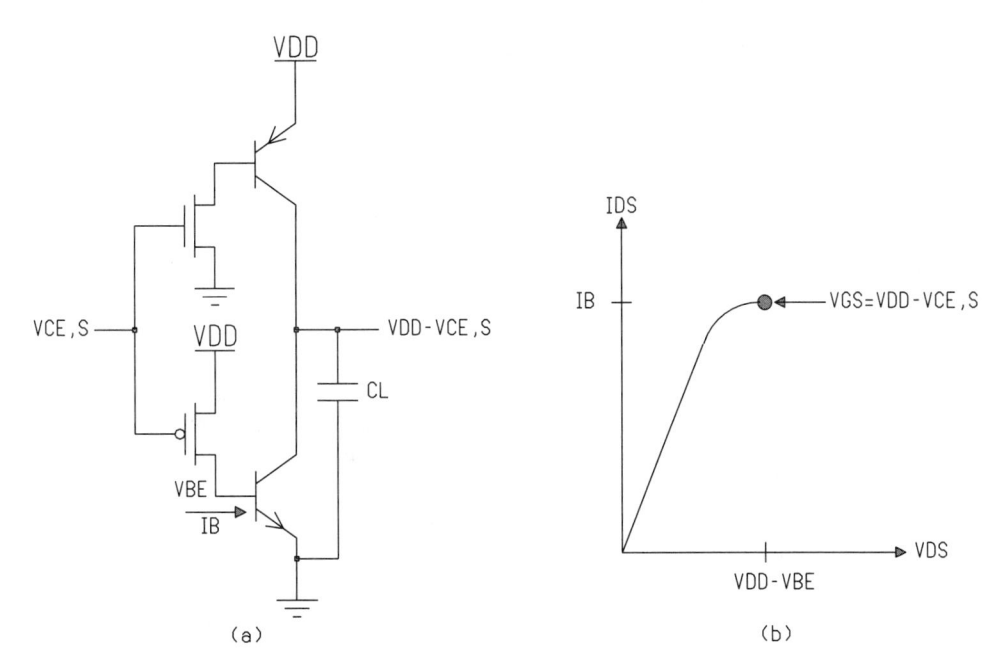

Figure 4–17 (a) Common-emitter (CE) BiCMOS driver configuration; (b) its base current during pull-down.

4.3.2.2 Gated-Diode Driver Configuration

The Gated-Diode (GD) driver configuration is shown in Figure 4–18(a). When the input is high (V_{IH}), the nMOS device turns on and the base node is charged up by a current from the output node until the BJT becomes active. As the base potential increases, the output voltage starts to drop, and as a result, both the gate-source and drain-source voltages of the nMOS transistor decrease. As shown in Figure 4–18(b), the MOSFET current decreases quite rapidly. After the BJT is biased into the active region, its base potential becomes fairly constant and the gate-source bias of the nMOS device maintains at ($V_{IH} - V_{BE}$). As the output drops, eventually the drain potential reaches V_{BE} and the base current stops flowing. Hence, the major problem of this circuit is that its output voltage swing is restricted from ($V_{DD} - V_{BE}$) to V_{BE}. This limitation results in degraded speed performance with the scaling down of the power supply voltage. Since this circuit is complementary, it suffers from the imbalance of the rise and fall times due to the unsymmetrical pull-up and pull-down sections. The pull-down circuitry, consisting of an nMOSFET and an npn BJT, is relatively faster than the pull-up circuitry, which consists of a pMOSFET and a pnp BJT. Another disadvantage of this configuration is its sensitivity to the body effect and the parasitic resistance at the base and emitter.

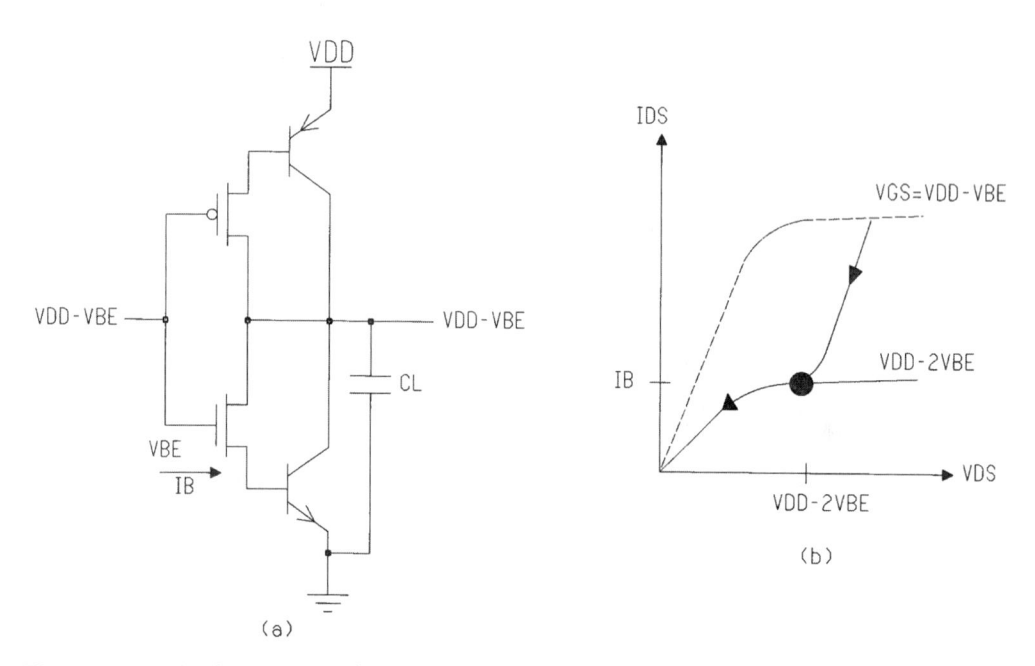

Figure 4–18 (a) Gated diode (GD) driver configuration; (b) its base current during the pull-down phase.

4.3.2.3 Emitter-Follower Driver

The Emitter-Follower (EF) driver configuration is shown in Figure 4–19(a). The nMOSFET in the pull-down circuitry operates in the common-source mode and drives the EF at the output. Because of the EF output stage, this circuit is constrained by a partial logic voltage swing like the GD driver. However, this circuit has a larger gate-source voltage and hence a better base drive because the drain node of the pull-down nMOSFET is not tied to the output node. Therefore, this configuration is more suited for scaled power supplies. Furthermore, the circuit is insensitive to the body effect and parasitic resistance. Some other advantages are, first, it is possible to physically merge the MOSFET and BJT into a compact structure to save on the area and hence reduce the parasitic capacitance. This structure is realized by having the drain of the nMOSFET and the base of a pnp BJT sharing a common n-type diffusion area. Second, this configuration has good rise and fall times balance because the pull-up circuitry consists of a pMOSFET and an npn BJT while the pull-down circuitry consists of an nMOSFET and a pnp BJT. Third, the speed performance of the circuit is not degraded since the BJTs driving the output node do not saturate.

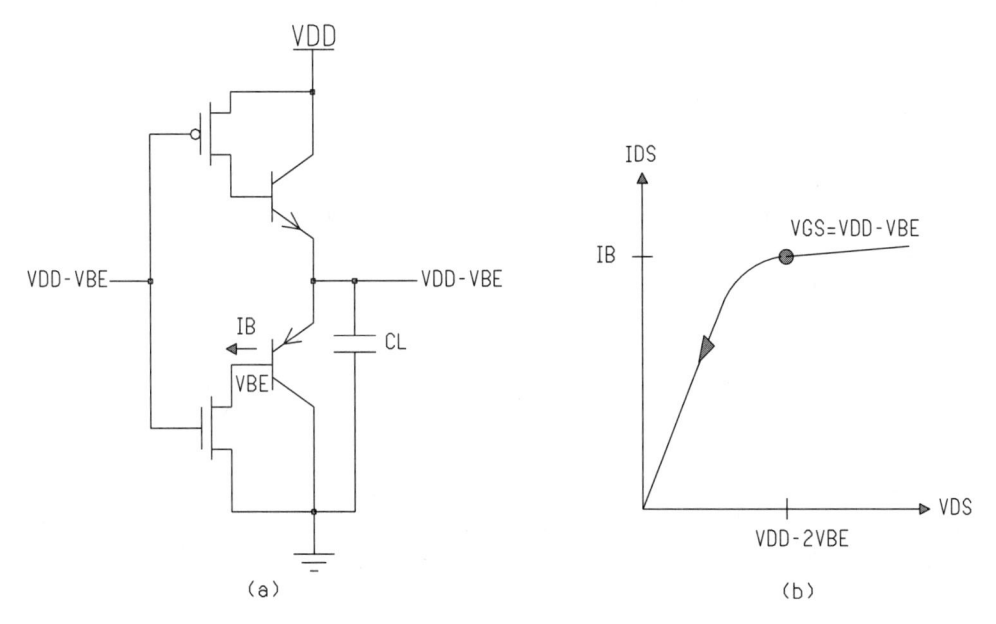

Figure 4–19 (a) Emitter-follower (EF) driver configuration; (b) its base current during the pull-down phase.

4.3.3 Full-Swing Techniques

In the previous sections, the GD and the EF driver configurations are restricted to partial output voltage swings due to the small gate-source voltages. Such restriction will, in turn, affect the speed of these circuits. The first technique to achieve the full-swing is to use a resistive shunt network between the collector and emitter of the BJT in the pull-down section for the GD configuration and the BJT in the pull-up section for the EF configuration (Figure 4–20(a)). The shunt network can be a simple resistor, a MOSFET, or a CMOS positive feedback circuit. With the shunt element adding current to the output, the final stage of the transient response is improved. However, this technique introduces a crossover current that flows through the shunt element when, for example, the output begins to fall from a high voltage level. This current will cause an increase in the power dissipation and also slows down the pull-down transient response.

Although these problems can be solved by increasing the resistance value of the shunt element, doing so increases the total time required to achieve a full-swing operation. The second technique, shown in Figure 4–20(b), is a resistive network to shunt the base and emitter of the respective BJTs. This technique does not produce a crossover current through the shunt element. Also, the shunt element clamps the base-emitter junction of the BJT, when the MOSFET is off, for the opposite output transition. The major drawback of this technique is that if a passive device is used for shunting, it diverts part of the MOSFET current flowing into the base during the main transient period, and this diversion will degrade the speed of the transistor. Again this problem can be solved by increasing the resistance value at the expense of a longer time to achieve a full output voltage swing.

4.3.4 Comparison Between the Three Different Driver Configurations

For comparison purposes, the MOSFET transconductance is set to be equal for the different circuits, and the substrate bias effect on the MOSFET threshold voltage is ignored. Identical BJTs with zero parasitics are assumed for the different circuits. Table 4–2 shows the key parameters and their values used in the simulations. For the gated-diode and emitter-follower circuits, both partial-swing (PS) and full-swing (FS) cases are considered. Also, for the GD driver, a case of intermixing the BiCMOS circuit with pure CMOS (Figure 4–21(a)) is compared with that without intermixing.

The relative maximum base currents of the various configurations as a function of the supply voltage are plotted in Figure 4–22. With the absence of FS techniques, the Common-Emitter (CE) provides the largest base current for all supply voltages. The GD with intermixing shows much less current than the one without intermixing. The base current decreases as the supply voltage decreases and eventually, the circuit fails to operate below 2.3 V. By implementing FS techniques, the base currents of the EF and GD configurations increase substantially. The base current of the FS EF circuit is the largest and that of the FS GD is comparable to the current of the CE driver.

In summary, although the CE type has a good driving capability, it may not be suitable for VLSI circuits due to the saturation of bipolar transistors and the relatively high power dissipation. On the other hand, the EF shows the best performance with reduced power supply voltages, and it does not experience any of the above mentioned problems. Techniques to achieve a full-swing operation have also been discussed and the base-emitter shunting is more favorable than collector-emitter shunting due to less crossover current.

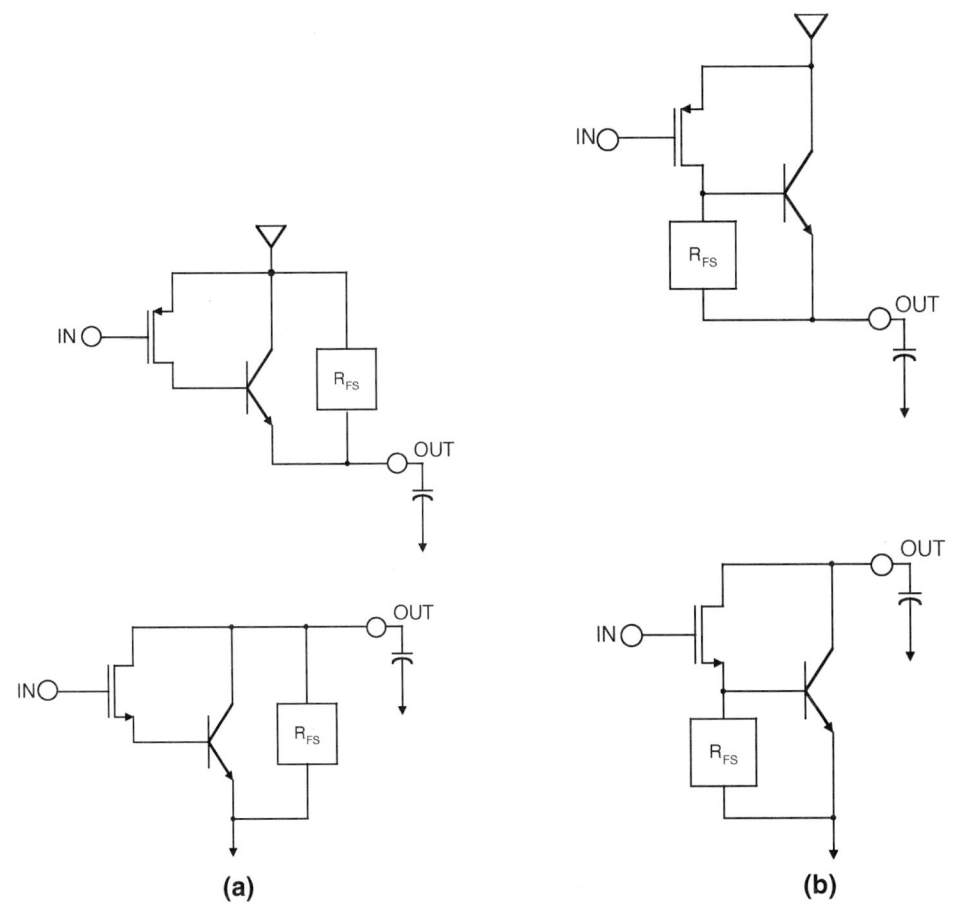

Full-Swing BiCMOS Logic Circuits with Complementary Emitter-Follower Driver Configuration   151

Figure 4–20 (a) Collector-emitter shunt (b) base-emitter shunt full-swing techniques.

Table 4–2 Key parameters and their values used in the simulation.

- Zero-bias threshold voltage: $V_{TO} \cong 0.2\, V_{DD}$.
- Minimum turn-off margin for MOSFETs: 0.2 V.
- BJT turn-on voltage: $V_{BE} = 0.7$ V.
- Clamped collector voltage before saturation: $V_{CE,M} = 0.3$ V.

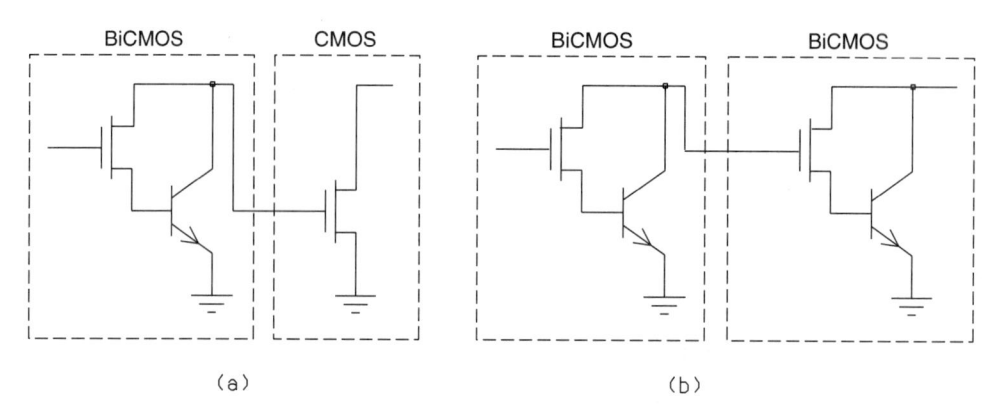

Figure 4–21 (a) BiCMOS-to-CMOS interface when BiCMOS circuits are intermixed with pure CMOS (b) BiCMOS-to-BiCMOS interface without intermixing.

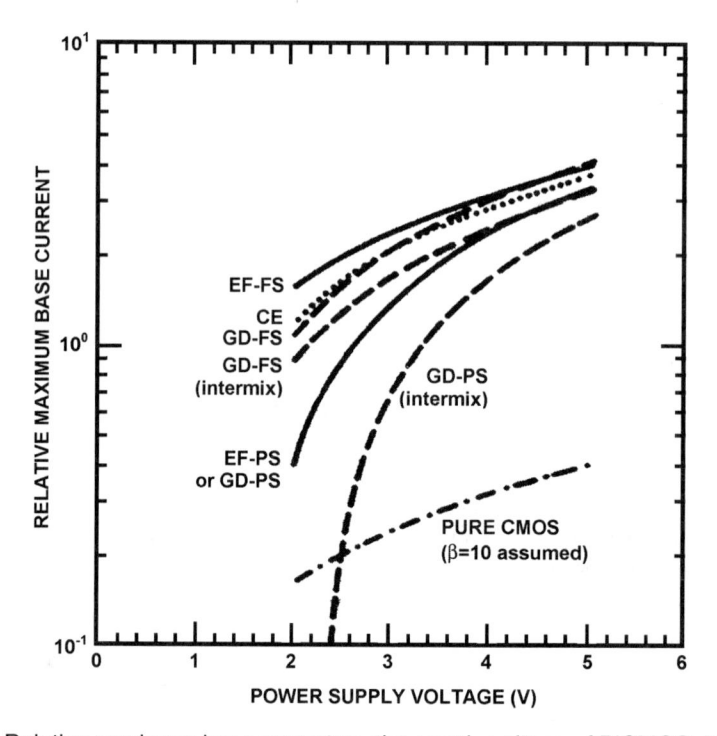

Figure 4–22 Relative maximum base current vs. the supply voltage of BiCMOS drivers.

4.3.5 Full-Swing Complementary MOS/Bipolar Logic (FS-CMBL) Circuits

This section describes two high-speed FS BiCMOS circuits that utilize a complementary EF configuration for efficient driving, switched base-emitter shunting to achieve an FS, and CMOS diodes for base-to-base clamping. The performance of these circuits will be demonstrated experimentally using a 0.8 µm BiCMOS process.

4.3.5.1 FS-CMBL Circuit without Feedback

Figure 4–23 shows the basic FS-CMBL circuit realizing a two-input NAND function [10]. In the initial state, with both inputs (A and B) being high, the node Y is fully discharged to ground via transistors MN1 and MN2. Because transistors MP3 and MN3 have ground and V_{DD} as their gate voltages, respectively, they will turn on once their source-gate voltage (or gate-source voltage) exceeds their threshold voltage. With the node Y discharged to ground, the gate-source voltage of nMOSFET MN3 exceeds V_{TN} and it turns on. Without MN3, the node O would stay at V_{BE}. However, with MN3, nodes Y and O are shorted through MN3, and the output is pulled down to low. Thus QP1 turns off. The CMOS diode, made up of a parallel pair of nMOS diode (MN4) and pMOS diode (MP4), acts as a transmission gate. As the node Y is at ground level, and the gate of MP4 is connected to Y, MP4 will turn on, shorting the node X to Y. This will ensure that QN1 remains off so that the output node is brought successfully to a low value.

If the logic at input A changes to low, MP2 turns on and its drain current starts to charge up the node X, which will eventually be pulled up to V_{DD}. Because the node X is connected to the base of QN1, QN1 will be biased into the active region. Hence the output node will be pulled up to $(V_{DD} - V_{BEon})$. This is the upper bound of the output high voltage without MP3. But once the node X reaches V_{TP}, the pMOS device MP3 turns on providing an alternative path to charge the output up to V_{DD}. Meanwhile, part of the MOSFET current from MP2 flows via MN4 to charge up the node Y. As MN2 is off, this current will flow into the base of QP1 and will keep QP1 off. Furthermore, MN3 will be on to keep QP1 off.

A major drawback of this circuit is that MP3 (or MN3) starts to turn on as soon as the source-gate (or gate-source) voltage is greater than V_{TP} (or V_{TN}) during pull-up (or pull-down). This will result in a premature bypass of the base drive current, which must actually go through QN1 or QP1 for current amplification. This will effectively degrade the circuit performance and result in a slower switching speed. A solution of this problem is presented in the next section, which discusses the use of positive feedback to enhance the speed [10].

4.3.5.2 FS-CMBL Circuit with Feedback

Figure 4–24 shows the improved FS-CMBL circuit utilizing positive feedback. The circuit is basically similar to the one shown in Figure 4–23 except that the gates of MN3 and MP3 are driven by the output of a CMOS inverter, made up of a pMOSFET MP5 and an nMOSFET MN5, which inverts the output signal. The purpose of the inverter is to delay the turning on of the device MP3 (or MN3) during the pull-up (or pull-down) cycle until the output reaches the

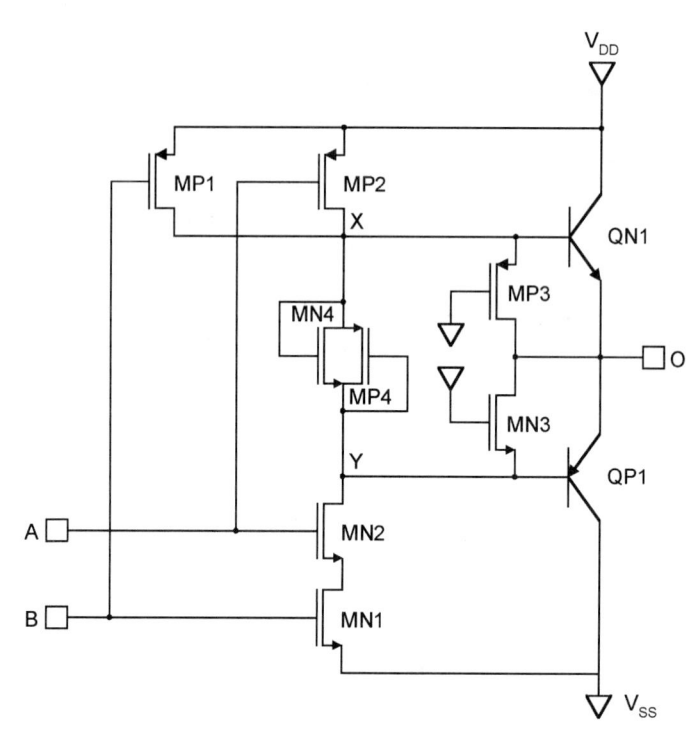

Figure 4–23 Basic FS-CMBL two-input NAND circuit with emitter-follower pull-up and gated-diode pull-down circuitry.

valid output high- or low-voltage level, respectively. When both inputs are high, the node Y is discharged to GND and, assuming that the output is initially high, transistor MN3 turns off. As a result, the transistor QP1 turns on and the output voltage starts to discharge to a low value. When the output reaches V_{BEon}, which is lower than the logic threshold voltage of the CMOS inverter (~ half V_{DD}), the output of the inverter becomes high and the device MN3 turns on. The conduction of MN3 shorts the node O to Y and pulls down the output voltage to ground. At the same time, QP1 turns off. Because MN3 turns on only when the output reaches the valid low logic, no premature bypass of current occurs. Furthermore, below V_{BEon}, QP1 starts to turn off and, therefore, it does not matter whether the current flows into the pnp transistor or the nMOS device MN3 since QP1 no longer provides a current amplification.

Similarly for the case when the input A changes to low, MP3 remains off until the output reaches the valid logic high level. Then MP3 turns on to bypass a fraction of the base current of QN1. Although the conduction of MP3 causes the final transient response to be slower, because the output has already changed its logical state, the circuit speed will not be degraded. This circuit performs better than the circuit in Figure 4–23 due to the latency in the shunting provided through the feedback inverter. However, the propagation delay of the inverter will add to the overall circuit delay.

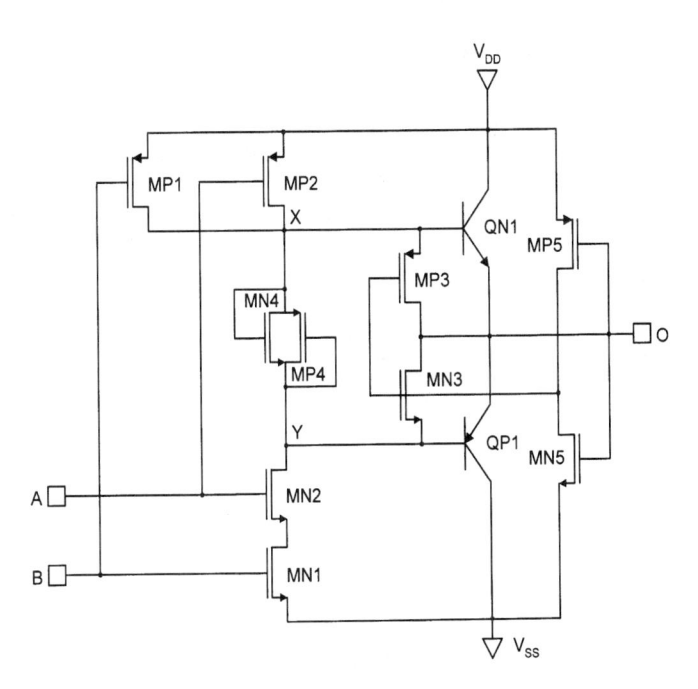

Full-Swing BiCMOS Logic Circuits with Complementary Emitter-Follower Driver Configuration 155

Figure 4–24 FS-CMBL two-input NAND circuit with positive feedback.

4.3.6 Experimental Results and Analysis

To evaluate the performance of the FS-CMBL circuits, they have been fabricated using the 0.8 μm BiCMOS technology [12]. For the pnp devices, two different pnp processes were used. These were the n-well base lateral pnp BJTs and the substrate pnp BJTs. Table 4–3 summarizes the key parameters of the technology.

Table 4–3 Key BiCMOS parameters for the 0.8 μm technology.

Threshold voltage (nMOSFET / pMOSFET)	0.8 V / – 0.7 V
Channel Length (nMOSFET / pMOSFET)	0.6 μm / 0.6μm
BJT f_T's (npn / substrate-pnp)	15 GHz / ≈0.5 GHz

The circuit of Figure 4–23 was fabricated using only the lateral pnp process, whereas the circuit of Figure 4–24 was fabricated using both the substrate pnp and lateral pnp process. The gate width of each MOSFET used for the logic function (MN1, MN2, MP1, and MP2) is 20 μm, and the emitter sizes of the npn and the pnp are 8 μm by 0.8 μm and 10 μm by 2.4 μm, respectively. The gate delays have been measured, using 19-stage ring oscillator circuits, as a function of the load capacitance, and the results are illustrated in Figure 4–25.

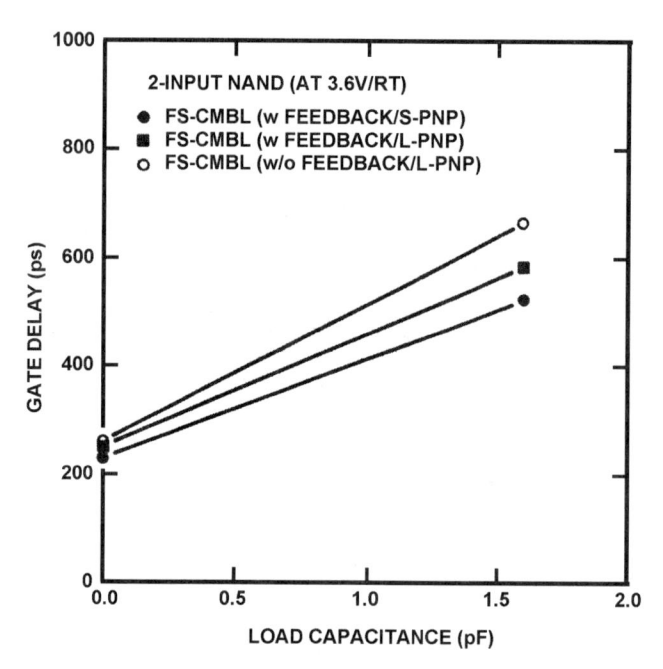

Figure 4–25 Measured gate delay as a function of the load capacitance.

The FS-CMBL circuit with feedback utilizing the substrate pnp shows the lowest gate delay for the entire range of load capacitance. This technique proves, first that the feedback mechanism is effective, and second that substrate pnp gives a better performance than lateral pnp. Figure 4–26 compares the measured gate delays of the FS-CMBL with those of the CMOS and the conventional BiCMOS circuits.

The results show that although the characteristics of the pnp device are poor, the FS-CMBL circuit shows a better driving capability than does the CMOS for load capacitances greater than 0.3 pF. It is also slightly faster than the conventional BiCMOS for loads below 0.5 pF. Figure 4–26 further illustrates that if a better pnp (with $f_T = 8$ GHz) is used (shown in dashed lines), the delay and driving capability of the FS-CMBL circuit will be improved further. Figure 4–27 demonstrates that the FS-CMBL circuit maintains its performance leverage over the conventional BiCMOS below 3 V. The FS-CMBL circuit operates for supply voltages down to 1.4 V, whereas the conventional BiCMOS circuit only operates down to 2.2 V.

4.3.7 Circuit Variations of the FS-CMBL with Feedback

Variations of the basic FS-CMBL circuit [11], shown in Figure 4–24, are presented in this section to evaluate the dependency of the delay and power consumption on the characteristics of the clamping diode, parasitic capacitance at the two base nodes, and circuits for full-swing.

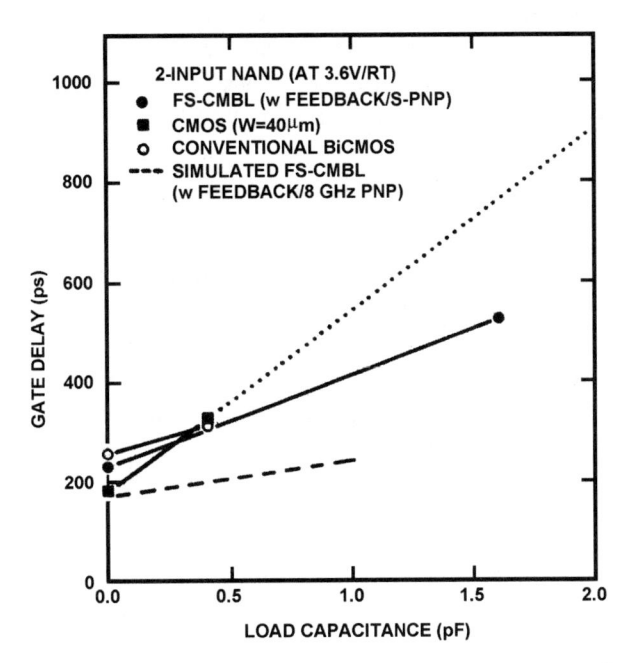

Figure 4–26 Measured delay comparison among the FS-CMBL circuit, CMOS (gate width = 40 μm), and conventional BiCMOS circuits.

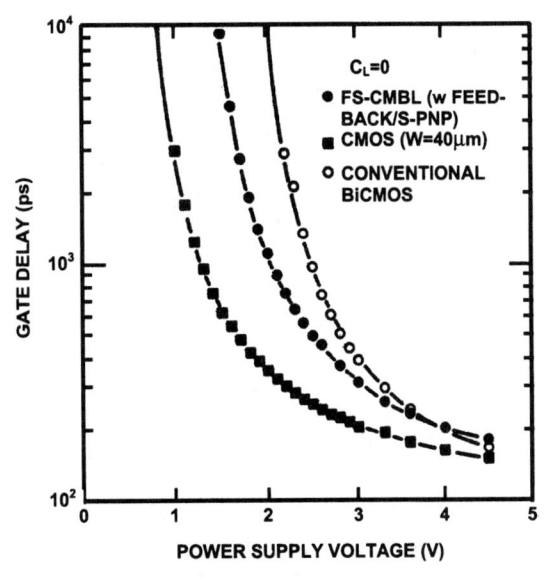

Figure 4–27 Measured unloaded-delay comparison among the FS-CMBL, pure CMOS and conventional BiCMOS circuits.

4.3.7.1 FS-CMBL with Low-Threshold Clamping Diode

The advantage of using a CMOS clamping diode is that the diode drop may be altered by changing the MOSFET threshold voltage. For instance in Figure 4–24, the diode drop of the CMOS diode MN4 can be lowered by using a lower threshold voltage MOSFET. However, the range of the base-emitter voltage of the BJT QN1 between the clamped and turn-on states is expected to increase. (Note that this range is defined as a dead band.) Increasing this range will result in a lower speed but less power consumption.

4.3.7.2 FS-CMBL with BJT Clamping Diode

The diode drop of a BJT diode is the same as the V_{BEon} of the BJT. This ideal dc characteristic provides an optimum bias and clamping level that reduces both the crossover current and the dead band. However, a BJT diode using a base-collector junction with the base-emitter shorted or a base-emitter junction with the base-collector shorted introduces a large parasitic capacitance associated with the collector-substrate junction. A BJT diode configuration constructed using double emitter transistors can reduce the parasitic and is shown in Figure 4–28.

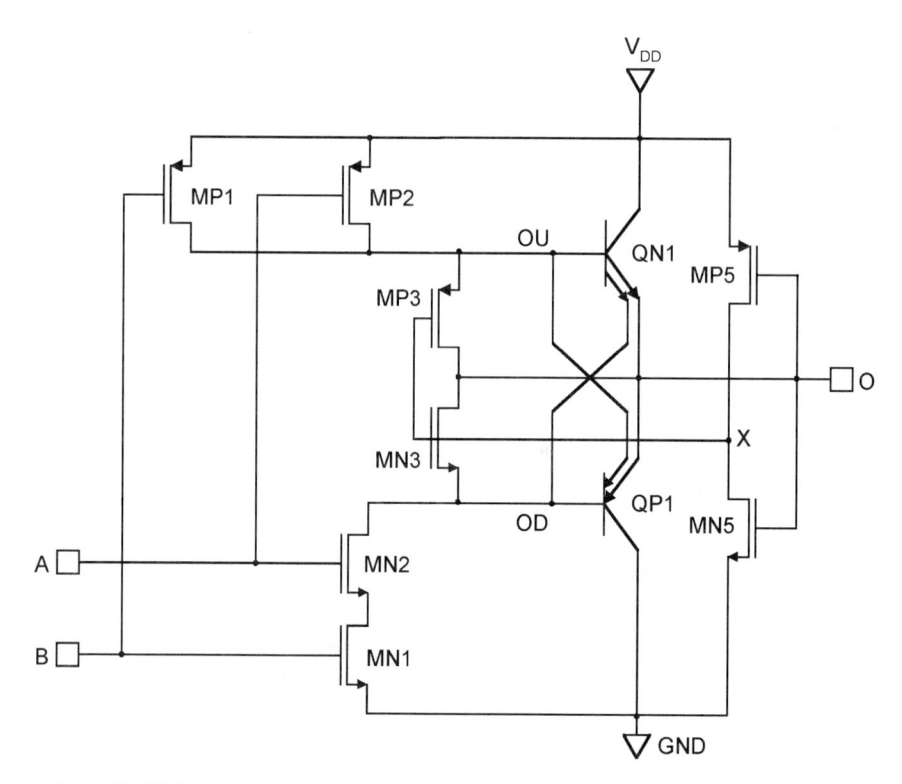

Figure 4–28 FS-CMBL circuit with a parallel complementary BJT clamping diode.

In this circuit, when the output becomes high, the node OD is clamped by the base-emitter diode of QN1. On the other hand, for the output high-to-low transition, the node OU is clamped by the base-emitter diode of QP1. Since the parallel complementary BJT diode is merged with the driver transistors by combining collectors, the parasitic capacitances at nodes OU and OD, added by the diode, is the only difference in the base-collector junction capacitance between double emitter BJT and the single emitter BJT, which is minimal. The parasitic capacitance present at the node OD may cause the diode drop to become excessive during transient and therefore turns on the secondary base-emitter junction of QP1. The conduction of QP1 will lead to an increased crossover current and a decrease in the speed.

4.3.7.3 FS-CMBL with Low-Parasitic Capacitance (Cp) Clamping Diode

The parasitic capacitance at the critical nodes OU and OD can be reduced by modifying the placement of the CMOS diode, shown in Figure 4–29. By comparing this circuit to the one in Figure 4–24, we realize that the drain terminals of the clamping CMOS diodes MP4 and MN4 are connected to the output node, instead of nodes OD and OU, respectively. This arrangement helps to eliminate the drain junction capacitance of MP4 and MN4 at nodes OD and OU, respectively. The operation of the CMOS diode is different from the earlier one presented in section 4.3.5.2 and it is described as follows: When both inputs are high, the output node O is shorted to the node OD to ground. The pMOS MP4 will turn on, biasing the node OU at V_{TP} (threshold voltage of MP4). In order to clamp QN1 off, this bias voltage V_{TP} must be smaller than V_{BEon} of QN1. The same requirement applies to MN4 due to the complementary structure of the circuit. MN4 will clamp the node OD to $(V_{DD} - V_{TN})$ during the pull-up transient and prevents QP1 from turning on as long as V_{TN} is smaller than V_{BEon} of QP1. A major drawback of this circuit is that its speed and power depend heavily on the availability of MOSFETs with low-threshold voltages.

4.3.7.4 FS-CMBL Without Feedback Control and Clamping

In this configuration, shown in Figure 4–30, both the clamping diode and the feedback controlled full-swing technique are absent. Instead, MP3 and MN3 serve a dual function of clamping and providing full-swing. When compared to the previous circuits shown in Figures 4–24 and 4–29, there is a dramatic decrease in the junction capacitance at the critical nodes, and this reduction in the capacitance will be translated into a speed advantage. However, if the circuit is implemented in standard technologies, it may suffer from a crossover current due to the body effect, especially for high-voltage operation.

The circuit operation can be explained as follows: When both inputs are high, the node OD is discharged to ground and transistor MN3 turns on, shorting the output node O to the node OD, causing a full output voltage swing. The node OU is clamped by MP3 at a voltage of V_{TP}. If V_{TP} is made smaller than V_{BEon} of QN1, QN1 will be cut off. During the pull-up period, MP3 will turn on and provide a strong shunting to pull the output node up to V_{DD}. The node OD will be clamped by MN3 at $(V_{DD} - V_{TN})$. As discussed earlier in section 4.3.5.2, MP3 and MN3 implemented in such a configuration (shown in Figure 4–30) will cause a premature bypass of part of the base currents during transient periods. Hence, this configuration is not recommended for use.

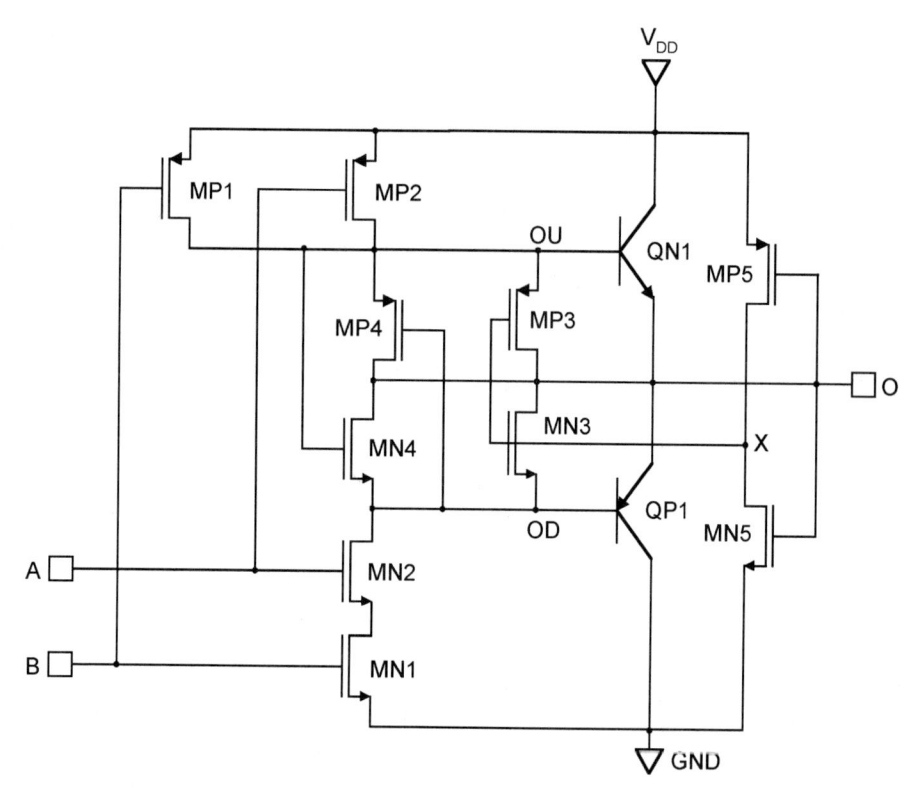

Figure 4–29 FS-CMBL circuit with a low-parasitic capacitance CMOS diode.

4.3.7.5 FS-CMBL with a Collector-Emitter Shunt

In Figure 4–31, the collector-emitter (CE) shunting technique is used to achieve a full-swing output voltage. This technique helps to minimize the parasitic capacitance at nodes OU and OD. Although the source junction capacitance of MP3 and MN3 are eliminated from nodes OU and OD respectively, this circuit may consume more power and have higher delays because of crossover currents flowing through the driver and shunting devices [8].

4.3.8 Experimental Results and Analysis

The basic FS-CMBL circuit is shown in Figure 4–24 and its variations (see Figures 4–29 to 4–31) and the conventional npn BiCMOS circuits have been implemented in a 1.2 μm complementary BiCMOS technology for mixed digital/analog applications. The key device parameters are summarized in Table 4–4.

Note that there are two threshold voltages: 0.85 V (standard) and 0.35 V (low) are available for the nMOSFET. The low threshold voltage nMOSFET is used to implement a low-diode drop CMOS diode MN4 in the low-V_D variation. All other circuits use the standard threshold

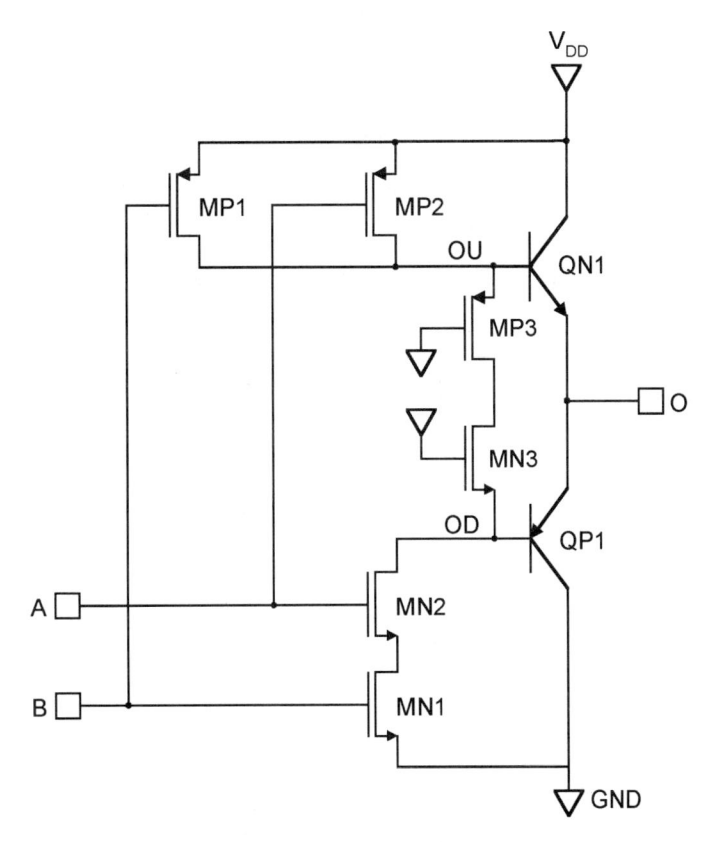

Figure 4–30 FS-CMBL circuit without feedback control and clamping diode.

voltages. This technology features 1.1 μm (on wafer) gate-length LDD MOSFETs as well as a 6 GHz (f_T) npn BJT and a 2 GHz (f_T) isolated vertical pnp BJT. The device cross-section is shown in Figure 4–32.

Figure 4–33 shows the gate delays (fan-in = 2, fan-out = 1) for circuits with and without additional loading of 0.5 pF. For the basic FS-CMBL circuit, the unloaded delay is 366 ps and the slope of the delay versus capacitance plot is 288 ps/pF. On the other hand, the conventional BiCMOS circuit has an unloaded delay of 432 ps with a slope of 278 ps/pF. This difference translates to more than 15% improvement of the basic FS-CMBL circuit over the conventional BiCMOS for the unloaded case. This leverage at high supply voltage is due to the presence of the fast pnp BJT. It has been demonstrated that there is little leverage even for 3.6 V if poor pnp transistors are employed [10]. Figure 4–33 also reveals that the FS-CMBL with a CE shunt gives the worst delay performance, whereas the FS-CMBL without feedback control and clamping gives the best delay timings.

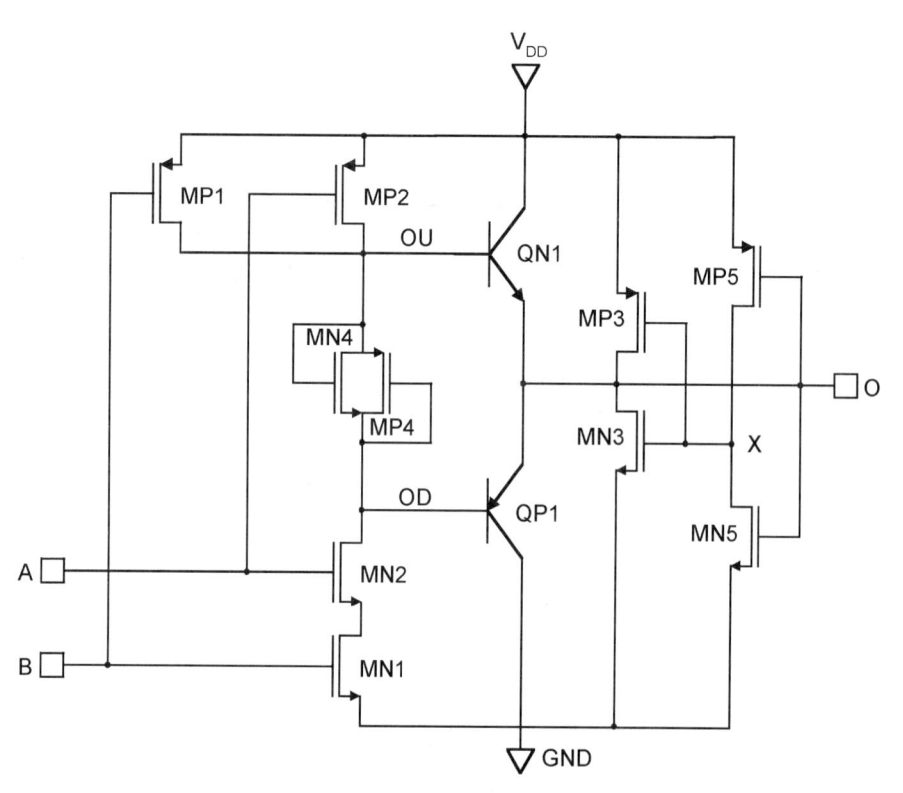

Figure 4–31 FS-CMBL circuit with collector-emitter (CE) shunt.

Table 4–4 Key MOSFET and BJT parameters for the 1.2 μm process.

MOSFET	nMOSFET / pMOSFET
Threshold voltage (V_T)	0.85 V(0.35 V) / –0.75 V
Effective channel length (L_{eff})	0.90 μm / 0.85 μm
BJT	**npn / pnp**
Current gain (β)	120 / 15
Cutoff frequency (f_T)	6 GHz / 2 GHz
Breakdown voltage (BV_{CEO})	9 V / 8 V

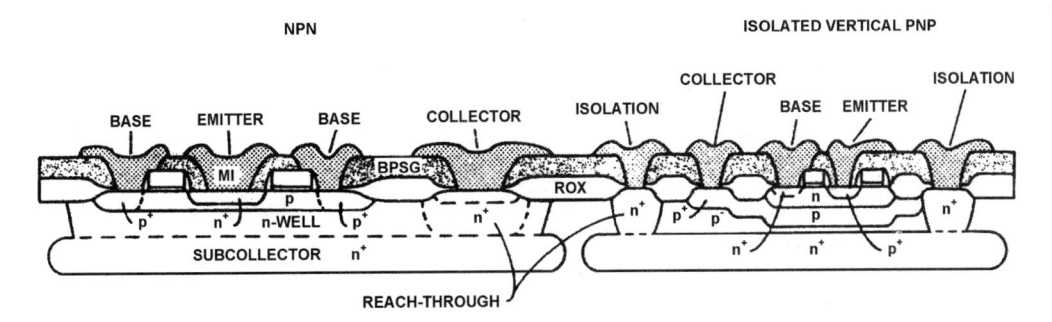

Full-Swing BiCMOS Logic Circuits with Complementary Emitter-Follower Driver Configuration

Figure 4–32 Device cross-section of the npn and pnp transistors.

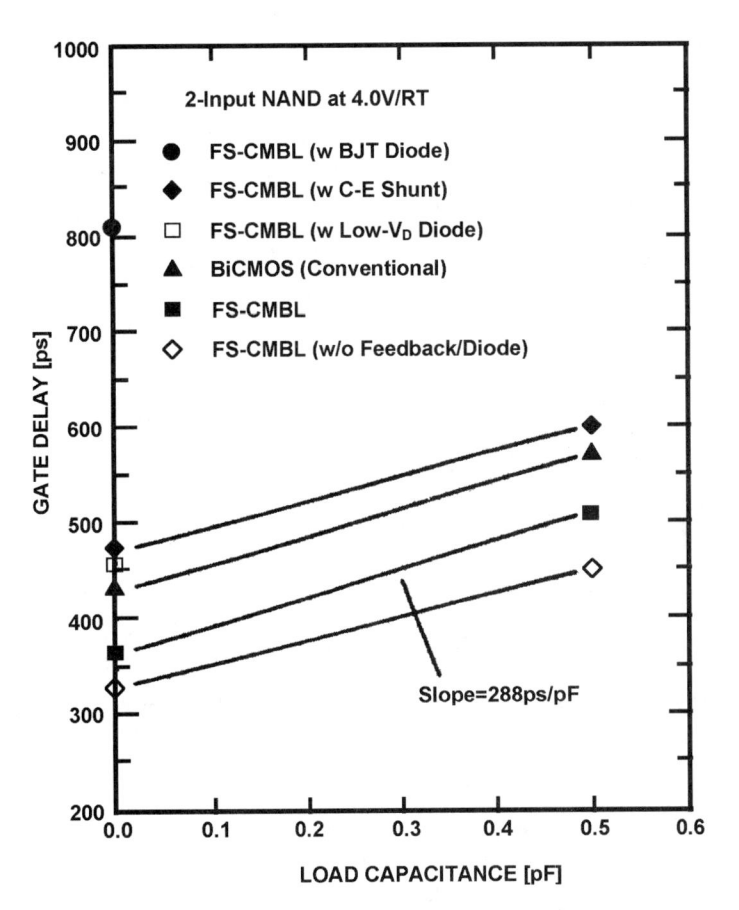

Figure 4–33 Measured gate delays vs. load capacitance at 4 V and room temperature.

Figure 4-34 illustrates the gate delays as a function of the load capacitance for a lower supply voltage of 3.3 V. For the basic FS-CMBL circuit, the unloaded delay is 495 ps with a driving capability of 310 ps/pF. Comparatively, the unloaded delay of the conventional BiCMOS circuit is 666 ps with a driving capability of 352 ps/pF. This difference translates to a delay improvement of over 26%, which is more than the previous case at a power supply voltage of 4 V.

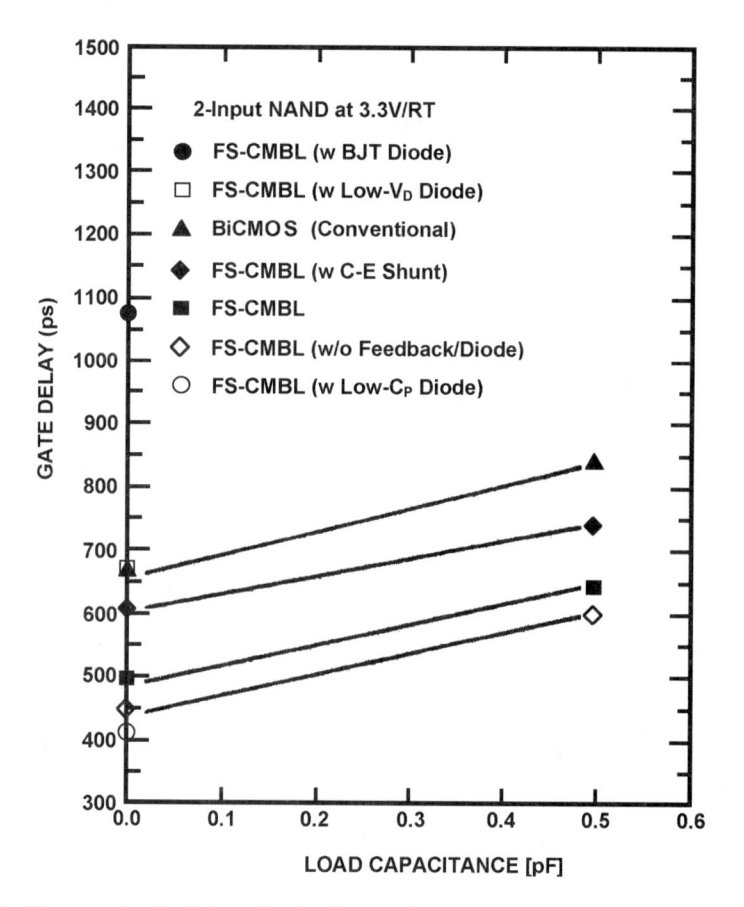

Figure 4-34 Measured gate delays vs. load capacitance at 3.3 V and room temperature.

Figure 4-35(a and b) shows the unloaded and loaded delays of each circuit as a function of the supply voltage. It can be observed that the BiCMOS circuit shows a surge in the delay below approximately 2.7 V whereas the FS-CMBL circuits operate down to 1.6 V. This disparity implies that the performance leverage of the basic FS-CMBL circuit over the conventional BiCMOS increases substantially as the supply voltage is scaled down.

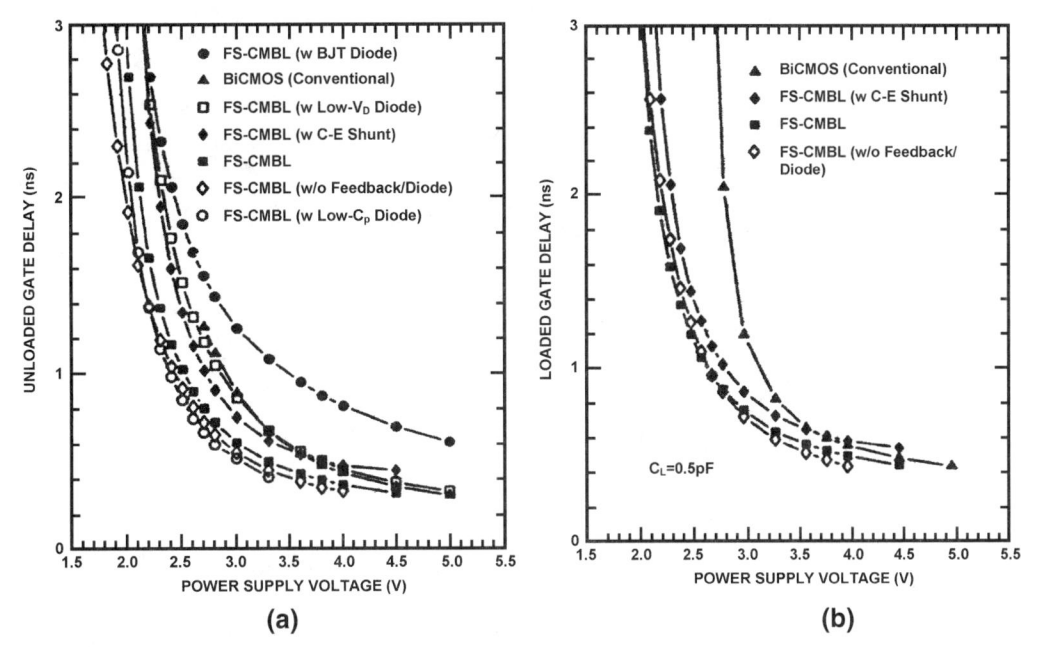

Figure 4–35 (a) Unloaded (b) loaded gate delays vs. the supply voltage.

Figure 4–36(a and b) shows the unloaded and loaded power consumption at 100 MHz as a function of the power supply voltage. It can be seen that the FS-CMBL circuits consume more power than the conventional BiCMOS circuit. This is due to the improperly high-drop clamping diodes consisting of MOSFETs with high-threshold voltages and body-effect constants.

For the loaded case, shown in Figures 4–35(b) and 4–36(b), there is a distinct separation of the BiCMOS curve from the FS-CMBL curves. This separation is due to the degradation of the driving capability of the conventional BiCMOS circuit as the power supply voltage is reduced. In Figure 4–36(b), it is clear that the power consumption of the FS-CMBL circuits decreases as the supply voltage is scaled down. This is because at higher supply voltages, dc power dissipation occurs due to the improperly high-drop clamping diodes. But as the supply voltage is lowered, the dc power dissipation is negligible, and the dynamic power forms the significant part of the total power. For the unloaded case, illustrated in Figures 4–35(a) and 4–36(a), compared with the basic FS-CMBL, the circuit with a BJT diode is much slower due to the larger parasitics and insufficient clamping, resulting in base-current bypass and crossover current. On the other hand, the circuit with a low parasitic CMOS diode and that without the feedback and clamping diode show better delay performance than does the basic FS-CMBL circuit. This is due to the reduced parasitics. Also, the circuit with the CE shunt is slower due to the crossover current and, as shown in Figure 4–36(a), the circuit has a relatively high power consumption.

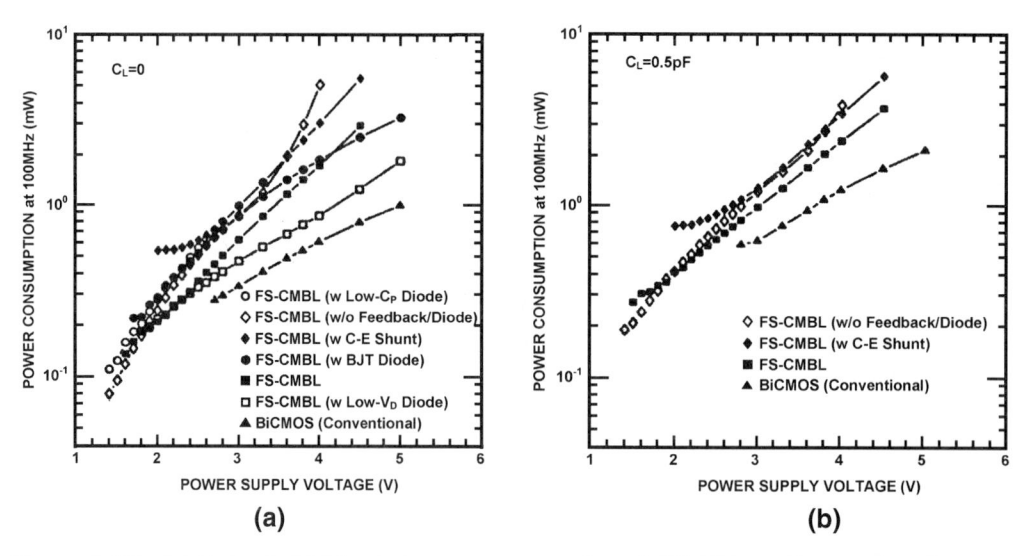

Figure 4–36 (a) Unloaded (b) loaded power consumption at 100 MHz vs. the supply voltage.

In conclusion, this section covers various FS-CMBL circuits that have been implemented and compared in a 1.2 μm complementary BiCMOS technology. For these circuits, the delay-power trade-offs depend on the clamping diode and the parasitic capacitance. In general, the circuit delay decreases with a higher drop-clamping diode while the power consumption decreases with a lower drop-clamping diode. Good performance has been obtained from the feedback controlled base-emitter shunting technique (in terms of the delay and power dissipation), as compared to the collector-emitter shunting technique.

4.4 Merged BiCMOS (MBiCMOS) Logic Gates

4.4.1 Introduction

In order to improve the BiCMOS gate performance at low-supply voltages, several techniques have been proposed and published in the literature [8,10,11,13–18]. These include the complementary BiCMOS with emitter-follower drivers [11], complementary BiCMOS with inverter drivers [14] and the BiNMOS circuits [17]. These circuits have been successful in extending the operating range of the BiCMOS to lower supply voltages, but at the expense of added process complexity (pnp devices), higher power and circuit complexity (due to additional devices per gate), and the absence of performance symmetry (as in the case of BiNMOS).

The merged BiCMOS was proposed recently [3]. As compared to those previously published techniques, MBiCMOS achieves performance leverage over CMOS in the sub-2 V regime without any additional gates or additional processing and yet consumes approximately the same amount of power as a standard BiCMOS gate. Furthermore, merging devices within a BiCMOS gate has been shown to reduce the area and increase the speed by reducing the internal capacitance.

4.4.2 MBiCMOS Gate

4.4.2.1 Circuit Design

The schematic diagram of the conventional BiCMOS inverter [19] and the MBiCMOS gate are shown in Figures 4–37(a) and (b), respectively. Figure 4–37(c) shows the cross-sectional view of the merged pMOS/npn device. Note that the pMOS drain is merged with the npn base while the pMOS n-well is merged with the npn collector. Figure 4–38 shows a microphotograph of the conventional BiCMOS and MBiCMOS gates in a 1 μm process. From Figure 4–38, it has been found that the BiCMOS gate occupies 876 μm^2, while the MBiCMOS occupies only 582 μm^2, giving an area saving of 34%.

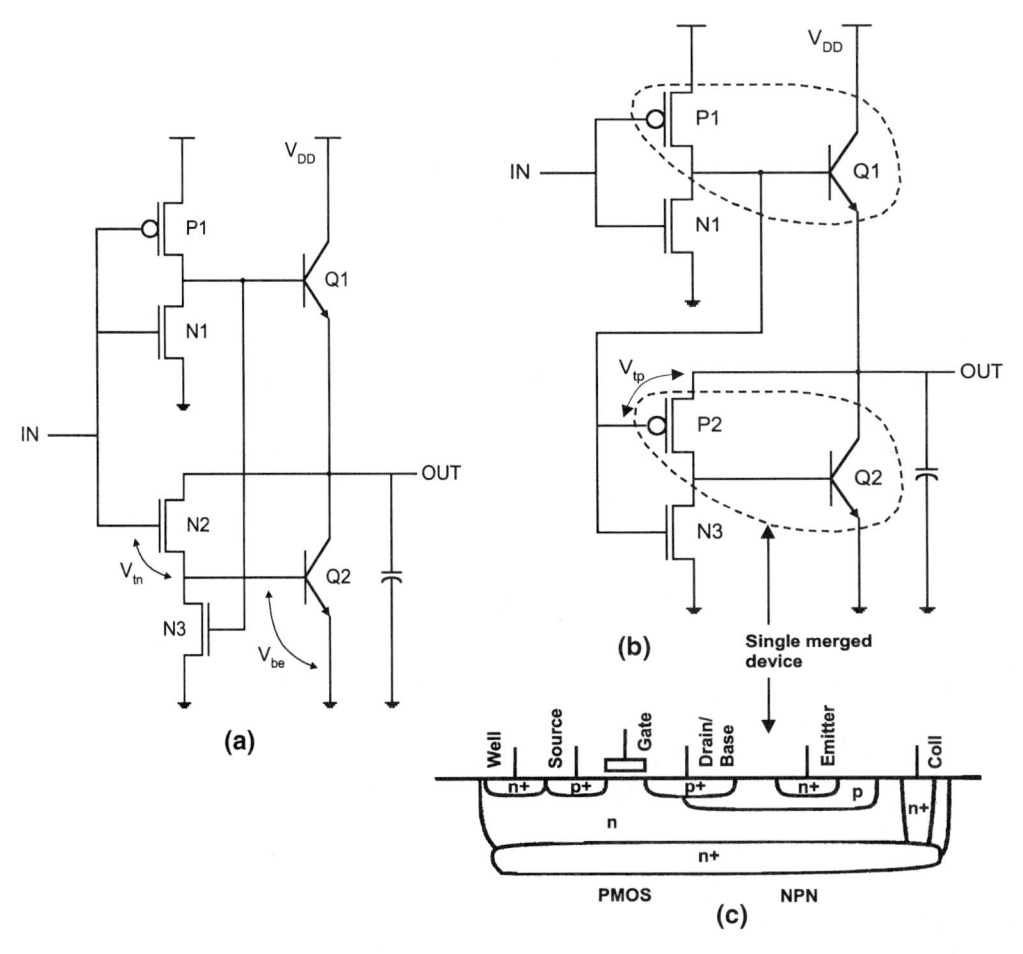

Figure 4–37 (a) Conventional BiCMOS gate circuit with six discrete devices (b) MBiCMOS gate circuit with only four discrete devices (c) cross-section of the merged pMOS/npn device.

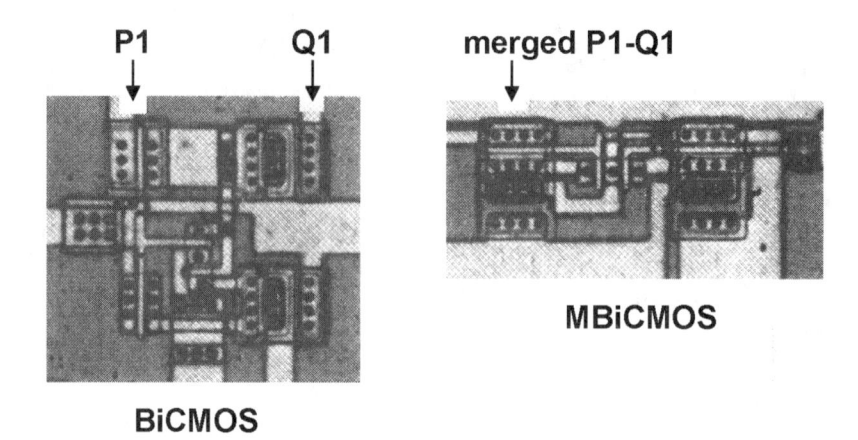

Figure 4–38 Microphotograph of the BiCMOS and MBiCMOS gates in the 1 μm technology.

The susceptibility of the merged pMOS/npn device to latch-up is greater than its non-merged counterpart. The latch-up phenomenon in the merged pMOS/npn device is illustrated in Figure 4–39(a) where the parasitic pnp device is formed by the source, body, and drain of the pMOS; and the parasitic npn device is formed by the base, emitter, and collector of the bipolar device. With a proper layout design, the direction of the current flow can be controlled [17] so as to avoid triggering the low impedance path, which causes latch-up. Figure 4–39(b) shows that by removing the external contact to the source or well, the source current will be supplied by the collector contact (indicated by the arrows in Figure 4–39(b)), thereby preventing latch-up since the emitter-base junction of the parasitic lateral pnp device is reverse biased.

4.4.2.2   Circuit Operation and Analysis

As shown in Figure 4–37(b), during the pull-up transition, the input voltage switches from high to low. As such, P1 turns on and N1 turns off, pulling the node B1 to V_{DD} and subsequently the output node to ($V_{DD} - V_{BEon}$). The high voltage at node B1 turns N3 on and keeps P2 off. This voltage will keep the base of Q2 at ground potential and ensure that Q2 is not conducting. For the pull-down transition, the input voltage goes from low to high. Now P1 is off and N1 conducts, so that node B1 is at ground level. The transistors Q1 and N3 are now off, while P2 conducts to supply the base current to turn on Q2. Hence, the output node is pulled down to a low level. However, the output low voltage depends on the magnitude of the threshold voltage of P2 relative to the base-emitter voltage of Q2. V_{OL} is the higher of these two voltages.

In order to achieve the optimized performance, the threshold voltage of P2 should be low to increase the base drive of Q2. By modifying P2 to be a zero-threshold device, the current drive of P2 will increase, but this modification is at the expense of a single threshold-adjusting implant.

Merged BiCMOS (MBiCMOS) Logic Gates

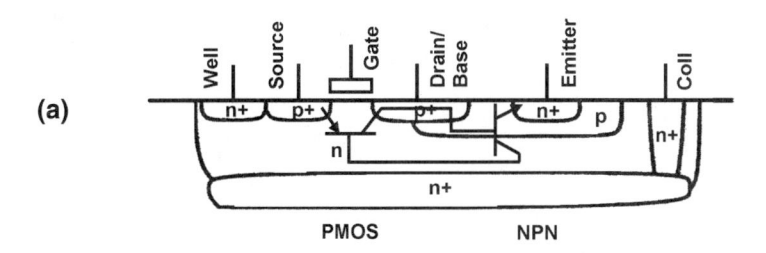

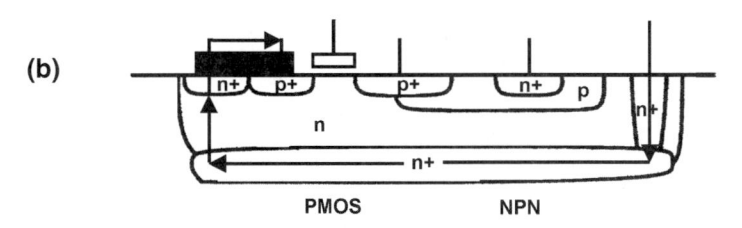

Figure 4-39 Latch-up design of a merged pMOS/npn device: (a) latch-up path of concern (b) current flow path to avoid latch-up.

4.4.3 Circuit Performance and Comparison

The pull-up portion of the MBiCMOS gate is essentially the same as that of a standard BiCMOS gate, which has been well documented in [20,21]. The major differences occur in the pull-down portion of the circuit. In the BiCMOS gate, a low-to-high step at the input turns on N2 (see Figure 3.37(a)), which then supplies the base current to Q2. For a similar input transition in the MBiCMOS, the node B1 must be pulled low through N1 before P2 switches on and N3 switches off. However, this delay can be minimized by appropriate sizing of the transistor N1 in order to achieve a fast pull-down at node B1.

Before comparing the pull-down responses of the MBiCMOS and BiCMOS, we first compare the pull-up and pull-down responses of the BiCMOS gate itself. First, for the pull-up phase, N1 shuts off quickly when the input goes low, whereas in the pull-down phase N3 remains on until the voltage at the node connecting the gate of N3 and the base of Q1 drops to zero. Consequently, the total base current being supplied to the base of Q2 is reduced. Second, the body effect must be considered for the transistor N2 during the pull-down cycle, unlike P1 during the pull-up phase. Third, the gate-to-source voltage of N2 during the pull-down is reduced by a diode drop at the input and another diode drop between the source and ground. Hence, it is smaller than the gate-to-source voltage of P1 during the pull-up cycle.

The MBiCMOS gate is designed to eliminate the above mentioned differences. Figure 4-40 illustrates the pull-down cycle for both the BiCMOS and MBiCMOS circuits. First, the

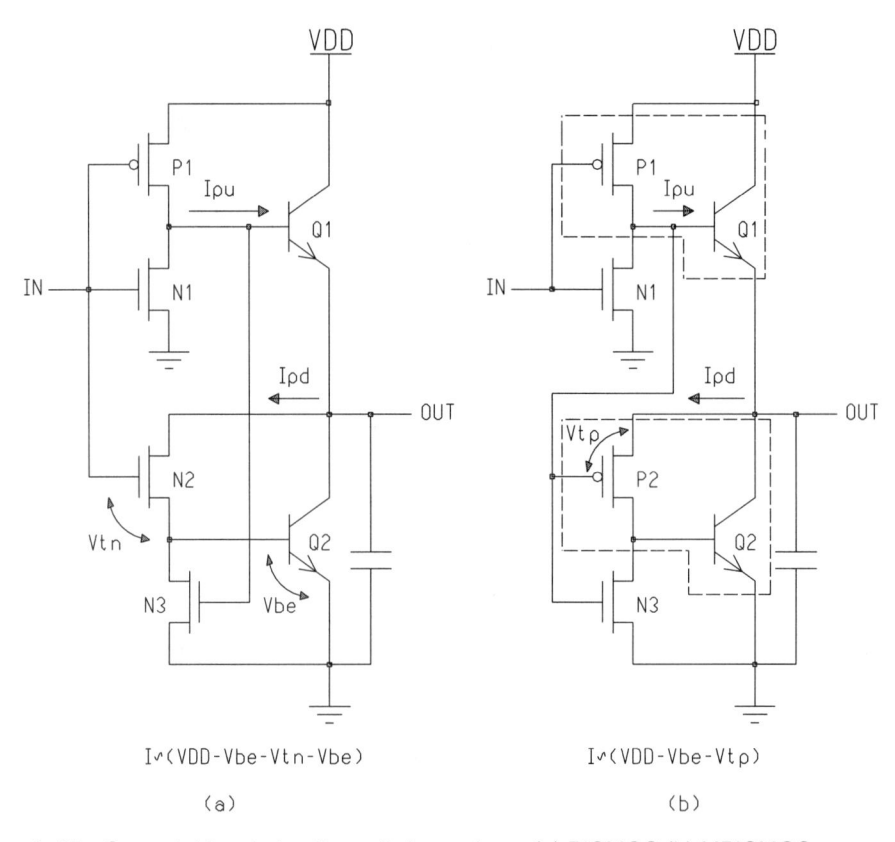

170   Chapter 4 • Low-Voltage BiCMOS Digital Circuits

Figure 4–40 Current drive during the pull-down phase (a) BiCMOS (b) MBiCMOS.

circuit exhibits symmetry, thus eliminating the body effect of the transistor P2 during the pull-down phase. Second, the gate-to-source voltage of P2 is reduced neither by a diode drop at the input nor by a diode drop between the source and ground. Hence, the base drive of Q2 increases leading to better performance, especially at low-supply voltages. Third, the input capacitance of the standard BiCMOS gate consists of the capacitance of three transistors, P1, N1, and N2, whereas, that for the MBiCMOS gate originates from only two devices, P1 and N1. Thus, by making N1 larger in the MBiCMOS gate than its counterpart in the BiCMOS gate, we can achieve a fast pull-down at the base of Q1, thus improving the switching speeds. Another noticeable difference between the MBiCMOS and BiCMOS operation is the state of Q2 near the end of the output low transition. In the conventional BiCMOS circuit, the base of Q2 is held at the same voltage as the output node through N2. But in MBiCMOS, the output voltage will continue to drop after the pMOS device P2 turns off due to the charge stored in the base of Q2, which keeps it conducting. Hence, Q2 will go into the saturation state. However, Q2 will just experience shallow saturation, which does not significantly slow the following output rising transition.

Merged BiCMOS (MBiCMOS) Logic Gates

During the next output rising transition, the excess minority charges in the base of Q2 will be depleted quickly by the nMOS device N3. Moreover, P2 also begins to conduct in the opposite direction after the output voltage drops below the base voltage of Q2. This change in the current direction will help to relieve the base of Q2 of stored charges. The charge removal is especially efficient in the MBiCMOS gate where P2 is modified to be a zero-threshold device. Figure 4–41 shows the voltage and current against the time characteristics of the BiCMOS, MBiCMOS and the modified MBiCMOS (i.e., P2 is a zero-threshold device) during the pull-down and pull-up transitions.

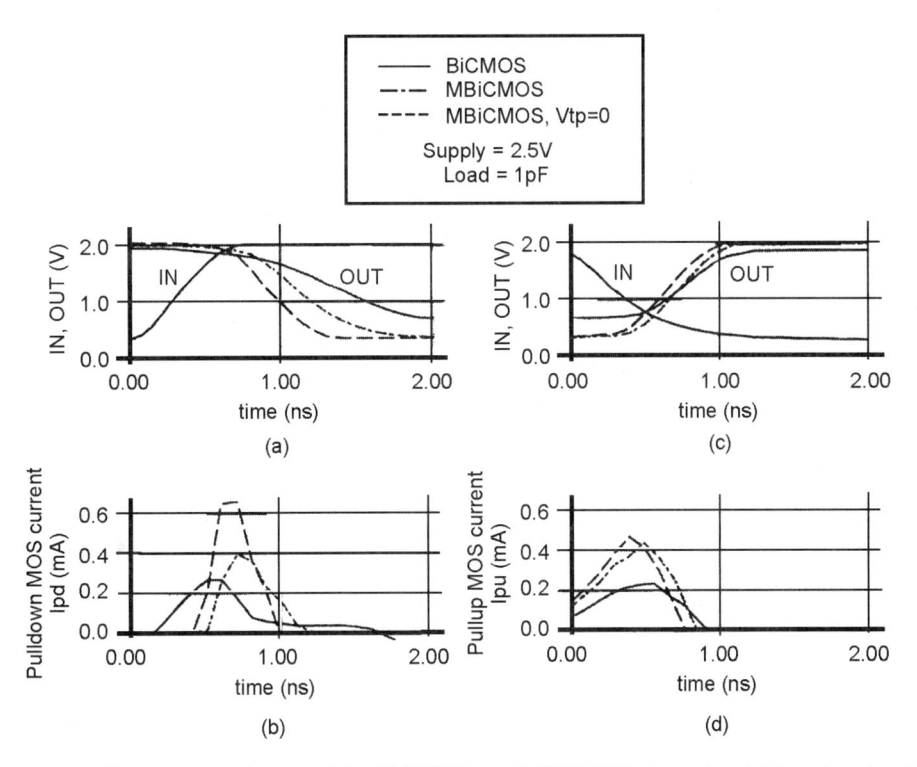

Figure 4–41 Transient response of the BiCMOS and MBiCMOS gates for: (a) input and output voltages for an output falling transition; (b) pull-down current during the output falling transition; (c) input and output voltages for an output rising transition; (d) pull-up current during the output rising transition.

From Figure 4–41(a), it is evident that the pull-down operation of the MBiCMOS is faster than that of the BiCMOS under the same load and power supply voltage conditions. This difference is due to the increase in the current drive to the base of Q2 during the pull-down cycle (see Figure 4–41(b)). As for the pull-up operation, shown in Figure 4–40(c), similar results are obtained for the three gates.

4.4.4 Experimental Tests and Analysis

Two different technologies were used to fabricate the BiCMOS and MBiCMOS gates. The first is a 1 μm conventional buried-layer/epi process with 15 GHz npn bipolar devices, and the second is a 2 μm process with a triple-diffused npn bipolar transistors [22]. Figure 4–42 illustrates and describes the gate fabricated using the two different technologies. For the 1 μm technology, the BiCMOS and MBiCMOS have the same device size, but the MBiCMOS has both a smaller gate area and a smaller input capacitance than those of the BiCMOS. As for the 2 μm technology, all gates have equal gate area and input capacitance.

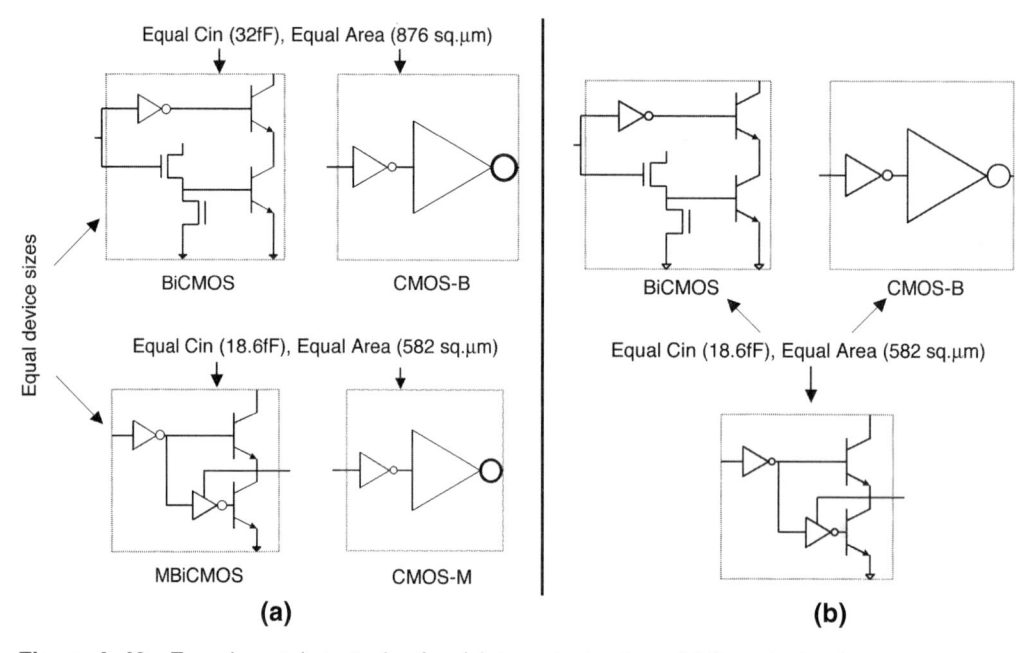

Figure 4–42 Experimental strategies for: (a) 1 μm technology (b) 2 μm technology.

Notice that two CMOS gates were built for the 1 μm technology; one is normalized to the BiCMOS gate (CMOS-B), and the other is normalized to the MBiCMOS gate (CMOS-M). For the 2 μm technology, all gates were normalized. From Figure 4–43, we can see that for this technology, the size of the MBiCMOS N1 is twice that of the BiCMOS N1. This is done to achieve equal input capacitance and improve the pull-down response of the MBiCMOS. Experiments involving those gates, shown in Figure 4–42, were carried out and the results displayed in Figures 4–44 through 4–47. Figure 4–44 shows the measured delay versus the power supply voltage for gates in the 1 μm technology with a capacitive load of 1 pF.

Merged BiCMOS (MBiCMOS) Logic Gates

1 μm		2 μm	
MBiCMOS		**MBiCMOS**	
P1	7.8/1	P1	8/2
N1	3/1	N1	8/2
P2	7.8/1	P2	8/2
N3	3/1	N3	6/2
Q1	1x5	Q1	2x8
Q2	1x5	Q2	2x8
BiCMOS		**BiCMOS**	
P1	7.8/1	P1	6/2
N1	3/1	N1	4/2
P2	7.8/1	P2	6/2
N3	3/1	N3	6/2
Q1	1x5	Q1	2x8
Q2	1x5	Q2	2x8
CMOS-B		**CMOS**	
P1	11/1	P1	10/2
N1	7.6/1	N1	6/2
P2	66/1	P2	72/2
N2	45.6/1	N2	38/2
CMOS-M		**BiCMOS-M**	
P1	6.5/1	P1	6/2
N1	4.3/1	N1	4/2
P2	32.5/1	N2	6/2
N2	21.5/1	N3	4/2
		Q1	2x8
		Q2	2x10

Figure 4–43 Size of devices used by the various gates for: (a) 1 μm technology (b) 2 μm technology.

The BiCMOS gate shows better delay performance than does the CMOS-B gate until the power supply voltage is reduced to 3 V, whereas the MBiCMOS shows a measured delay advantage over its normalized gate, CMOS-M down to a supply of 2.4 V. Notice that the BiCMOS is slightly faster than the MBiCMOS between supply voltages of 3 to 5 V. This improvement is due to the fact that for the 1 μm technology, the size of N1 in the BiCMOS device is the same as its counterpart in the MBiCMOS device. However, if the size of N1 in the MBiCMOS device is

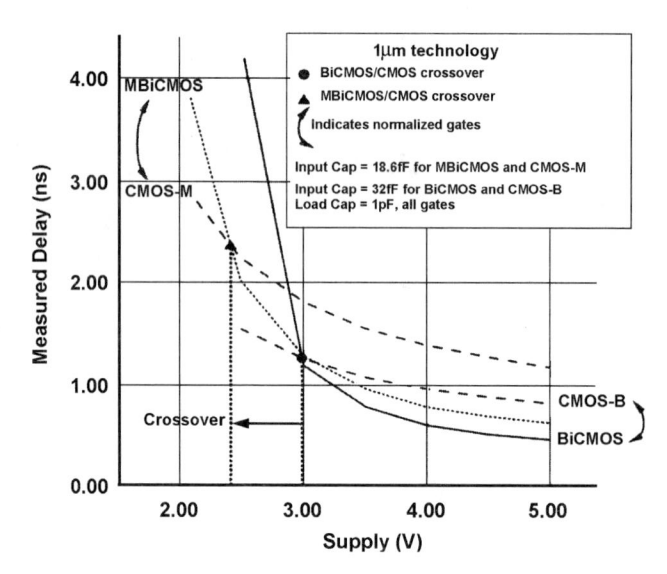

Figure 4–44 Measured delay vs. the power supply voltage for the 1 μm technology.

increased, and the input capacitance of both gates are made equal, the MBiCMOS will show a better speed performance than will the BiCMOS (as demonstrated in the 2 μm technology case in Figure 4–45). This increase in size will be necessary in the MBiCMOS to drive the increased internal capacitance at the base of Q1.

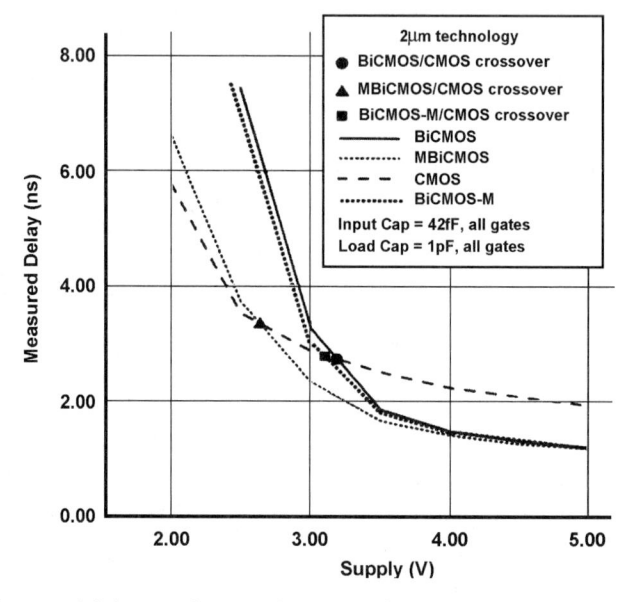

Figure 4–45 Measured delay vs. the supply voltage for the 2 μm technology.

Merged BiCMOS (MBiCMOS) Logic Gates 175

For the 2 μm technology, the size of N1 in the MBiCMOS is made twice that of N1 in BiCMOS. As such, the MBiCMOS device outperforms the BiCMOS device for all supply voltages. Figure 4–46 shows the measured delay versus the load capacitance for the 2 μm BiCMOS and MBiCMOS gates. At 5 V supply, both the MBiCMOS and BiCMOS outperform the CMOS for the entire range of capacitance tested. As for a supply voltage of 3.3 V, the MBiCMOS outperforms the CMOS for capacitance higher than 0.5 pF.

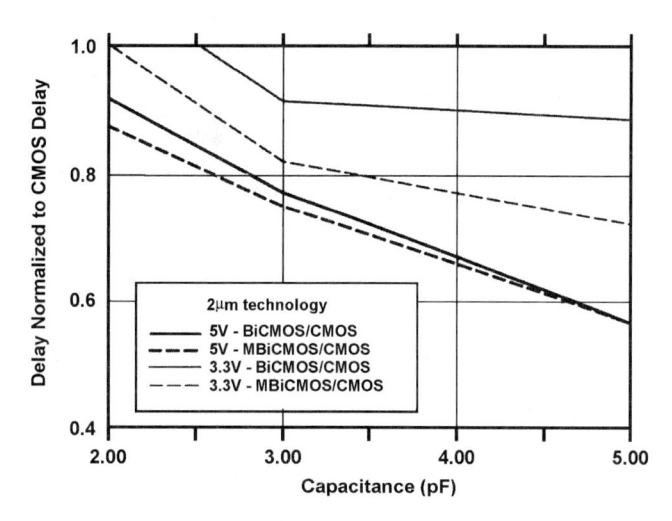

Figure 4–46 Measured delay vs. the load capacitance for the 2 μm technology.

Figure 4–47 shows the measured power-delay products as a function of the supply voltage for the 1 μm technology. For high-supply voltages, the power-delay products of the MBiCMOS and BiCMOS are comparable. Below 3 V supply, the power-delay products of the BiCMOS increase due to the static power dissipation resulting from an incomplete switching, whereas those of the MBiCMOS continue to decrease.

4.4.5 Simulation Results and Analysis

A SPICE simulation was performed to evaluate the MBiCMOS gate based on the 0.5 μm CMOS technology and using the 18 GHz npn bipolar transistors. Figure 4–48 shows the simulated delay as a function of the supply voltages for the normalized CMOS, BiCMOS and modified MBiCMOS gates.

The sizes of the BJT in both the BiCMOS and the modified MBiCMOS gates, and the sizes of the MOS devices in the second stage of the CMOS gate were optimized to produce the lowest possible delay at the respective load and supply voltage. Hence, the crossover points in Figure 4–48 are independent of the loading conditions. The MBiCMOS delay crosses over the CMOS delay curve at approximately 1.9 V supply, whereas the crossover point for the BiCMOS

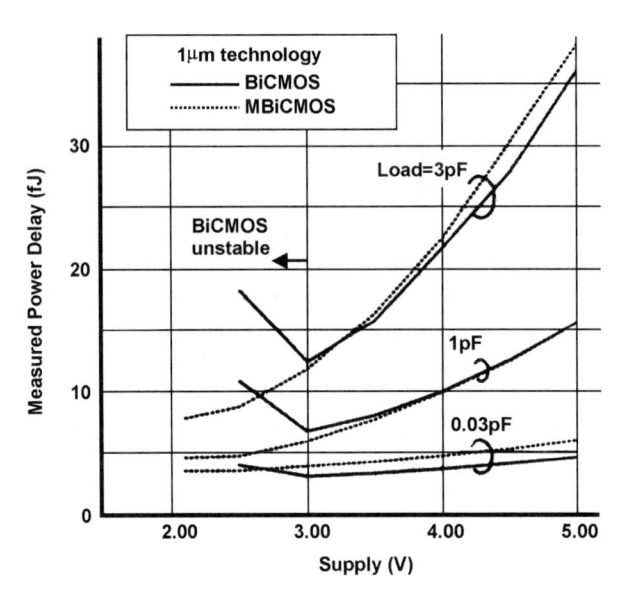

Figure 4–47 Measured power-delay product vs. the power supply voltage for the 1 μm BiCMOS and MBiCMOS circuits.

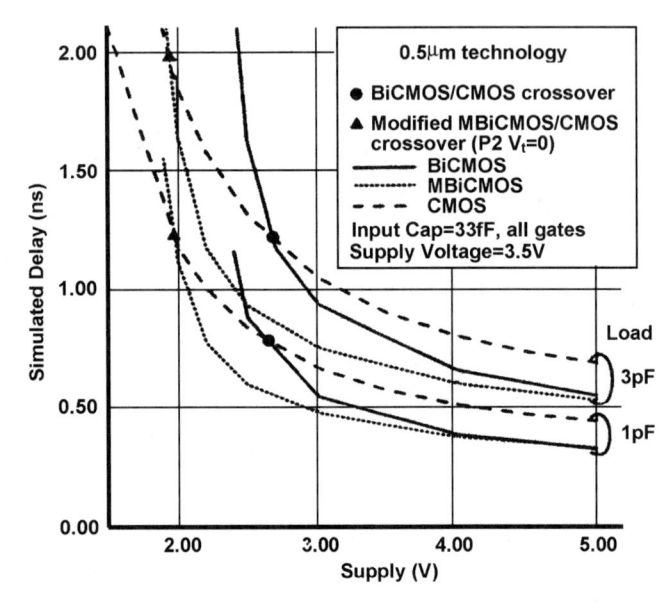

Figure 4–48 Simulated delay vs. the supply voltage for the modified MBiCMOS gate for the 0.5 μm technology.

gate is at a higher supply voltage of 2.63 V. Therefore, with the modified MBiCMOS gate, the performance advantage over CMOS is extended downwards below the 2 V supply.

4.4.6 Full-Swing MBiCMOS Gate

The full-swing MBiCMOS (FS-MBiCMOS) gate is an adaptation of the full-swing complementary BiCMOS (FS-CMBL) reported by Shin [10]. Figure 4–49 shows the schematic diagram of the FS-MBiCMOS gate. In the FS-MBiCMOS gate, the pnp pull-down of the FS-CMBL is replaced by a merged pMOS/npn device. Figure 4–50 shows the simulated delay versus the power supply voltage for the FS-MBiCMOS, MBiCMOS, BiCMOS and CMOS gates. As shown in Figure 4–50, the crossover voltage of the FS-MBiCMOS (CMOS for the 2 μm technology) can be reduced further to 2.2 V as compared to 2.4 V in Figure 4–45.

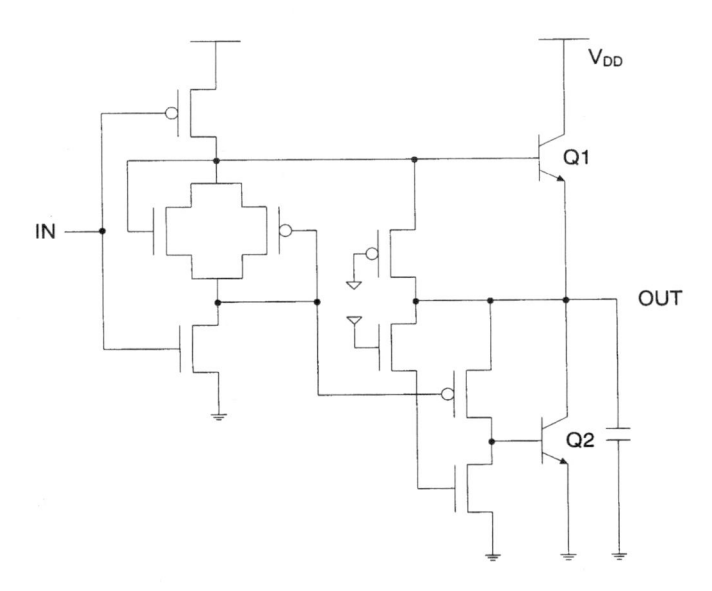

Figure 4–49 FS-MBiCMOS gate structure.

4.5 Full Voltage Swing Multi-Drain/Multi-Collector Complementary BiCMOS Buffers

4.5.1 Introduction

Scaling of the supply voltage (< 3 V) in conjunction with scaled BiCMOS buffers degrades the circuit performance [23,24]. Attempts to overcome this problem include shunting the base to the emitter or the collector to the emitter of the BJTs by a MOSFET or a resistor [8,25]. Due to the passive nature of the pull-up (pull-down) circuit path as the voltage increases

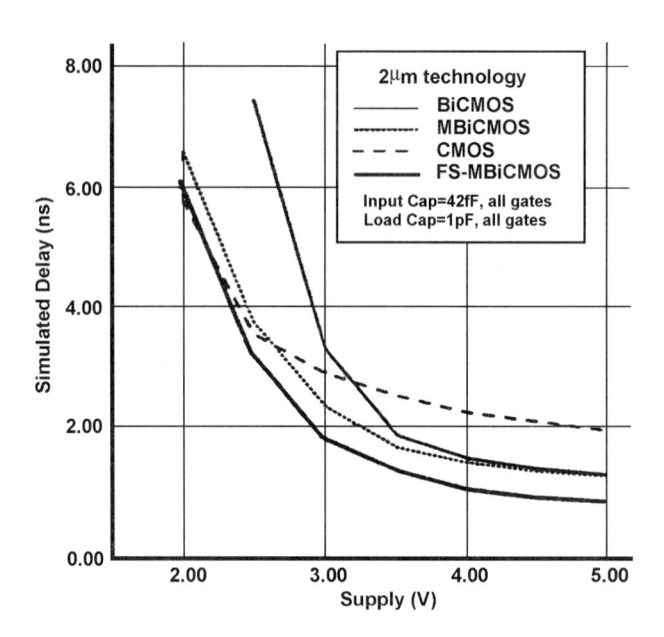

Figure 4–50 Simulated delay vs. the supply voltage for the BiCMOS, MBiCMOS, CMOS and FS-MBiCMOS gates for the 2 μm technology.

from ($V_{DD} - V_{BEon}$) to V_{DD} (or decreases from V_{BEon} to zero) through the shunting elements, this method fails to achieve full swing under high operating frequency and heavy loading conditions. Another attempt is the use of the CBiCMOS (see Figure 4–51(b)) [26]. It has been shown that the CBiCMOS circuits have the best speed performance at reduced supply voltages [26–29]. However, the major drawbacks are the increased process complexity due to the fabrication of high performance pnp BJT and the circuit complexity of extra devices per gate. In this section, new CBiCMOS circuits employing multi-drain/multi-collector structures will be introduced. These circuits can provide near rail-to-rail output voltage and high performance at low supply voltages.

4.5.2 Conventional BiCMOS and CBiCMOS Buffers

The conventional BiCMOS and CBiCMOS circuits are shown in Figure 4–51. We will be looking into the circuit operation of the conventional BiCMOS. The conventional CBiCMOS circuit will be used to make comparisons with other circuits in the later section.

The voltage swing of the conventional BiCMOS buffer suffers from a voltage loss of 2 V_{BEon}. But under low-capacitance loading conditions, the output voltage was observed to swing above ($V_{DD} - V_{BEon}$) and below V_{BEon}. During the low-to-high transition at the input, the output voltage drops until it reaches V_{BEon}. Consequently, the current flowing into the base of Q2 becomes zero. But as the excess minority charge stored in the base of the output transistor is not

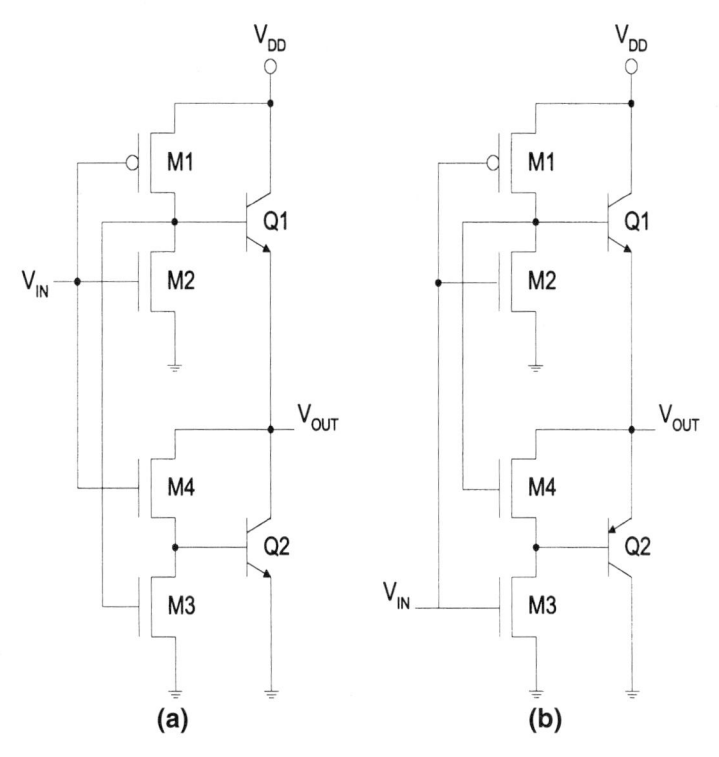

Figure 4–51 The structure of (a) conventional BiCMOS (b) conventional CBiCMOS buffers.

depleted, Q2 remains on and the load capacitor is further discharged, causing the output voltage to drop below V_{BEon}. Thus the main idea, utilized by the circuits in subsequent sections, is to trap the excess minority charge in the base of the bipolar transistors of the output driver. As a result, the bipolar transistors will keep on charging or discharging the output node until they saturate.

4.5.3 Multi-Drain/Multi-Collector BiCMOS Buffers

Figure 4–52 shows two circuit implementations that employ the concept of driving the output drivers into saturation using base current pulses. The advantage of using a base current pulse over a dc base current to drive the output drivers into saturation is that there is no dc power consumption and the turn-off time of the BJT is substantially reduced [8].

4.5.3.1 First Implementation

The first implementation [27] uses a multi-drain [30] CMOS input stage (referred to as transistors P and N in Figure 4–52(a)). Transistors Pd and Nd function as discharging transistors for the bases of Q1 and Q2, respectively. Transistors Q11 and Q22 act as input/output controlled switches. The complementary pair Q1 and Q2 serves as output drivers.

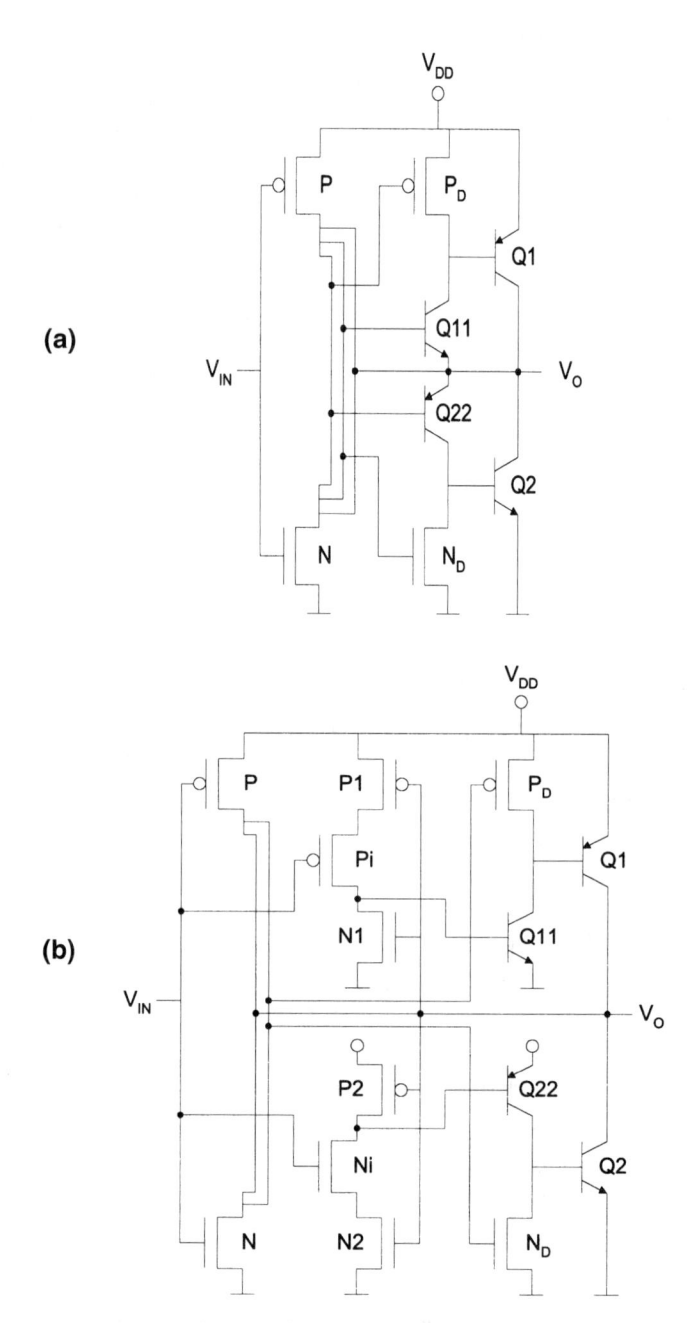

Figure 4–52 Circuit diagrams of (a) first (b) second implementations of the complementary BiCMOS buffers.

For the input falling transition and with the output being initially low, the transistor P turns on, driving Q11 and subsequently Q1 into the active region. The output node starts to pull up to high and when it reaches ($V_{DD} - V_{BEon}$), Q11 turns off. The base of Q1 is now isolated, preventing any leakage of its charge, thus allowing the output voltage to increase until Q1 saturates. The pull-down process, involving transistors N, Q22 and Q2, is similar to the pull-up process. The major drawback of this circuit is that the differential swing ($\Delta V = V_{DD} - V_{CEsat} - V_H$, where V_H is the output voltage at the time when the base current is cut off) is fixed at the value of V_{BEon}. Thus at very low supply voltages (< 2.5 V), the full-swing condition cannot be satisfied unless the areas of Q1 and Q2 are increased significantly.

4.5.3.2 Second Implementation

Figure 4–52(b) shows the second implementation [27] of the CBiCMOS buffer. It incorporates a multi-drain CMOS at the input stage. The first drain is used to compensate for any leakage of the charge stored in the load capacitance during the steady state. The second drain drives the discharging transistors Pd and Nd. The transistors P2, Ni, and N2 control the base current of Q22, which in turn drives the output npn, Q2. The pull-up section is the complement of the pull-down section.

During the low-to-high transition, transistors Ni and N2 turn on and consequently Q22 and Q2 switch on. The output node is pulled down. As the output voltage drops below the threshold voltage of N2, it turns off and the base voltage of Q22 starts to drop below V_{BEon}, turning it off. When Q22 cuts off, the excess minority charge trapped in the base of Q2 enables it to remain on to discharge the output node below V_{BEon} until Q2 saturates. The conduction period of Q22 can be controlled by sizing the transistors P2, Ni, and N2 in order to allow the output to get close to the ground level before the base current of Q2 falls to zero. Hence, the differential voltage swing (ΔV) can be reduced. In fact, simulations have shown that the circuit in Figure 4–52(b) can achieve a rail-to-rail output voltage swing while that of Figure 4–52(a) achieved voltage swing within 0.2 V from the rail voltages.

These circuits are recommended for driving large capacitive loads and not for implementation of logic gates. However, this is not a serious drawback because even the conventional BiCMOS circuits are mainly used as buffers, and logic circuits are usually implemented using CMOS logic.

4.5.4 Simulation Results and Discussion

The 1 μm BiCMOS technology was used for the simulations. The circuits in Figure 4–52 are compared to the conventional BiCMOS buffer (see Figure 4–51(a)), a three-stage CMOS buffer chain, and a full-swing CBiCMOS buffer (without feedback and clamping diode) proposed by Shin [25]. The simulations were performed using HSPICE [31]. The delay per buffer was extracted from the frequencies of a loaded seven-stage ring oscillator. All five circuits have the same input capacitance. The area ratios of the CMOS buffer, the conventional BiCMOS, the full-swing BiCMOS buffer proposed by Shin [25], and the first and second implementations in Figure 4–52 are 1:0.3:0.3:0.7:1, respectively. Figure 4–53 illustrates the normalized delay of the five different circuits as a function of the supply voltage for a load capacitance of 5 pF.

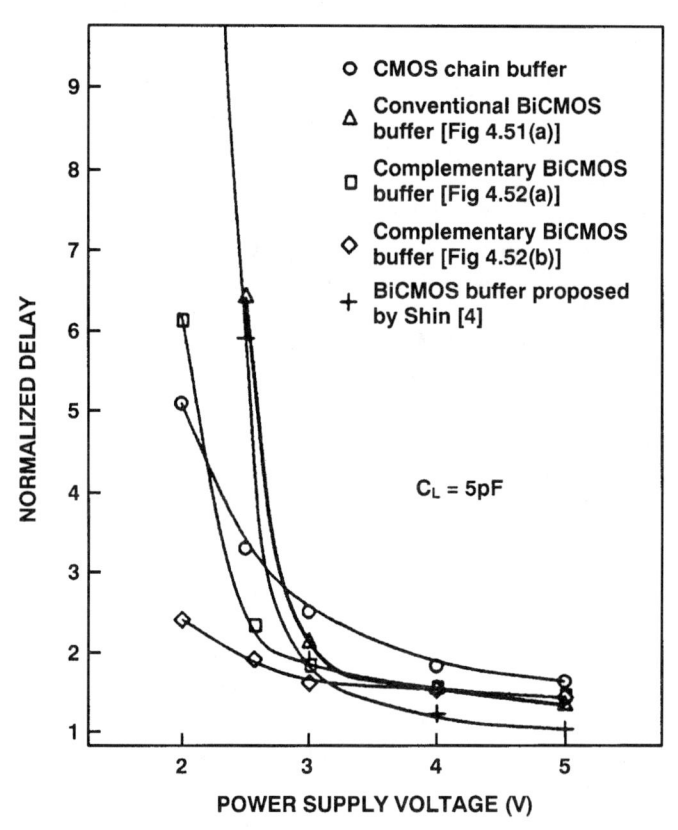

Figure 4–53 Comparison among CMOS chain, conventional BiCMOS, and complementary BiCMOS buffers for different supply voltages.

Above 3 V supply voltages, the BiCMOS circuit proposed by Shin [25] has the least delay, whereas the CMOS buffer has the highest delay. However, below 3 V the speed performance of the BiCMOS circuit proposed by Shin [25] degrades. Also the speed of the conventional BiCMOS buffer decreases below 3 V. The second implementation in Figure 4–52(b) shows the best delay to supply voltage sensitivity below 3 V. In fact, its output voltage swings rail-to-rail even at a power supply voltage of 2 V.

Figure 4–54 shows the normalized delay of different circuits as a function of the load capacitance for a V_{DD} of 2.5 V. Again the new CBiCMOS buffers, shown in Figure 4–52, demonstrate better delay to load sensitivity as compared to that of the CMOS, the conventional BiCMOS and the BiCMOS circuit reported in [25]. The major drawback of these circuits, shown in Figure 4–52, is that by allowing the base currents of Q1 and Q2 to flow for longer time periods, their turn-off times are increased. To solve this problem, the size of the discharging transistors (Pd and Nd) needs to be increased. Another group of multi-drain/multi-collector CBiCMOS

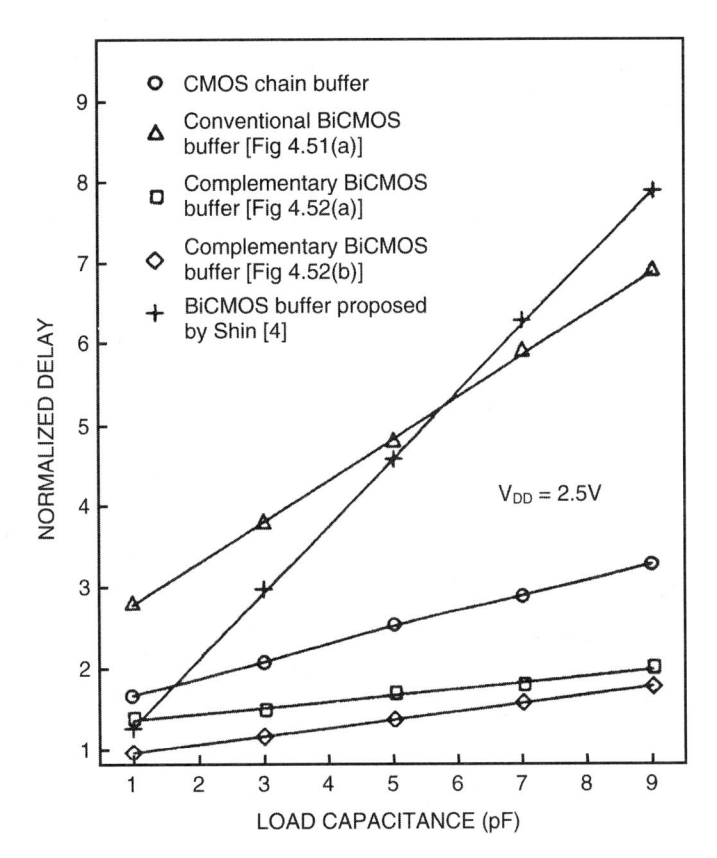

Figure 4–54 Comparison among CMOS chain, conventional BiCMOS, and complementary BiCMOS buffers for different load capacitance.

buffers is reported [32], which not only can reduce the BJTs turn-off time for better speed performance but also offers less circuit and process complexities and utilizes less silicon area. These circuits are presented in the next section.

4.5.5 Full-Swing Multi-Drain/Multi-Collector BiCMOS Buffers

Three CBiCMOS circuits [32] are shown in Figure 4–55. These circuits have been designed such that the pnp BJTs are implemented internally between the inputs and the output. The output drivers employ only npn BJTs. The advantage of having such a design is that the effect of the collector resistance, R_{Cp}, of the internal pnp transistor on the propagation delay is negligible. The collector resistance, R_{Cn}, of the npn output driver will have a significant impact on the propagation delay. This allows the implementation of nonoptimized pnp BJTs with little additional process complexity [33].

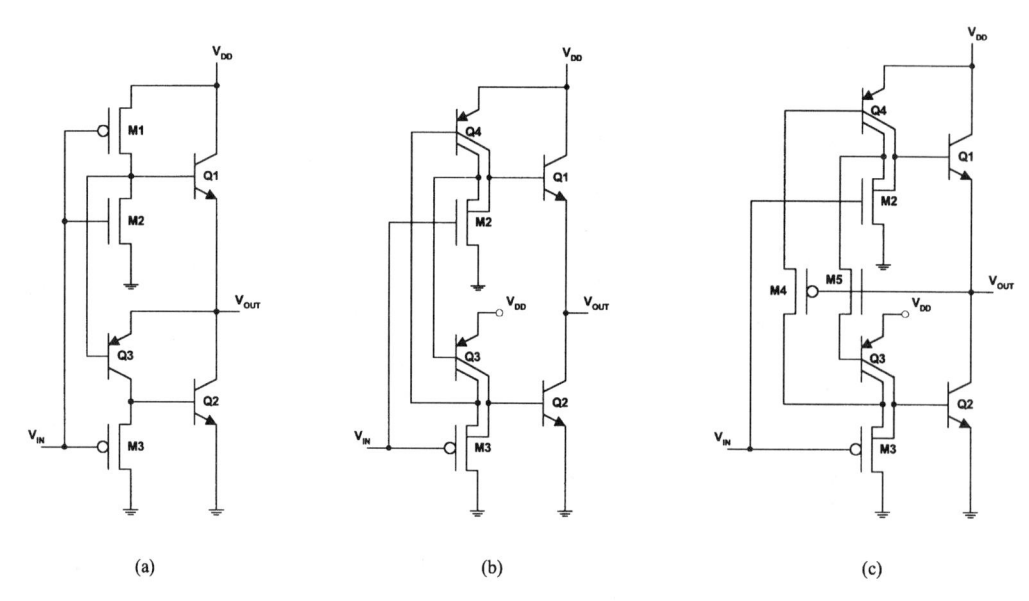

Figure 4–55 Configurations of (a) CBiCMOS-A (b) CBiCMOS-B (c) CBiCMOS-C buffers.

4.5.6 Circuit Implementation and Operation

4.5.6.1 First Implementation

Figure 4–55(a) shows the CBiCMOS-A circuit. During the pull-up operation, the transistor M1 drives the output npn transistor Q1 on and discharges the base of the pnp transistor Q3. Meanwhile, M3 conducts and depletes the base charge of Q2. During the pull-down phase, transistor M2 discharges the base of transistor Q1 and drives transistor Q3 and subsequently turns on transistor Q2. As a result, the output node is pulled low and when it reaches V_{BEon}, Q3 turns off, isolating the base of Q2. The excess minority charge trapped in the base of Q2 keeps it on so that the output voltage can continue to decrease below V_{BEon} until Q2 saturates. However, when the supply voltage is low (less than 2.5 V), the situation described above might not occur. Hence, the CBiCMOS-A suffers from a partial output voltage swing at low-supply voltages.

4.5.6.2 Second Implementation

The CBiCMOS-B circuit shown in Figure 4–55(b) consists of a multi-drain/multi-collector structure implemented to isolate the bases of the BJTs Q1 and Q3 (or Q2 and Q4) [27,30,33]. This isolation will reduce the turn-off time of Q1 and Q3 (or Q2 and Q4), and ensure a full-swing operation. The circuit has a simple structure and a good speed performance at lower supply voltages (< 2 V).

In Figure 4–55(b), note that the function of the transistor M3 is to discharge the base of Q2 and trigger Q4 to conduct during the pull-up phase. The collector current of Q4 will charge the base of Q1 and deplete the carriers at the base of Q3 during the same phase. In the pull-down cycle, transistor M2 is there to discharge the base of Q1 and help Q3 to turn on. Meanwhile, the collector current of Q3 charges the base of Q2 and depletes the carriers at the base of Q4. Consequently, the output node will be pulled down close to V_{CEsat}. The CBiCMOS-B uses a dc base current [8] to saturate the BJTs in order to achieve a full voltage swing. The consequence is an increase in both the dc power consumption and the turn-off time of the BJTs.

4.5.6.3 Third Implementation

Figure 4–55(c) shows the CBiCMOS-C circuit, which is similar to the CBiCMOS-B circuit except that the pMOS M4 and the nMOS M5 are incorporated to control the base currents of Q4 and Q3, respectively. The gates of M4 and M5 are controlled by the output voltage. The sizing of transistors M4 and M5 can be designed to allow the output voltage to get as close as possible to ($V_{DD} - V_{CEsat}$) or V_{CEsat} before the base current of Q4 (or Q3) drops to zero, thus ensuring a full-swing operation [27].

During the input low-to-high transition, the transistor M2 turns on. Since the output voltage is initially high, the nMOS transistor M5 remains in conduction. The transistors Q3 and, subsequently, Q2 will be driven into the active region. When the output node is pulled down to a low value, transistor M5 starts to turn off. When this happens, the base of Q3 is isolated and the trapped minority charge ensures that Q3 remains on to drive the base of Q2 until both of them saturate, hence allowing the output voltage to drop to V_{CEsat} [27]. The major advantage of this circuit is that full-swing operation is achieved without any dc power consumption, and the turn-off time of the BJTs is reduced.

By comparison, the CBiCMOS-A circuit has the simplest structure, but both the CBiCMOS-B and CBiCMOS-C circuits employed multi-drain/multi-collector structure to ensure a full-swing operation at low-supply voltages (< 2 V). Unlike the CBiCMOS-B circuit, CBiCMOS-C has no dc power consumption and the turn-off time of the BJTs is shorter.

4.5.7 Simulation Results and Discussion

SPICE circuit simulations were performed for the conventional BiCMOS, CBiCMOS (see Figure 4–51), and the three CBiCMOS buffers discussed in section 4.5.6. A 0.5 μm BiCMOS technology was used and Tables 4–5 and 4–6 indicate the key CMOS and BJT device parameters. After the circuit optimization, the area ratios of the BiCMOS, CBiCMOS, CBiCMOS-A, CBiCMOS-B, and CBiCMOS-C are 0.67:0.67:0.75:0.83:1, respectively. The delay per buffer was extracted from the frequencies of a seven-stage loaded ring oscillator with all circuits having the same input capacitance. The simulated cut-off frequencies for the npn and pnp BJTs are 9 and 3.6 GHz, respectively.

Figure 4–56 shows the propagation delay (t_{pd}) of the five circuits as a function of the power supply voltage (V_{DD}) for a load capacitance of 5 pF. The CBiCMOS buffers (Figure 4–55) show a better delay-to-supply voltage sensitivity, and hence a better speed performance as compared to those of the conventional BiCMOS and CBiCMOS circuits. Notice that CBiCMOS-B and CBiCMOS-C operate efficiently for supply voltages less than 2 V whereas

Table 4–5 Key CMOS device parameters used in SPICE simulations.

Device parameter	nMOS	pMOS
Oxide thickness, T_{OX}	16 nm	16 nm
Effective channel length, L_{eff}	0.43 µm	0.45 µm
Channel width, W	10 µm	10 µm
Threshold voltage, V_T	0.5 V	–0.6 V

Table 4–6 Key BJT device parameters used in SPICE simulations.

Device parameter	npn BJT	pnp BJT
Emitter area, A_E	1.2×5 µm^2	1.2×5 µm^2
Current amplification factor (max), β_O	100	50
Base transit time, τ_{fo}	20 ps	50 ps
Collector resistance, R_C	92 Ω	190 Ω
Emitter resistance, R_E	6 Ω	15 Ω
Base resistance, R_B	220 Ω	550 Ω
Collector junction capacitance, C_{BC}	29 fF	29 fF
Emitter junction capacitance, C_{BE}	29 fF	29 fF
Substrate junction capacitance, C_{SUB}	110 fF	110 fF
Knee current density, J_K	11 mA/µm^2	11 mA/µm^2

CBiCMOS-A operates efficiently only for supply voltages more than 2.5 V. Figures 4–57(a) and 4–57(b) show the propagation delay of five different buffers as a function of the load capacitance (C_L) for V_{DD} of 5 V and 2.5 V, respectively. The CBiCMOS-A, -B, and -C buffers show better delay-to-load sensitivities as compared to that of the conventional BiCMOS and CBiCMOS buffers. For lower supply voltages (see Figure 4–57(b)), the delay-to-load sensitivities for the BiCMOS and CBiCMOS buffers increase significantly, whereas those of the CBiCMOS-A, -B, and -C buffers deviate slightly from those in Figure 4–57(a). Figures 4–53 and 4–54(a) demonstrate that the CBiCMOS-A buffer shows a better speed performance and better delay-to-load sensitivity than the CBiCMOS-B buffer for V_{DD} > 3.6 V. The opposite is true for supply voltages lower than 3.6 V (see Figures 4–57(a) and 4–57(b)). This difference is due to the degradation in speed performance of the CBiCMOS-B buffer, since the effect of saturating the BJTs is

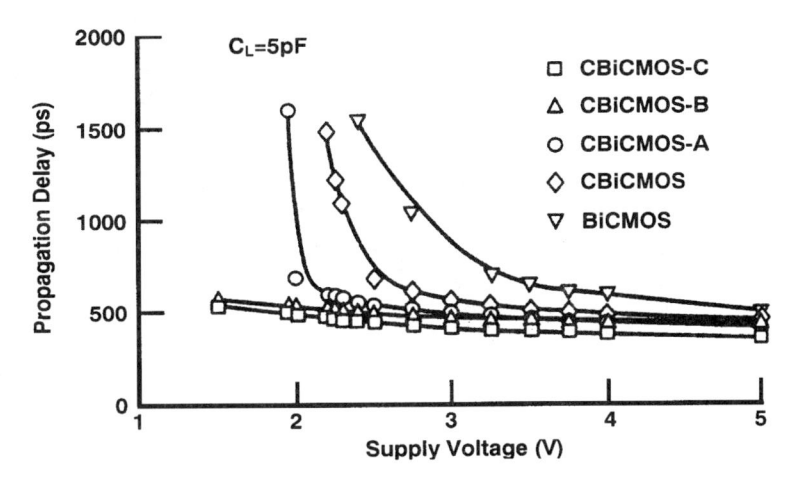

Figure 4–56 Propagation delay as a function of the supply voltage for C_L = 5 pF.

more effective at supply voltages higher than 3.6 V. Figure 4–57(c) compares the propagation delay of the TS-FS BiCMOS [29], CBiCMOS-B, and CBiCMOS-C buffers as a function of the load capacitance (C_L) for a supply voltage of 1.5 V. Both the CBiCMOS-B and CBiCMOS-C buffers show better speed performance and better delay-to-load sensitivity as compared to that of the TS-FS-BiCMOS buffer [29] (see section 4.10).

Figure 4–58 shows the propagation delays (t_{pd}) of the CBiCMOS buffers as a function of the collector resistance R_{Cn} of the npn BJT Q2 and the collector resistance R_{Cp} of the pnp BJT Q3 for a supply voltage of 5 V and a load capacitance of 5 pF. It can be seen that the effect of the collector resistance R_{Cp} of the pnp BJT Q3 is insignificant, as compared to the effect of the collector resistance R_{Cn} of the npn BJT Q2, on the propagation delay. This disparity occurs because the device parameters of Q2 affect the collector current I_{Cn} of Q2. On the other hand, the device parameters of the internal pnp BJT Q3 mainly affect the collector current I_{Cp} of Q3, which is approximately equal to the base current I_{Bn} of the output npn BJT Q2. When driving large capacitive loads, the discharging time of these loads, which is inversely proportional to I_{Cn}, is the dominant term of the propagation delay. Therefore, the effect of the device parameters (I_K, β_f, τ_f, R_C, R_e, C_{BE}, C_{BC}, etc.) of Q3 on the delay-time is less significant as compared to that of Q2. Furthermore, the effect of the base resistance R_{bp} of Q3 on the propagation delay is negligible because the voltage drop across it is typically too small to affect the drain current of M2 [34]. This small voltage drop allows the implementation of the nonoptimized pnp BJTs (using lateral BJT in MOSFET structure) in the CBiCMOS circuits, thus reducing the process complexity [26–29,33].

The CBiCMOS-B and CBiCMOS-C buffers, as compared to CBiCMOS-A, conventional BiCMOS and CBiCMOS buffers, have symmetrical pull-up and pull-down structures and hence do not suffer from speed degradation due to the V_{BE} loss in the pull-up and/or pull-down transi-

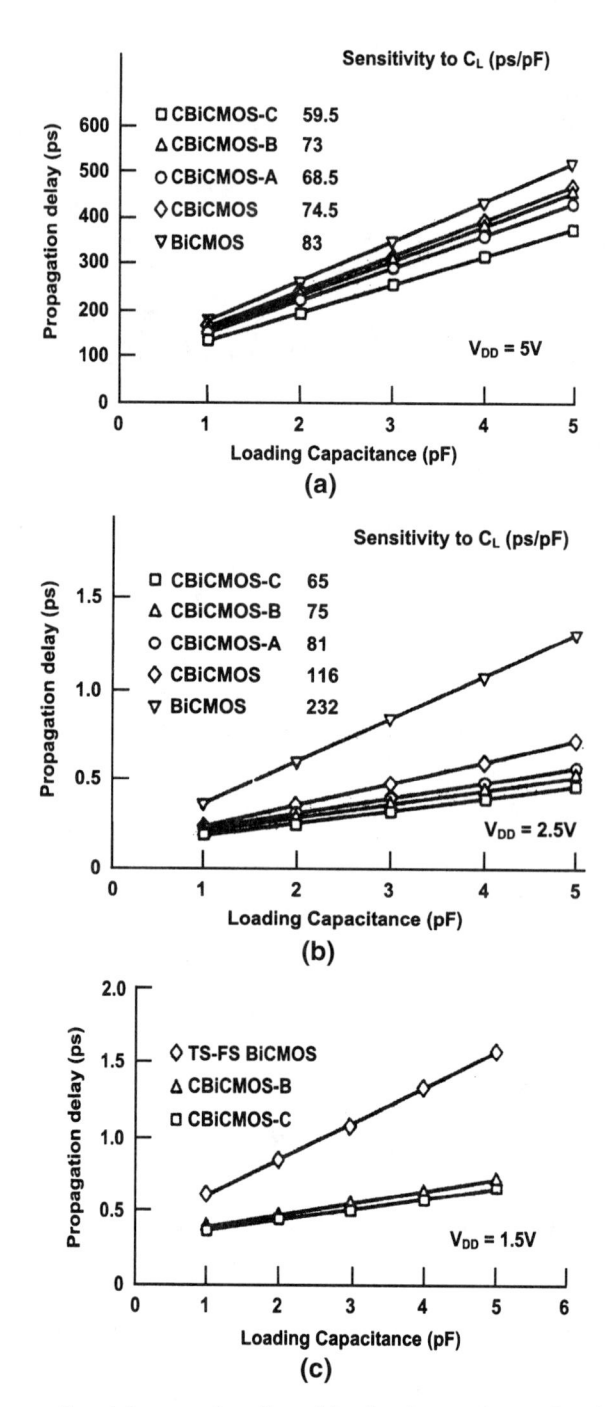

Figure 4–57 Propagation delay as a function of the load capacitance for: (a) V_{DD} = 5 V (b) V_{DD} = 2.5 V (c) V_{DD} = 1.5 V.

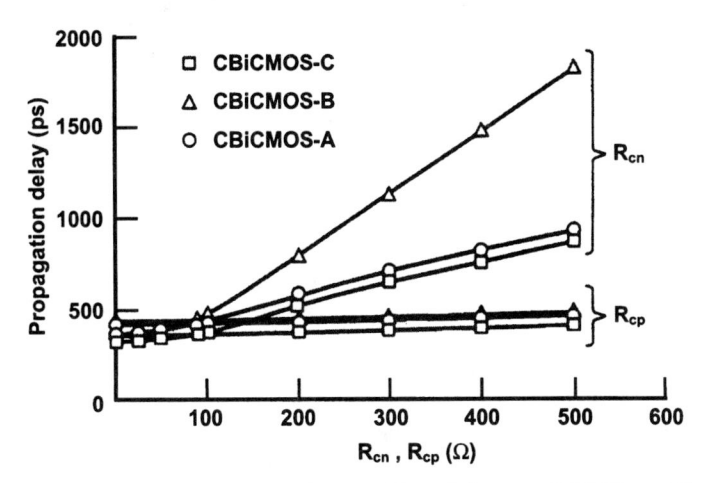

Figure 4–58 The effect of the collector resistance R_{Cn} of the npn BJT Q2 and the collector resistance R_{Cp} of the pnp BJT Q3 on the delay of the three CBiCMOS buffers.

tions [23,26–29]. The CBiCMOS-C buffer has an added advantage over the CBiCMOS-B buffer since it has no dc power consumption, and the turn-off time of the BJTs is shorter. Furthermore, the CBiCMOS-C buffer is superior to the TS-FS-BiCMOS (see section 4.10), especially when driving large capacitive loads because it has a high output collector current driving capacity and uses a much simpler process [27,29]. The CBiCMOS-C buffer has also more advantages than has the full-swing CBiCMOS buffer [27]. These include less circuit complexity (8 instead of 14 devices), less area coverage (about 30% saving in area), less process complexity, symmetrical pull-up and pull-down structures, and better efficiency even for supply voltages less than 2 V.

In conclusion, three different implementations of the CBiCMOS circuits (see Figure 4–55) have been presented in this section. They show better delay-to-supply voltage sensitivity, better delay-to-load sensitivity and high-speed performance at low-supply voltages compared to those of the conventional BiCMOS and CBiCMOS buffers. They are also better than the complementary emitter-follower driver circuits reported in section 4.3 [27], which also employ the same technique, in terms of the circuits' complexity and the turn-off time of the BJTs.

4.6 Quasi-Complementary BiCMOS Logic Circuits

4.6.1 Introduction

The quasi-complementary BiCMOS (QC-BiCMOS) [35] was developed to overcome the problem of BiCMOS performance degradation at low-supply voltages and in the deep submicrometer regime. Two key circuit techniques are used: (1) a separation between the base of the pull-up bipolar and the base of the quasi-pnp, and (2) the carefully designed base discharging circuit enabling the QC-BiCMOS to exhibit a better low-voltage performance than that of other conventional BiCMOS circuits. These techniques will be described in the following sections.

4.6.2 Circuit Concept of the QC-BiCMOS and Its Advantages

First, let us examine the cause for the degraded performance of the conventional BiCMOS circuits below a power supply voltage of 4 V. Figure 4–59 shows the conventional BiCMOS inverter (also called the "Darlington type" BiCMOS inverter [35]) and the QC-BiCMOS inverter.

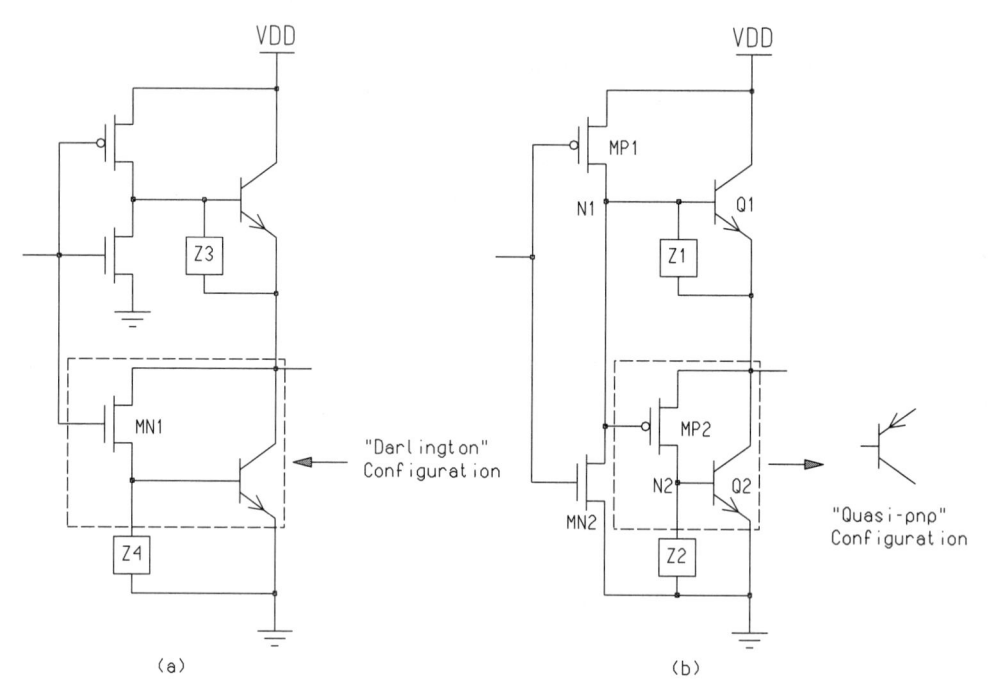

Figure 4–59 Schematic diagrams of (a) conventional (Darlington-type) BiCMOS inverter and (b) QC-BiCMOS inverter.

During the pull-down operation, the maximum gate-source voltage that can be applied to the nMOS transistor MN1 is:

$$V_{GS} = V_{DD} - 2V_{BE} \qquad 4.1$$

where V_{GS} is the gate-source voltage of MN1, V_{DD} is the supply voltage, and V_{BE} is the base-emitter voltage when the bipolar transistor turns on. From Eq. 4.1, when the supply voltage is scaled down, V_{GS} will be significantly reduced since V_{BE} remains constant. Figure 4–60(a) shows that when V_{GS} of MN1 is reduced as in the conventional BiCMOS circuit, its drain current decreases tremendously.

Quasi-Complementary BiCMOS Logic Circuits

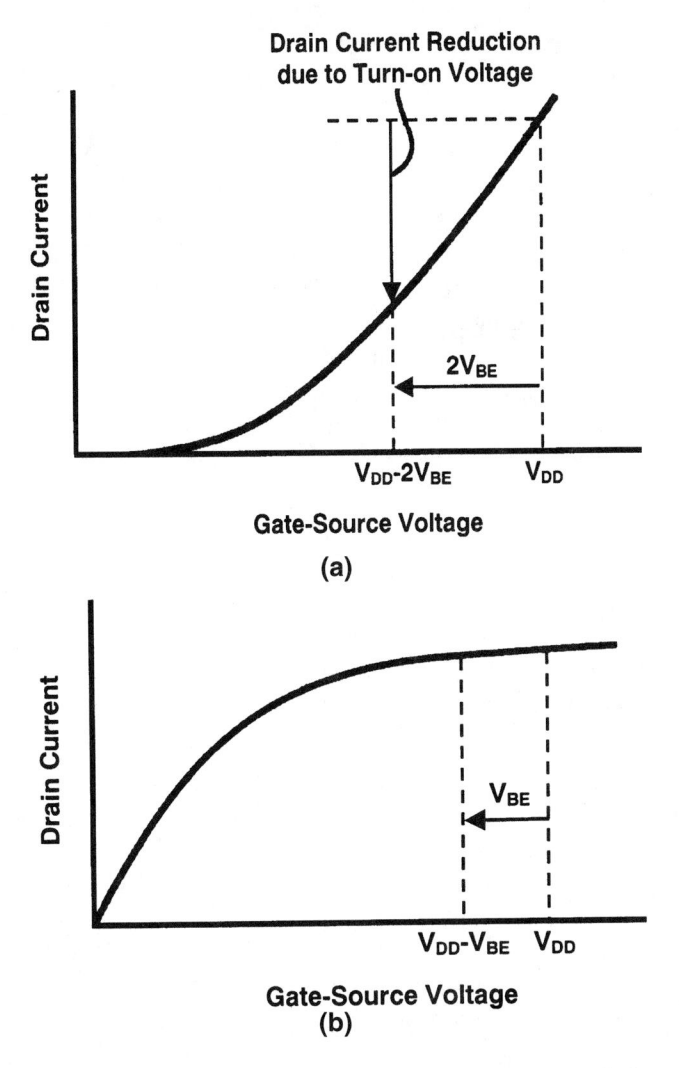

Figure 4-60 Current-voltage characteristics of MOSFETs that show the drain current reduction due to base-emitter turn-on voltage in (a) conventional Darlington-type BiCMOS and (b) QC-BiCMOS.

The QC-BiCMOS circuit overcomes this problem by employing the "quasi-pnp" connection, which consists of a pMOS transistor (MP2) driving the output npn bipolar transistor (Q2) (see Figure 4–59(b)). The idea of quasi-pnp, where the poor lateral pnp is reinforced by adding the high-performance vertical npn, is already well-known and resembles class-B amplifiers in analog ICs [36]. This concept has been adopted in the design of the QC-BiCMOS inverter, with the poor pMOS being reinforced by the high-performance npn.

During the pull-down operation, the base of the npn bipolar transistor Q2 (see Figure 4–59(b)) is driven by the pMOS transistor MP2. The drain-source voltage of MP2 is reduced by the base-emitter voltage V_{BE} of Q2; however, the gate-source voltage remains unaffected. Figure 4–60(b) shows that in the saturation region, the drain current dependency on the drain-source voltage is small. This means that the influence of V_{BE} on the delay is small, and the delay dependence on the supply voltage is smaller than that of the conventional BiCMOS inverter. Figure 4–61 shows the simulated voltage and current waveforms of the QC-BiCMOS in the pull-down operation. When the output voltage drops below the threshold voltage of MP2, the transistor MP2 switches off. The carriers in the base of the npn transistor Q2 can only be depleted through the resistor Z2 and the intrinsic base current component. Hence by selecting appropriate values for Z2, these base carriers can be retained until the output is driven to ground level by Q2. This phenomenon is illustrated in Figure 4–61(a) where the voltage of the node N2 does not drop to ground until the output voltage is completely pulled down to 0 V.

The bipolar transistor Q2 will be temporarily saturated just after the pull-down operation. However, this saturation is relatively shallow because, as shown in Figure 4–61(b), the base-current driving through MP2 is already terminated when Q2 is saturated. The shallow saturation of Q2 is relieved promptly by the resistor Z2 after the pull-down operation. On the other hand, for conventional BiCMOS circuits such as the C-BiCMOS and the Darlington-type BiCMOS circuits, saturation of the output transistors is avoided. Hence they lose the bipolar drive when the output is lower than V_{BE}. Although small MOSFETs or resistors can be employed to lower the output voltage below V_{BE}, the discharging speed is much slower than that of bipolar transistors.

Two circuit techniques are employed in the design of QC-BiCMOS. The first involves a separation between the base of the pull-up bipolar (Q1) and the base of the quasi-pnp (the gate of MP2). This separation is expected to improve the QC-BiCMOS performance. The second technique involves a carefully designed base discharging circuit (Z2), which enables the quasi full-swing operation. The discharging circuit Z2 needs to be designed correctly because it provides a base charge retention and a relief from shallow saturation. Four configurations of QC-BiCMOS will be analyzed in the next section and their schematic diagrams are shown in Figure 4–62, which indicates the gate widths of the MOSFETs and the emitter lengths of the bipolar transistors.

The QC-BiCMOS-a connects a simple CMOS logic to a quasi-complementary buffer. The QC-BiCMOS-b has the nodes N3 and N4 separated by diode-connected MOSFETs. The QC-BiCMOS-c features an independent MOS logic for pull-up and pull-down. It can construct a large fan-in logic without suffering short-circuit current through Q1 and Q2 due to the large degree of freedom in its MOS size design. The QC-BiCMOS-d has a feedback circuit, which reduces the leakage of the base charge during the pull-down transient. This reduction is achieved by using the output signal to determine the timing of extracting the charge from Q2 via MN4.

Quasi-Complementary BiCMOS Logic Circuits 193

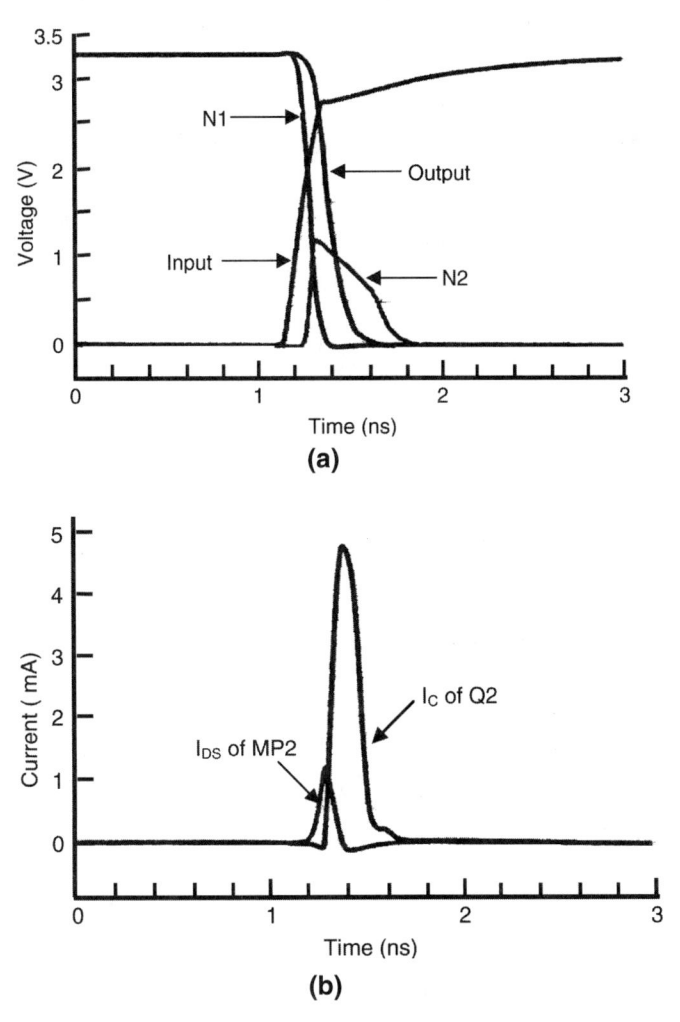

Figure 4–61 Simulated (a) voltage waveforms (b) current waveforms of the QC-BiCMOS in the pull-down phase (three-input NAND with 5-fan-out load).

4.6.3 Circuit Performance Comparison and Analysis

The performance of the QC-BiCMOS is compared with the conventional BiCMOS circuit performance. A 0.3 μm BiCMOS technology is used in the circuit simulation and key MOS and bipolar parameters are listed in Table 4–7. To ensure a fair comparison between the CMOS and BiCMOS circuits, a certain degree of standardization is required. First, the threshold voltage is defined here as the gate voltage that drives 1 nA drain current per 1 μm gate width. Second, the output fan-out load is set to 5, which is the middle of the optimum loaded condition for the CMOS and the BiCMOS. Also since the circuit delay is heavily dependent on the gate-width

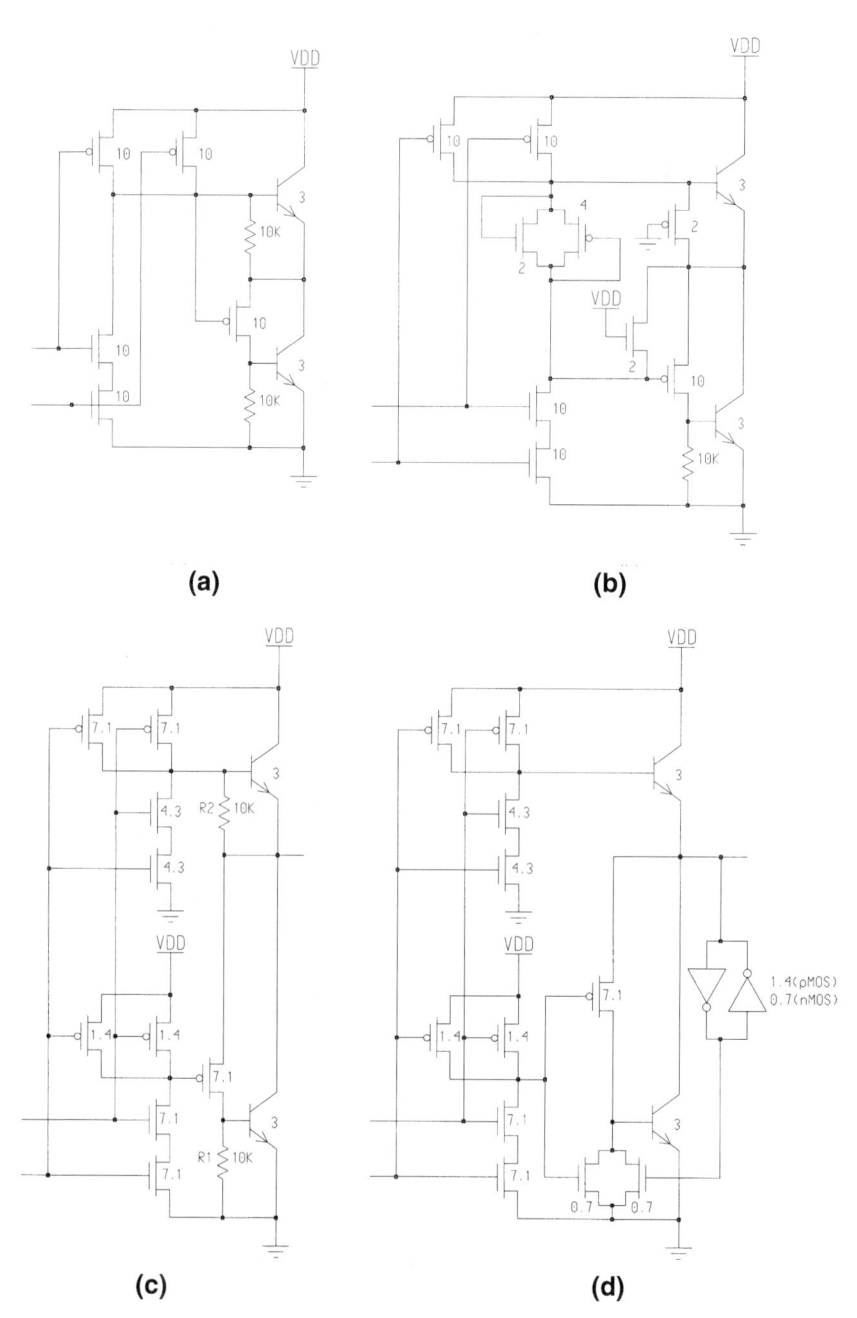

Figure 4–62 Typical QC-BiCMOS two-input NAND configuration (a) QC-BiCMOS-a (direct-connected type) (b) QC-BiCMOS-b (diode-connected type) (c) QC-BiCMOS-c (separate type) (d) QC-BiCMOS-d (feedback MOS discharging type).

Quasi-Complementary BiCMOS Logic Circuits

design of the base discharging MOSFETs or resistance, it is therefore essential that these parameters are carefully set. Figure 4–63 shows the different implementations of the 2-input NAND gate. Note that the numbers shown in Figure 4–63 are the gate widths of MOSFETs and emitter lengths of the bipolar transistors. Figure 4–64 and Table 4–8 compare the various QC-BiCMOS circuits in terms of the device count, gate area, delay, power dissipation, and delay dependence on the supply voltage.

Table 4–7 Key BiCMOS device parameters used in circuit simulations.

MOSFET	Gate oxide thickness		8 nm
	Drain current		2.8 mA (nMOS), 2.0 mA (pMOS)
	Drain-substrate capacitance		10 fF
Bipolar Transistor	Cutoff frequency		24 GHz
	Knee current		5 mA
	Junction capacitance	C_{TE}	6.5 fF
		C_{TC}	5.0 fF
	Base resistance		250 Ω

The QC-BiCMOS-a circuit is the slowest among the four types of QC-BiCMOS configurations because the time taken to charge and discharge the node N1, which has a larger parasitic capacitance than those of N3 or N4, is longer. In the QC-BiCMOS-b circuit, the node N4 is discharged in a short time without being affected by the parasitic capacitance of the node N3 in the pull-down operation. Similarly, QC-BiCMOS-c and -d, which have small MOSFETs (MN3 and MP3), operate quickly without interference from those large parasitic capacitances. From Figure 4–64, we can see that the QC-BiCMOS-d circuit shows the best delay performance and is suitable for operation below a power supply voltage of 2.5 V, because it has a steep pull-down waveform due to the good charge retention capability of the bipolar transistor Q2. Furthermore, in the pull-up operation, the charge outflow from the base of Q1 is small due to the output latch. The QC-BiCMOS-b and -c circuits, which have a resistance between their base and emitter, lose some of the stored base charges in the pull-down transients. In future comparisons, unless otherwise stated, the QC-BiCMOS-d circuit configuration will be used. But the conclusions drawn will be also valid for the QC-BiCMOS-b and -c circuits. The only drawback of the QC-BiCMOS-d circuit (see Table 4–8) is the large number of device count and gate area. Figure 4–65 shows the delay dependencies of the QC-BiCMOS (QC-BiCMOS-d circuit), BiNMOS, C-BiCMOS, conventional BiCMOS, and CMOS inverters on the supply voltage.

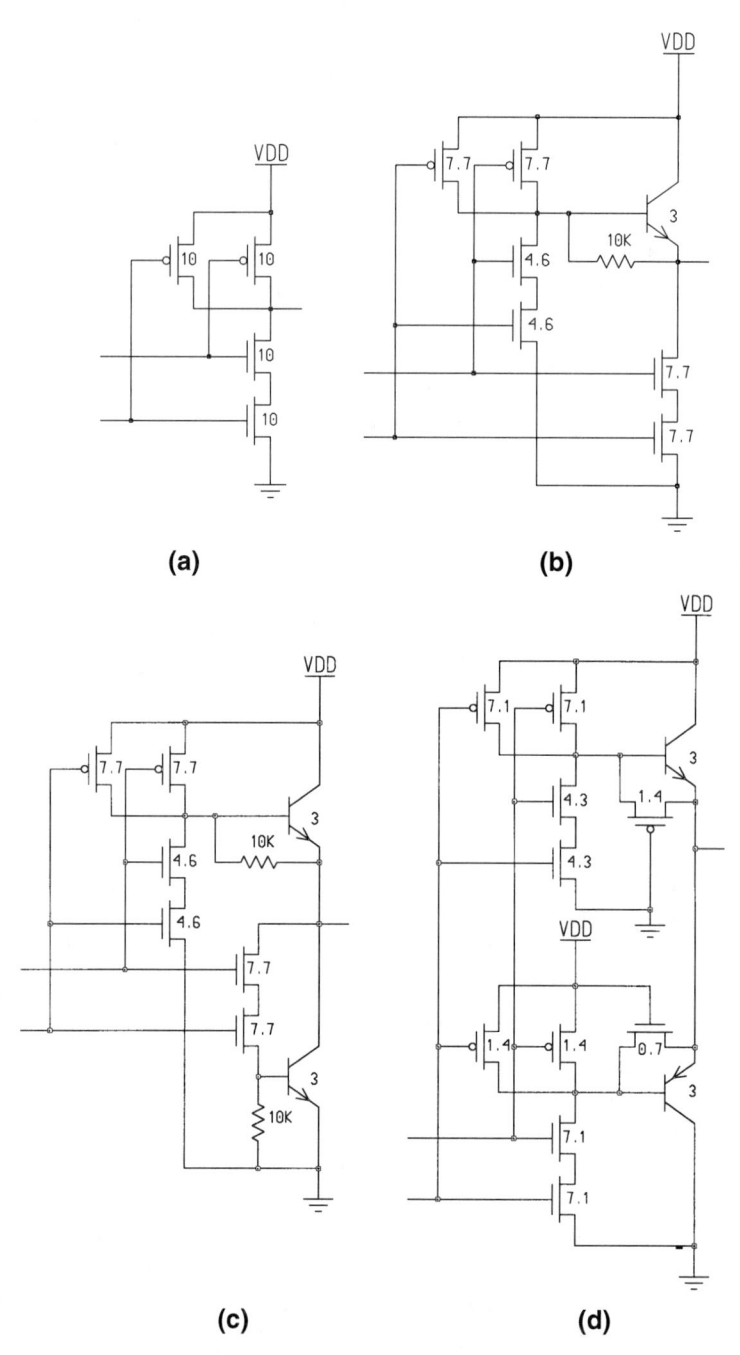

Figure 4–63 Two-input NAND gate implemented using (a) CMOS (b) BiNMOS (c) Darlington (d) C-BiCMOS.

Quasi-Complementary BiCMOS Logic Circuits

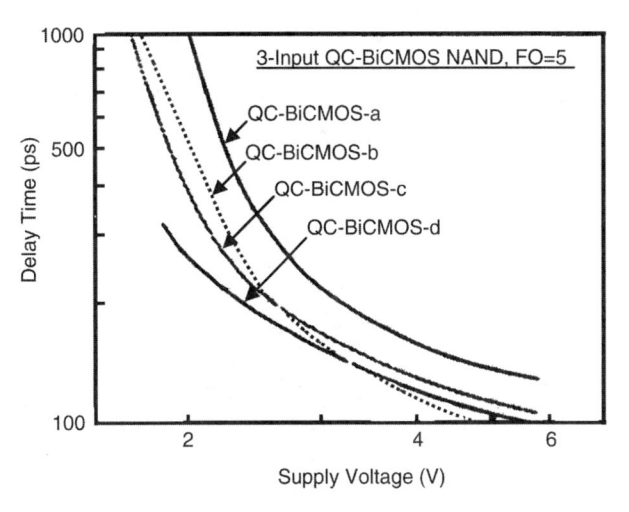

Figure 4–64 Delay dependencies on the supply voltage for various QC-BiCMOS circuits.

Table 4–8 Comparison among circuits (three-input NAND with 5 fan-out load at 3.3 V).

	Conventional				QC-BiCMOS			
	CMOS	BiNMOS	Darlington	C-BiCMOS	(a)	(b)	(c)	(d)
No. of MOSFETs	6	10	11	14	9	11	15	19
No. of Bipolars	0	1	2	2	2	2	2	2
Gate Area[i]	0.5	0.81	1	1.23	0.98	1.06	1.30	1.39
Delay (ps)	277	198	192	152	191	139	144	138
Power[ii] (μW)	146	151	150	151	174	179	151	165

i. Relative area with respect to Darlington type BiCMOS.
ii. 100 MHz operation.

The QC-BiCMOS has the lowest delay time as compared to that of the other circuits. At 2.5 V supply, the QC-BiCMOS is twice as fast as the CMOS. Even at a power supply voltage of 2 V, it is 1.5 times faster than the CMOS. By contrast, the conventional BiCMOS, BiNMOS, and C-BiCMOS lose their performance leverage over the CMOS at 2.9 V, 2.4 V and 2.1 V, respec-

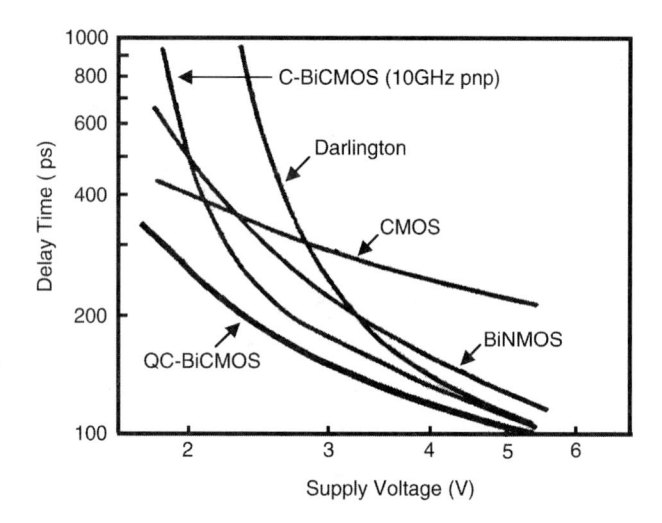

Figure 4–65 Delay dependencies on supply voltage for various circuits.

tively. Next the QC-BiCMOS is compared with the C-BiCMOS in greater detail. But first to clarify the performance correspondence between the quasi-pnp and the real pnp, the equivalent cutoff frequency of the quasi-pnp is estimated as follows: The cut-off frequency of the real pnp transistor, which provides the same pull-down delay time by replacing the quasi-pnp, is defined as the equivalent cut-off frequency of the quasi-pnp. Figure 4–66 shows the simulation results for the dependence of the equivalent cut-off frequency of the quasi-pnp on the pMOS gate length and cut-off frequency of the npn transistor.

The results reveal that the quasi-pnp generally has a higher cut-off frequency than the real pnp transistors in the deep submicrometer regime. For instance, using a 0.3 μm pMOS gate length at 20 GHz cut-off frequency of the npn bipolar transistor at 3.3 V supply gives a cut-off frequency of approximately 10 GHz, which is higher than that of any pnp transistor fabricated in the BiCMOS process. Figure 4–67 shows the simulated input and output waveforms of the C-BiCMOS and QC-BiCMOS circuits at different supply voltages.

The pull-down waveform of the QC-BiCMOS is very steep and is a quasi full-swing. On the other hand, the C-BiCMOS achieves a full-swing but has a slow transient response. Also notice that the transition from steep discharging (due to bipolar devices) to slow discharging (due to small MOSFETs) occurs at lower supply voltages. The reduced stored charge in the base of the pnp transistor results in an early termination of the bipolar current drive. Figure 4–68 shows the delay dependence of the various circuits on the fan-out.

The dependence of the QC-BiCMOS delay on the load capacitance is stronger than that of other conventional circuits because the output current is amplified by the nMOS (MN2), the

Quasi-Complementary BiCMOS Logic Circuits

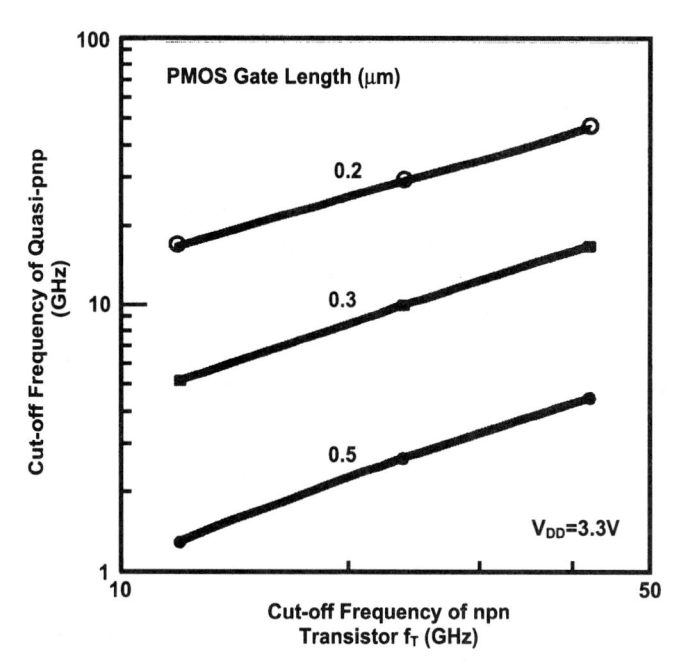

Figure 4-66 Dependence of equivalent cut-off frequency of quasi-pnp on pMOS gate length and cut-off frequency of npn transistor.

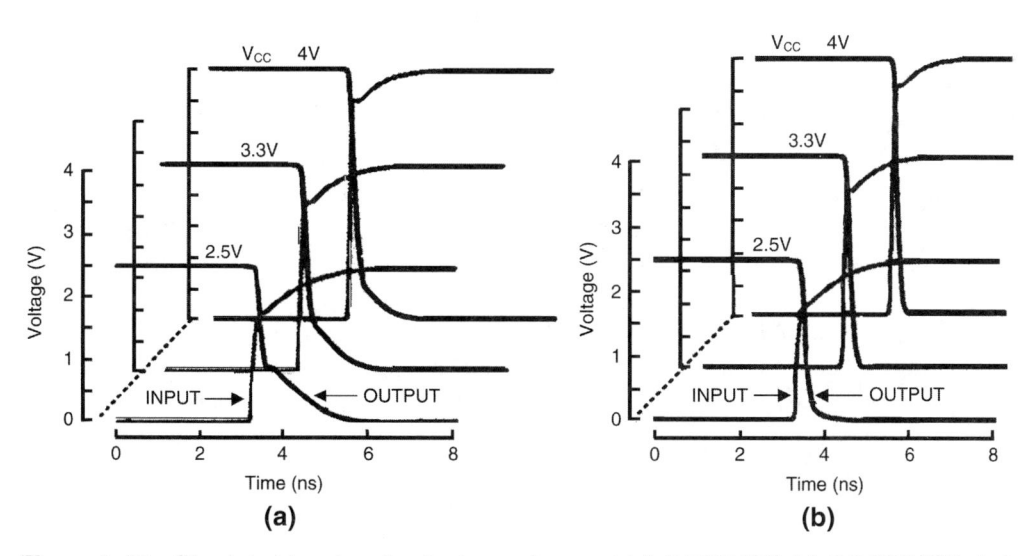

Figure 4-67 Simulated input and output waveforms of (a) C-BiCMOS (b) QC-BiCMOS-d at various supply voltages.

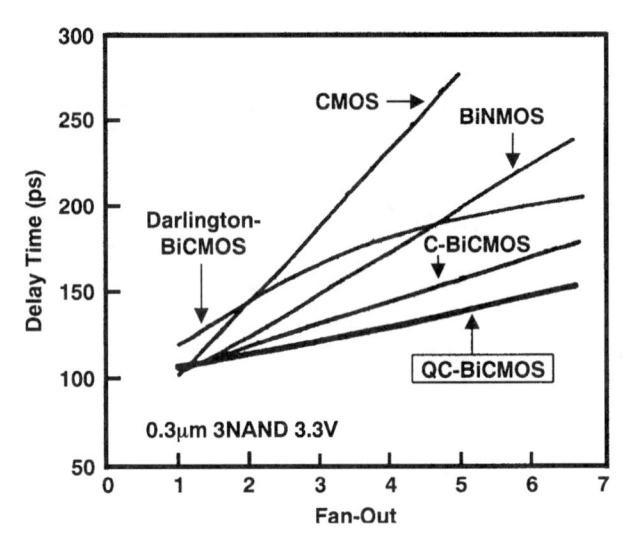

Figure 4–68 Dependence of delay time on the fan-out load.

pMOS (MP2), and the npn (Q2) in the pull-down operation (Figure 4–59(b)). In the pull-up operation, the npn (Q1) in the QC-BiCMOS drives a larger current than does the npn (Q1) in the conventional BiCMOS because the pMOS (MP1) receives a quasi-full-input swing (Figure 4–59(b)).

From Table 4–8, we notice that the gate area of the QC-BiCMOS is larger than that of the C-BiCMOS due to the additional area needed for the pMOS (MP2). This larger area may pose a problem in the IC design layout where the area is limited. But this area requirement is tolerable when the QC-BiCMOS is used in the sea-of-gates gate array where it can be used as the writing area. Also the power dissipation of the QC-BiCMOS is slightly more than that of the CBiCMOS circuit due to the additional power required for the pMOS (MP2) drive.

Next, the impact of the excellent low-voltage performance of the QC-BiCMOS on the performance of the deep submicrometer ULSIs is assessed. Figure 4–69 shows the delay time dependence on the channel length for three different circuit configurations.

The plots take the J-shaped curves because the delay increase caused by the supply voltage reduction is more significant than the delay reduction that arises from the gate-length reduction at the characteristic length [37]. The QC-BiCMOS maintained its performance leverage over CMOS even for a channel length of 0.13 μm at 2 V supply voltage. In contrast, the conventional Darlington BiCMOS loses its advantage over the CMOS at channel lengths less than 0.3 μm. Therefore, the QC-BiCMOS extends the scaling limit of BiCMOS by two generations.

Quasi-Complementary BiCMOS Logic Circuits

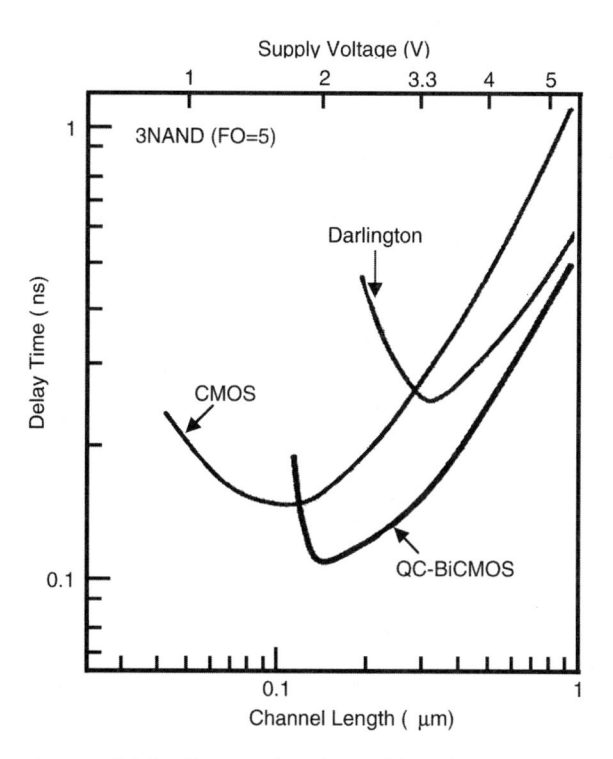

Figure 4-69 Dependence of delay time on the channel length.

4.6.4 Delay Analysis

In this section, the delay expression for the QC-BiCMOS is derived and compared to the delay expression of the conventional BiCMOS circuit. The simplified circuit shown in Figure 4–70 is used to derive the delay expressions.

4.6.4.1 Pull-Down Delay

From section 4.6.7, we can write the pull-down delay time t_{phl} of the QC-BiCMOS as

$$t_{phl} = \left(\frac{3\tau\tau_p C_L V_{DD}}{I_{DN}}\right)^{\frac{1}{3}} \qquad 4.2$$

where τ denotes the transit time of the bipolar transistor (Q2), τ_p is the transit time of the pMOS (MP2), C_L is the load capacitance, V_{DD} is the supply voltage, and I_{DN} is the drain current of the nMOS (MN2). For the conventional BiCMOS circuit [38], the pull-down delay time is given by:

$$t_{phl} = \left(\frac{\tau C_L V_{DD}}{I_{DN}}\right)^{\frac{1}{2}} \qquad 4.3$$

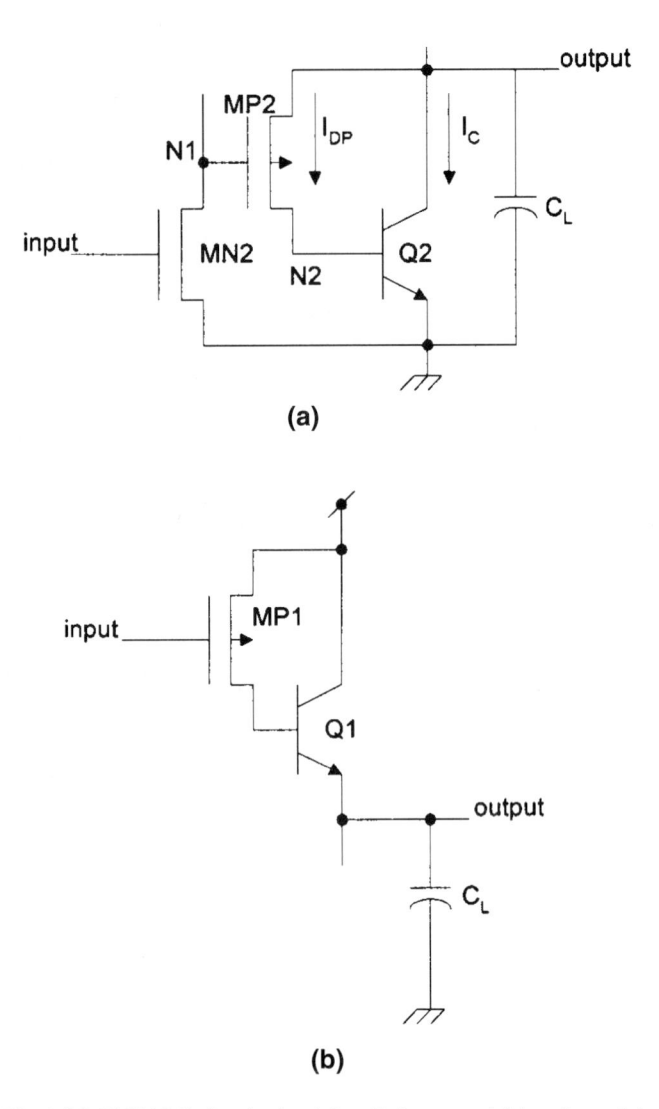

Figure 4–70 Simplified QC-BiCMOS circuits for (a) pull-down and (b) pull-up delay analysis.

Comparing Eqs. 4.2 and 4.3 demonstrates that the QC-BiCMOS delay has a weaker dependence on the load capacitance, since t_{phl} of the QC-BiCMOS is proportional to $C_L^{1/3}$ and that of the conventional BiCMOS is proportional to $C_L^{1/2}$. This weaker dependence is due to the current drive amplification by MN2, MP2, and Q2. Also the delay expression for the QC-BiCMOS is the product average of the performance of nMOS, pMOS, and npn bipolar device, whereas in the conventional BiCMOS, the performances of the nMOS and npn bipolar transistors determine the pull-down delay. Hence, the QC-BiCMOS delay is less sensitive to the bipolar cut-off frequency because three device performances contribute equally to the delay. By solving Eq. 4.2 with the MOSFET drain current relation in the velocity-saturation limit:

$$I_{DS} = qWv_{sat}C_{OX}(V_{GS} - V_T) \qquad 4.4$$

The supply voltage dependent delay expression for the QC-BiCMOS is:

$$t_{phl} \propto \left(\frac{V_{DD}}{V_{DD} - V_{BE} - V_T}\right)^{\frac{1}{3}} \qquad 4.5$$

By contrast, for the conventional BiCMOS gate, the delay expression is given by:

$$t_{phl} \propto \left(\frac{V_{DD}}{V_{DD} - 2V_{BE} - V_T}\right)^{\frac{1}{2}} \qquad 4.6$$

These two equations show that the QC-BiCMOS gate has weaker supply-voltage dependence than have the conventional BiCMOS gates.

4.6.4.2 Pull-Up Delay

The pull-up delay expression for the QC-BiCMOS, which is the same as the conventional BiCMOS, can be written as:

$$t_{plh} = \left(\frac{\tau C_L V_{DD}}{I_{DP}}\right)^{\frac{1}{2}} \qquad 4.7$$

The drain current expression for the pMOS (I_{DP}) is, however, different from that of the conventional BiCMOS because the pull-down waveform is quasi full-swing in the QC-BiCMOS. The drain current for the transistor MP2 can be expressed as:

$$I_{DP} \propto (V_{DD} - V_T)^B \text{ for the QC-BiCMOS}$$

$$I_{DP} \propto (V_{DD} - V_{BE} - V_T)^B \text{ for the Darlington BiCMOS}$$

where B is a fitting parameter and is between 1 and 2. From the above expressions, we can deduce that the QC-BiCMOS pull-up delay is smaller than that of the conventional BiCMOS, especially at low supply voltages, although the pull-up circuit itself is essentially the same as the conventional BiCMOS.

4.6.4.3 Low-Voltage Limitation

The QC-BiCMOS has a low-voltage limit and is explained as follows. Eq. 4.2 is based on the assumption that the pMOS (MP2) operates in the saturation region. But from Figure 4–60(b), we see that the operating point moves to the linear region at low supply voltages due to V_{BE}. In the linear region, the drain current degrades severely with decreasing supply voltage. Therefore, the condition for MP2 entering the linear region is given as:

$$V_{DD} = V_{BE} + V_{DS(sat)} \qquad 4.8$$

where $V_{DS(sat)}$ denotes the drain-source voltage for the saturation region. We can further define $V_{DS(sat)} = \gamma_a V_{DD}$ where γ_a is the scaling coefficient. Eq. 4.8 can be rewritten as:

$$V_{DD} = \left[\frac{V_{BE}}{1 - \gamma_a} \right] \qquad 4.9$$

Assuming V_{BE} to be around 0.8 to 1 V, the transition voltage from saturation to the linear region is about 1.6 to 2 V. Below this voltage, the current drive into the Q2 base reduces rapidly, degrading the circuit performance.

4.6.5 Design Issues

One of the main design issues of Figure 4–59(b) is the discharging circuit for the base charge of Q2 because it should be retained until the pull-down transient is completed. Hence the resistance Z2 ideally should be high. However, the saturated Q2 must be discharged shortly after the pull-down transition; otherwise, the next pull-up transition will cause a large short-circuit current through Q1 and Q2. Therefore, the resistance Z2 should not be too large. Figure 4–71 shows the simulation results of the output waveforms for the pull-down operation of the two-input NAND QC-BiCMOS (implemented using the QC-BiCMOS-c circuit) for different values of R1.

Notice that for values of R1 less than 3 kΩ, a transition from steep transient to slow transient far above the ground level is observed. This transition shows that the base charge is discharged through the resistance R1 before the pull-down is completed. An undershoot below ground level is observed for values of R1 between 10 kΩ and 30 kΩ because the electrons stored in the base flow to the collector and emitter when the holes in the base are discharged through the resistance R1. This phenomenon can be viewed as a capacitive coupling between the base and collector through the diffusion capacitance, thereby lowering the output voltage when the base voltage is discharged. The undershoot is gradually diminished by the current flowing through the resistance R2.

Another design issue is that the bipolar transistor must have high current gain. This condition is much more critical in QC-BiCMOS than in conventional BiCMOS for the purpose of charge retention in the base. Figure 4–72 shows the simulated pull-down waveform for different current gains. For current gains below 30, the waveforms start to degrade. Therefore, in QC-BiCMOS, bipolar transistors with high current gain (> 30) have been used so that performance degradation due to low current gain does not occur.

4.6.6 Experimental Results

The QC-BiCMOS gate chains have been fabricated using a 0.3 µm double polysilicon-self-aligned BiCMOS technology. The delays from the input signal to the output signal have been measured relative to those of the CMOS circuit and compared against those of the conven-

Quasi-Complementary BiCMOS Logic Circuits

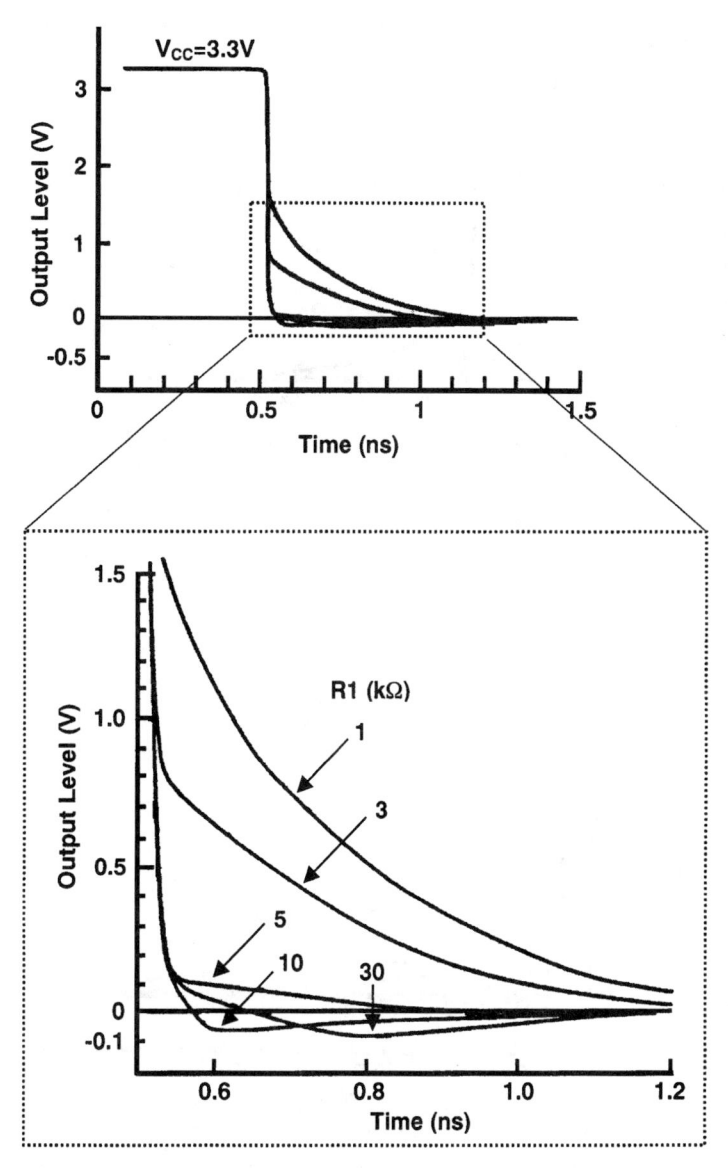

Figure 4–71 Influence of resistance R1 on the output waveform in the pull-down operation of the two-input QC-BiCMOS NAND gate using QC-BiCMOS-c configuration.

tional BiCMOS (Figure 4–73). Figure 4–73 shows that the fully loaded (C_L = 0.16 pF, fan-out = 5) three-input QC-BiCMOS NAND gate maintains a performance leverage over the CMOS even down to below 3 V supply. In contrast, the conventional BiCMOS loses its performance leverage over CMOS at about 3 V supply.

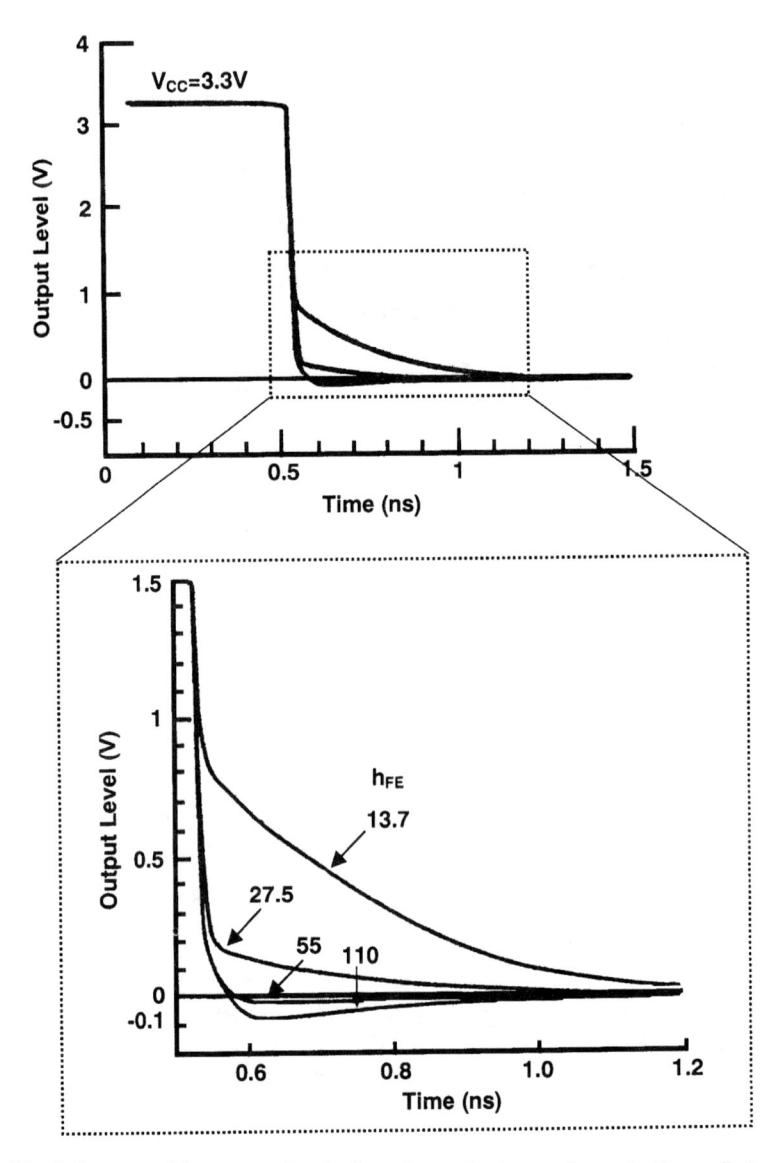

Figure 4–72 Influence of the current gain β on the output waveforms in the pull-down operation of two-input QC-BiCMOS NAND gate using QC-BiCMOS-c configuration.

4.6.7 Appendix

The input pulse is assumed to be a step function and the parasitic capacitance is neglected for simplicity. If the drain current through the nMOS transistor MN2 is a constant (denoted by I_{DN}), we can write:

Quasi-Complementary BiCMOS Logic Circuits

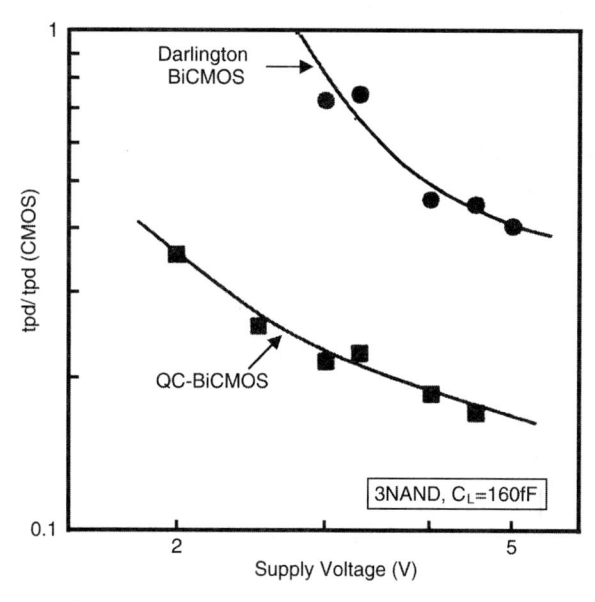

Figure 4-73 Measured performance leverage of the conventional and QC-BiCMOS-d gate over the CMOS.

$$\frac{C_L V_{DD}}{2} = \int_0^{t_{phl}} I_C \, dt$$

$$I_C = \frac{\int_0^t I_{DP} \, dt}{\tau}$$

$$I_{DP} = \frac{\int_0^t I_{DN} \, dt}{\tau_p}$$

where I_C denotes the collector current, I_{DP} is the drain current of the pMOS (MP2), τ is the transit time of the npn bipolar transistor, and τ_p is the time constant of MP2. Integrating, we get:

$$\frac{C_L V_{DD}}{2} = \frac{I_{DN} \times t_{phl}^3}{6 \tau \tau_p}$$

Therefore, the pull-down delay time can be written as:

$$t_{phl} = \left(\frac{3 \tau \tau_p C_L V_{DD}}{I_{DN}}\right)^{\frac{1}{3}}$$

where the input has a ramp waveform defined by:

$$\frac{V_{DD}}{I_{in}} = \frac{t_r}{0.8C_{in}}$$

I_{in} is the current that drives the input capacitance C_{in} of the nMOS (MN2), and t_r is the input rising time. The drain current of MN2 is given by:

$$I_{DN} = \frac{\int I_{in} dt}{\tau_N}$$

The pull-down delay time for the input ramp waveform can be expressed as:

$$t_{phl} = \left(\frac{12\,\tau\tau_p\tau_N t_r C_L V_{DD}}{0.8 I_{DN}}\right)^{\frac{1}{4}}$$

4.7 Full-Swing Schottky BiCMOS/BiNMOS Logic Circuits

4.7.1 Introduction

The circuit performance of the conventional BiCMOS buffer, illustrated in Figure 4–74 for convenience, degrades with scaling down of the power supply voltage. The problems associated with its speed degradation can be explained as follows: The gate voltage of the nMOS driver in the pull-down section is equal to ($V_{DD} - 2V_{BE}$). As the power supply voltage is reduced, the gate-source voltage of the pull-down nMOS driver will also be reduced. This results in a reduction of the drain current and consequently a decrease in the circuit speed. Many alternative circuits such as the BiNMOS [18,39–41], CBiCMOS [10], and quasi-complementary BiCMOS (QC-BiCMOS) [35] were proposed. All these circuits use shunting elements (resistors or MOS) between the collector-emitter or base-emitter junctions of the output bipolar transistors. Although these circuits have a full output voltage swing at the end of the pull-up cycle, their transient rise responses are poor. Furthermore, the QC-BiCMOS circuit, which has been reported to be the fastest compared to the BiNMOS and CBiCMOS, consumes a large area. To overcome these shortcomings, a novel full-swing BiCMOS/BiNMOS logic circuit utilizing Schottky diodes in the pull-up section was developed [42]. Figure 4–75(a) shows the proposed structure. A Schottky diode was added to the pull-up section of the conventional BiNMOS gate.

4.7.2 Basic Concept

The operation of the Schottky BiNMOS (SBiNMOS) gate is based on the concept of driving the output npn transistor into saturation with a base current pulse applied to its base. At the end of the pulse, both the base of Q1 and the anode of the Schottky Diode (SD) will have reached V_{DD}. Hence the Schottky diode turns off and the base of Q1 is isolated, thereby confin-

Full-Swing Schottky BiCMOS/BiNMOS Logic Circuits 209

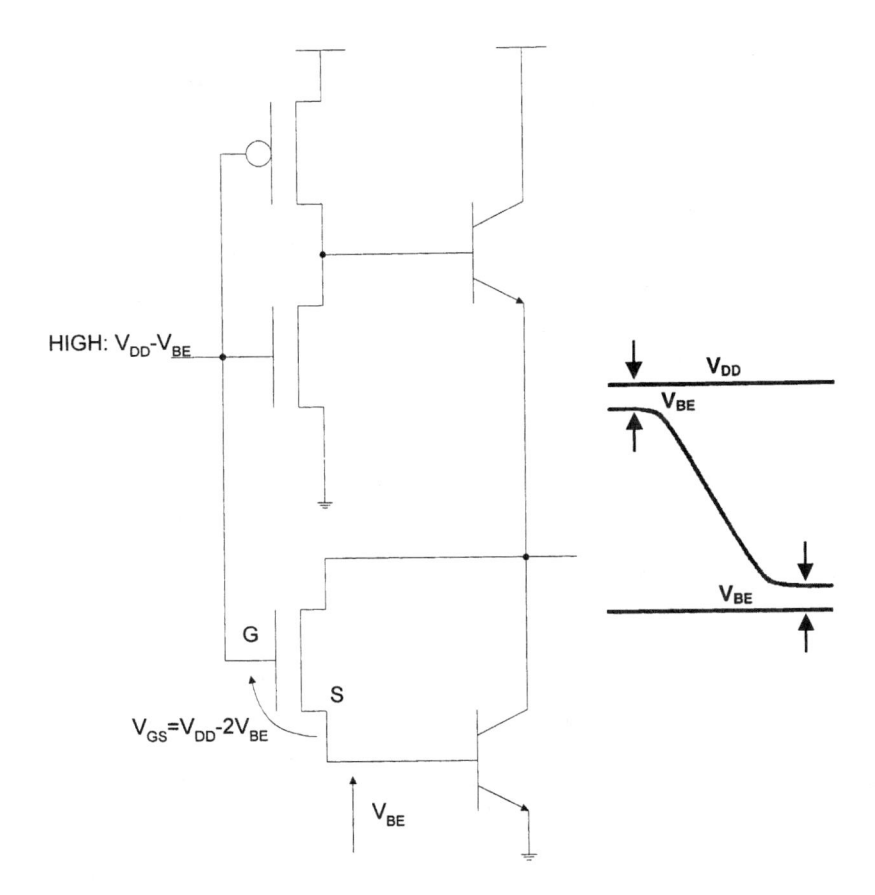

Figure 4–74 Conventional BiCMOS buffer.

ing the excess minority charge to the base. The base charge will keep Q1 on until it saturates. Figure 4–76(a and b) shows the voltage and current waveforms of the SBiNMOS as a function of time.

Figure 4–76(a), illustrates that the base voltage of Q1 is pulled up beyond V_{DD} as a result of the capacitive coupling between the base of Q1 and the output. It can also be seen that when the base voltage reaches V_{DD}, the base-collector voltage of Q1 is equal to 0.3 V. Consequently, at the beginning of the bootstrapping period, the bipolar transistor remains in the forward mode. At the end of the bootstrapping cycle, the internal base-collector voltage reaches the turn-on voltage of the base-collector junction of Q1 causing it to saturate. For the conventional BiNMOS gate, when the output reaches ($V_{DD} - V_{BEon}$), the coupling capacitance tends to pull-up the base of Q1 beyond V_{DD}, causing a reverse drain current in the pMOS. However, this reversal of current will not occur in the SBiNMOS circuit because the Schottky diode will turn off at the end of

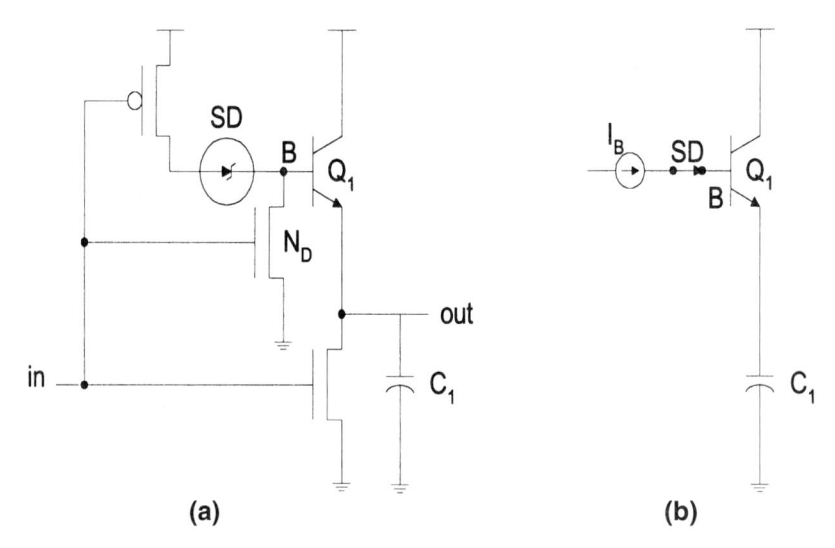

Figure 4–75 Schottky BiNMOS (SBiNMOS) inverter (a) schematic diagram and (b) pull-up equivalent circuit.

the pull-up phase. Hence, there will be no reverse drain current flowing in the pMOS transistor (see Figure 4–76(b)).

The parasitic pnp transistor (Q_{pnp}) residing in the npn transistor Q1, shown in Figure 4–76(c), has no significant impact on the bootstrapping effect for two reasons. First, the output would have reached its steady-state value by the time it turns on. Second, the collector current of Q_{pnp} is negligible due to its low current gain and high forward base transit time. The Schottky BiNMOS gate was simulated with and without the parasitic pnp, and no significant effect on the bootstrapping of the base and performance of the circuit was observed.

4.7.3 Circuit Implementations

Figure 4–77 shows two implementations of the SBiCMOS circuit. Their pull-up section consists of a pMOS device, an nMOS device Nd, a Schottky Diode SD, and a bipolar device Q1. When the input goes from low to high, Nd turns on to discharge the base of Q1. The charge accumulated at the drain of the pMOS transistor is also discharged through SD and transistor Nd to ground. For the pull-down section, two configurations are proposed (Figure 4–77). The pull-down section employs the principle of transient saturation [27,43] (see section 4.10). For the circuit in Figure 4–77(a), when the input changes from low to high, the transistor Ni turns on. Since the output is initially high, the transistor P1 will remain on. Together, they drive Q2 into the active region and, consequently, the output node is pulled low. When the output voltage approaches zero, the inverted output signal through I1 turns P1 off and N1 on. Hence, the transistor Q2 discharges through N1 and turns off. The operation of the

Full-Swing Schottky BiCMOS/BiNMOS Logic Circuits 211

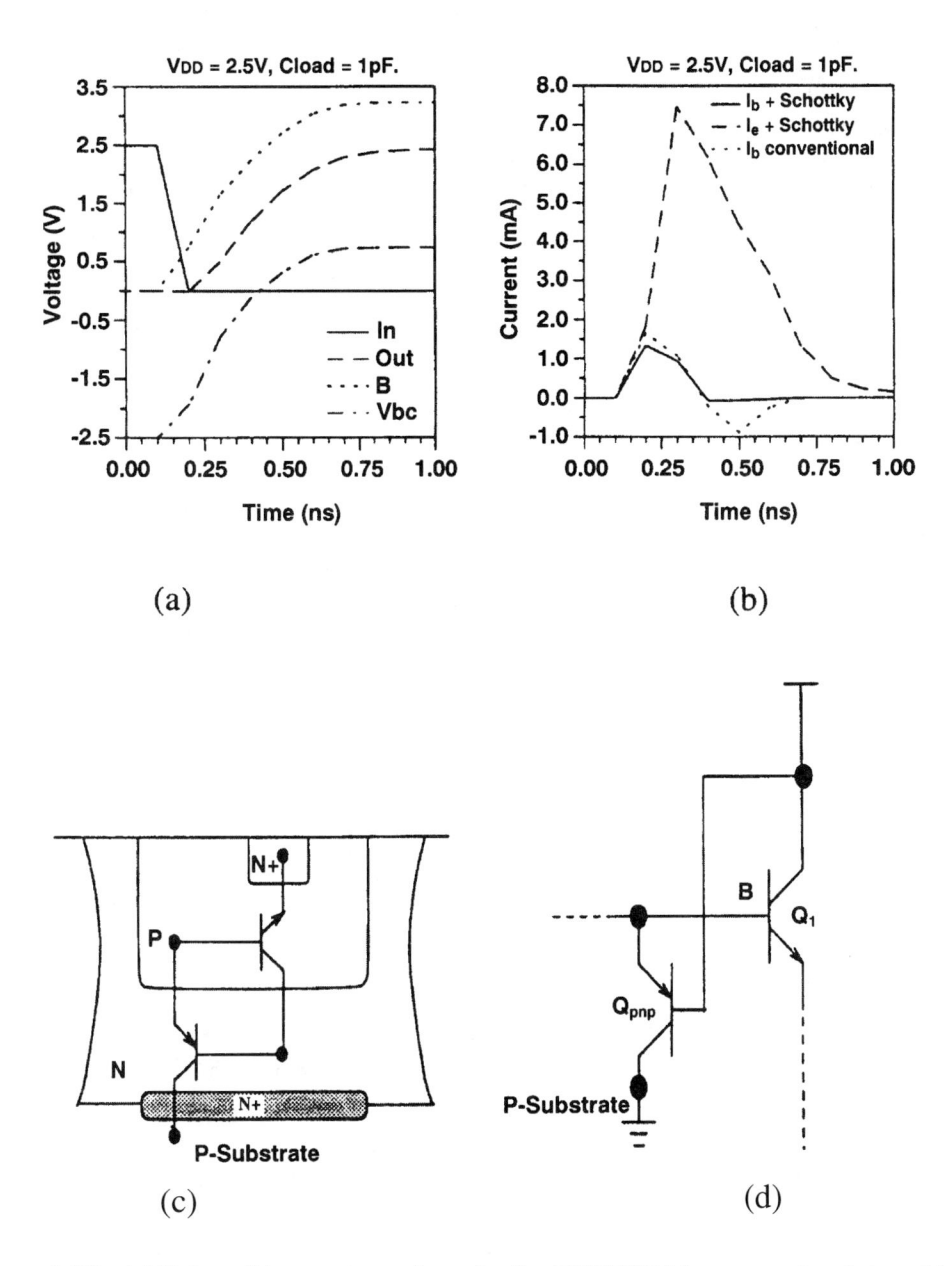

Figure 4–76 (a) Voltage (b) current waveforms for the SBiNMOS (c) cross-sectional view of Q1 showing the parasitic pnp and (d) the equivalent circuit of Q1 with the parasitic pnp.

circuit shown in Figure 4–77(b) is similar to that of Figure 4–77(a) except that the nMOS device Ni is replaced by a pMOS device Pi, which is driven by the input through an inverter I2. For both circuits of Figure 4–77, the output floats when the input is high. To overcome this problem, a small size cross-coupled inverter is added to inverter I1.

(a)

(b)

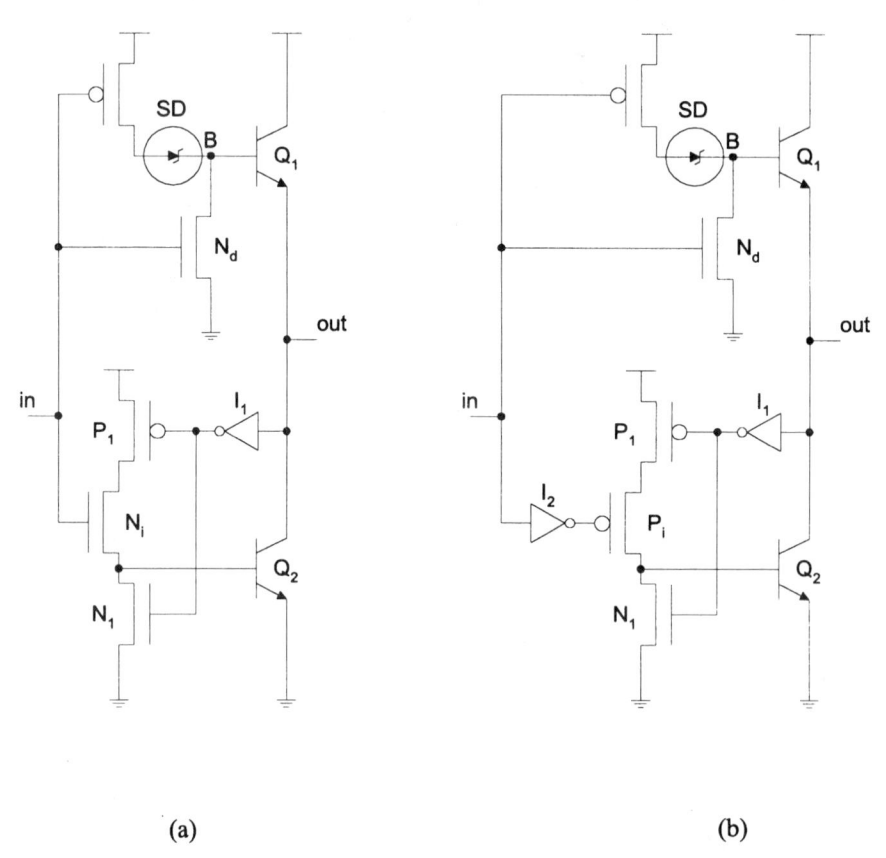

Figure 4–77 BiCMOS inverter configurations (a) Schottky BiCMOS (SBiCMOS) (b) Second SBiCMOS (SSBiCMOS).

4.7.4 Circuit Simulation and Discussion

The circuit performances of the SBiNMOS/BiCMOS and the Second SBiNMOS (SSBiCMOS) were simulated using HSPICE [44] and compared to that of the CMOS, the conventional BiCMOS, BiNMOS, QC-BiCMOS, TS-FS-BiCMOS, and CBiCMOS circuits. Figure 4–78 shows the different circuit implementations of the two-input NAND gate, which is chosen for evaluation and performance comparison. The simulation was carried out using the arrangement shown in Figure 4–79. The gate under test is between nodes 2 and 3. The input of the first gate is

Full-Swing Schottky BiCMOS/BiNMOS Logic Circuits

driven by a signal with a variable frequency and sharp rise and fall times. The slope and the shape of the voltage at node 2 will affect the delay of the gate under test. Hence, the response of the second gate is different from the first gate. This method of having a chain of the same gate is used to estimate the worst case delay when the input/output signals are degraded. The simulation was performed using the 0.5 µm BiCMOS technology. Table 4–9 lists the key device parameters for the 0.5 µm BiCMOS process.

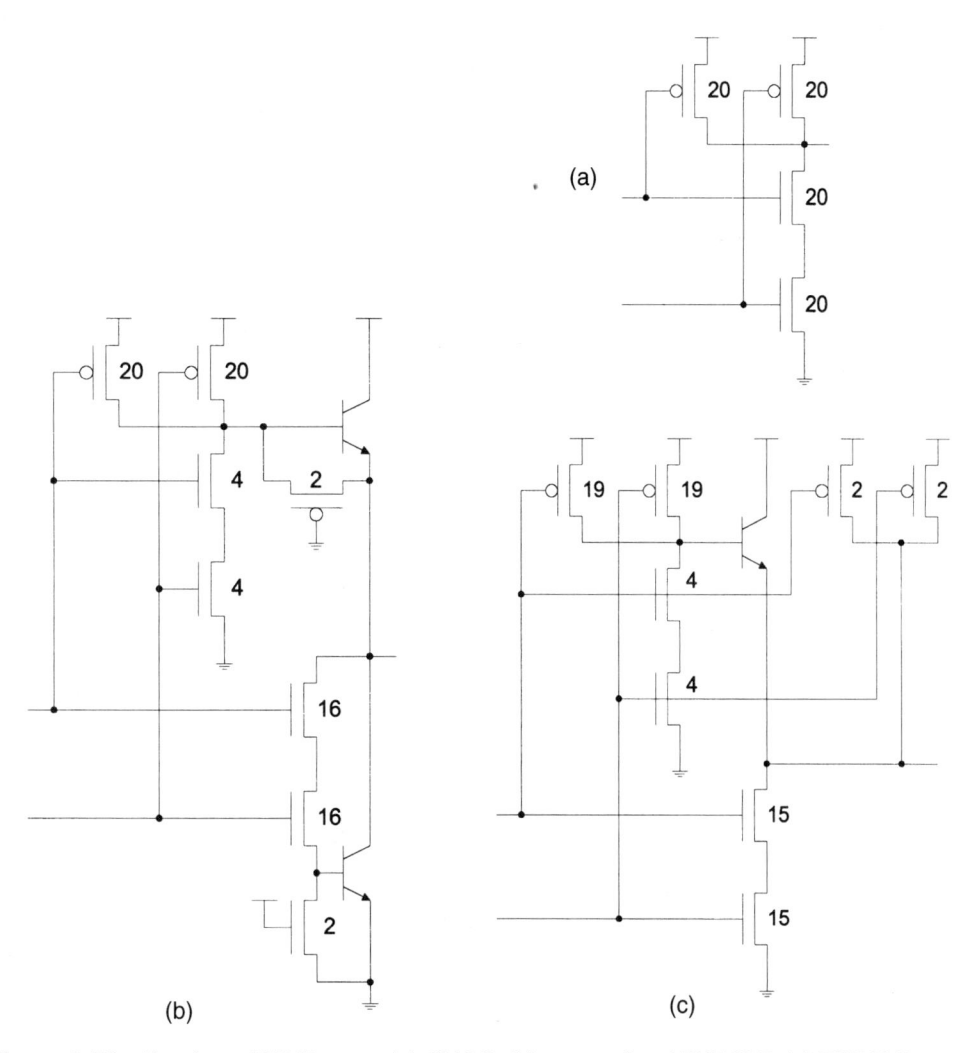

Figure 4–78 Two-input NAND gates (a) CMOS; (b) conventional BiCMOS; (c) BiNMOS.

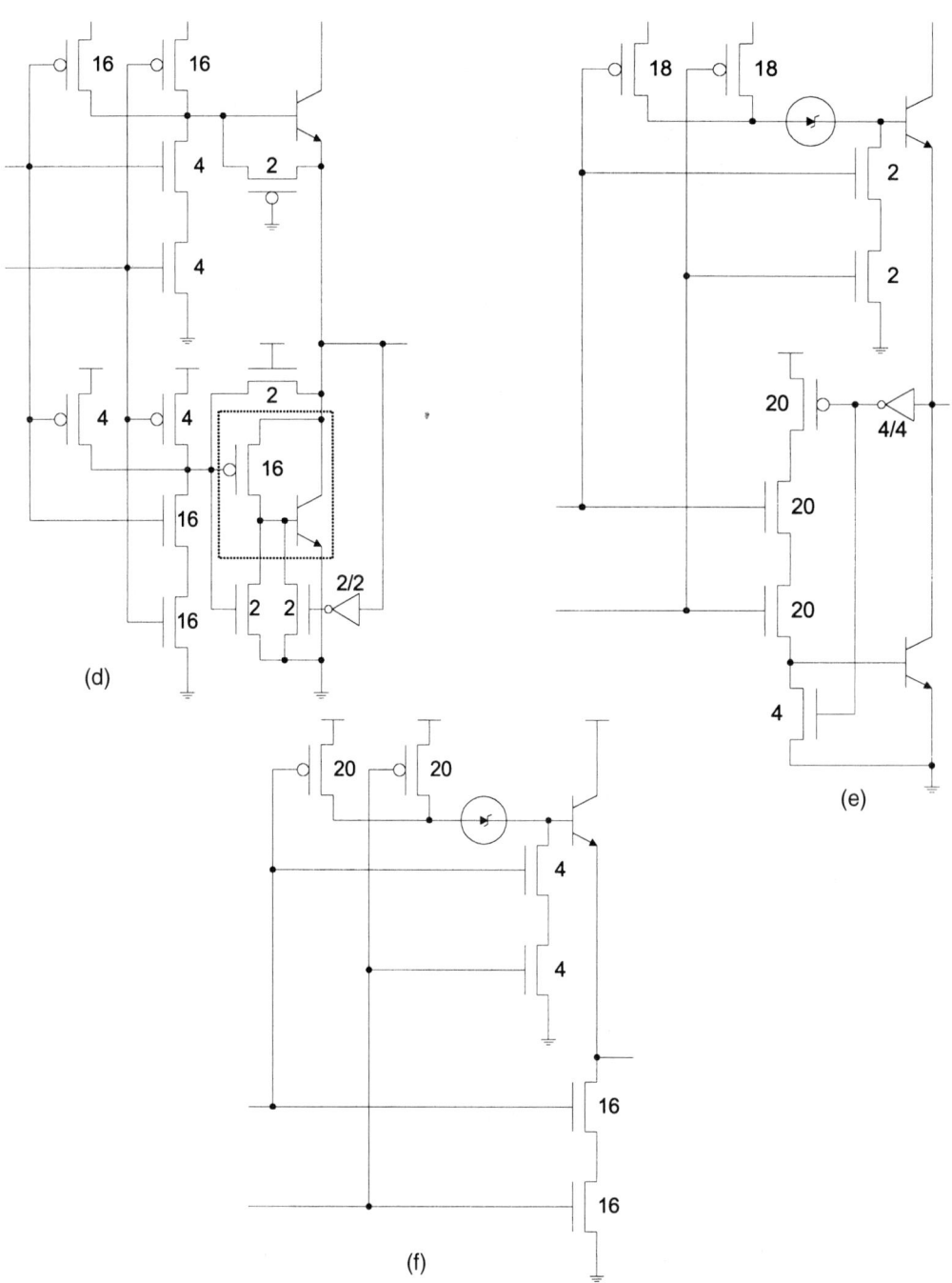

Figure 4–78 (cont.) Two-input NAND gates (d) QC-BiCMOS; (e) SBiCMOS; (f) SBiNMOS.

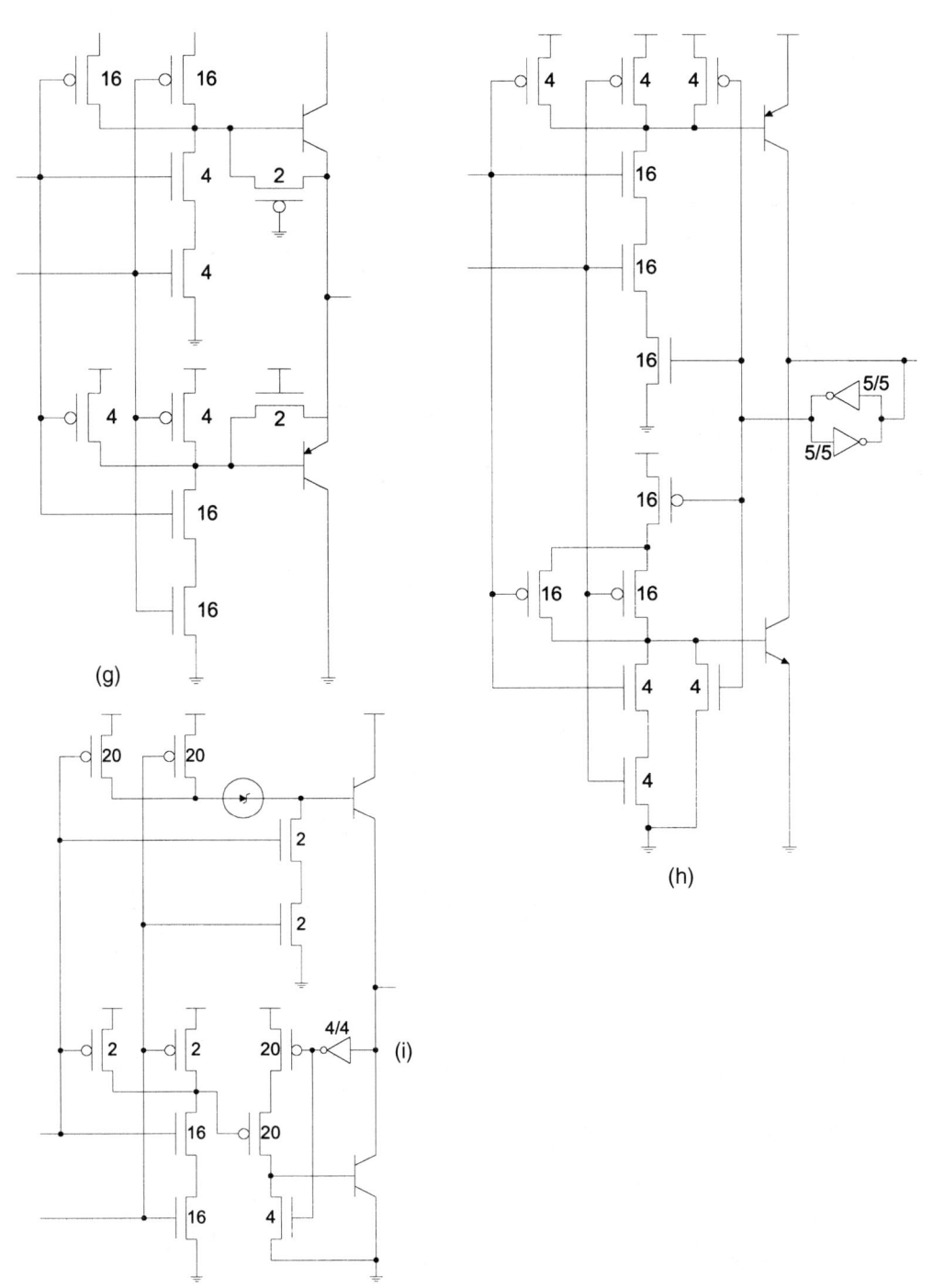

Full-Swing Schottky BiCMOS/BiNMOS Logic Circuits

Figure 4–78 (cont.) Two-input NAND gates (g) CBiCMOS; (h) TS-FS-BiCMOS; (i) SSBiCMOS.

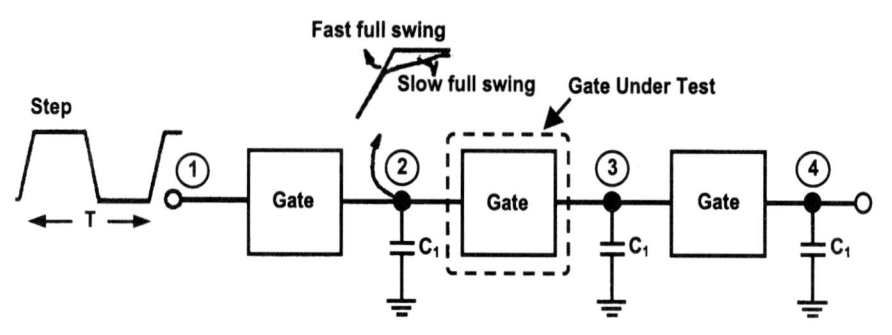

Figure 4–79 Delay simulation setup.

Table 4–9 Key device parameters for the 0.5 μm BiCMOS process.

		nMOS	pMOS
MOS	L_g	0.5 μm	0.5 μm
	L_{eff}	0.4 μm	0.4 μm
	I_{DS}	4.6 mA @ $V_{DS} = V_{GS} = 3.3$ V, W = 10 μm	2.4 mA
	T_{OX}	10 nm	10 nm
	V_T	0.60 V	−0.60 V
		npn	pnp
Bipolar	A_E	1×2 μm^2	1×2 μm^2
	β_f	90	90
	τ_f	9 ps	27 ps
	C_{JE}	7 fF	7 fF
	C_{JC}	9 fF	14 fF
	R_E	75 Ω	75 Ω
	R_C	60 Ω	60 Ω
	R_B	520 Ω	520 Ω
Schottky Diode	Area	6 μm^2	
	V_f	0.2 V @ 1mA/μm^2	
	C_j	17.5 fF	

Full-Swing Schottky BiCMOS/BiNMOS Logic Circuits

4.7.5 Scaling the Supply Voltage

Figure 4–80 illustrates the percentage delay reduction of the various circuit configurations as compared to the delay reduction of the CMOS for three different supply voltages and three different input frequencies. The percentage delay reduction is a measure of a circuit delay to that of the CMOS and is given as:

$$(\frac{\text{Delay}_{\text{CMOS}} - \text{Delay}_{\text{CKT}}}{\text{Delay}_{\text{CMOS}}}) \times 100\% \qquad 4.10$$

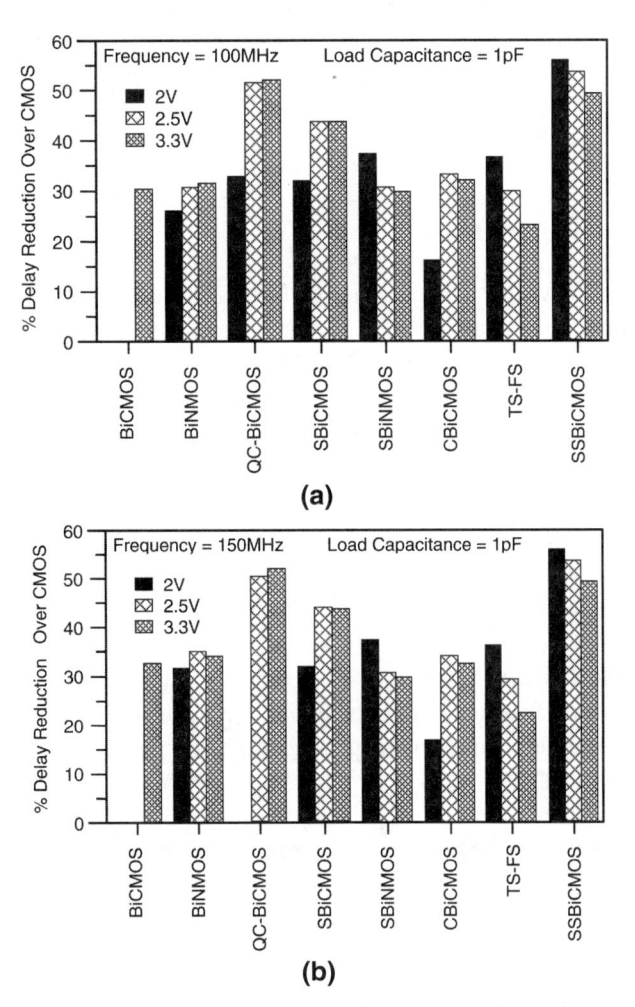

Figure 4–80 Bar plot of percentage delay reduction over CMOS at 3.3 V, 2.5 V and 2 V supply voltage for (a) 100 MHz (b) 150 MHz.

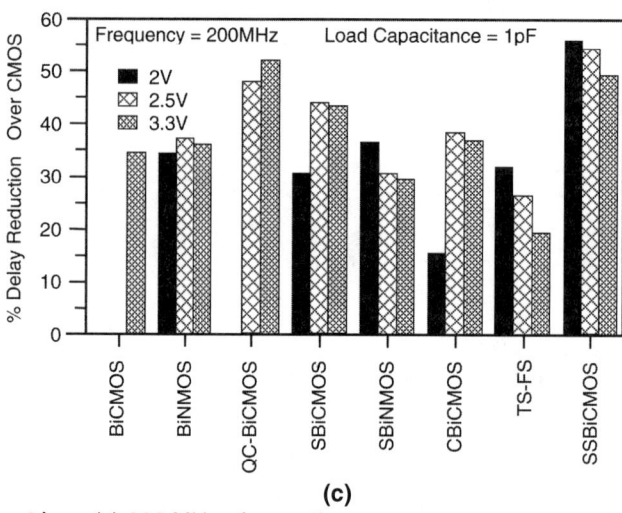

Figure 4–80 *(cont.)* (c) 200 MHz of operation.

Figure 4–80(a) shows that at a frequency of 100 MHz and at a supply voltage of 3.3 V, the QC-BiCMOS circuit has the better percentage delay reduction over that of the CMOS by 52%. This percentage is followed closely by the SSBiCMOS and SBiCMOS circuits with delay reductions of 47% and 43% respectively. The TS-FS-BiCMOS circuit has the worst performance at 3.3 V as predicted [43]. Figure 4–80 shows that as the input frequency increases, the delays of the Schottky circuits remain relatively unchanged, whereas the delays of the BiCMOS and BiNMOS circuits decrease and those of the TS-FS-BiCMOS and QC-BiCMOS increase. At 2 V power supply voltage and for frequencies above 150 MHz, the QC-BiCMOS circuit fails to operate. On the other hand, the Schottky circuits maintain their performance at 2 V and provide considerable delay reduction over the CMOS. The SSBiCMOS and SBiNMOS give the best delay reduction of 56% and 37% at 2 V, respectively. A major drawback of the Schottky BiNMOS and BiCMOS circuits is that for power supply voltages below 1.8 V, the pull-up section of the circuit starts to lose its full rail operation because the charge stored in the base is not sufficient to drive the output to full-swing. To overcome this problem, deep submicron BiCMOS technology should be used.

4.7.6 Delay Dependence on the Frequency of Operation

The delay dependence on the frequency of operation of various full-swing circuits is examined in this section. The aim is to find out how the frequency of operation affects the gate delay and the output voltage swing of these circuits. Figure 4–81 shows the simulated results of the delay as a function of frequency for a power supply voltage of 2 V and a load capacitance of 1 pF. Both the conventional BiCMOS and BiNMOS circuit delays decrease as the frequency increases due to the reduced dynamic voltage swing. For the BiNMOS NAND gate, this phenomenon is shown in Figure 4–82. For frequencies of 50 MHz and 250 MHz, the gate has a slow full-swing and a fast reduced swing, respectively. Figures 4–82(b) and 4–82(c) show that at higher operating frequencies, the gate has lower delays.

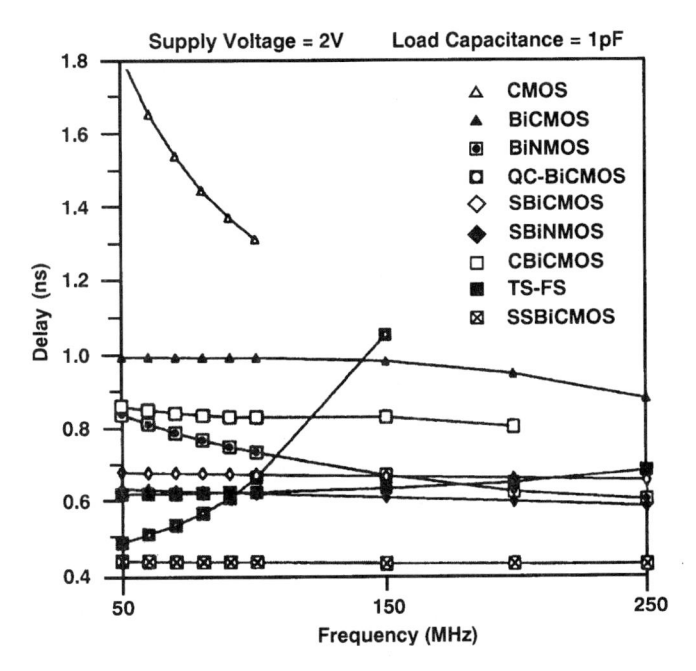

Figure 4–81 Delay dependence of various two-input BiCMOS NAND gates on the frequency of operation.

The QC-BiCMOS delay increases with frequency because the small size shunting transistors are incapable of delivering a sufficient current to rapidly charge/discharge the output. From Figure 4–81, we notice that the delay of the QC-BiCMOS approaches that of BiCMOS at high frequency, and both circuits fail to work above certain frequencies. On the other hand, the delays of the SBiCMOS, SBiNMOS, and SSBiCMOS circuits are unaffected by the frequency of operation due to the bootstrapping effect of the Schottky diode, which boosts the base voltage of the output npn transistor, and hence achieves fast full-swing operation at higher frequencies.

4.7.7 Power Dissipation

Figure 4–83 shows the power dissipation as a function of the supply voltage for the circuits shown in Figure 4–78. The TS-FS-BiCMOS circuit has the highest power dissipation. The power dissipation of SBiCMOS and SSBiCMOS are also relatively high compared to that of the CMOS. But this high power dissipation has been compensated for by the high speed they provide for power supply voltages between 2 V and 3.3 V.

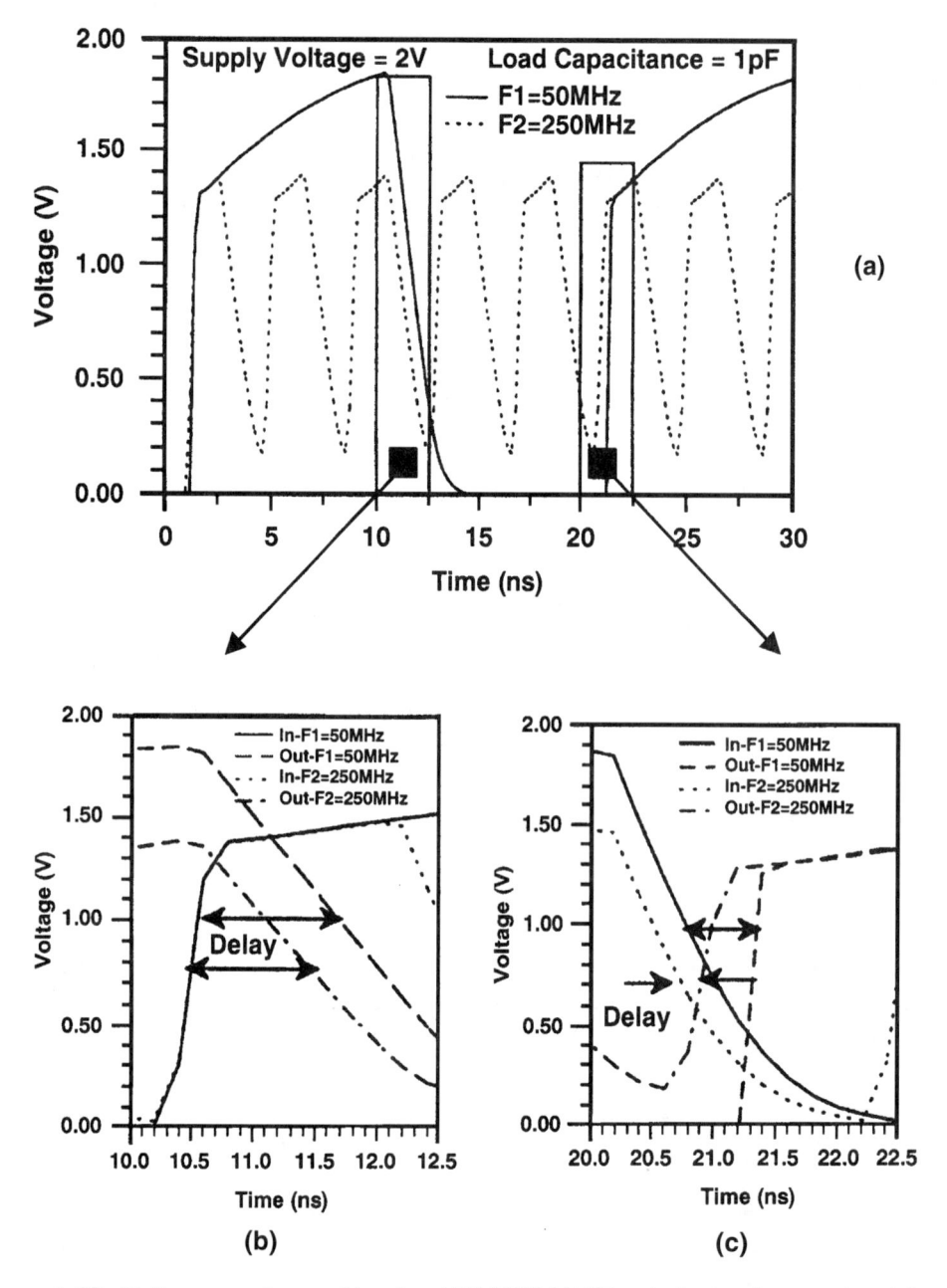

Figure 4–82 Voltage waveforms of two-input BiNMOS NAND gate for two frequencies: (a) output waveforms (b) input and output waveforms for falling transition (c) input and output waveforms for rising transition.

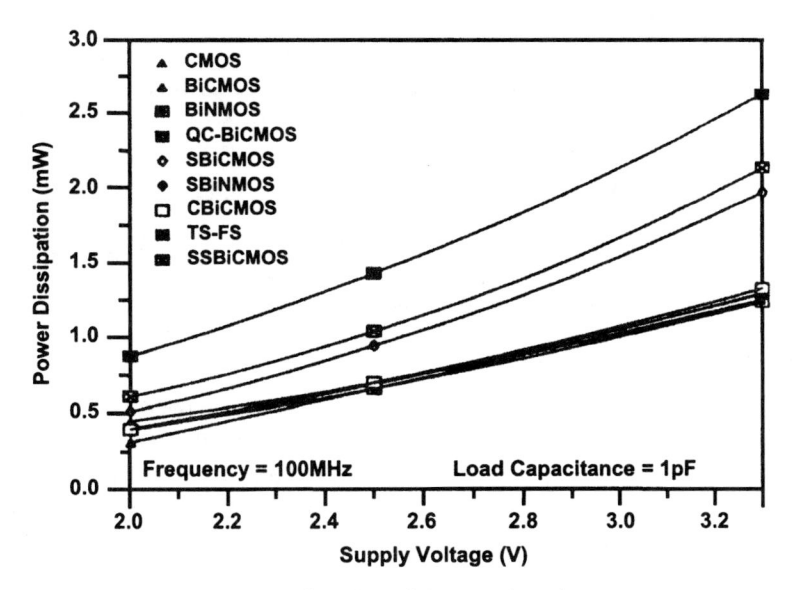

Full-Swing Schottky BiCMOS/BiNMOS Logic Circuits 221

Figure 4–83 Power dissipation as a function of the supply voltage.

4.7.8 Area Comparison

Table 4–10 gives the device count and area of the two-input NAND gates shown in Figure 4–78. The TS-FS-BiCMOS, SSBiCMOS, and QC-BiCMOS gates have the highest area ratio compared to the CMOS gate. A high area ratio from 2 to 3 is only tolerable if it is offset by the speed advantage. The SSBiCMOS, which has the second highest area ratio of 2.75, offers the best speed performance, particularly at 2 V. The SBiCMOS also shows a good speed performance with an area approximately 1.8 times smaller than that of the SSBiCMOS. Moreover, its power dissipation is lower than that of the SSBiCMOS.

Table 4–10 Device count and area of two-input NAND gates.

	Device Count		
Circuit Configuration	MOS	Bipolar	Area Ratio
CMOS	4	–	1
BiCMOS	8	2	1.8
BiNMOS	8	1	1.5
QC-BiCMOS	15	2	2.65
SBiCMOS	10	2	2.5

(continued)

Table 4–10 Device count and area of two-input NAND gates.

	Device Count		
SBiNMOS	6	1	1.5
CBiCMOS	10	2	2.05
TS-FS-BiCMOS	16	2	2.9
SSBiCMOS	13	2	2.75

4.7.9 Crossover Capacitance

Figure 4–84 illustrates the ratio of the crossover capacitance over the gate input capacitance for various BiCMOS/CMOS circuits as a function of the power supply voltage. This ratio represents the fan-out count where the BiCMOS circuits start to have a speed advantage over that of the CMOS. To outperform the CMOS, the BiNMOS and SBiNMOS circuits must have a fan-out of 1 and 2 at 3.3 V and 2 V, respectively. The SSBiCMOS circuit requires a fan-out of 3 in order to outperform the CMOS.

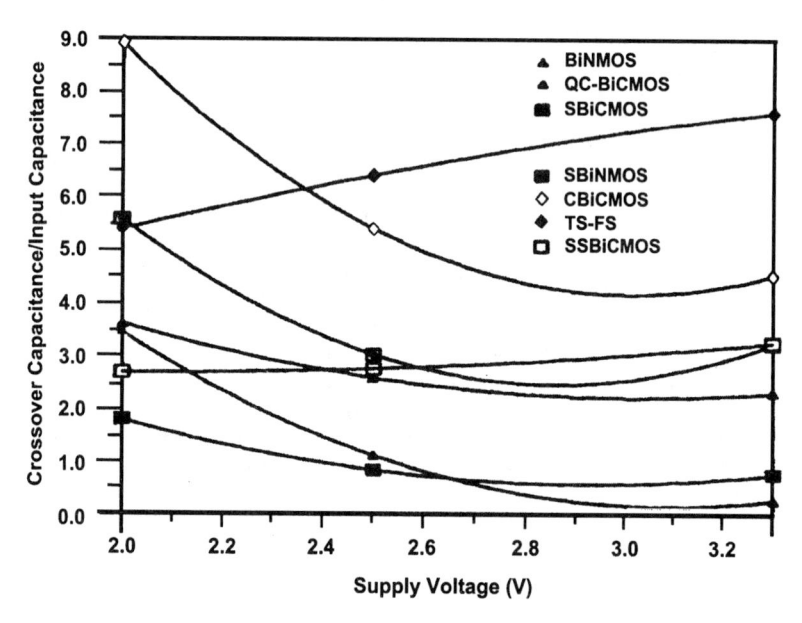

Figure 4–84 Ratio of the crossover capacitance/input capacitance as a function of the supply voltage.

In summary, this section covers full-swing BiCMOS/BiNMOS logic circuits employing Schottky diodes. These circuits are designed to operate with a low power supply voltage. The pull-up operation is achieved by saturating the bipolar transistor with a base current pulse. The Schottky diode can be easily fabricated using a standard noncomplementary BiCMOS process. The simulation results have shown that they outperform the conventional BiCMOS, BiNMOS, CBiCMOS, QC-BiCMOS, and TS-FS-BiCMOS circuits especially at a power supply voltage of 2 V. Also they are not affected by the frequency of operation, hence establishing their superiority over the above mentioned circuits.

4.8 Feedback-Type BiCMOS Logic Circuits

4.8.1 Introduction

The high-performance VLSI applications of the future will require high-performance, low-power consumption at low-supply voltages [45] along with high-packing density. A few BiCMOS circuits for a low-voltage, low-power environment have been suggested [46–48]. But they have yet to meet the demand because their performance features start to degrade as the power supply is scaled down to 2 V and below. A 1.5 V full-swing BiCMOS circuit [29] has been reported, but it too has not satisfied the above requirements because it consumes a large power, uses too many devices and has a large input capacitance. As the BiCMOS technology, together with the supply voltage, is scaled down for future VLSI applications, the threshold voltages of the MOS devices are also reduced. Hence, a full logic swing function is necessary to ensure that no dc current flows in the next gate and that the speed of the multi-input gate does not decrease. In this section, several BiCMOS circuits that are able to meet the requirements of high-speed, high-density, and low-power dissipation at low-supply voltages will be presented as alternatives to the conventional BiCMOS logic gates. Also a low-voltage BiCMOS dynamic logic circuit will be discussed. Its main advantage over the static circuit is its low power consumption.

4.8.2 Conventional BiCMOS Circuits and Their Limitations

The conventional BiCMOS logic gates [49–51] are shown in Figure 4–85. The difference between the two types lies in the technique used to discharge the base of the bipolar transistor. The R type is achieved through a resistor while the N type is done by nMOS transistors. But both types have their limitations. In the R type, a high resistance is required for high speed operation. However, during the pull-down phase, when the output level falls, the base potential of the pull-up BJT falls with the output level due to the large resistance of the bypass circuit. Consequently, the pull-up BJT cannot turn off and the power dissipation becomes twice as large as that of the CMOS logic gates. In the N type, the output high level voltage is ($V_{DD} - V_{BE}$), and the low-level voltage is V_{BE}. For the low-output level, the nMOS transistor nearest to the ground will have a V_{BE} as its gate voltage. If this transistor's threshold voltage is low, the V_{BE} input gate voltage will be sufficient to turn it on and a dc current will flow, hence dissipating power. Figure 4–85 (a and b) shows that these circuits utilize an emitter-follower configuration for the pull-up portion.

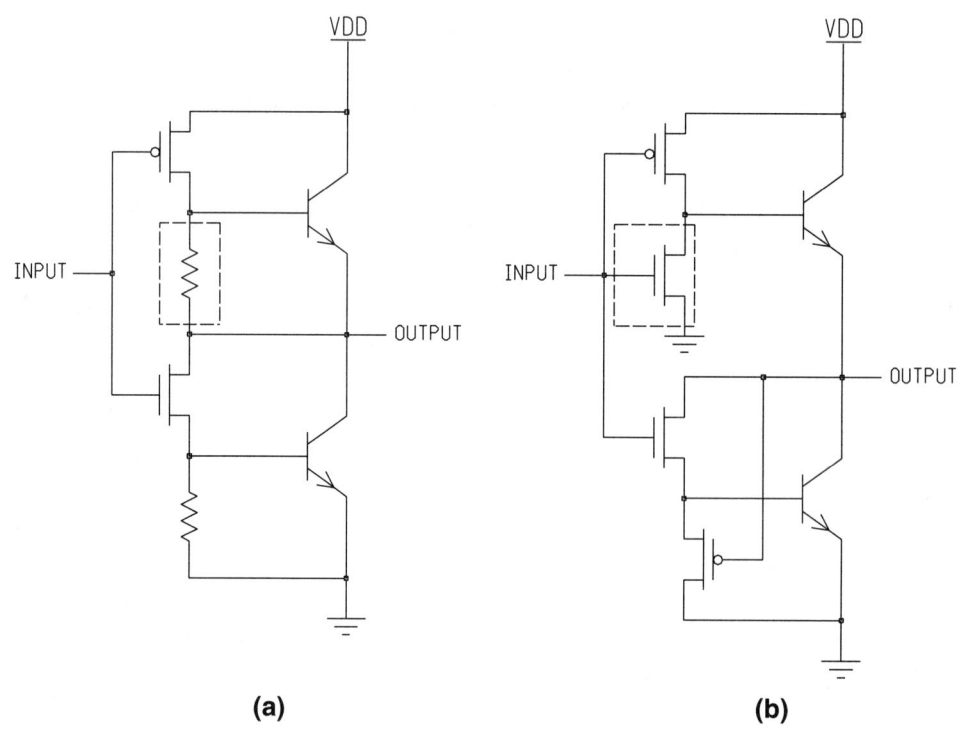

Figure 4–85 Conventional BiCMOS logic gates (a) R type and (b) N type.

Figure 4–86(a) shows the pull-up section of the conventional BiCMOS circuit and Figure 4–86(b and c) shows its transient response, during the pull-up cycle, using HSPICE [52] simulations. During the pull-up phase, when the input becomes low, the pMOS transistor turns on. Next the bipolar transistor starts to conduct pulling the output node to V_{DD}. As the output rises, the base voltage also increases and subsequently approaches V_{DD}. When the base voltage exceeds V_{DD}, the pMOS transistor starts to turn off and, consequently, the base current decreases. Due to the residual base (minority carriers) charge, there is still a small collector current. Hence the output voltage climbs further before leveling at a value slightly above ($V_{DD} - V_{BE}$). The base voltage, which is still following the output, increases above V_{DD}. However, the pMOS transistor starts to conduct in the reverse direction, depleting some of the residual base charge and hence lowering the base voltage back to V_{DD} (see Figure 4–86(b)).

4.8.3 R + N Type and Feedback-Type (FB) BiCMOS Logic Gates

Figure 4–87 shows two BiCMOS logic gates [53]. The R + N type gate is a modification of the R type (see Figure 4–85(a)). It adds a discharging nMOS, MN, to the R type, leading to improved power dissipation. Also, this gate has a full logic swing due to the resistance between

Feedback-Type BiCMOS Logic Circuits

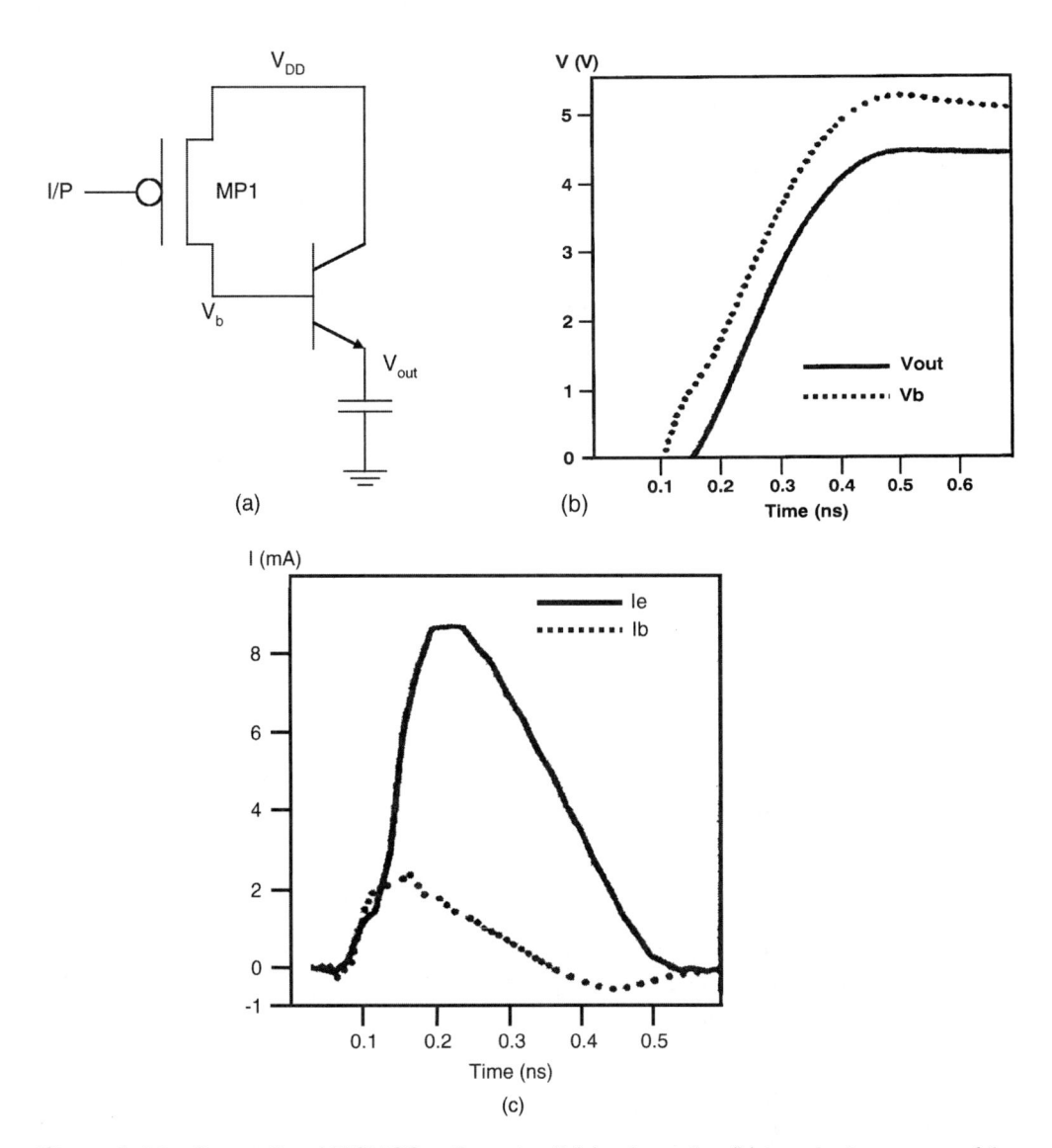

Figure 4-86 Conventional BiCMOS pull-up circuit (a) schematics (b) transient response of the base and output voltages (c) transient response of the base and emitter currents.

the base and the emitter. Its drawback is that when the input falls and the base current is supplied to the base of Q1, the driving pMOS current flows partly into the discharging nMOS the same way as the N type. This problem is solved by a Feedback-type (FB) BiCMOS circuit, which is shown in Figure 4-87(b), where the leakage current is terminated by an nMOS transistor N1 and two feedback CMOS inverters.

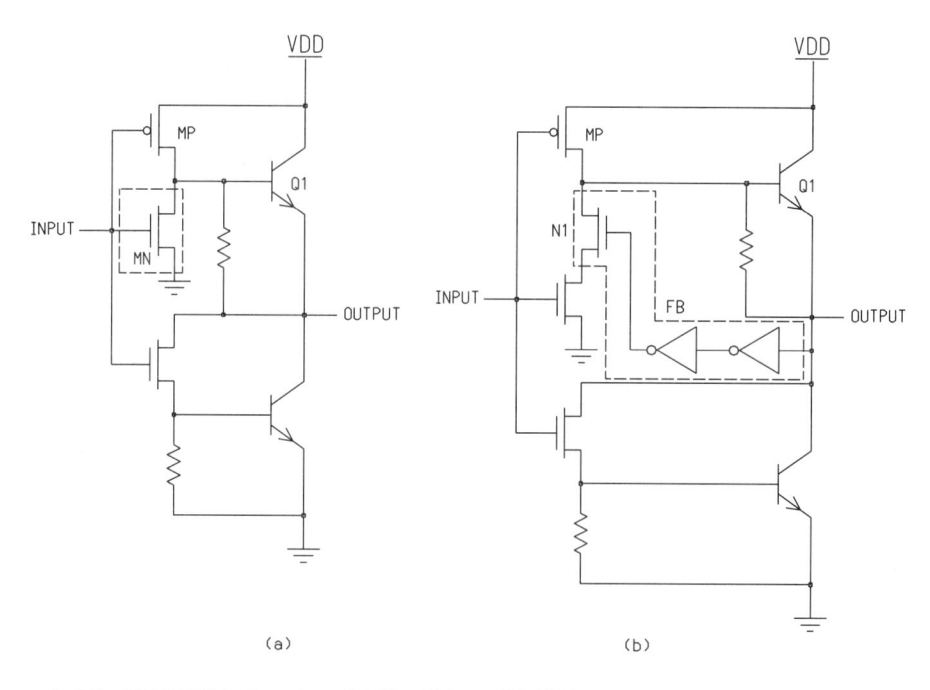

Figure 4-87 BiCMOS logic gates: (a) R + N type (b) FB type.

Figure 4–88 shows the circuit operation of the FB type inverter. When the input is low, the transistor MP turns on, supplying a current to the base of Q1 causing the output voltage to increase towards V_{DD}. The FB signal, which is a delayed signal of the output, is initially low and the MOS transistor N1 continues to be in the off state for a while. Consequently, there is no nMOS leakage current and a sufficient driving current from MP is supplied to the base of Q1, and high-speed performance can be achieved. When the input becomes high, the transistor Q1 turns off. Since the FB signal is initially high, N1 continues to be in the on state for a while due to the feedback inverters. The base of Q1 is able to discharge quickly through N1 and MN, thus reducing the penetration current and lowering the power dissipation.

4.8.3.1 Simulation Results

Figure 4–89 shows the simulation results of the propagation delay and power dissipation versus the load capacitance C_L for four different implementations of the three-input NAND gate. Figure 4–89(a) shows that the delay time of the N type is large and the CMOS type is faster than all the other circuits for load capacitances below 0.25 pF because the effect of the bipolar transistor of all the other circuits on the delay is not significant at very low load capacitance. The FB type circuit gives the best speed performance among the various BiCMOS types for all values of C_L. Figure 4–89(b) shows that the power dissipation of the R type is about twice as much as that of the other types of circuits since Q1 (see Figure 4–85(a)) takes a long time to turn off because of the large penetration current.

Feedback-Type BiCMOS Logic Circuits

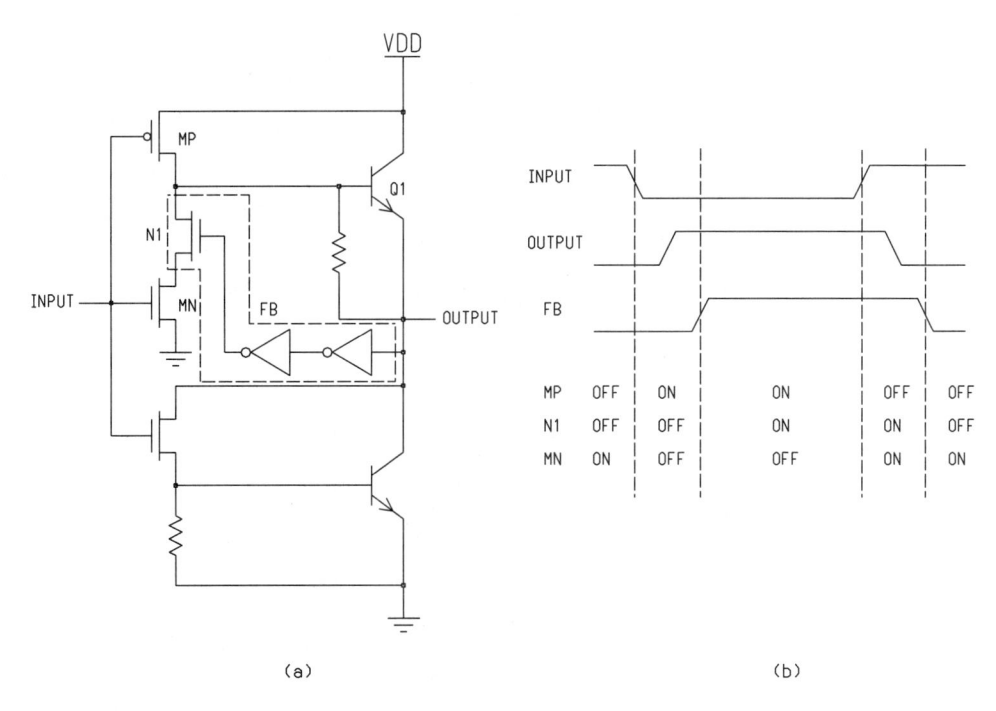

Figure 4–88 Circuit operation of the FB type (a) FB type inverter and (b) timing diagram.

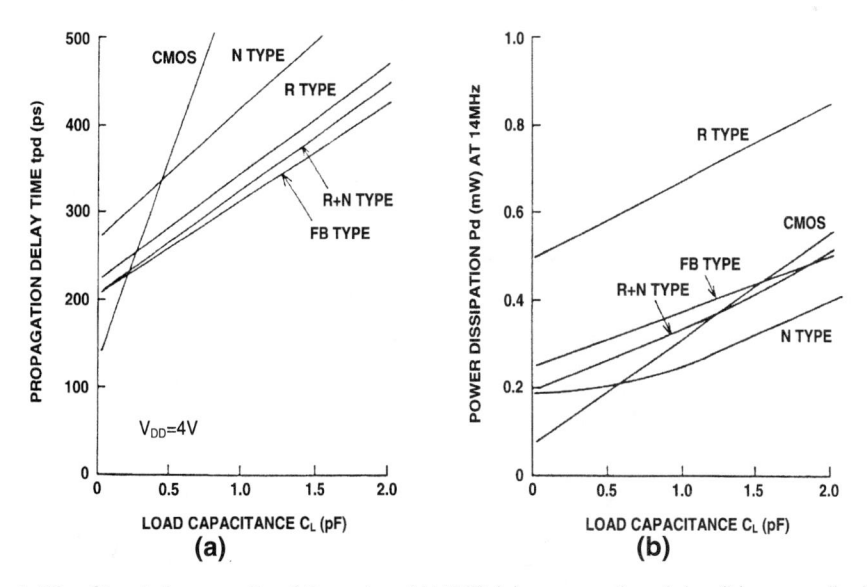

Figure 4–89 Simulation results of three-input NAND (a) propagation delay (b) power dissipation.

4.8.3.2 Experimental Results

The proposed BiCMOS logic gates were fabricated using a 0.5 μm BiCMOS technology. The performance of the three-input NAND gate was evaluated by measuring various ring oscillator frequencies and currents. Figure 4–90 shows part of the pattern of the 31-stage ring oscillator in the FB type. Figure 4–91 shows the experimental results of the propagation delay time versus the load capacitance for the three-input NAND at a power supply voltage of 4 V. The FB type is the fastest for C_L greater than 0.25 pF. Also, even in the submicron level, the speed performance of the BiCMOS gate is still superior to that of the CMOS for C_L greater than 0.25 pF. Figure 4–92 shows the experimental results of the power dissipation versus the load capacitance and Figure 4–93 shows the measured voltage dependence of the propagation delay for the FB type and three-input CMOS NAND gates.

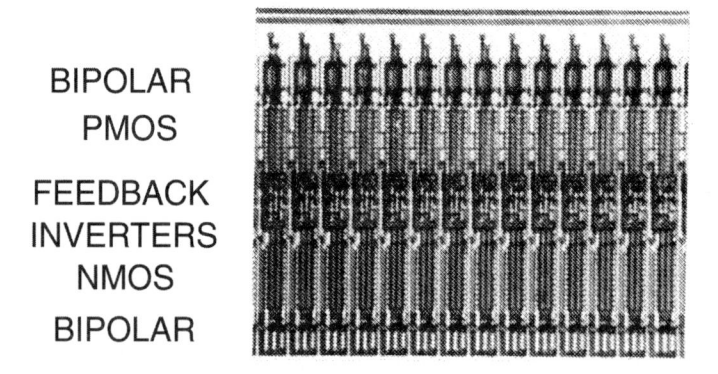

Figure 4–90 Pattern of a ring oscillator.

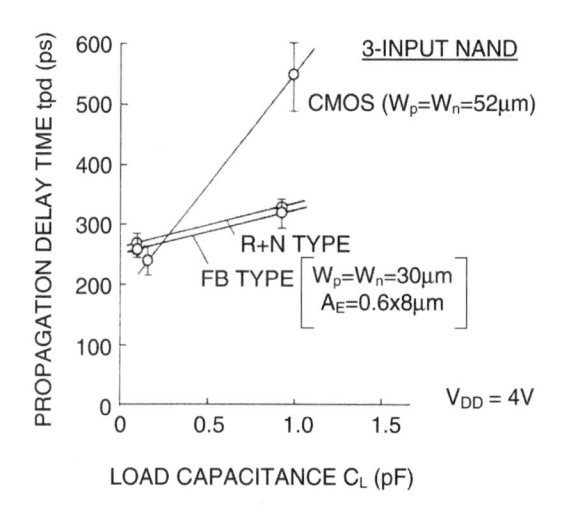

Figure 4–91 Experimental results for the propagation delay.

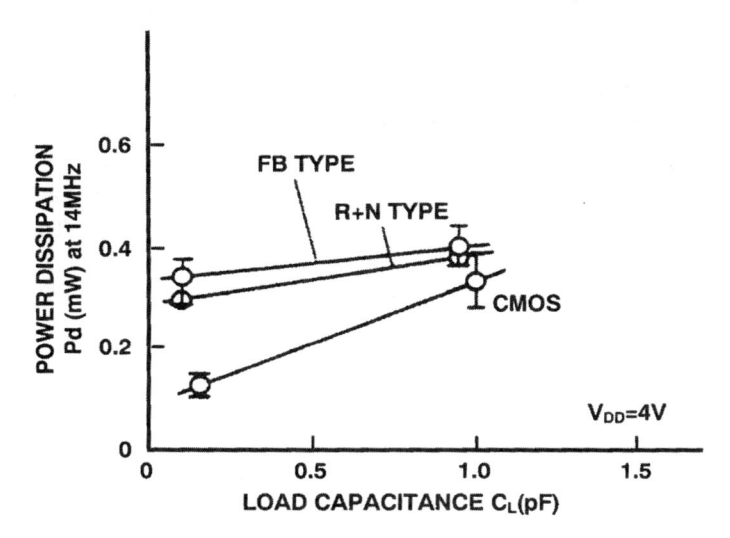

Figure 4–92 Experimental results for the power dissipation.

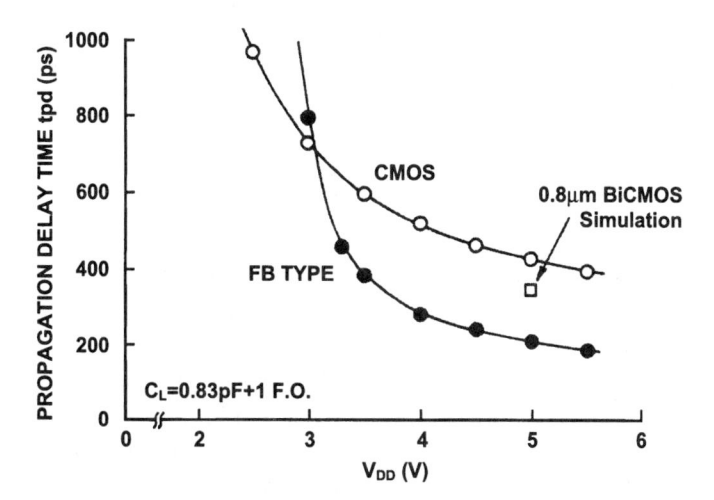

Figure 4–93 Voltage dependence for the propagation delay.

Figure 4–92 illustrates that the power consumption of the BiCMOS circuit is comparable to that of the CMOS at a load of 1 pF. Figure 4–93 shows that the propagation delay time of the BiCMOS circuit increases rapidly as the power supply voltage goes below 3.3 V. At 3 V, the speed performance of the BiCMOS circuit is almost equal to that of the CMOS. Also shown in Figure 4–93, the performance of the BiCMOS circuit using the 4 V/0.5 µm technology is superior to that of the 5 V/0.8 µm technology. The main disadvantage of the FB type BiCMOS circuit is that its speed performance degrades for power supply voltages below 3.3 V.

4.8.4 BiCMOS Full-Swing Circuits Utilizing a Positive Capacitively Coupled Feedback Technique

Different configurations of the BiCMOS circuits utilizing the positive dynamic feedback are shown in Figure 4–94 [48]. Note that the circuits in Figure 4–94(b and c) are noninverting, whereas that in Figure 4–94(a) is inverting. A feedback switch consisting of MP2 and MN1 is used to turn off the pMOS device (MP1) at the end of the output transition and a positive feedback capacitance C_{fb} is used to provide an additional source of base charge. The feedback switch and the feedback inverters In1 and In2 form the feedback circuit.

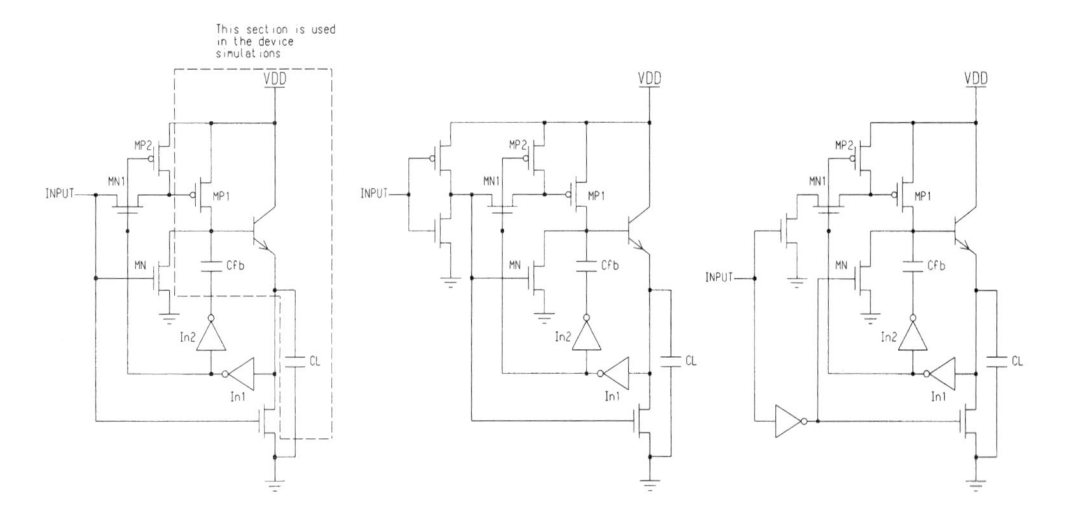

Figure 4–94 Different implementations of novel BiCMOS pull-up circuits utilizing the positive capacitively coupled feedback techniques.

The feedback circuit works as follows: As the output voltage increases, a charge is stored in C_{fb}. The feedback circuit then injects the charge stored in C_{fb} back into the BJT base when the output reaches the upperbound of $(V_{DD} - V_{BE})$. The base voltage will increase, thus turning on the BJT and finally the output voltage reaches V_{DD}. As C_{fb} begins to inject charges into the base, the feedback switch turns off the pMOS minimizing charge losses and injecting more charges into the base. Also with the pMOS device off, MN will have to discharge only the BJT base and hence the pull-down time reduces. Note that the feedback timing is controlled by the sizing of the feedback inverters and will be discussed in a later section.

A merged BipMOS device, combining the npn and MP1, was used in the simulations mainly to reduce the overall area and to ensure that any additional charge in the MP1 drain-substrate diode will not be dissipated when the base node voltage rises above the supply voltage. Figure 4–95 shows a cross-section of the merged device with its dimensions illustrated in Table 4–11.

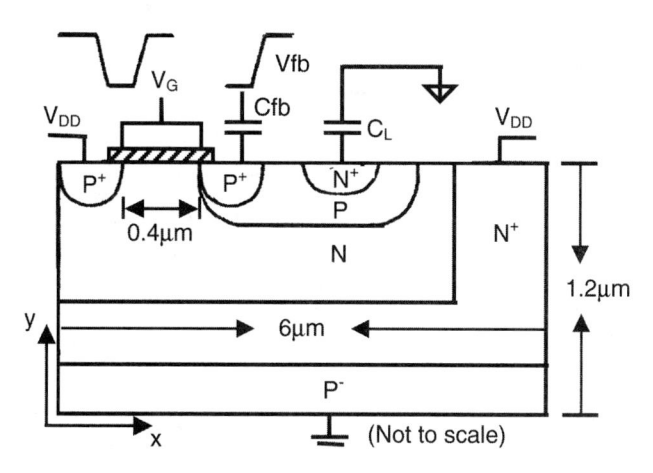

Figure 4–95 Cross-section of the merged BipMOS device used in the simulations.

Table 4–11 Dimensions of the merged BipMOS device.

Effective channel length	0.38 μm
Channel width	20 μm
Depth of emitter	0.01 μm
Depth of base	0.11 μm
Depth of buried layer	0.6 μm
Depth of p substrate	1.0 μm
Drain depth	0.8 μm

Device and circuit simulations were also performed on the novel pull-up circuits, as shown in Figure 4–94. Table 4–12 gives the device parameters of three generic BiCMOS technologies used in the simulations. Figures 4–96 and 4–97 present the simulation results. In Figure 4–96, the solid lines represent the novel circuit of Figure 4–94(a) while the dashed lines represent the circuit reported in [54].

Figure 4–96 shows that the output of the novel pull-up circuit is fast for the entire transient response, unlike the circuit reported in Hara et al. [54], where there is a slow period. Figure 4–97 shows that during the pull-down transients, the emitter current is zero, implying that the additional minority carriers trapped in the collector due to saturation do not affect the pull-down transients.

Table 4–12 Device parameters of three generic BiCMOS technologies.

| \multicolumn{5}{c|}{MOS} | \multicolumn{5}{c}{Bipolar (minimum size)} |

L_{eff} (μm)	V_T (V) nMOS	V_T (V) pMOS	T_{OX} (nm)	C_J (fF/μm)	β	τ_f (ps)	I_K (mA)	C_C (fF)	C_S (fF)
0.8	0.8	−0.8	15	0.4	100	10	2.5	9.5	25
0.5	0.56	−0.6	12	0.6	100	7	1.8	7.5	18
0.2	0.35	−0.35	7	1	100	4	1.0	3.5	8.5

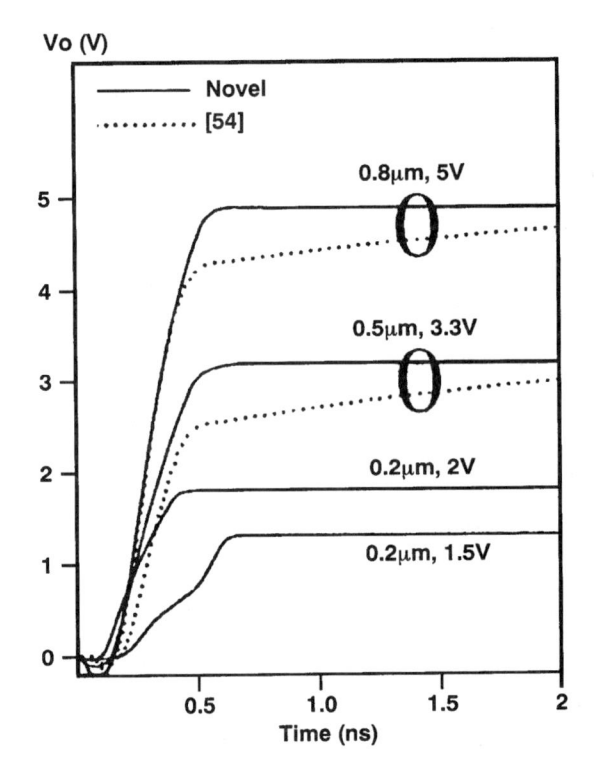

Figure 4–96 Pull-up transient response of the novel circuit and the circuit reported in [54] for various technologies and supply voltages for a load of 0.3 pF.

A two-dimensional transient device simulator TRASIM [55] was used to check the output voltage V_O and the effects of turning off the pMOS and the feedback capacitance on its performance. The results, in Figure 4–98, show that turning off the pMOS increases the swing significantly, and with the feedback capacitance, the output voltage swing increases further. The feedback capacitance, C_{fb}, helps to speed up the pulling up of the output node to a high value.

Feedback-Type BiCMOS Logic Circuits

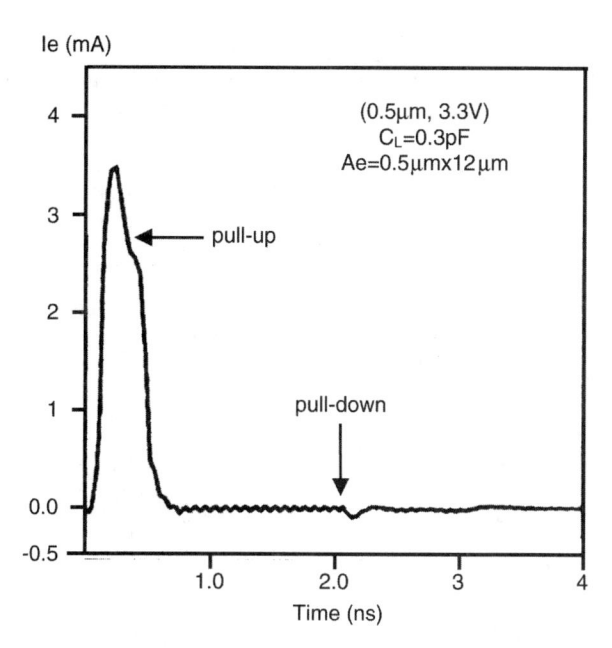

Figure 4–97 Emitter current of the pull-up BJT during pull-up and pull-down at 250 MHz frequency.

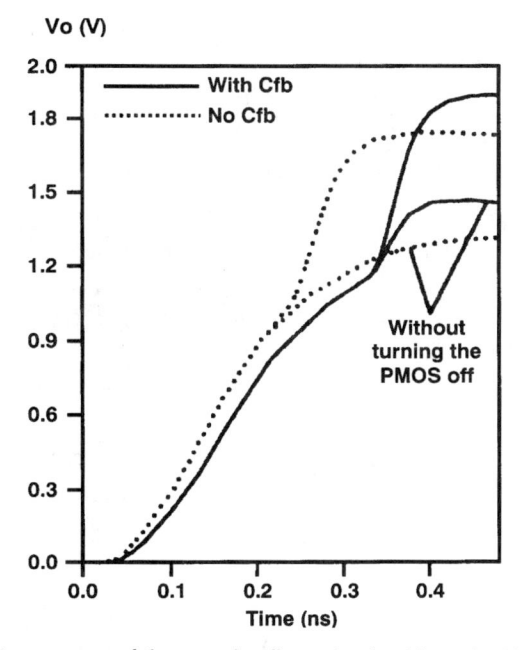

Figure 4–98 Transient response of the novel pull-up circuit with and without turning off the pMOS, as well as with and without C_{fb}.

Device simulations were also used to check for latch-up. It was reported [17] that the two parasitic pnp BJTs in the merged BipMOS devices would be a potential source of latch-up. The two parasitic pnp BJTs to be considered are first, under the pMOS gate between the base and the source, and second, under the base between the base and the substrate. Figure 4–99 shows the 2-D holes distribution at different points of time during transients.

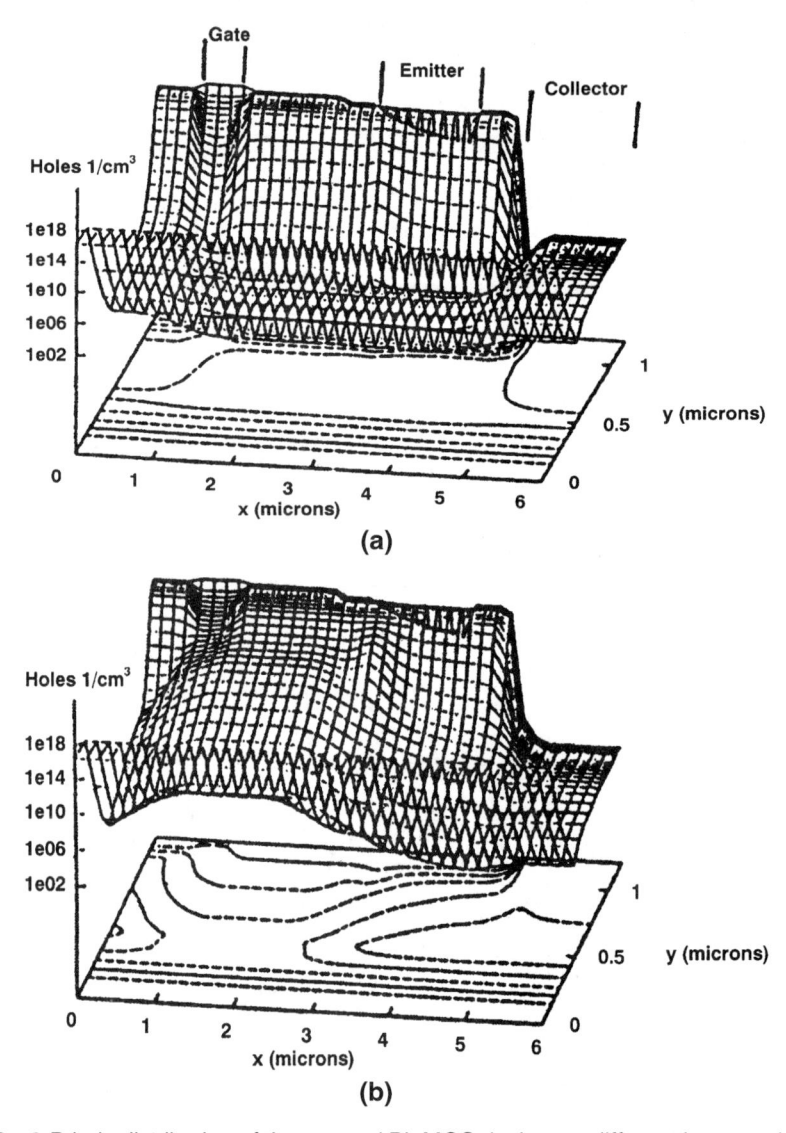

Figure 4–99 2-D hole distribution of the merged BipMOS devices at different instants during the pull-up transient (a) before the feedback (b) just after the start of the feedback.

Feedback-Type BiCMOS Logic Circuits

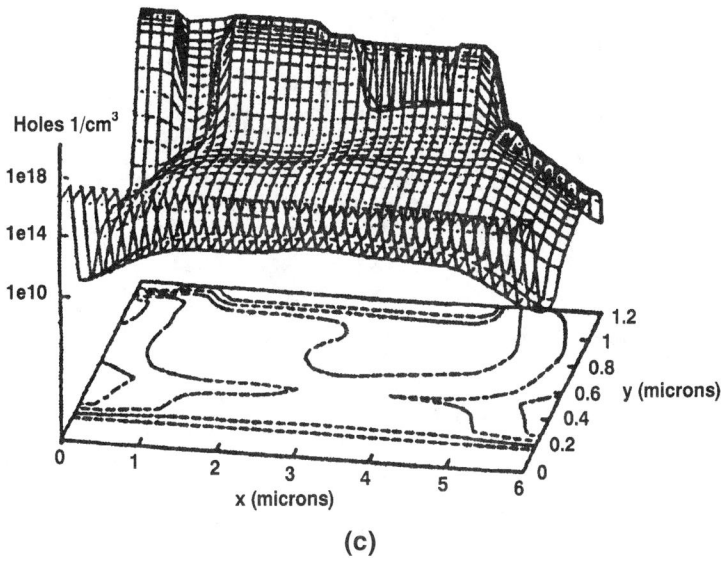

(c)

Figure 4–99 *(cont.)* (c) at the end of transition.

From Figure 4–99, we can deduce that the parasitic pnp between the source and the base improves the performance by increasing the base current before feedback. With the feedback in action, the base current reverses the direction (see Figure 4–99(b)) and depletes some of the base charge, and therefore decreases the output voltage swing. This pnp can be eliminated by using special layout techniques for the merged structure [17]. At the end of transition, the hole concentration in the n-well, under the gate and base, falls by a few orders of magnitude (see Figure 4–99(c)). This means that there is no longer any parasitic pnp BJTs.

Figure 4–100 shows the 2-D electron concentration in the base. The BJT continues to conduct after the start of the feedback, as indicated by the large concentration gradient in the base of Figure 4–100(a). At the end of the transition, the electron concentration drops by several orders of magnitude, as indicated in Figure 4–100(b), implying that the pull-down circuit will not be affected by the excess minority carriers trapped in the collector due to saturation.

4.8.4.1 Circuit Performance Comparison

The performance of the three circuits shown in Figure 4–94 will be evaluated and compared against the CMOS circuit. They will be implemented as simple buffers and AND gates.

4.8.4.1.1 Simple Buffers

The delay time, T_D, is defined as the average of the rise and fall times, measured from the input at $V_{DD}/2$ to the output at $V_{DD}/2$. HSPICE simulations were carried out and the results are shown in Figure 4–101. The novel circuits outperformed the CMOS circuits down to 0.2 pF load capacitance. As the load capacitance increases, T_D of CMOS increases initially rapidly and then gradually when the load capacitance exceeds 2 pF.

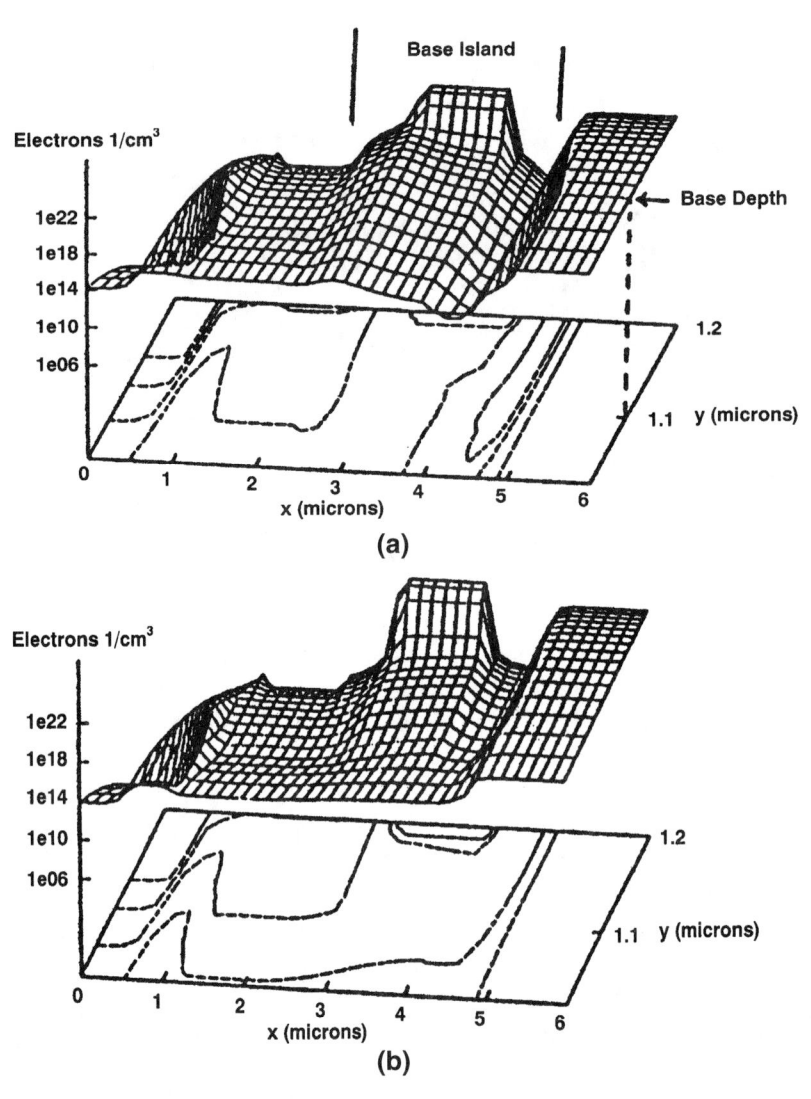

Figure 4–100 2-D electron concentration in the base for two instants after the start of the feedback (a) just after the start of the feedback (b) at the end of transition.

Although for large loads, the CMOS becomes faster than the first novel BiCMOS circuit, its buffer chain at that point will have become excessively large.

Two other pull-down circuits, together with the pull-up circuit shown in Figure 4–94(a), were implemented and evaluated. Figure 4–102 shows the two pull-up/pull-down circuits. Figure 4–103 shows the propagation delay of the two circuits shown in Figure 4–102, the circuit in Figure 4–94(a), and the optimized CMOS buffer. The pull-up/pull-down circuits achieved the

Feedback-Type BiCMOS Logic Circuits

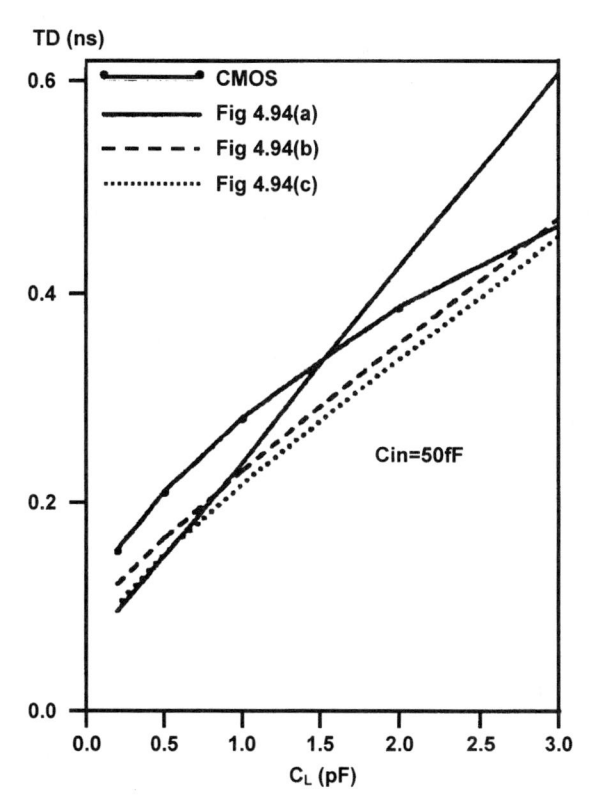

Figure 4–101 Average delay of the circuits shown in Figure 4–94 as compared to that of the CMOS using the 0.2 μm, 2 V technology.

best speed performance over the CMOS for load capacitance greater than 0.5 pF. The circuit with the nMOS pull-down achieved good speed-up at lower load capacitance and had the least area among the three BiCMOS circuits.

Figure 4–104 shows the propagation delay versus the supply voltage of the circuits shown in Figure 4–102(a) and the CMOS buffer for three BiCMOS technologies. The areas of the circuit shown in Figure 4–102(a) and the CMOS buffer are kept approximately equal and the value of the load capacitance is approximately equivalent to a fan-out of 4. For the 0.8 μm technology, the novel BiCMOS circuit did not operate below 2 V. Above 2 V, it outperformed the CMOS buffer. For channel length of 0.5 μm, these circuits outperformed the CMOS down to a power supply voltage of about 1.7 V. If the gate length of the MOSFET is reduced to 0.2 μm, the novel circuit will outperform the CMOS down to a power supply voltage of 1.5 V. This is indeed an achievement as compared to the circuits described previously in other sections.

The average power dissipation of the three BiCMOS circuits in Figures 4–94(a), 4–102(a), and 4–102(b), and the CMOS buffer is illustrated in Figure 4–105 at a frequency of 100 MHz. It

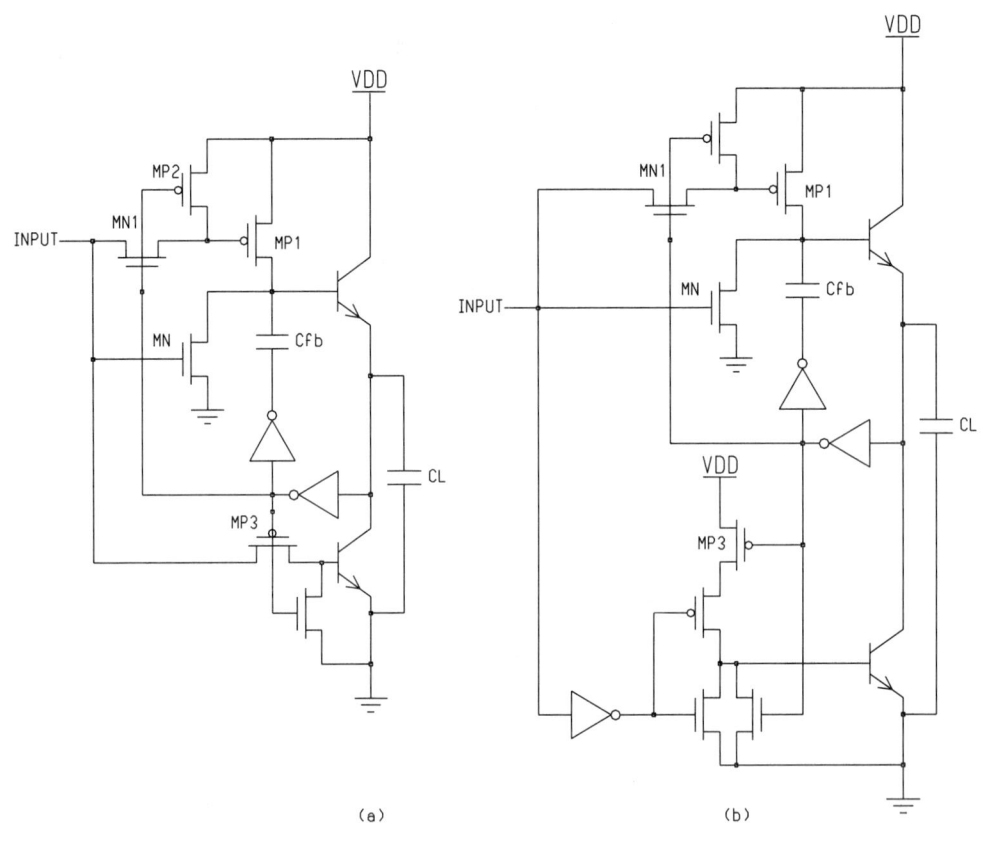

Figure 4–102 Pull-up circuit combined with (a) a novel pull-down circuit (b) a pull-down circuit similar to the one reported in [8].

is found that the circuit of Figure 4–94(a) has the least power dissipation. The other two BiCMOS circuits have power dissipation close to that of the CMOS buffer.

4.8.4.1.2 AND Gates A multi-input AND gate was implemented using the novel circuit techniques with slight modifications to increase the speed. Figure 4–106 illustrates the AND gate implementation. The construction of the circuit is such that the transistor MP2 is connected as a load and the nMOS device MN1 is placed at the bottom of the nMOS logic block. This arrangement allows the pMOS logic block in the pull-up section to be removed. Hence, the circuit layout area is reduced. Furthermore, this configuration reduces the parasitic capacitance of the BipMOS gate and enhances the speed. The feedback switch will turn MN1 off before the end of the transition, as such there is no static power dissipation.

Feedback-Type BiCMOS Logic Circuits

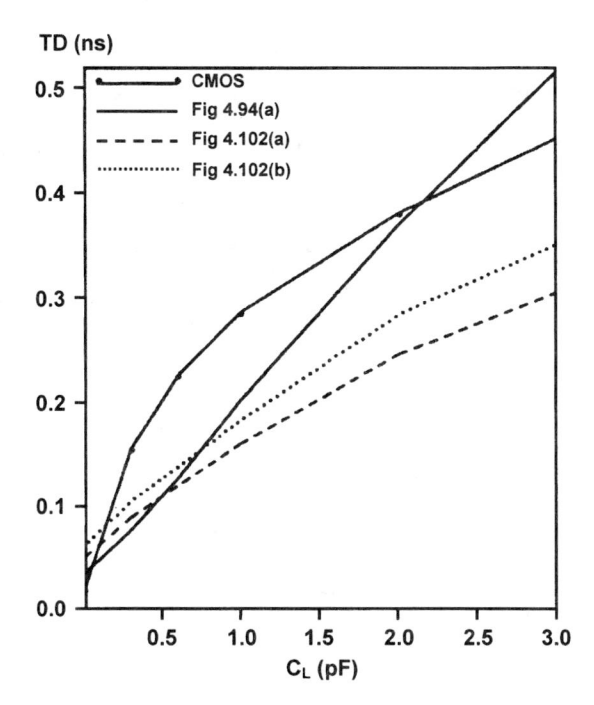

Figure 4–103 Average delay of the circuits in Figures 4–102 and 4–94(a) when compared to the CMOS for the 0.2 µm, 2 V technology.

The feedback capacitance was left out from the circuit because the circuit was intended for the 0.5 µm, 3.3 V technology. The output voltage can reach 3.1 V with the pMOS (MP1) turned off. An nMOS base-emitter shunt is used to discharge the base of the BJT. As the threshold voltage of the nMOS is less than V_{BEon}, the BJT will be turned off at the end of pull-up cycle and remain off during the pull-down. The pull-down section consists of the diode connected nMOS device, MN2, and the circuit operation can be explained as follows: During the pull-down phase, the output is initially high. When one or more inputs switch to low, the pMOS (MP3) device in the pull-down section conducts which, in turn, turns on the bipolar transistor. The output voltage starts to discharge and when it reaches a low value, the feedback inverter will send a high signal to turn off the transistor MP3. Meanwhile, MN2 starts to discharge the base of the bipolar transistor until it turns off completely. During this period, the shunting nMOS in the pull-up section keeps the pull-up BJT off. At the end of the transition, the feedback switch, consisting of MP2 and MN1, in the pull-up portion is turned on and the circuit is prepared for the pull-up phase. The pull-up operation is the same as that explained in section 4.8.2 except that the nMOS shunting the BJT is switched on at the end of the pull-up transition and the circuit is ready for the next pull-down transition.

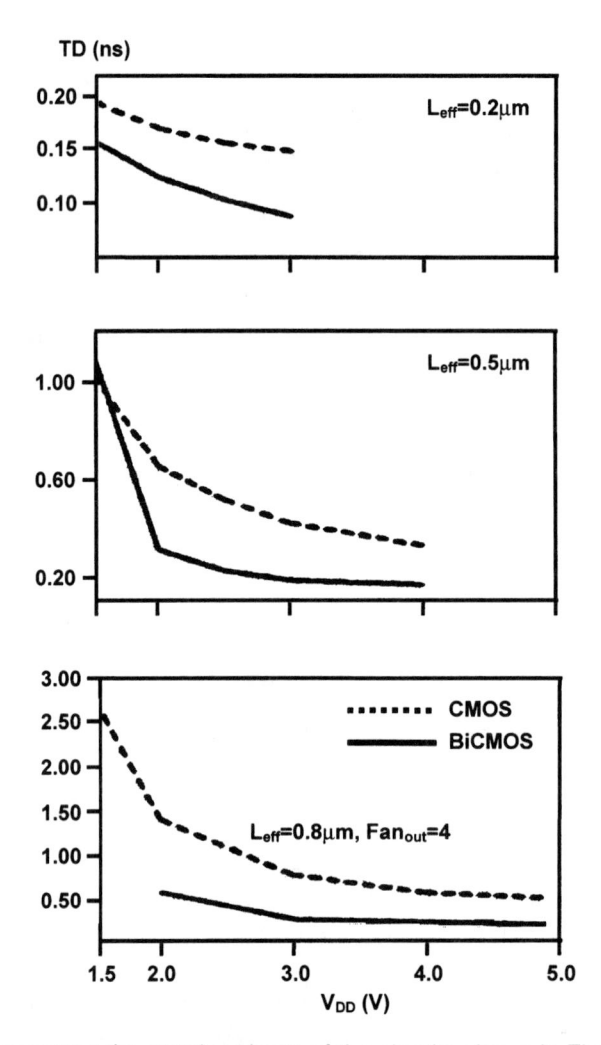

Figure 4–104 Delay versus the supply voltage of the circuits shown in Figure 4–102(a) and the CMOS buffer for three BiCMOS technologies.

The sizing of transistors MP2 and MN2 is crucial. It must be done such that MP2 is small enough not to slow down the nMOS logic chain, yet large enough to reduce or prevent glitches at the BipMOS gate node. Also, MN2 must be small enough to speed up the parallel pMOS logic block from turning on the BJT, yet large enough to promptly discharge the BJT. Figure 4–107 shows the speed up and area ratio between the novel BiCMOS AND gate and the CMOS NAND gate for the 3.3 V/0.5 μm technology.

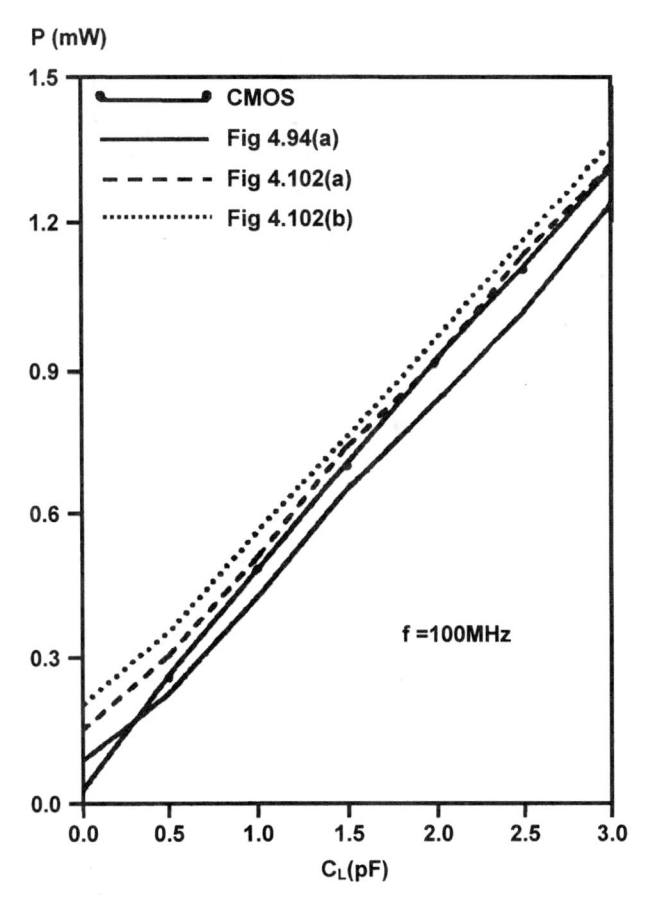

Figure 4–105 Power dissipation vs. the load capacitance for circuits shown in Figures 4–94(a) and 4–102 and the CMOS buffer using the 0.2 μm, 2 V technology.

The speed-up factor is defined as the ratio of the delay of the CMOS gate to the delay of the novel BiCMOS gate. Figure 4–107 shows that the speed-up factor increases as the number of inputs increases, whereas the area ratio drops. This means that for macro cells implemented using this BiCMOS circuit, as the number of inputs per logic gate increases, the speed performance over the CMOS implementation increases and the circuit layout area decreases.

4.8.4.2 Design Issues Regarding the Feedback Circuit

In Figure 4–94, we see that the feedback circuit consists of the feedback switch, MN1 and MP2, the feedback inverters, In1 and In2, and the feedback capacitance, C_{fb}. The size of the device MP2 is strongly coupled to that of the feedback inverter In1. Thus, the nMOS transistor in In1 and MP2 are sized simultaneously such that the feedback switch is turned off when the output is about $(V_{DD} - V_{BEon})$ but not earlier or later than that point. And if In1 is also used to

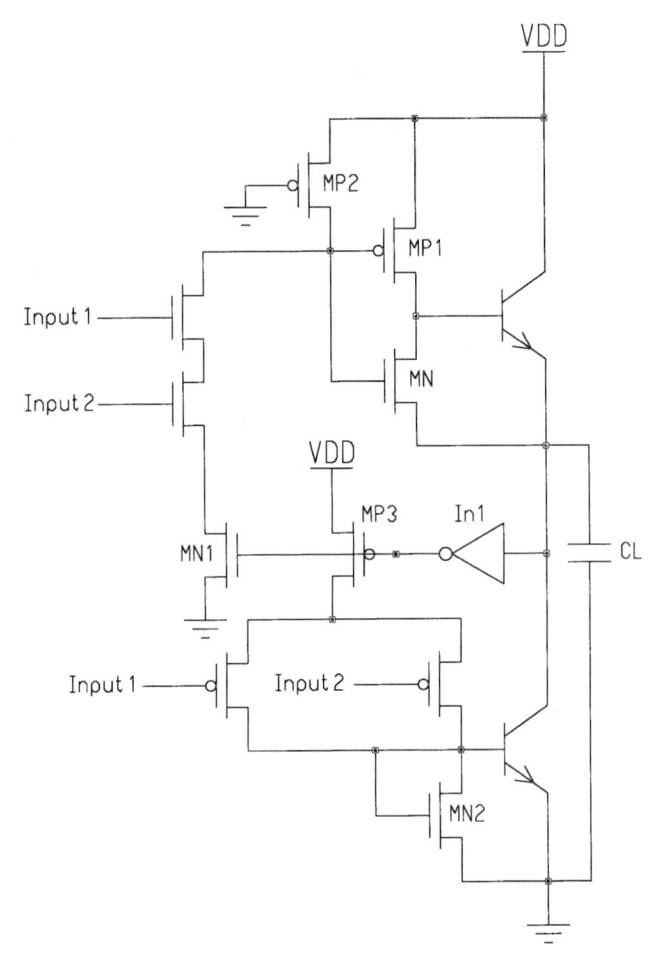

Figure 4–106 An AND gate implemented using the novel circuit technique.

control the pull-down section, as in Figure 4–106, its pMOS (MP3) device must be sized such that the pull-down feedback switch is turned off at the correct time. The nMOS transistor, MN, used to discharge the base of the BJT should be sized minimally and according to the operation frequency. It must be sized such that it can discharge the base of the BJT in adequate time, yet not load the circuit significantly.

The sizing of inverter In2 (Figure 4–94) is dependent on the value of C_{fb}. The larger the value of C_{fb}, the larger the sizing of the pMOS device in In2 should be. However, the width of the nMOS device in In2 is set approximately to the minimum width. The setting of the value of C_{fb} depends on several factors, such as the technology used, the power supply voltage used and the output loading. C_{fb} should be able to store sufficient charge to charge up the base of the bipo-

Feedback-Type BiCMOS Logic Circuits

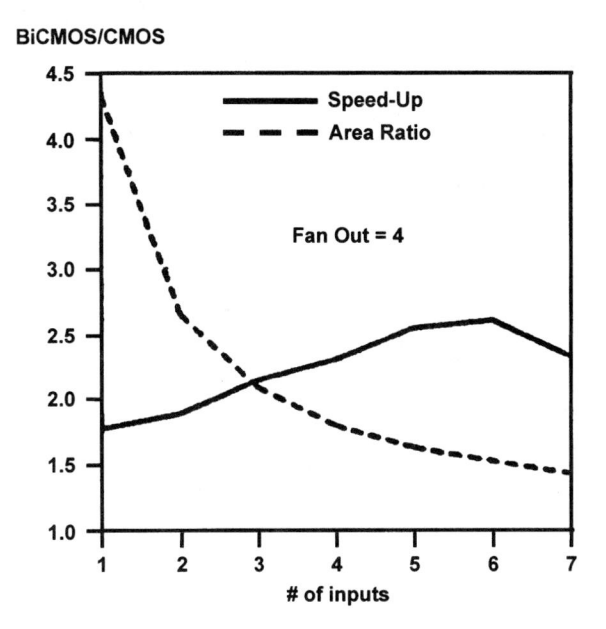

Figure 4–107 Speed-up and area ratio between the novel BiCMOS AND and the CMOS NAND for the 3.3 V, 0.5 μm technology.

lar to about 0.7 V above V_{DD} and continue to supply the base charge required for the bipolar to continue conducting until the output reaches V_{DD}. Hence, we can write:

$$C_{fb}(\min) = \frac{0.7 C_{BC}}{V_{DD}} \qquad 4.11$$

where C_{BC} is the base-collector junction capacitance. Since C_{fb} is assisted by the charge injected by turning off the pMOS device MP1, the value of C_{fb} does not need to be much greater than $C_{fb}(\min)$. The serious repercussion of increasing the value of C_{fb} is that the output slew rate decreases and, consequently, the circuit delay increases, as shown in Figure 4–98. This means that for a power supply voltage of 3 V, a C_{BC} value of about 20 fF, and a C_{fb} value of 15 fF would be enough for the circuit to achieve a full-swing without compromising on the speed. For very high loads, the value of C_{fb} needs to be increased above $C_{fb}(\min)$, especially if the size of the charging pMOS (MP1) device is small. The amount of charge supplied by the pMOS, when it turns off, to the bipolar base, ΔQ, can be estimated as half of the total channel charge under the gate and given by:

$$\Delta Q = \frac{C_{ox} V_{DD}}{2} \qquad 4.12$$

where C_{OX} is the gate oxide capacitance of transistor MP1. However, as shown in Figure 4–98, at a low-supply voltage, the circuit designer should not only rely on ΔQ to achieve full-swing.

In this section, full-swing BiCMOS circuits using positive capacitively coupled feedback techniques are presented. These circuits have superior performance over CMOS down to a power supply voltage of 1.5 V. Since scaling of BiCMOS is inevitable for future VLSI applications, a new generation of BiCMOS logic gates suitable for scaled sub-half micrometer and supply voltages below 1.5 V is required. The next section will describe an example of this class of circuits.

4.8.5 1.2 V Complementary Feedback BiCMOS Logic Gates

It was reported [56] that a complementary feedback BiCMOS (FB-BiCMOS) digital circuit has been designed to give a full voltage swing and relatively high speed at 1.2 V supply. The circuit, shown in Figure 4–108, is an improved version of a similar circuit reported in [57].

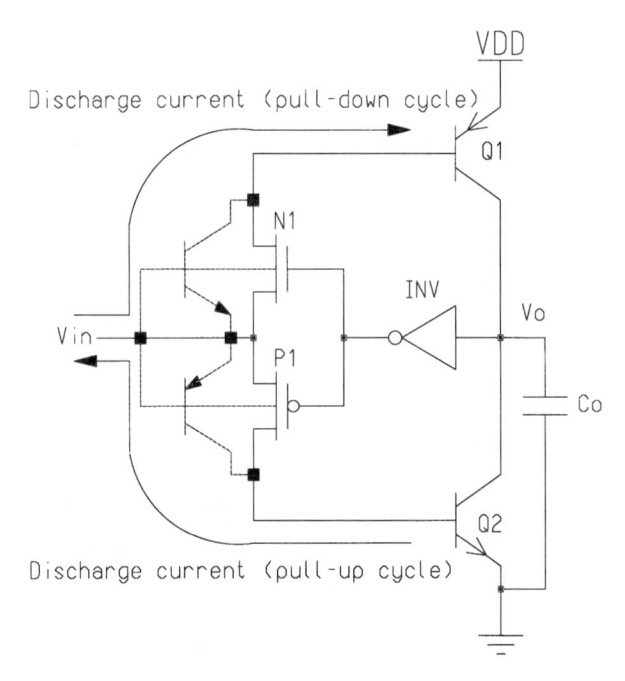

Figure 4–108 Feedback BiCMOS (FB-BiCMOS) inverting circuit.

The CMOS input stage drives the complementary bipolar transistors, Q1 and Q2, of the output stage. The gate of the CMOS input stage is determined by the feedback CMOS inverter, INV. For the pull-up transition, an initial low-output voltage would feedback a high voltage to the gate of the input stage, hence turning on the transistor N1. A low input at Vin causes N1 to

Feedback-Type BiCMOS Logic Circuits

conduct, drawing its current from the conducting pnp, Q1. As the output voltage rises, the feedback signal via the inverter will eventually turn off the transistors N1 and Q1. The excess charge stored in the base of the saturated transistor, Q1, would get to the output and the output voltage, V_O, would increase further. Upon the start of the pull-down phase, the remaining base charge is entirely discharged through the substrate of N1. This can be explained as follows: During the pull-down cycle, Vin goes from low to high, causing the base-collector junction of the parasitic npn associated with the device N1 to be forward-biased, hence injecting the discharging current into the base of Q1. The discharge path during the pull-down phase is shown in Figure 4–108.

HSPICE simulations of the circuit during the pull-up transition were carried out and the results are presented in Figure 4–109. The figure shows that Q1 starts conducting in the active region before it saturates. The FB-BiCMOS circuit has been modified and later fabricated using the 0.8 µm, n-well/p-substrate (single well), double polysilicon, double metal, standard BiCMOS process. The modification made was to connect the body of N1 to ground rather than to the input.

4.8.5.1 An Analytical Transient Model

An analytical transient model is developed to describe the transient performance of this BiCMOS circuit. During the pull-up phase at $t = 0$, $V_O = 0$, $I_C = 0$, the output of INV is high. As the input changes from high to low, V_O increases, the gate voltage of N_1, $V_{G(N1)}$, decreases with time, and N_1 is in the triode region for most of the output rise time. Therefore, I_{DS} of N_1 can be written as:

$$I_{DS(N1)} = K_{t(N1)}[(V_{GS(N1)}(t) - V_{T(N1)})V_{DS(N1)}] \qquad 4.13$$

where:

$$K_{t(N1)} = \frac{k'_{N1} \mu_{neff(N1)} C_{OX(N1)} W_{N1}}{L_{eff(N1)}} \qquad 4.13a$$

In Eq. 4.13, the term $0.5(V_{DS(N1)})^2$ has been assumed to be negligibly small and k'_{N1} represents the short-channel effects factor in the triode region [58], $\mu_{neff(N1)}$ is the effective electron mobility, $C_{OX(N1)}$ is the gate oxide capacitance, W_{N1} is the channel width, and $L_{eff(N1)}$ is the effective channel length of N_1. $V_G(t)$ is obtained by solving the differential equation governing the discharging of the output capacitance seen by INV, C_{inv}, which is given by:

$$-C_{inv} \frac{dV_{GS(N1)}(t)}{dt} = K_{S(N2)}(V_O - V_{T(N2)}) \qquad 4.14$$

where:

$$K_{S(N2)} = k_{N2} C_{OX(N2)} W_{N2} \upsilon_{sat} \qquad 4.14a$$

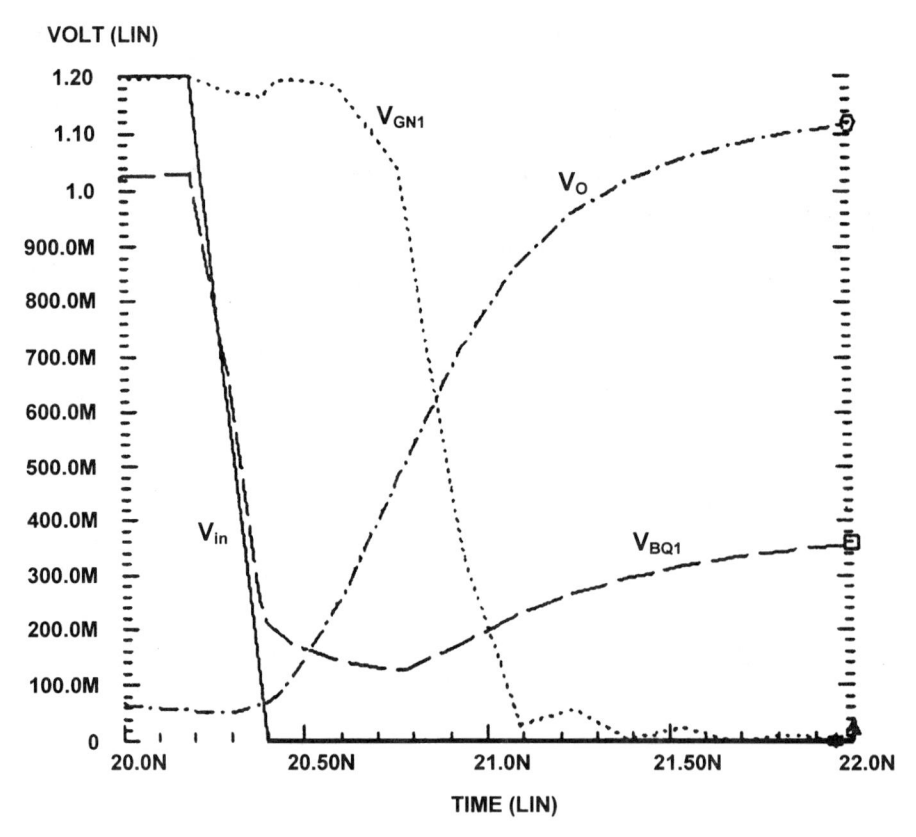

Figure 4–109 Input voltage (Vin), the gate voltage of N1 (V_{GN1}), the base voltage of Q1 (V_{BQ1}) and the output voltage (V_O), during the pull-up cycle of the FB-BiCMOS inverting circuit vs. time at a supply voltage of 1.2 V and a load of 1 pF.

$$V_O = st \qquad 4.14b$$

k_{N2} represents the short-channel effects factor of N_2 in the saturation region [58], υ_{sat} is the saturation velocity of the nMOS transistor, and s is the slope of $V_O(t) \sim t$. Solving Eq. 4.14 gives:

$$V_{GS(N1)}(t) = V_{DD} - \frac{K_{S(N2)}}{C_{inv}}(0.5st^2 - V_{T(N2)}(t)) \qquad 4.15$$

The general form of $I_{DS(N1)}$ governing the charging of the output load, C_O, [59] can be obtained by substituting Eq. 4.15 into Eq. 4.13, which gives:

$$I_{B(Q1)} = I_{DS(N1)} = \frac{I_{C(Q1)}}{\beta} + \frac{d}{dt}(\tau_f I_{C(Q1)}) \qquad 4.16$$

Feedback-Type BiCMOS Logic Circuits

Considering high level injection [59] where $\tau_f \sim B\,\tau_{fo}\,I_C/I_K$ and $\beta \sim \beta_o\,I_K/I_C$, B is the linear proportionality constant in the τ_f dependence on I_C [59], and τ_{fo} and β_o are the forward transit time and current gain at low level injection, respectively. Eq. 4.16 can be integrated to solve for I_C, and it is given by:

$$I_{C(Q1)} = \sqrt{A\left[t - (\tau + \tau^*)\left(1 - \exp\left[-\frac{t}{\tau}\right]\right)\right]} \qquad 4.17$$

where:

$$A = \frac{I_K K_{t(N1)} K_{S(N2)} \beta_o V_{DS(N1)}(av)}{C_{inv}} (V_{T(N2)} + s\tau) \qquad 4.17a$$

$$\tau = B\tau_{fo}\beta_o \qquad 4.17c$$

$$\tau^* = \frac{1}{V_{T(N2)} + s\tau}\left[\frac{st^2}{2\left(1 - \exp\left[\frac{-t}{\tau}\right]\right)} - \frac{C_{inv}(V_{DD} - V_{T(N1)})}{K_{S(N2)}}\right] \qquad 4.17c$$

Figure 4–110 shows an excellent agreement between the analytical and HSPICE results. Eq. 4.17 indicates that the parameters I_K and $K_{t(N1)}$ (which is proportional to $W_{(N1)}$) affect the value of I_C and thus the circuit propagation delay. Hence, the higher the channel width of N_1 and the knee current of Q_1, the higher the I_C will be and, subsequently, the circuit speed. The channel width, $W_{(N2)}$, has a very slight influence on the transient performance because an increase in $W_{(N2)}$ increases both τ^* and the factor A (through $K_{S(N2)}$ in Eq. 4.17).

4.8.5.2 Performance Evaluation

The FB-BiCMOS circuit will be compared to an optimized two-stage CMOS buffer [60] and the BiCMOS circuits [48,57] in terms of the output voltage swing, the power dissipation, the propagation delay, the fabrication area, and the sensitivity of the output voltage to the input voltage. The comparison is made for a ramp input of 200 ps rise time and for an area ratio of 1:1.5:1:1 for the FB-BiCMOS, the circuits [48,57], and the CMOS circuits, respectively. Table 4–13 gives the key SPICE parameters used in the HSPICE simulation.

Figure 4–111 shows the results of the output response versus time. As shown in the figure, the voltage swing of the BiCMOS circuit [57] decreases steadily with the increasing number of input transitions. This decrease is attributed to the premature turning off of the pMOS [57] at high load capacitance and/or low supply voltages resulting in a reduced voltage swing and an increased propagation delay. Second, V_{GS} of the pMOS driver [57] is smaller by at least 0.7 V

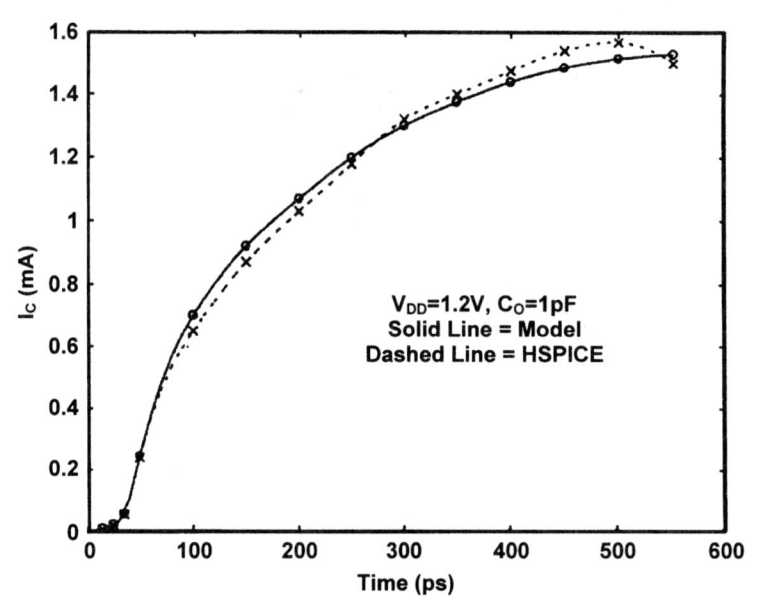

Figure 4–110 Analytical and HSPICE results for $I_C(t)$.

Table 4–13 Key SPICE parameters used for simulation performed on the FB-BiCMOS, BiCMOS circuits [48,57] and the CMOS circuits.

pMOS	nMOS	pnp	npn
T_{OX} = 6 nm	T_{OX} = 6 nm	β_f = 40	β_f = 80
V_T = –0.27 V	V_T = 0.2594 V	I_K = 0.666 mA	I_K = 1.2 mA
C_{jw} = 245 pF/m	C_{jw} = 200 pF/m	τ_f = 16.392 ps	τ_f = 3.5 ps
C_{jm} = 678.7 µF/m²	C_{jm} = 580 µF/m²	R_C = 150 Ω	R_C = 150 Ω
		C_{je} = 4.31 fF	C_{je} = 5.5 fF
		C_{jc} = 10 fF	C_{jc} = 5 fF
		C_{js} = 23.81 fF	C_{js} = 12 fF

than the nMOS driver in the FB-BiCMOS circuit, hence the supply of base current [57] is less than that of the FB-BiCMOS circuit which, in turn, leads to a smaller output voltage swing.

Figure 4–112 compares the performance of the FB-BiCMOS circuit to the CMOS and the BiCMOS circuit in [48] (same as the circuits described in section 4.8.4) at 1.2 V, 0.2 µm technology in terms of the voltage swing, propagation delay, average power dissipation versus the load capacitance, and the sensitivity of the output voltage to the input voltage transition.

Feedback-Type BiCMOS Logic Circuits

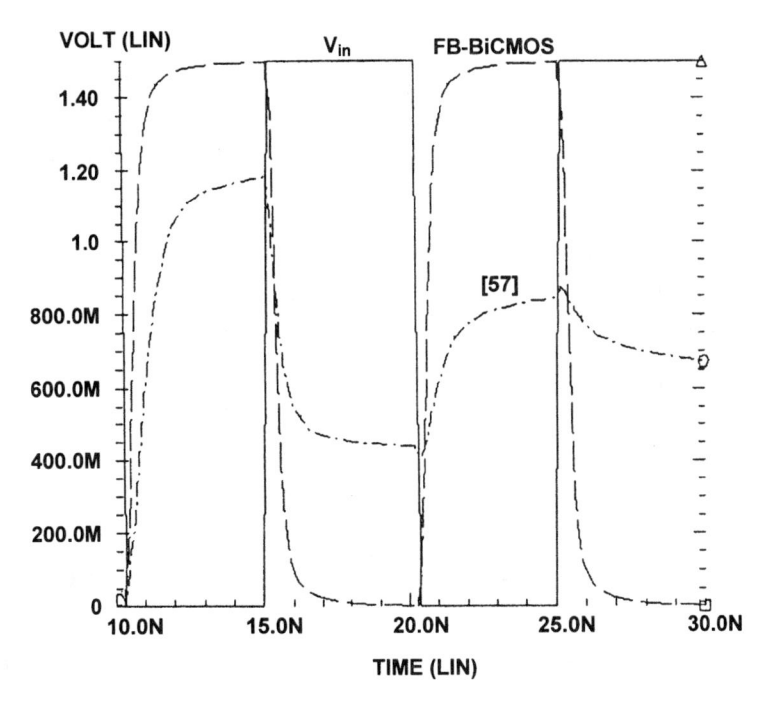

Figure 4–111 Output response of the FB-BiCMOS circuit and [57] at 1.5 V, 0.2 μm technology and a load of 0.6 pF.

From Figure 4–112(a), it can be seen that for the FB-BiCMOS circuit, full output voltage swing is maintained over a wide range of capacitive loading as compared to the poor swing performance of the BiCMOS circuit in [48]. From Figure 4–112(b and c), the slightly higher levels of propagation delay and average power dissipation of the FB-BiCMOS circuit are compensated by their excellent voltage swing performance. Also, the FB-BiCMOS circuit compares favorably with the CMOS circuit. For instance at V_{DD} of 1.2 V and a load of 1 pF, it only consumes 14% more power but about 20% faster than the CMOS. Figure 4–112(d) shows that the FB-BiCMOS circuit maintains its full-swing performance even at low load capacitance, thereby leading to high noise margins. The low sensitivity of the circuit as compared to the BiCMOS circuit in [48] is due to the fact that the input gate voltage to N1 and P1 is determined primarily by the CMOS inverter, INV and is almost independent of the input voltage, Vin.

The FB-BiCMOS circuit also compares favorably with the BiCMOS circuits in [48] and [29] in terms of fabrication area. It uses only 6 devices as compared to 11 and 16 devices in [48] and [29], respectively.

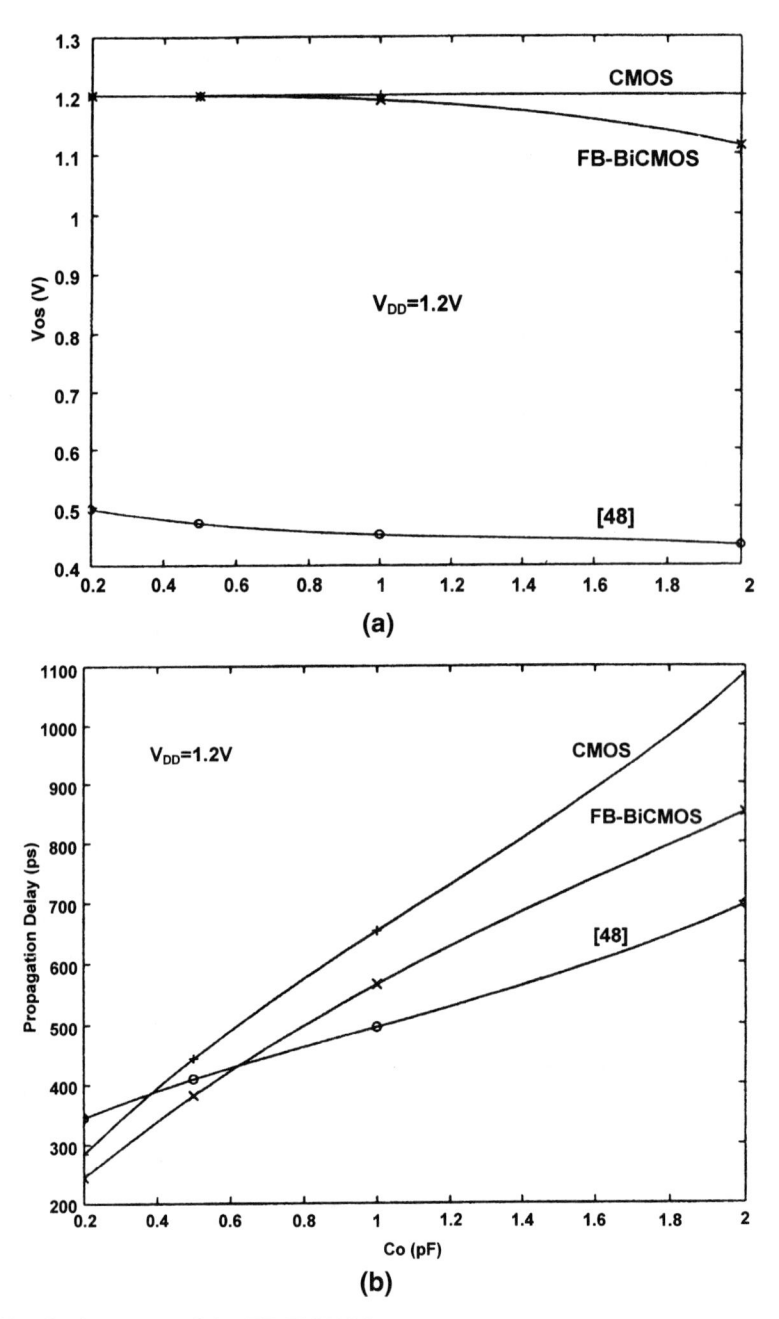

Figure 4–112 Performance of the FB-BiCMOS circuit as compared to the CMOS and the BiCMOS circuit in [48] at 1.2 V, 0.2 μm technology (a) voltage swing (b) propagation delay.

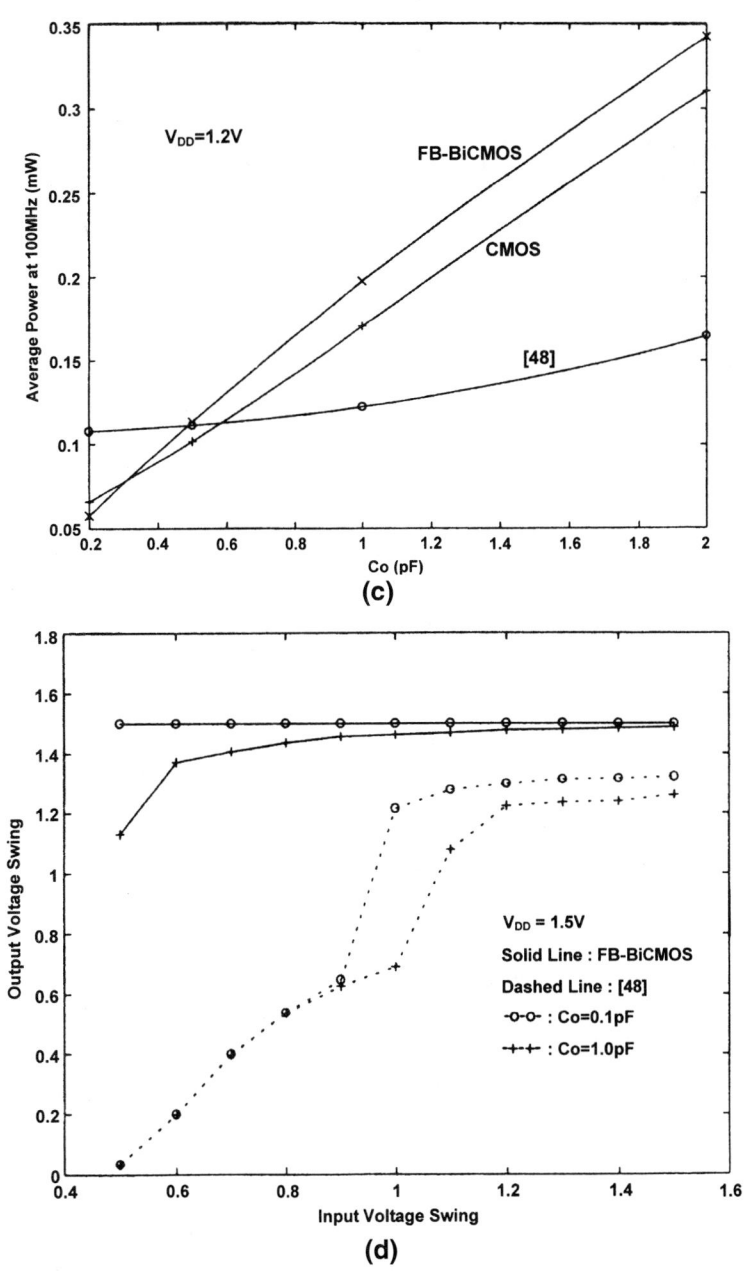

Figure 4–112 (cont.) (c) average power dissipation (d) sensitivity of output voltage to input voltage.

4.8.5.3 Two-Input FB-BiCMOS AND Gate

The FB-BiCMOS inverting circuit has been used to implement a two-input AND gate, as shown in Figure 4–113. The device channel widths and emitter sizes are as follows: P1 (4 μm), P2 (4 μm), P3 (7 μm), N1 (4 μm), N2 (4 μm), N3 (4 μm), pMOS of INV (2 μm), nMOS of INV (1 μm), Q1 (0.3 μm by 1.3 μm), and Q2 (0.3 μm by 1.3 μm). The FB-BiCMOS was compared to an optimized two-stage CMOS AND gate, as shown in Figure 4–114. As demonstrated in this figure, the FB-BiCMOS AND gate is faster than the CMOS AND gate for power supply voltages down to 1 V.

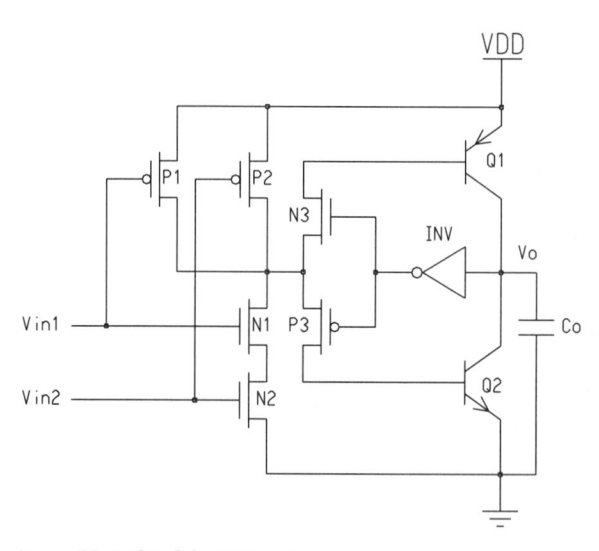

Figure 4–113 Two-input FB-BiCMOS AND gate.

In summary, high-performance complementary BiCMOS logic gates suitable for low-voltage operation were presented. They were designed for the scaled sub-half micrometer BiCMOS generation, and are able to give full output voltage swings at relatively high speeds. So far, the circuits described are all BiCMOS static circuits. In the next section, a low-voltage BiCMOS dynamic logic circuit for high-speed VLSI applications will be discussed.

4.8.6 1.5 V BiCMOS Dynamic Logic Circuit

Dynamic circuit techniques have been used to speed up the performance of the CMOS VLSI [61,62]. However, the n/p-type CMOS dynamic circuits may suffer speed degradation due to the n/p-channel device under/above the logic block. A 1.5 V BiCMOS dynamic logic circuit using a BipMOS pull-down structure that is free from race problems was reported [63]. Figure 4–115 shows the circuit that has been configured as a logic "OR" circuit.

Feedback-Type BiCMOS Logic Circuits 253

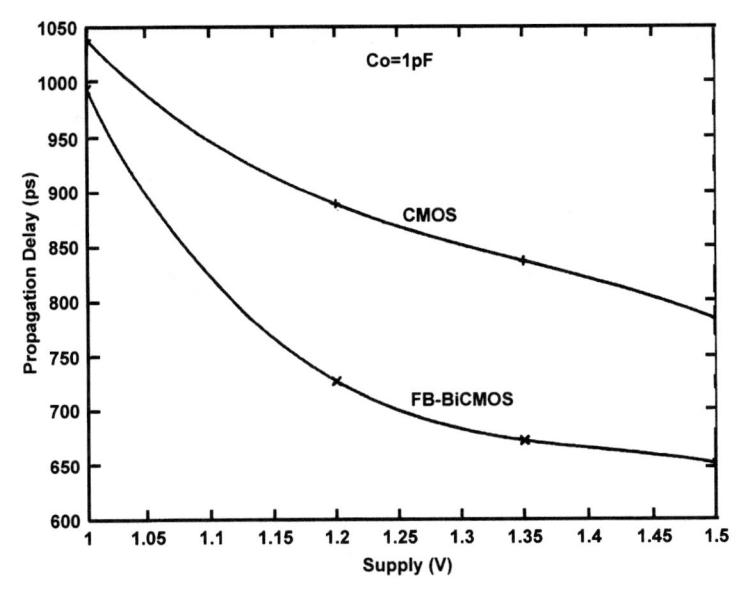

Figure 4–114 Propagation delay of the two-input FB-BiCMOS AND gate as compared to the standard CMOS AND gate as a function of the supply voltage for a load of 1 pF.

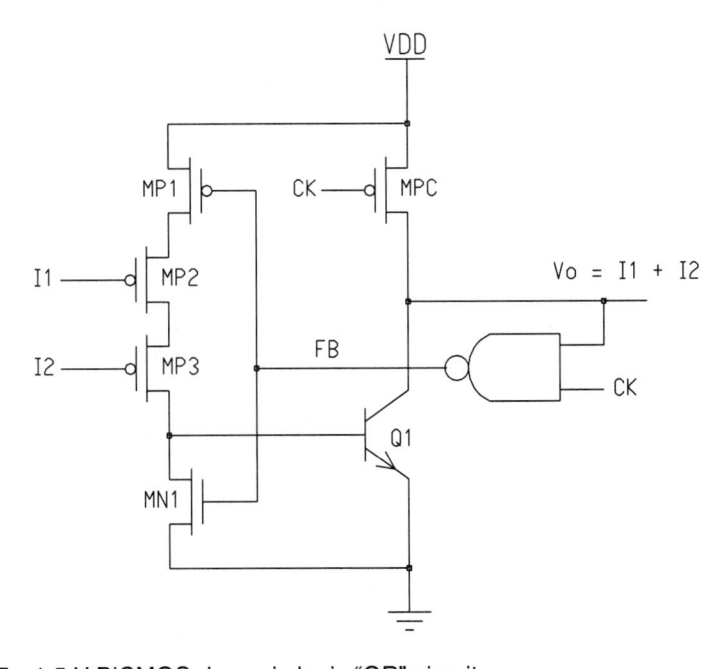

Figure 4–115 1.5 V BiCMOS dynamic logic "OR" circuit.

4.8.6.1 Circuit Operation

The BiCMOS dynamic logic OR circuit consists of the CMOS NAND feedback gate, the pMOS precharge switch MPC, two control switches MN1, MP1, and the BipMOS pull-down structure. The circuit operation is divided into two periods — the precharge period and the logic evaluation period. During the precharge period, the CK input is low and the transistor MPC turns on. Consequently, the output is pulled up to V_{DD}. The feedback signal, FB, is high and causes the transistor MN1 to conduct, pulling the base of Q1 to ground, and consequently turns off the transistor. During the logic evaluation period, the CK input becomes high. Hence, both the transistors MPC and MN1 turn off while MP1 switches on. If both inputs are low, both MP2 and MP3 conduct. Then current flows from V_{DD} via MP1, MP2, and MP3 into the base of Q1, turning on the bipolar device. As a result, the output voltage is pulled down to ground level. However, when the output is pulled down to ground with the CK set to a high level, the feedback signal, FB, switches to high. Consequently, MN1 turns on again and MP1 turns off. Since there is no current flowing in MP1, MP2, and MP3, the bipolar device Q1 shuts off. Also, the base charge of Q1 is now being discharged via MN1. Hence, the bipolar device is on only during switching, and as such reduces the power consumption. In addition, it was reported [63] that a full-swing of 1.5 V at the output of the BiCMOS dynamic logic gate can be obtained.

The BipMOS pull-down structure (consists of MP2, MP3, and the bipolar device in the BiCMOS dynamic logic circuit) allows the output to be pulled close to 0 V, despite the fact that the bipolar device may be in saturation during the pull-down transient. Once the output voltage goes down, the bipolar device will be turned off with the setting of the base to ground by MN1 and the NAND gate. Consequently, no dc power dissipation is consumed.

4.8.6.2 1.5 V BiCMOS Dynamic Full Adder

Using the 1.5 V BiCMOS dynamic logic circuit, a 16-bit full adder circuit, which is composed of half adders (see Figure 4–116) and a carry look-ahead circuit (see Figure 4–117) can be designed. The carry look-ahead circuit is used to process the propagate and generate signals produced by the half adders to generate the carry signals: $C_i = G_i$ OR C_{i-1} AND P_i, for i = 1 to n, where n is the bit number. G_i and P_i represent the generate ($G_i = X_i$ AND Y_i) and propagate ($P_i = X_i$ EXOR Y_i) signals produced from the two inputs (X_i, Y_i) to the half adder. In the BiCMOS dynamic circuits with the BipMOS pull-down structure, N-type and P-type BiCMOS dynamic cells have been placed alternatively to avoid race problems [64–70]. Moreover, as the BipMOS pull-down structure acts as a "buffer" instead of an "inverter", no race problems exist and, therefore, it becomes much easier to design than the N- and P-type BiCMOS dynamic cells.

4.8.6.3 Circuit Performance Comparison and Discussion

To evaluate the performance of the 1.5 V BiCMOS dynamic logic circuit, a test chip consisting of a single logic circuit and a 16-bit full adder using the BiCMOS dynamic logic circuit with the BipMOS pull-down structure was designed according to the parameters' specification, shown in Table 4–14.

Figure 4–118 shows the propagation delay time and delay improvement factor versus the output load of the OR logic circuit, using the CMOS static logic circuit and the BiCMOS

Feedback-Type BiCMOS Logic Circuits

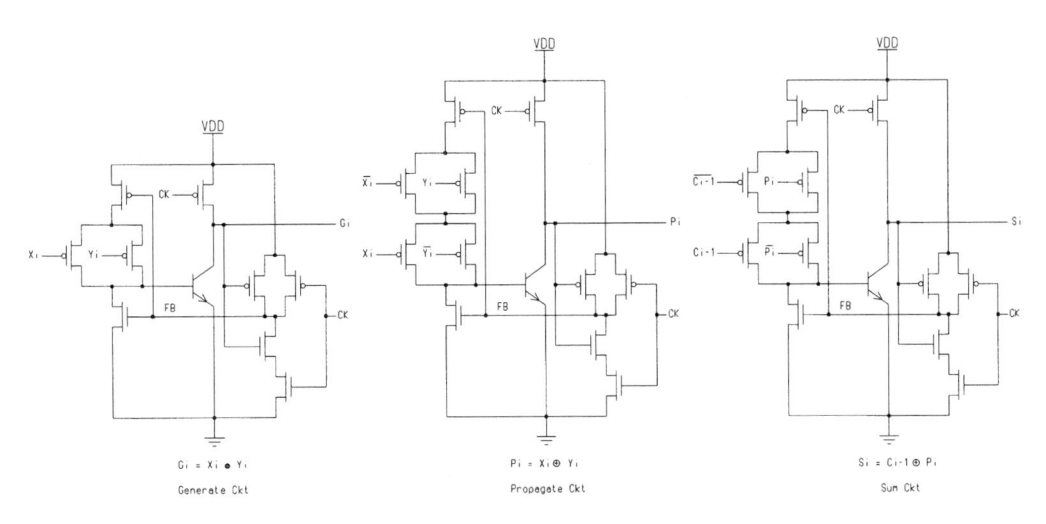

Figure 4–116 Half adder circuit using the BiCMOS dynamic logic circuit.

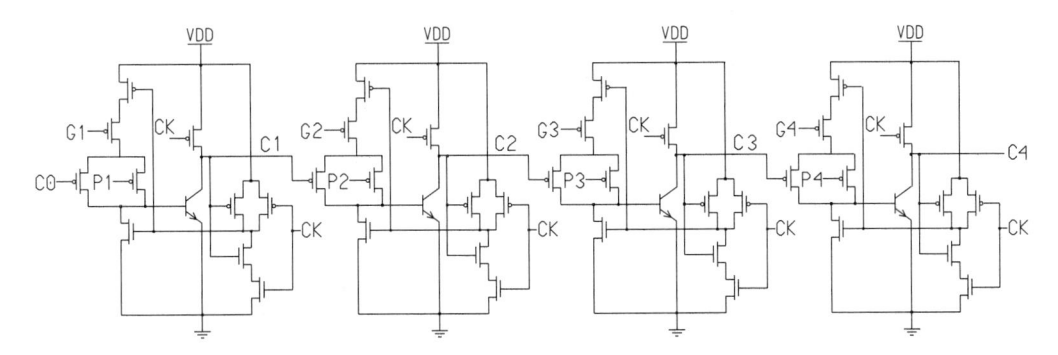

Figure 4–117 1.5 V 4-bit carry look-ahead circuit using the BiCMOS dynamic logic circuit.

Table 4–14 Parameters' specification for the 1.5 V BiCMOS dynamic logic circuit using a 1 µm BiCMOS technology.

Technology	1 µm BiCMOS
Gate oxide thickness	18 nm
Threshold voltage	± 0.7 V
Unity gain frequency of bipolar device	1 GHz
Aspect ratio of nMOS	36/1
Aspect ratio of pMOS	64/1

dynamic circuit. The delay improvement factor is defined as the ratio of the switching time of the CMOS static circuit to that of the BiCMOS dynamic counterpart. From Figure 4–118, we can see that the improvement factor increases with the output load. Overall, the BiCMOS dynamic circuit shows consistent improvement in the propagation delay, which means that it has a faster switching speed compared to the CMOS static circuit's. This is due to the enhanced pull-down capability from the bipolar device. Figure 4–119 shows the simulated transient waveforms of the carry signals: C4, C8, and C16 in the 1.5 V 16-bit CMOS static and 1.5 V BiCMOS dynamic carry look-ahead (CLA) circuits with a load capacitance of 0.2 pF.

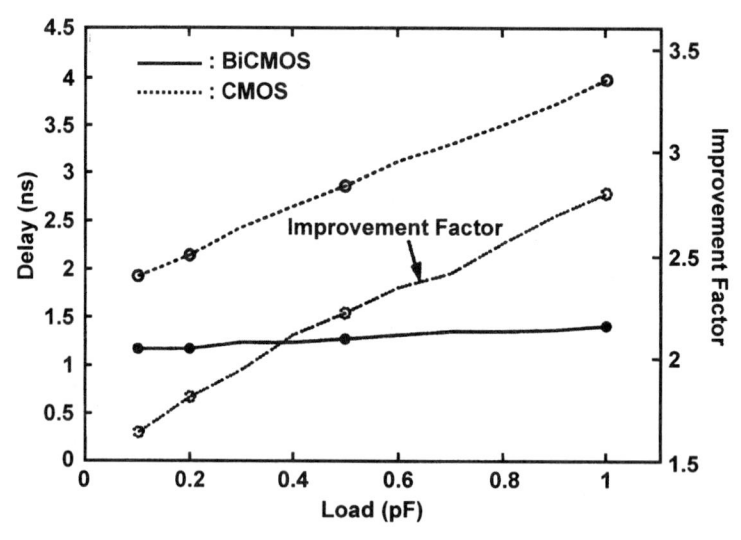

Figure 4–118 Propagation delay and the improvement factor of the OR logic circuit and the BiCMOS dynamic circuit.

In dynamic circuits, the speed performance is characterized by the cycle time or maximum operating frequency, which is determined mainly by the propagation delay in the logic evaluation period. The precharge period is much shorter, about 3 ns, and therefore is not significant. As shown in Figure 4–119, the improvement in the propagation delay for the longest critical path, that is, the path associated with C16, in the BiCMOS dynamic CLA circuit is 1.7 times that of the CMOS static circuit. Figure 4–120 shows the propagation delay of a 16-bit CLA using the BiCMOS dynamic circuit and the CMOS static circuit versus the supply voltage based on the 1 μm technology. Figure 4–120 also compares the BiCMOS dynamic circuit [64,65], which is denoted by "BiCMOS*". The graph shows that for a power supply voltage less than 2.3 V, the speed of the CLA, implemented using the "BiCMOS*" dynamic circuit, is not better than that of the CMOS circuit. However, the BiCMOS dynamic circuit described earlier shows an improvement factor of approximately 1.7 times to 2.5 times that of the CMOS circuit. The improvement

Feedback-Type BiCMOS Logic Circuits

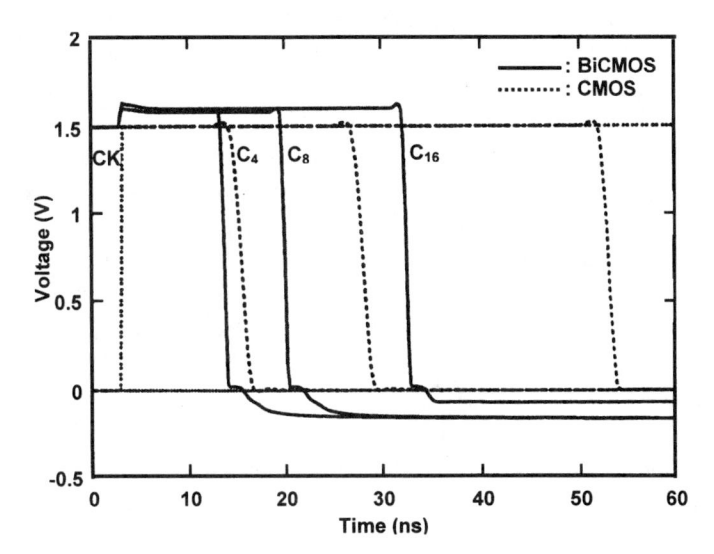

Figure 4–119 Transients of the 16-bit carry look-ahead circuit, using the 1.5 V BiCMOS dynamic and the 1.5 V CMOS static circuit techniques based on SPICE simulation results.

of the CLA using the BiCMOS dynamic circuit over that of the CMOS circuit generally becomes better for a larger supply voltage.

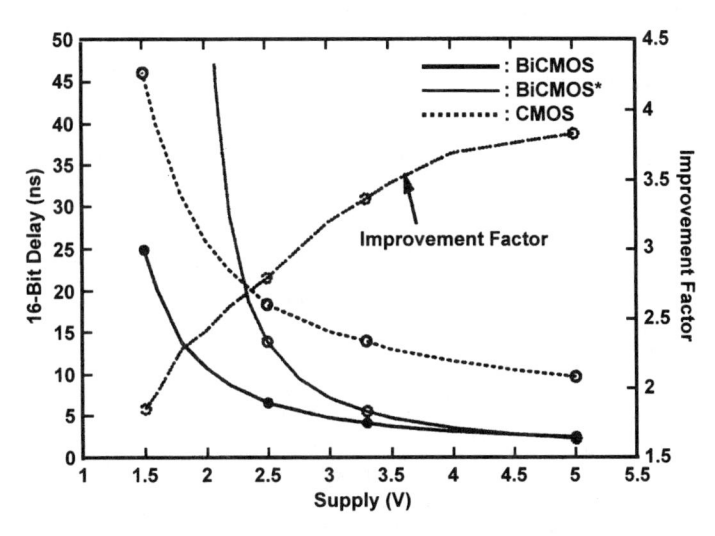

Figure 4–120 Propagation delay time vs. the supply voltage of the 16-bit carry look-ahead circuit using the BiCMOS dynamic and the CMOS static techniques and the propagation delay improvement factor.

The BiCMOS dynamic circuits provide speed advantages over the CMOS dynamic circuits [61,62] due to the following reasons. First, the n/p-type CMOS dynamic circuits may suffer from speed degradation due to the n/p-channel device under/above the logic block. For the BiCMOS dynamic circuits, since the BJT device is the major pull-up/pull-down device during the logic evaluation period, speed degradation will not occur. Hence, the BiCMOS dynamic circuits are more suitable for advanced low-voltage VLSI systems as compared to the CMOS dynamic circuits. Also, in the 1.5 V BiCMOS dynamic logic circuit, when the bipolar device is on, the output can be pulled down close to 0 V, where the bipolar device eventually saturates. Usually, in a typical bipolar circuit, the turn-off time increases if the device is saturated. In this case, the saturation of the bipolar device does not affect the speed performance due to the feedback CMOS NAND gate as shown in Figure 4–115. As the output is pulled down to low, the FB signal produced by the NAND gate turns high. Consequently, the device MN1 turns on, the bipolar device turns off, and the saturation of the bipolar device does not affect the switching time.

Second, during the precharge period, the pMOS device (MPC) is used. As a result, the BiCMOS dynamic circuit functions similarly to the CMOS circuits during the precharge period. The speed performance during the precharge period is not instrumental in determining the overall speed performance, which is determined mainly by the propagation delay during the logic evaluation period. This delay has been proven to be lower than their CMOS circuit counterparts. In retrospect, the BiCMOS dynamic circuit is applicable not just to the full adder circuit, its application is wider ranging, and it can enhance the speed performance of a large system at a cost of an extra bipolar transistor per stage.

4.9 High-Beta BiCMOS (Hβ-BiCMOS) Logic Circuits

4.9.1 Introduction

A high-beta BiCMOS (Hβ-BiCMOS) logic circuit with a low-input capacitance has been developed to achieve high-speed operation at sub-2 V power supply voltage [71]. It uses a base bias clamping circuit technique to maintain the base bias of a pull-down npn bipolar transistor at around V_{BE}. This technique will help to reduce the delay-time to charge the base parasitic capacitance of the bipolar transistor. Next, the Hβ-BiCMOS is combined with a BiCMOS charge pump to further lower the minimum required supply voltage.

4.9.2 Hβ-BiCMOS Logic Circuit

The Hβ-BiCMOS circuit is shown in Figure 4–121. The base bias clamping circuit is indicated by the dotted boundary lines. The key function of the clamping circuit is to ensure that the base node B2 of the output pull-down npn bipolar transistor Q_{DN} remains biased at around V_{BE}. Two important criteria of a base bias clamping circuit are: (1) to clamp the base bias voltage at node B2 slightly below V_{BEon} and (2) to provide a high impedance at the base node B2 of Q_{DN} so that only a small MOS current for MN1 and MN2 is required to produce a 150 mV voltage swing at its base, which in turn produces a large driving current of about 5 mA at the output node.

High-Beta BiCMOS (Hβ-BiCMOS) Logic Circuits

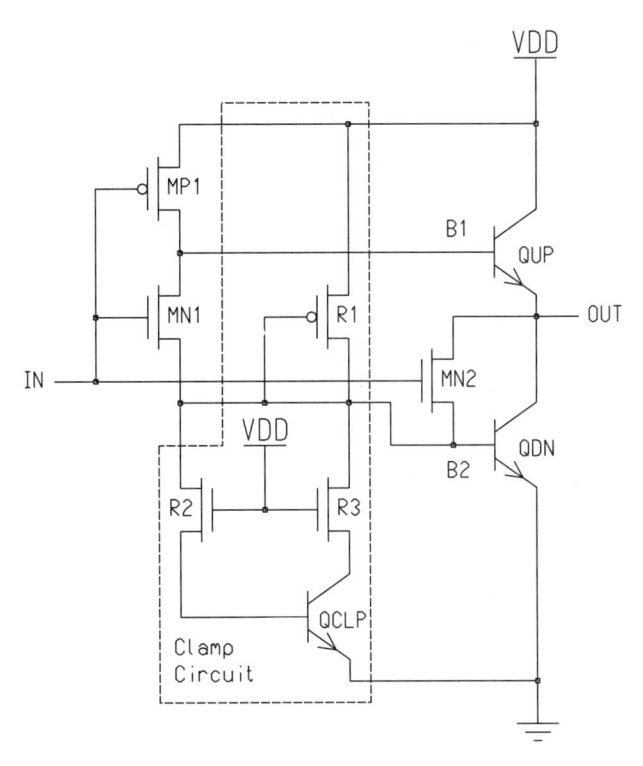

Figure 4–121 Hβ-BiCMOS logic circuit.

The local clamping circuit shown in Figure 4–121 utilizes two nMOS implemented resistors R2 and R3 to make the impedance of the base bias clamping circuit high. Hence a small MOS current through MN1 and MN2 is able to turn on Q_{DN}. Furthermore, as transistors Q_{DN} and Q_{CLP} form a current mirror circuit, the base of Q_{DN} is biased at a voltage such that its collector current is equal to that of Q_{CLP} given by $(V_{DD} - V_{BE})/R1$. Figure 4–122 shows the delay and the current variations of the Hβ-BiCMOS gate circuit as functions of the resistance R1, R2, and R3. These results were obtained by SPICE simulations using the 0.4 μm BiCMOS technology, a load capacitance of 2 pF, and a supply voltage of 3.3 V. The results show that the variations of the delay time and the current with the three resistances are very small.

The concept of the base bias clamping circuit technique was previously adopted in the BP-CBiCMOS/BiNMOS [72] as illustrated in Figure 4–123. But it was used only in the pull-up section, which does not strongly require this circuit technique. Moreover, because low impedance PN diodes are used as clamping devices in these circuits, the pull-up npn bipolar transistor is supplied with enough base current to generate the collector current only after the nMOS transistors are turned off. Figure 4–124 compares the gate delay (when the output voltage goes from low to high) for a BP-BiNMOS, a conventional BiNMOS, and a BiNMOS circuit with the base bias clamping circuit of the Hβ-BiCMOS added.

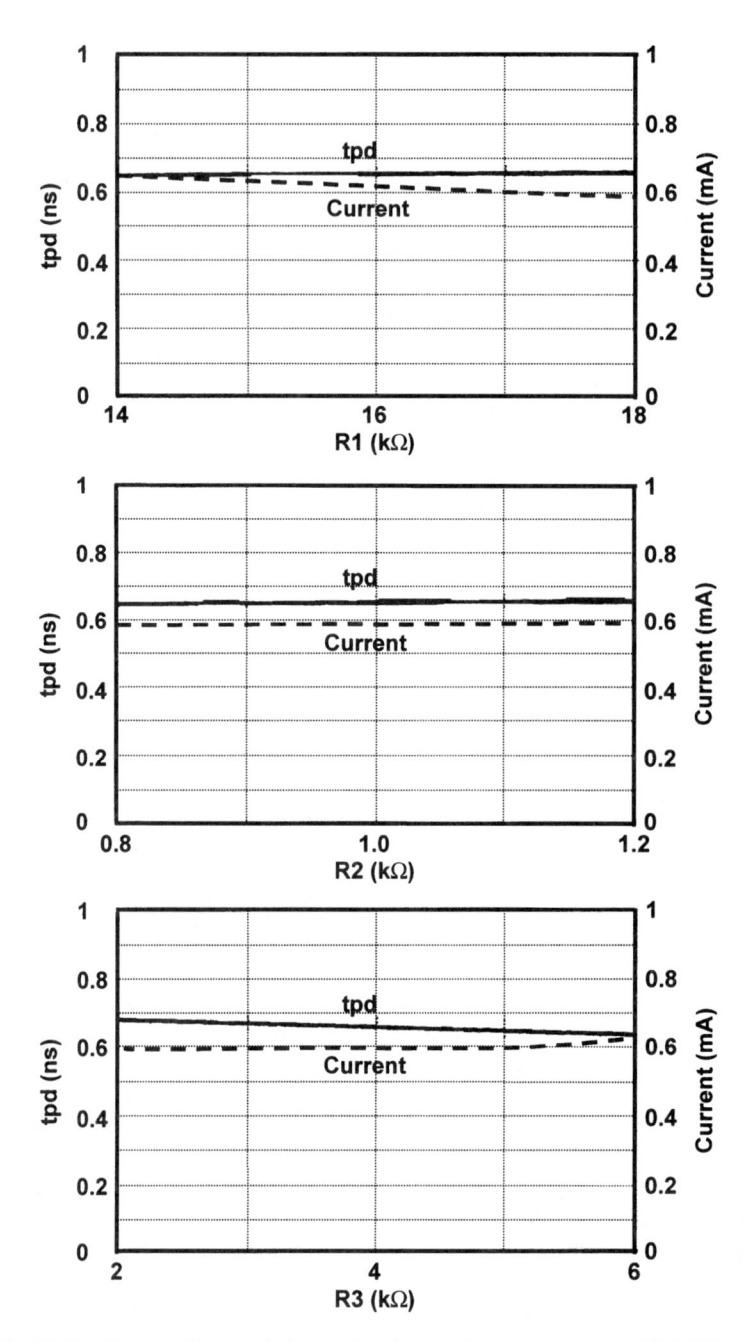

Figure 4–122 Delay-time and current dependencies on the resistances of R1, R2, and R3 in the Hβ-BiCMOS gate circuit.

High-Beta BiCMOS (Hβ-BiCMOS) Logic Circuits

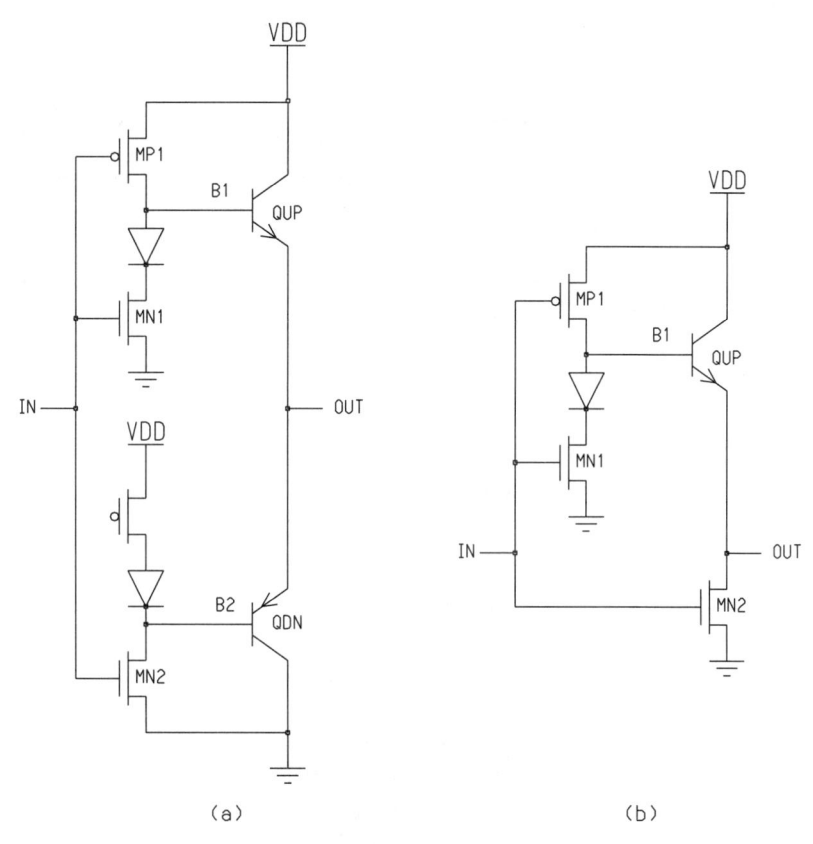

Figure 4–123 (a) BP-CBiCMOS (b) BP-BiNMOS.

As shown in Figure 4–124, the BiNMOS with a clamping circuit of Hβ-BiCMOS has the lowest gate delay. It was also found that the delay of the BP-BiNMOS is longer than that of a conventional BiNMOS gate circuit due to the parasitic capacitance of the clamping diodes. The BP-BiNMOS/BP-CBiCMOS is effective only when higher supply voltages are used.

4.9.3 Circuit Operation of the Hβ-BiCMOS Circuit

As shown in Figure 4–121, during the input low-to-high transition, both transistors MN1 and MN2 switch on immediately and start supplying a base current to Q_{DN} without any delay in charging its base parasitic capacitance. This process enables a faster pull-down operation as compared to that of the conventional BiCMOS circuits. The current through R1, MN2, and Q_{DN} will clamp the output voltage level to a value above ground such that Q_{DN} never enters the saturation region. Furthermore, the base bias of Q_{DN} will be kept at around V_{BE}, which is an important feature of the Hβ-BiCMOS circuit.

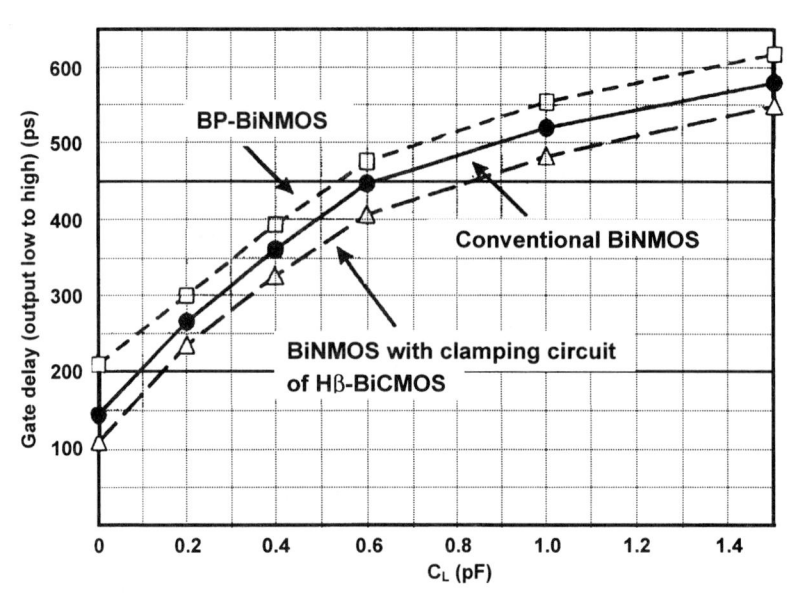

Figure 4–124 Gate delay (pull-up cycle) of BP-BiNMOS, conventional BiNMOS and BiNMOS with the base-bias clamping circuit used in the Hβ-BiNMOS.

During the input high-to-low transition, the pMOS transistor MP1 turns on to drive the base of Q_{UP}. Although MP1 needs to supply not only the base current of Q_{UP} but also enough current to pull up the base node B1 from V_{BE} to V_{DD}, the required on-current of MP1 is still small because the rise time at node B1 is equal to the rise time at the heavily loaded output node. Hence the output pull-up operation of Hβ-BiCMOS is fast even at low-supply voltages. Figure 4–125 compares the transconductance (g_m) of bipolar transistors and MOS transistors. The Hβ-BiCMOS circuit needs only sub-mA MOS drain current to produce a 150 mV voltage swing at the base of Q_{DN}, which, in turn, is sufficient to produce a 5 mA output load driving current. This is possible because of the Q_{DN}'s high current gain ($\beta = 25$ under high-level injection) and large transconductance, which is over 20 times better than that of an nMOS transistor with gate width/length of 10 μm/0.4 μm respectively. As illustrated in Figure 4–125, a transconductance of 33 mS can be achieved for an emitter area of 0.8 μm by 6.4 μm. Hence, the smallest geometric gate width of around 1.1 μm can be employed for MN1 and MN2, thereby realizing very small gate input capacitance.

4.9.4 Circuit Simulations and Discussion

The Hβ-BiCMOS inverter has been simulated based on a 0.4 μm gate length and a self-aligned emitter base BiCMOS technology. The device parameters are listed in Table 4–15.

High-Beta BiCMOS (Hβ-BiCMOS) Logic Circuits

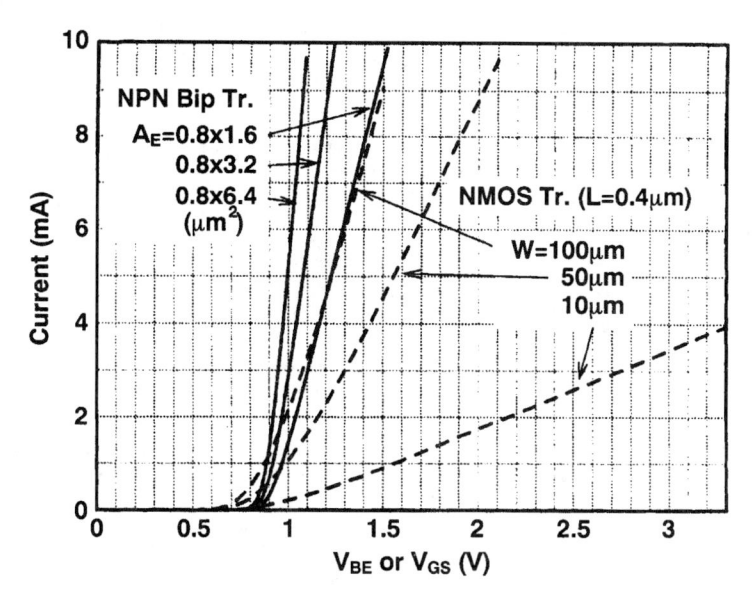

Figure 4–125 Drain current of nMOS transistors and collector current of npn transistors (the gradient gives the transconductance, g_m).

Table 4–15 Measured device parameters based on a 0.4 μm BiCMOS process.

nMOS : $L_N = 0.4$ μm $V_{TN} = 0.65$ V	
pMOS : $L_P = 0.5$ μm $V_{TP} = -0.7$ V	
npn (Emitter size of 0.8 μm × 6.4 μm)	
$f_{T(max)}$	21 GHz
β_f	100
R_C	14 Ω
R_B	150 Ω
R_E	13 Ω
C_{CB}	9.4 fF
C_{EB}	21.3 fF

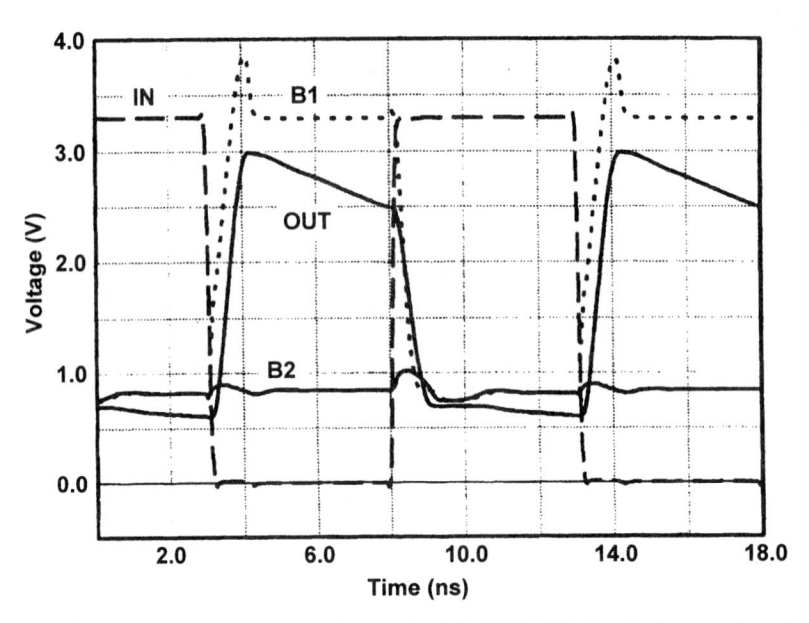

Figure 4–126 Simulated operating waveforms of a Hβ-BiCMOS circuit at a supply voltage of 3.3 V and without charge pumping circuit.

Figure 4–126 shows the voltage waveforms at various nodes against time. From the figure, we can see that the base node B2 of Q_{DN} is clamped at approximately V_{BE} (~ 0.8 V) throughout the entire duration. The voltage at B2 only rises above V_{BE} during the output pull-down operation. Figure 4–127 compares the delays of the Hβ-BiCMOS, CMOS, and BiNMOS inverters with the same input capacitance of 14 fF and a supply voltage of 3.3 V.

It can be seen that the Hβ-BiCMOS (without any charge pump) is approximately four times faster than the BiNMOS and 10 times faster than CMOS at a load of 1.5 pF. Also in order to achieve the same load capacitance dependent characteristics as the Hβ-BiCMOS, the CMOS needs to have 12 times the gate width (which means 12 times greater input capacitance) of the Hβ-BiCMOS circuit. This high input capacitance will cause a large fan-in delay, as shown in Figure 4–127. Figure 4–128 shows how the gate delays of the CMOS, conventional BiCMOS, and Hβ-BiCMOS inverters with the same input capacitance (14 fF) vary with the supply voltage.

The superior performance of the Hβ-BiCMOS inverter at low-supply voltage (< 2 V) has been achieved by the use of a base bias clamping circuit technique, which allows the full utilization of the bipolar transistors' characteristics.

4.9.5 A BiCMOS Charge Pump with Hβ-BiCMOS

A Hβ-BiCMOS gate circuit, which uses a base bias clamping circuit, significantly reduces the required on-current of the nMOS transistors needed to drive the pull-down bipolar transistor.

High-Beta BiCMOS (Hβ-BiCMOS) Logic Circuits

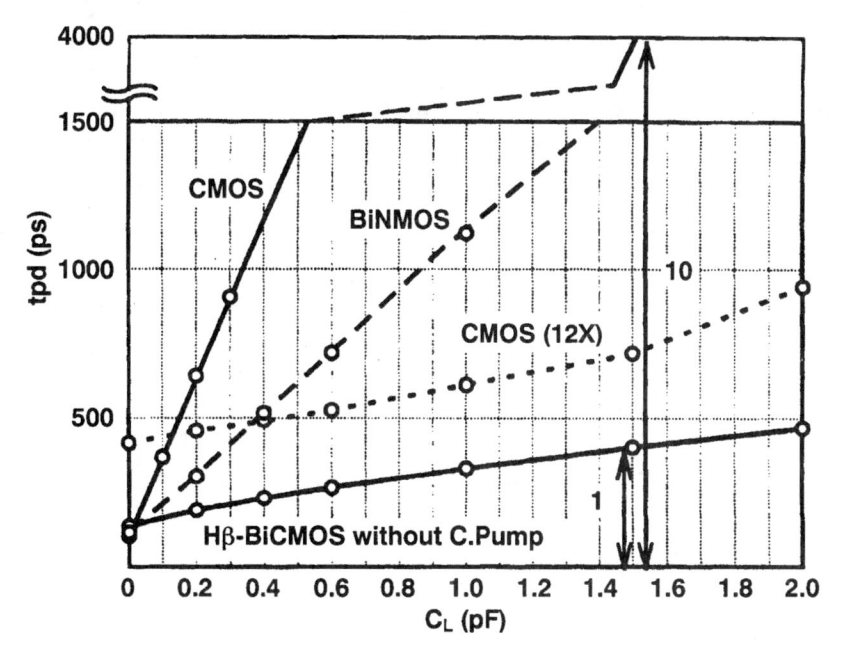

Figure 4–127 Simulated delay vs. load capacitance (V_{DD} = 3.3 V).

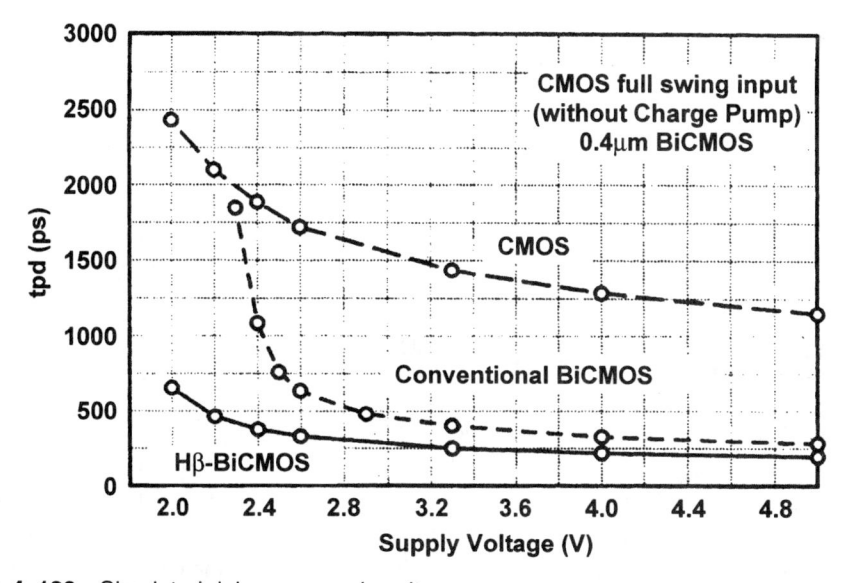

Figure 4–128 Simulated delay vs. supply voltage.

In order to further lower the minimum required supply voltage while maintaining sufficient output voltage swing, a charge pump circuit has been developed. It is tied to the source of MP1 of the Hβ-BiCMOS circuit to drive the base of Q_{UP}. This charge pump employs bipolar transistors and enables a maximum output voltage of (2 $V_{DD} - V_{BE}$) with high current drivability. Figure 4–129 shows the BiCMOS charge pump connected to the Hβ-BiCMOS gate circuit. In this figure, the MOS on-resistance has been used in the base bias clamping circuit of the Hβ-BiCMOS.

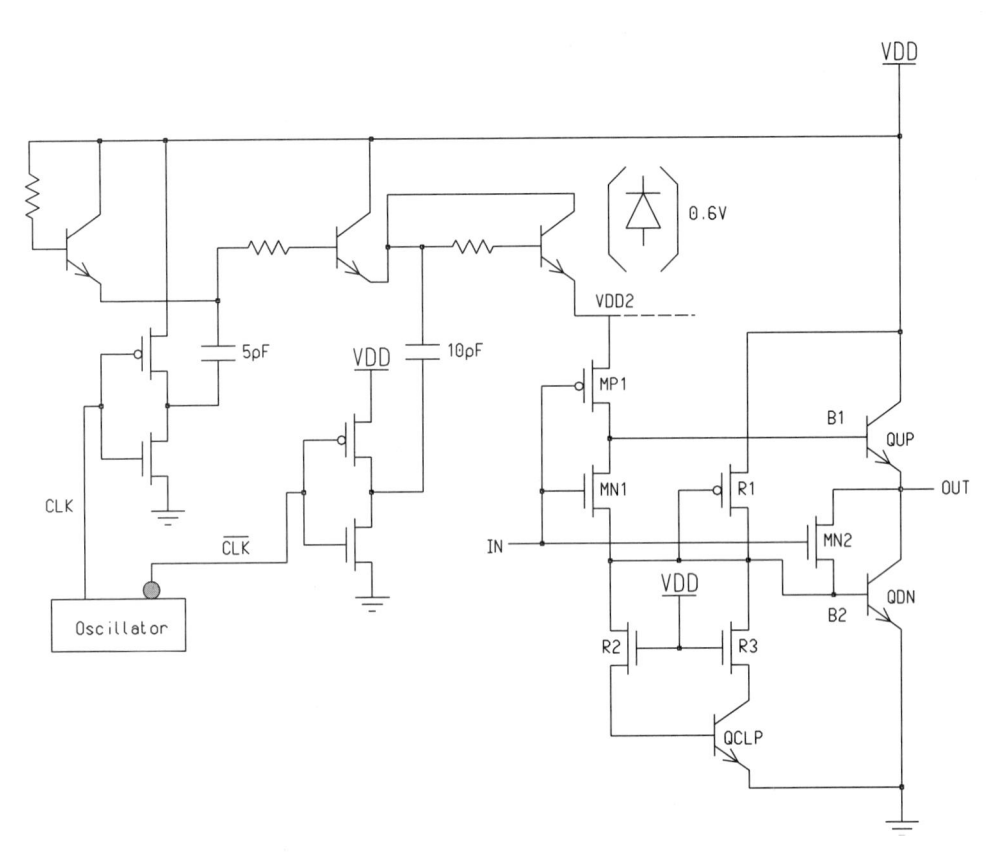

Figure 4–129 BiCMOS charge pump with Hβ-BiCMOS.

Figure 4–130 shows the variation of the output voltage (V_{DD2}) from the charge pump circuit with the supply voltage. Notice that the charge pump can operate at a supply voltage as low as 1.5 V. Since the charge pump is connected to the source of MP1 of the Hβ-BiCMOS, it needs to supply only the Q_{UP} base current and the current required to charge the parasitic capacitance of the base node to raise the output voltage to a high level. Hence, the current drivability of the charge pump in this circuit configuration is much smaller than in the configuration in which the

High-Beta BiCMOS (Hβ-BiCMOS) Logic Circuits

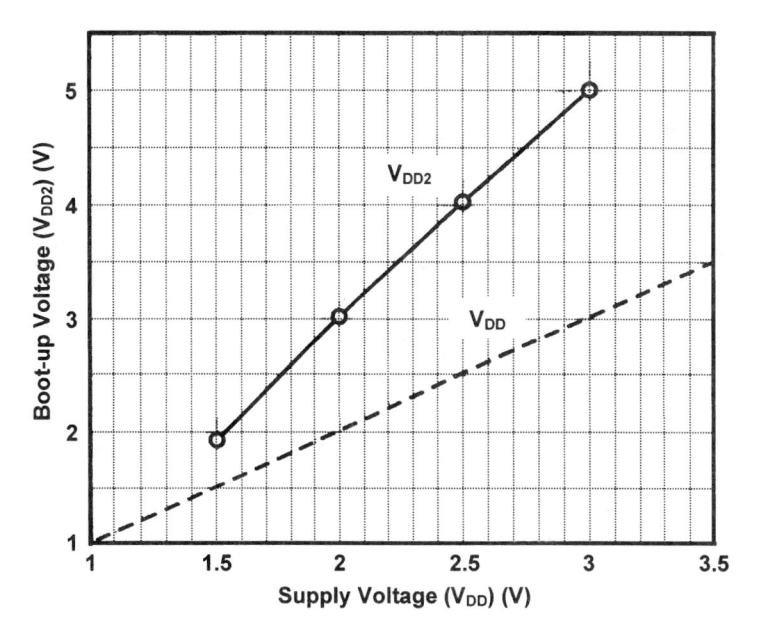

Figure 4–130 Simulated output voltage of the charge pump.

charge pump is also connected to the collector of Q_{UP}, as in a BiCMOS circuit configuration [73]. Figure 4–131 shows the relationship between the output voltage of the charge pump and the number of driven Hβ-BiCMOS gate circuits. The supply voltage and the synchronous clock frequency used in the simulation are 2.5 V and 200 MHz, respectively. As demonstrated in Figure 4–131, for a power supply voltage of 2.5 V, the BiCMOS charge pump can drive as many as 120 Hβ-BiCMOS gate circuits. This example shows that a small number of charge pumps can supply enough boot-up voltage for all the BiCMOS circuits on the chip. In order to avoid saturation of Q_{UP}, which may cause operating speed degradation, the boot-up voltage should be maintained at about (V_{DD} + 0.6 V). This can be achieved by either controlling the oscillator in the charge pump or inserting a diode between V_{DD} and the output.

Figure 4–132 shows the operating waveforms of the Hβ-BiCMOS with the charge pump at a load of 0.5 pF and a power supply voltage of 2 V. It is assumed that low-threshold (0.2 V) nMOS transistors are used for transistors MN1 and MN2. Note that Hβ-BiCMOS drives Hβ-BiCMOS in this case. By introducing the charge pump to increase the output voltage swing, direct connection of the Hβ-BiCMOS gate circuits becomes possible at a 2 V supply. In summary, the Hβ-BiCMOS gate circuit has a very small input capacitance and is able to achieve high speed at low-supply voltages by using a base bias clamping circuit technique. The circuit is able to achieve ten times the speed of a CMOS gate with the same input capacitance at a supply voltage of 3.3 V. In order to lower the minimum supply voltage of the Hβ-BiCMOS, a charge pump was developed to help increase the output voltage swing of the Hβ-BiCMOS circuit.

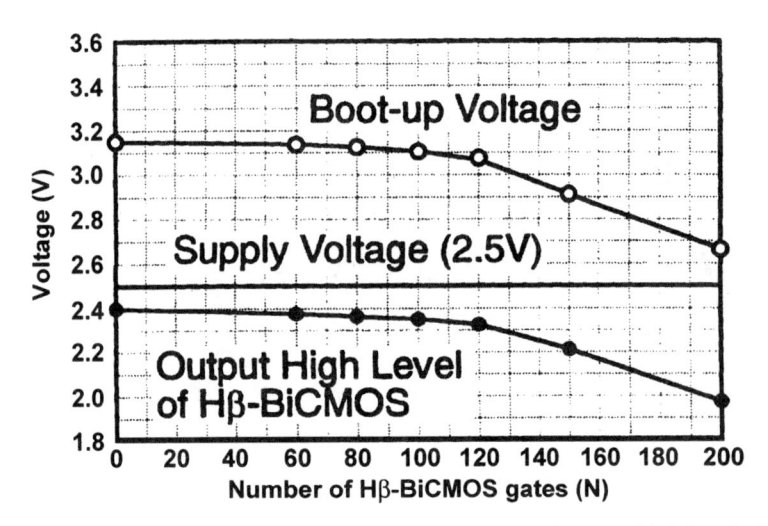

Figure 4–131 Simulated output voltage of the charge pump and output high level of the Hβ-BiCMOS versus the number of Hβ-BiCMOS inverters.

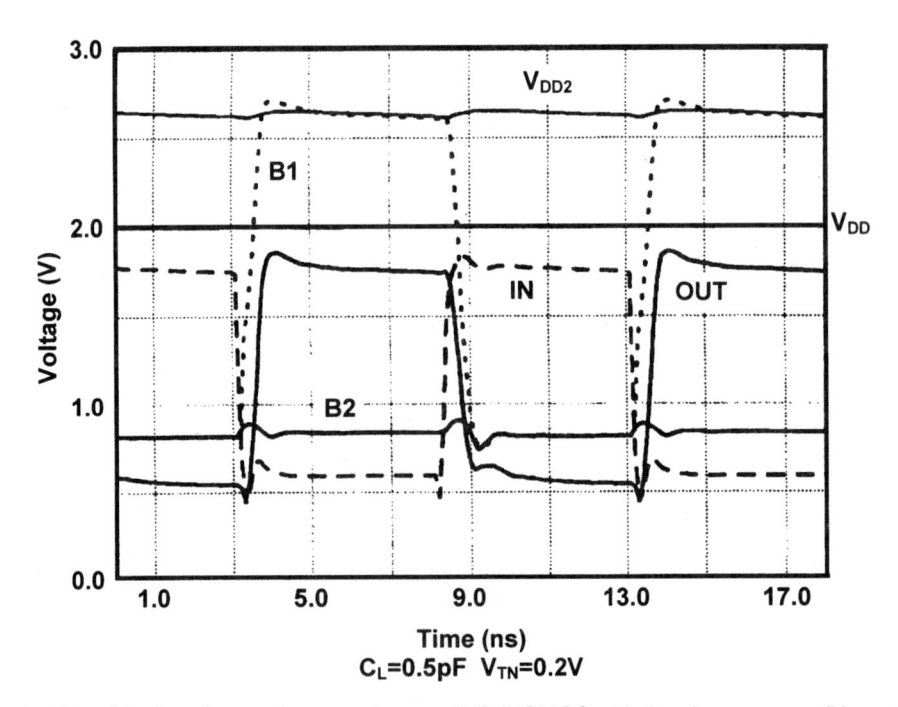

Figure 4–132 Simulated operating waveforms of Hβ-BiCMOS with the charge pump (V_{DD} = 2 V).

4.10 Transiently Saturated Full-Swing BiCMOS (TS-FS-BiCMOS) Logic Circuits

4.10.1 Introduction

Conventional BiCMOS circuits suffer from speed degradation as the supply voltage is reduced. This degradation is due to the reduction of the gate-source voltage applied to the nMOS transistor in the pull-down circuit by 2 V_{BE}. This voltage loss will seriously degrade the speed performance of the circuit when the supply voltage is reduced to about 3 V [74]. To improve the low-voltage performance of the BiCMOS circuits, several types of BiCMOS circuits have been proposed. In section 4.6, the QC-BiCMOS has been introduced and demonstrated to be better than the conventional BiCMOS, BiNMOS, C-BiCMOS, and CMOS circuits. Recently, a TS-FS-BiCMOS [43] has been reported, and it will be shown in the subsequent sections that it is by far the most promising circuit for BiCMOS ULSI applications in the sub-2 V regime. The key to its high performance is the transient saturation of the bipolar transistors only during the switching periods. This technique helps to remove the voltage loss due to the base-emitter turn-on voltage. Hence, while conventional BiCMOS, C-BiCMOS, QC-BiCMOS, MBiCMOS, and BiNMOS circuits still suffer from the one V_{BE} voltage loss in the output voltage transition, TS-FS-BiCMOS circuit suffers from none.

4.10.2 Circuit Concept and Operation

In order to achieve a high-speed full-swing operation in BiCMOS circuits, full-swing operation of bipolar transistors must be realized. Figure 4–133 shows two full-swing circuit techniques.

The saturation type circuit in Figure 4–133(a) achieves full-swing operation but at a slow switching speed because once the excess minority carriers are charged into the base, they stay there until the next transition [8,27]. On the other hand, the transient-saturation type circuit shown in Figure 4–133(b) enables the bipolar transistor to achieve high-speed full-swing operation because the excess minority carriers are initially charged into the base and subsequently discharged immediately after the output voltage transition. Figure 4–134(a) shows the circuit configuration of the TS-FS-BiCMOS circuit. This circuit is noninverting and the output voltage follows the input signal. The simulated waveforms as illustrated in Figure 4–134(b) aid to explain the transient behaviour of the circuit.

When the input goes from high to low, MP2 turns on. Since the output is initially high, the feedback signal from the output through the inverter INV1 will cause MP3 to remain on. The base of Q1 is charged through the current path from V_{DD} to the node B1. The transistor Q1 will continue to drive the load until the output voltage reaches nearly zero. Although Q1 saturates, the next pull-up transition is not affected because the excess minority carriers of Q1 are discharged immediately after the pull-down transition through MN4, which switches on as the delayed output voltage is fed back via INV1. Hence, Q1 achieves full-swing operation without any compromises on its speed performance. Note that the node voltage of B1 drops down to

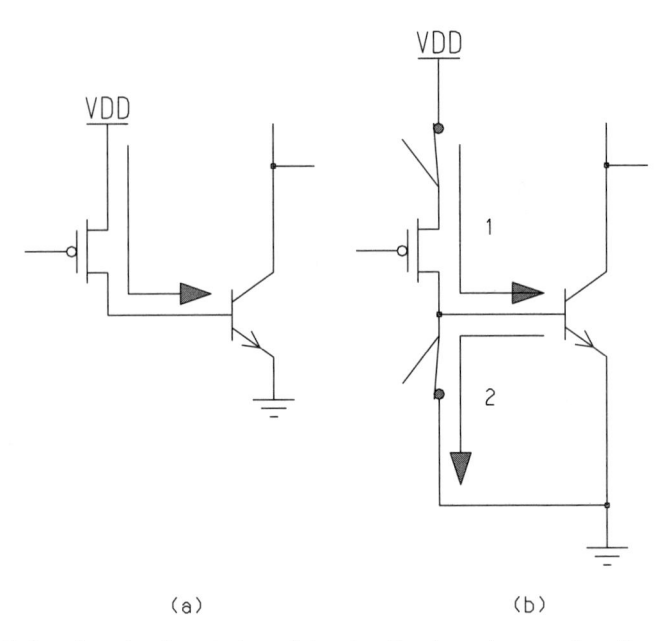

Figure 4–133 Full-swing circuit technique (a) saturation type (conventional) and (b) transient-saturation type.

almost zero immediately after the pull-down transition. Furthermore, static power dissipation is avoided as MN3 cuts off the dc current path after the pull-up transition, and MP3 cuts it off after the pull-down transition. Therefore, the bipolar transistors are saturated transiently in the switching periods only. Similar explanations apply for the pull-up transition due to the symmetry of the circuit.

The static output voltage is held by the small-size CMOS latch composed of INV1 and INV2. Although both Q1 and Q2 are off in steady state, this circuit suffers little from a switching noise at the output node, because if the switching noise is large enough to flip the CMOS latch, the bipolar transistor (Q1 or Q2) begins to drive the load again. As such, the output voltage is immediately restored back to the static level. If the switching noise is too small to flip the CMOS latch, it will not flip the next stage gate either.

The circuit has been simulated without NM4, as shown in Figure 4–134(a), to verify the advantage of the transient-saturation technique, and the results are shown in Figure 4–134(b) in hatched lines. We can see that Q1 continues to saturate even after the pull-down transition, thereby resulting in about 2.5 times longer delay time due to the inability to deplete the excess minority carriers at the base. Figure 4–135 shows the simulated output voltage waveforms of the C-BiCMOS and TS-FS-BiCMOS at a power supply voltage of 1.5 V when a step input is applied to the circuit.

Transiently Saturated Full-Swing BiCMOS (TS-FS-BiCMOS) Logic Circuits

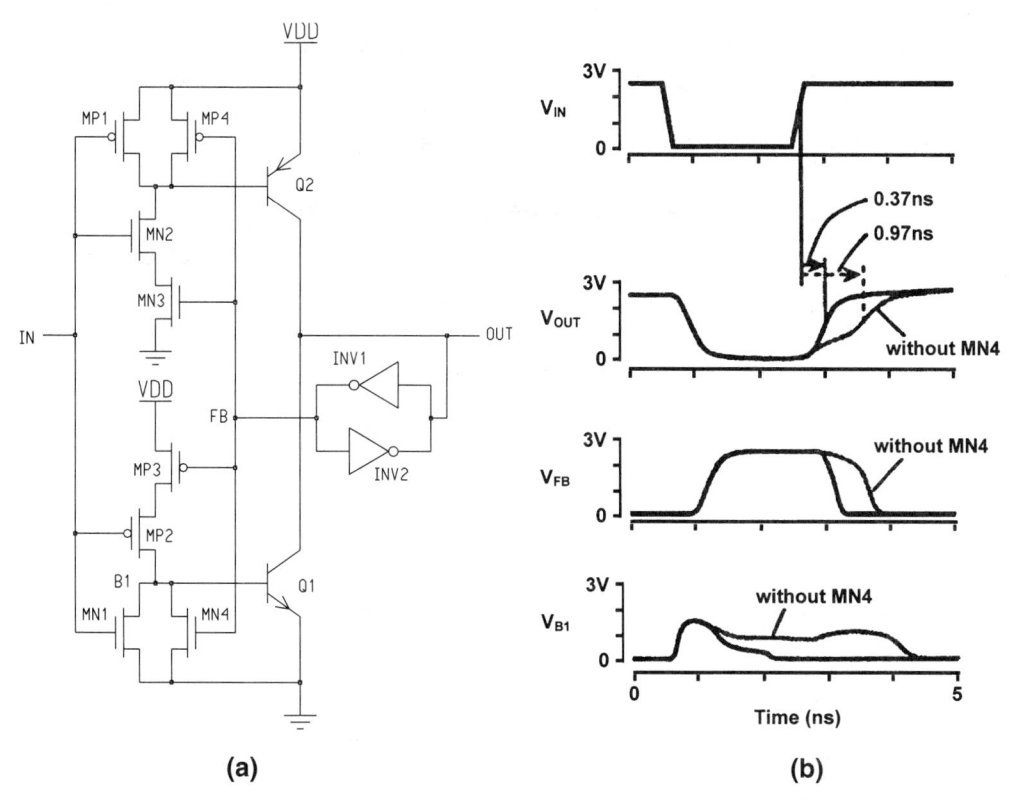

(a) **(b)**

Figure 4–134 (a) Configuration of TS-FS-BiCMOS buffer and (b) simulated voltage waveforms.

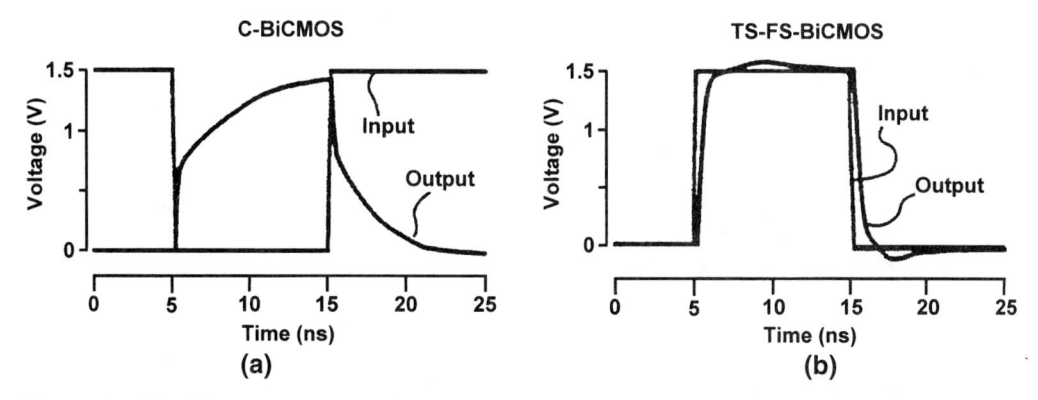

Figure 4–135 Simulated input and output voltage waveforms of (a) C-BiCMOS and (b) TS-FS-BiCMOS.

For the C-BiCMOS, the fast output voltage transitions are V_{BE} smaller than the supply voltage, and the following transitions by additional elements are slow. This reduction in the output voltage swing has resulted in degrading the speed performance and a large leakage current in the next stage gate, especially at sub-2 V supply voltages. In contrast, high-speed full-swing operation is achieved by the TS-FS-BiCMOS even at 1.5 V supply.

4.10.3 Circuit Performance Comparison and Analysis

Figure 4–136 shows the different implementations of the two-input NAND gate to be used in SPICE simulation for performance comparison. All circuits have an input capacitance of 60 fF, and the 0.3 μm technology was used for the simulation. Table 4–16 gives a list of all the key device parameters used in the simulations.

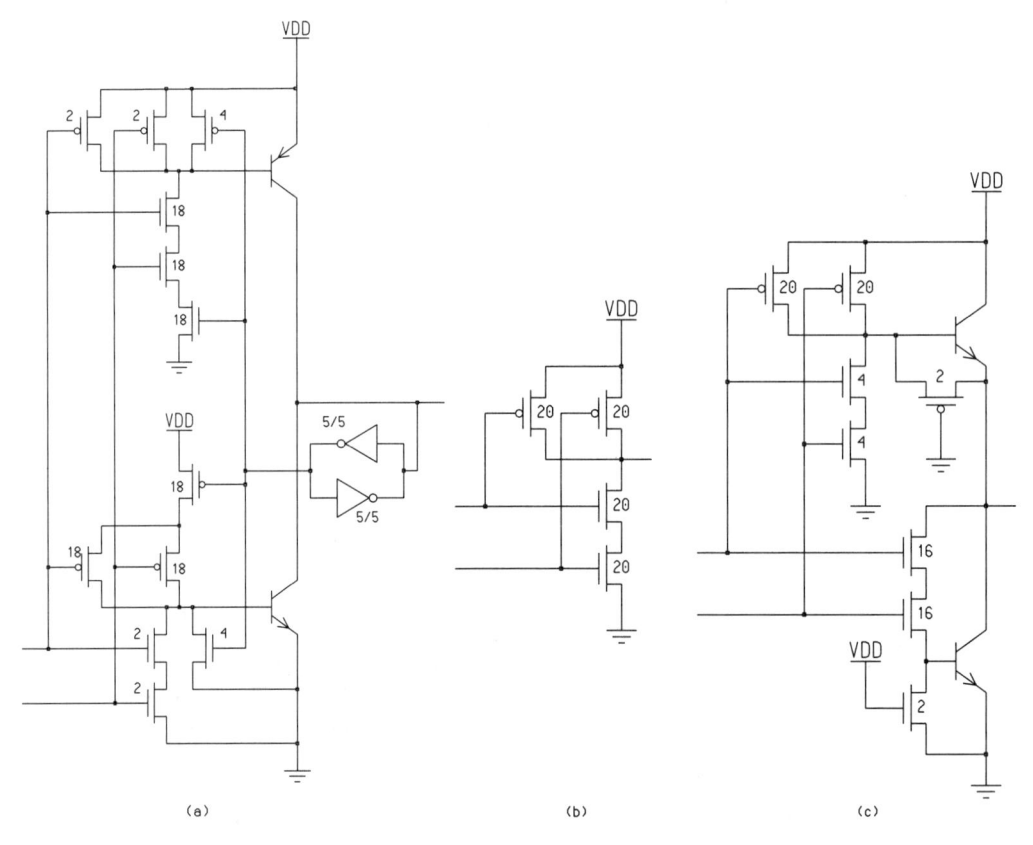

Figure 4–136 Comparison of two-input gates (a) TS-FS-BiCMOS; (b) CMOS; (c) conventional BiCMOS. (The numbers indicate gate widths of MOSFETs.)

Transiently Saturated Full-Swing BiCMOS (TS-FS-BiCMOS) Logic Circuits

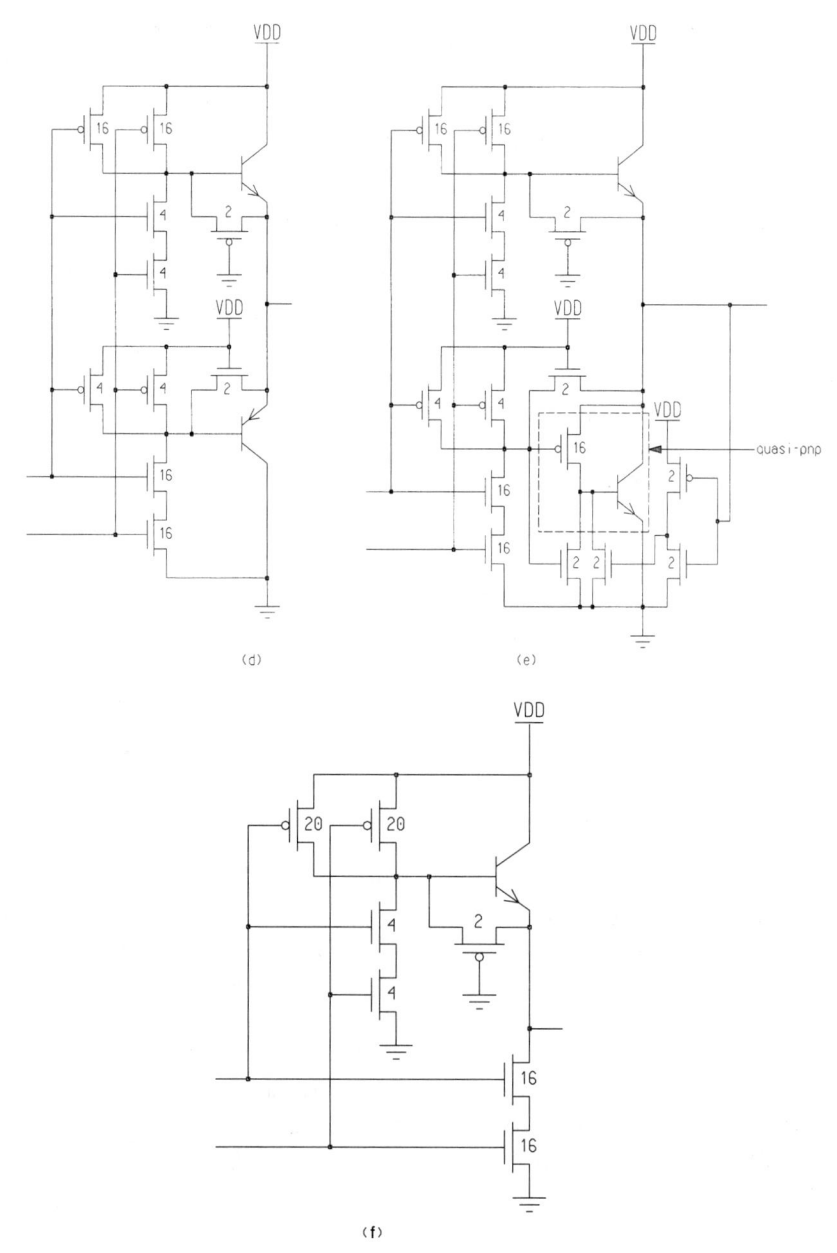

(d)

(e)

(f)

Figure 4–136 *(cont.)* (d) C-BiCMOS; (e) QC-BiCMOS; and (f) BiNMOS. (The numbers indicate gate widths of MOSFETs.)

Table 4–16 Device parameters used for SPICE simulation.

MOS	L_g	0.35 µm (nMOS), 0.35 µm (pMOS)
	L_{eff}	0.3 µm (nMOS), 0.3 µm (pMOS)
	I_{DS}	4.8 mA (nMOS), 2.4 mA (pMOS) ($V_{DS} = V_{GS} = 3.3$ V, $W = 10$ µm)
	T_{OX}	10 nm
	C_J	20 fF ($W = 10$ µm)
	V_T	0.3 V (nMOS), –0.3 V (pMOS)
Bipolar transistor	β_f	89 (npn), 89 (pnp)
	$f_{T(max)}$	15 GHz (npn), 6 GHz (pnp)
	C_{JE}	12.4 fF (npn), 23.8 fF (pnp)
	C_{JC}	10.6 fF (npn), 16.8 fF (pnp)
	R_E	25 Ω (npn), 25 Ω (pnp)
	$R_{bb'}$	300 Ω (npn), 250 Ω (pnp)
	R_C	42 Ω (npn), 42 Ω (pnp)

Figure 4–137 shows the simulated gate delay dependence on the supply voltage of a two-input TS-FS-BiCMOS AND gate, a two-input CMOS NAND gate, and other two-input BiCMOS NAND gates. Below 2 V supply, TS-FS-BiCMOS shows the best delay performance, and it is twice as fast as the CMOS at 1.5 V supply. In contrast, the QC-BiCMOS and BiNMOS lose their performance advantages over the CMOS at about 2 V owing to one V_{BE} voltage loss. The conventional BiCMOS circuit loses its performance advantage over the CMOS at about 3 V due to the 2 V_{BE} voltage loss. Figure 4–138 shows the simulated gate delay dependence on the load capacitance for a supply voltage of 1.5 V.

The dependence of the gate delay on the load capacitance at V_{DD} of 1.5 V is weaker for the TS-FS-BiCMOS than for all other BiCMOS circuits due to its high-speed full-swing operation. All the previously mentioned performances of the TS-FS-BiCMOS were based on a 0.3 µm technology. In order to investigate the circuit performance for other technologies, the performance improvement due to device scaling needs to be studied. Figure 4–139 shows the simulated gate delay time as a function of the technology as well as the supply voltage. The supply voltage must be scaled along with the devices to avoid hot-carrier effects [75]. The results reveal that the performance of the TS-FS-BiCMOS improves down to less than 0.1 µm, but the performance of the C-BiCMOS starts to degrade at less than 0.2 µm due to the loss of its base-emitter voltage.

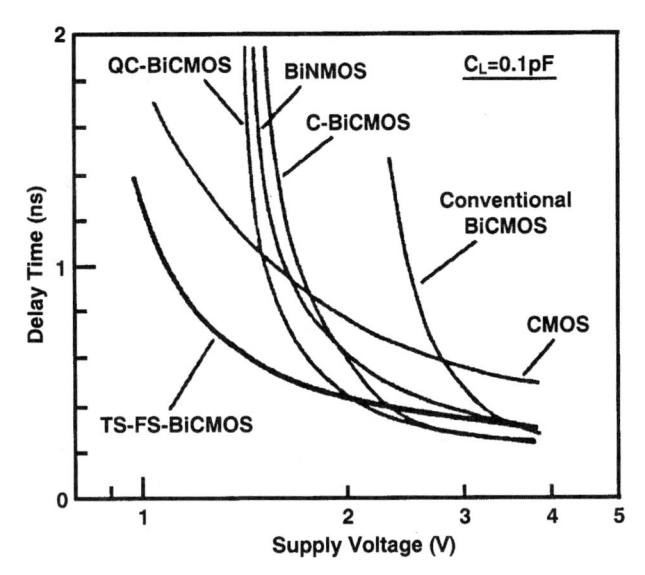

Figure 4–137 Simulated dependence of gate delay time on supply voltage of various two-input gates.

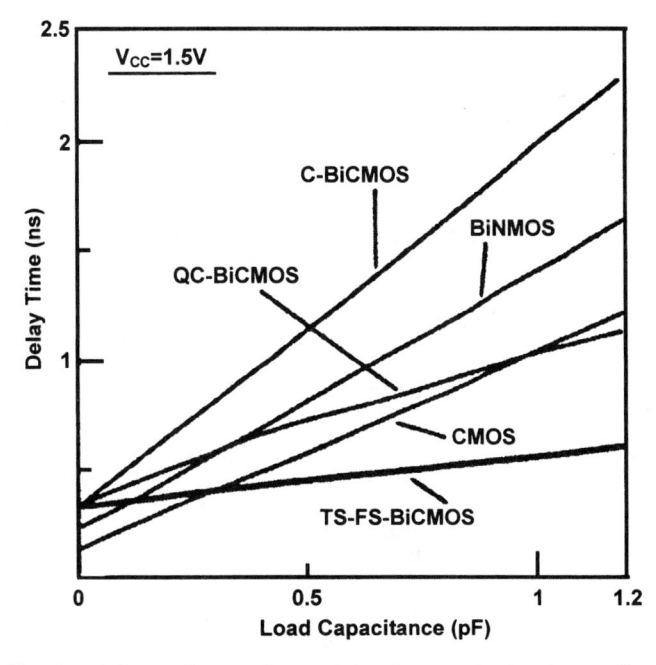

Figure 4–138 Simulated dependence of gate delay time on the load capacitance of various two-input gates.

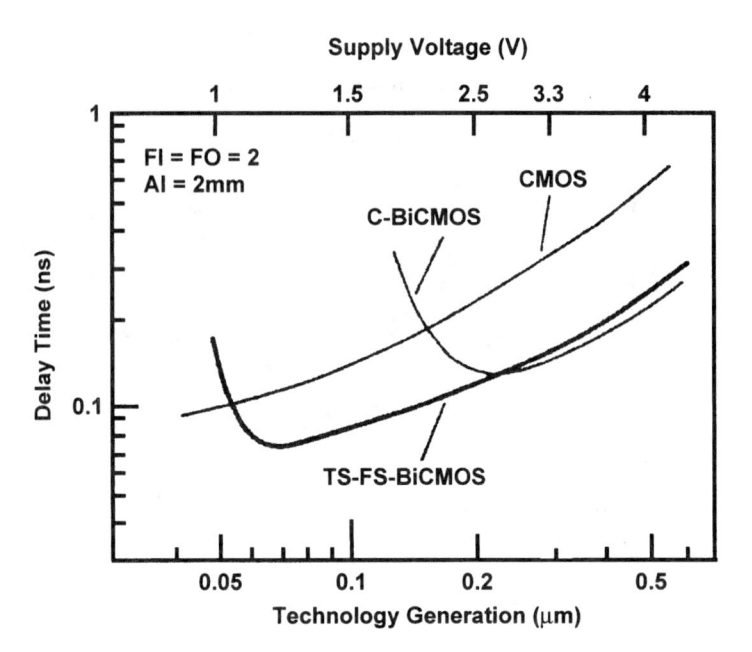

Figure 4–139 Simulated delay time for each technology generation.

4.10.4 Design Issues

The TS-FS-BiCMOS utilizes the feedback signal to discharge the excess minority carriers; therefore, the feedback CMOS inverter must be properly designed. Figure 4–140 illustrates the simulated gate delay time, Td, as a function of the input transition interval, Tp, for different gate width, Wn, of the nMOS in the feedback CMOS inverter.

Notice that when the gate width Wn is reduced to about 1 μm, a significant increase in the gate delay time is observed. This increase is due to the fact that the base-carrier discharging is not yet completed when the next transition starts. To rectify this problem, the feedback CMOS inverter must be designed so that it has enough driving capability. For instance, the increase in the delay-time can be reduced to only 10% by increasing Wn from 1 μm to 5 μm. Table 4–17 shows the comparison of the TS-FS-BiCMOS, C-BiCMOS and CMOS circuits in terms of their device count and relative gate area.

The gate area of the TS-FS-BiCMOS is 1.4 times larger than that of the C-BiCMOS and 3.5 times larger than that of the CMOS because of the larger number of devices in the TS-FS-BiCMOS. But this drawback is offset by its high performance. Figure 4–141 shows the simulated dependence of the gate delay time on the fan-in for the TS-FS-BiCMOS and CMOS.

The results demonstrate that the fan-in dependence of the TS-FS-BiCMOS delay is smaller than that of the CMOS. This feature of the TS-FS-BiCMOS allows it to achieve a very

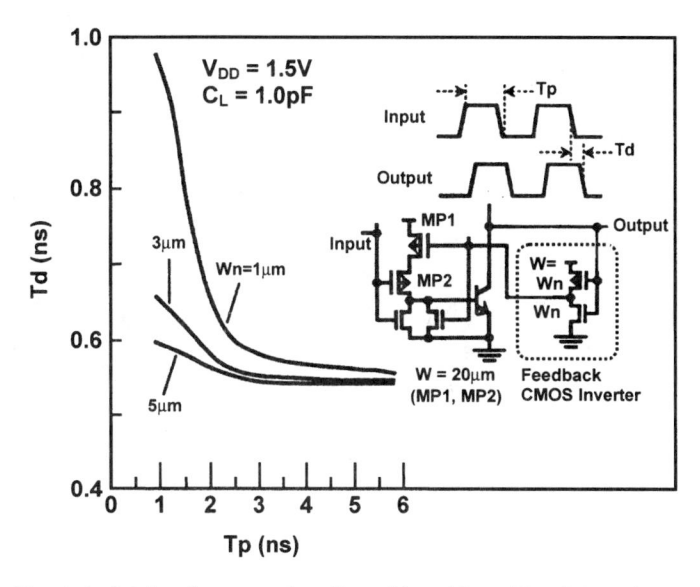

Figure 4–140 Simulated delay time as a function of input transition interval.

Table 4–17 Device count and gate area of 2-input gates.

	Device Count		
	MOS	**Bipolar**	**Relative Area**
TS-FS-BiCMOS	16	2	3.5
C-BiCMOS	10	2	2.5
CMOS	4	–	1

complex logic function with only a single-stage gate, whereas several stage gates are needed in the CMOS for the same logic function. This is another advantage of the TS-FS-BiCMOS, which will offset its gate area disadvantage. The reason why the fan-in dependence of the TS-FS-BiCMOS delay is smaller than that of CMOS can be explained with the aid of Figure 4–142.

The difference in the fan-in dependence originates from the difference of the voltage swing across the parasitic capacitance, Cd. As the fan-in increases, Cd for both circuits will also increase causing an increase in the gate delay time. However, the voltage swing across Cd of the TS-FS-BiCMOS is only one V_{BE}, whereas that of the CMOS can be as large as V_{DD}. This results in a smaller fan-in dependence for the TS-FS-BiCMOS than for CMOS.

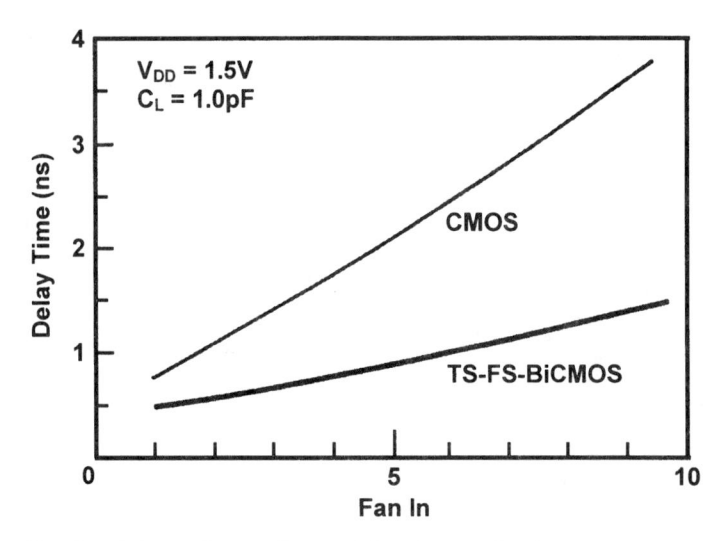

Figure 4–141 Simulated dependence of gate delay time on fan-in.

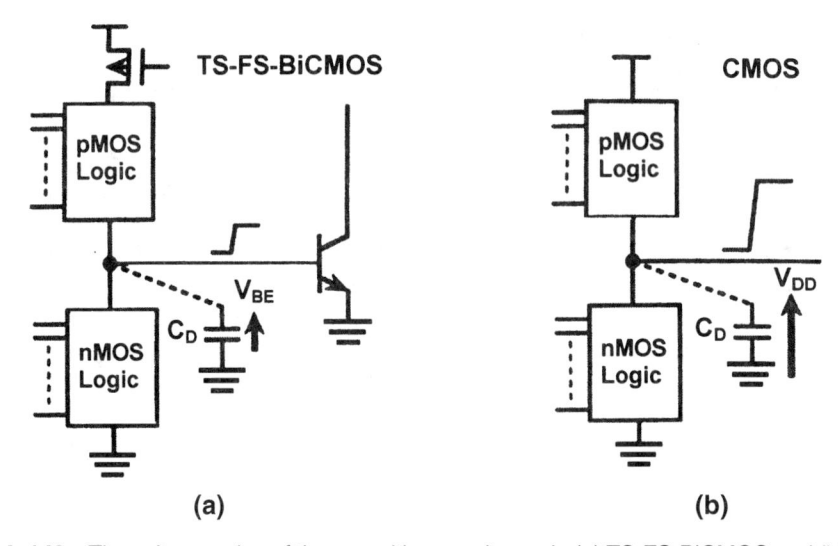

Figure 4–142 The voltage swing of the parasitic capacitance in (a) TS-FS-BiCMOS and (b) CMOS.

4.10.5 Experimental Results and Analysis

The TS-FS-BiCMOS, conventional BiCMOS, C-BiCMOS, BiNMOS, QC-BiCMOS, and CMOS were fabricated using a 0.3 μm BiCMOS technology with triple-poly-si and double-metal process. Figure 4–143 shows a microphotograph of a test chip used to measure the gate delay. The measurements were performed using a 17-stage gate chain.

Bootstrapped BiCMOS Logic Circuits 279

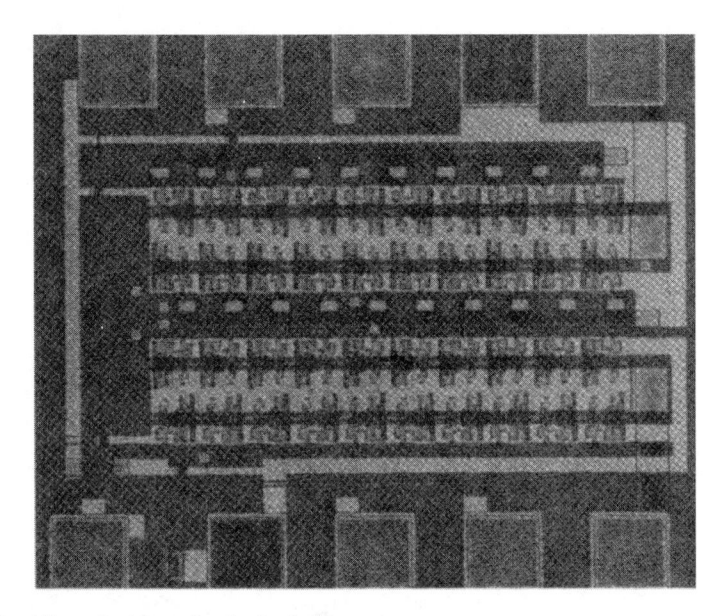

Figure 4-143 Microphotograph of a test chip.

Figure 4-144 shows the measured input and output waveforms of a loaded TS-FS-BiCMOS gate chain with a supply voltage of 1.5 V. Figure 4-145 shows the measured dependence of the gate delay-time on the supply voltage for the TS-FS-BiCMOS, conventional BiCMOS, C-BiCMOS, QC-BiCMOS, and BiNMOS. Below 2 V power supply, the TS-FS-BiCMOS has the lowest delay time. These measured results again demonstrate the performance advantage of TS-FS-BiCMOS over the performances of other BiCMOS circuits at sub-2 V supply. Figure 4-146 shows the measured dependence of the gate delay time on the load capacitance for the TS-FS-BiCMOS and CMOS operating at a power supply voltage of 1.5 V. The fabricated TS-FS-BiCMOS has shown to be twice as fast as the CMOS even at a supply voltage of 1.5 V. In conclusion, the TS-FS-BiCMOS logic circuit has been proven (through circuit simulation and experiment) to be far superior, in terms of speed, to BiCMOS and CMOS circuits at sub-2 V supply voltages.

4.11 Bootstrapped BiCMOS Logic Circuits

4.11.1 Introduction

As the supply voltage is scaled below 3.3 V, the conventional BiCMOS loses its performance advantage over that of the CMOS due to the 2 V_{BE} loss in its output voltage swing [34]. Several improvements to the conventional BiCMOS design have been reported to achieve good performance of BiCMOS circuits operating at 3.3 V power supply voltage and below [3,35].

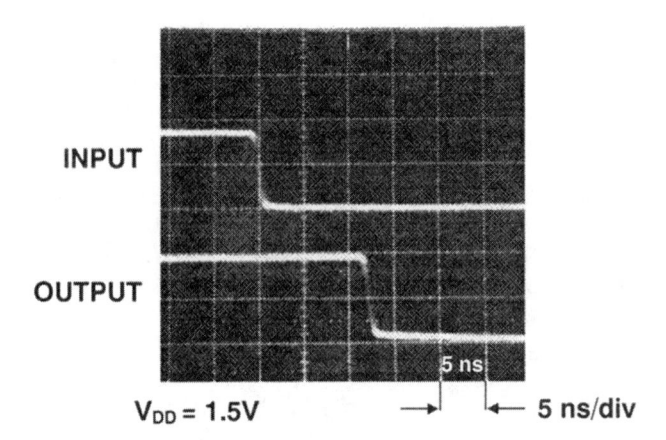

Figure 4-144 Measured voltage waveforms of a TS-FS-BiCMOS gate chain.

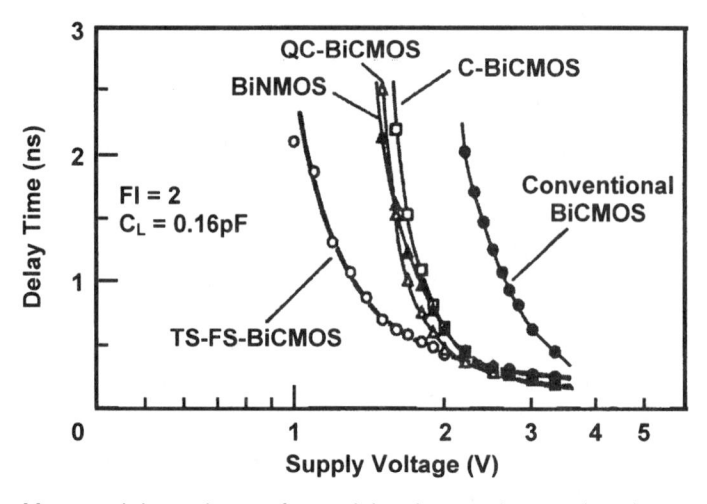

Figure 4-145 Measured dependence of gate delay time on the supply voltage.

Although the MBiCMOS [3] has been proven to improve the BiCMOS logic gate performance at sub-3 V supply, it depends on certain special processing steps to alter the threshold voltage of the MOS device to enable it to operate below 2 V supply. A 1.5 V full-swing BiCMOS circuit using complementary bipolar devices was reported [29]. However, due to both the high cost of complementary BiCMOS technology and the fact that to obtain high performance pnp devices out of a BiCMOS process is difficult, alternative designs were proposed [76–79]. Basically, these circuits employed a common technique, bootstrapping, to enhance their circuits' performance down to a power supply voltage of 1.1 V. Each design will be analyzed in greater detail in the subsequent sections.

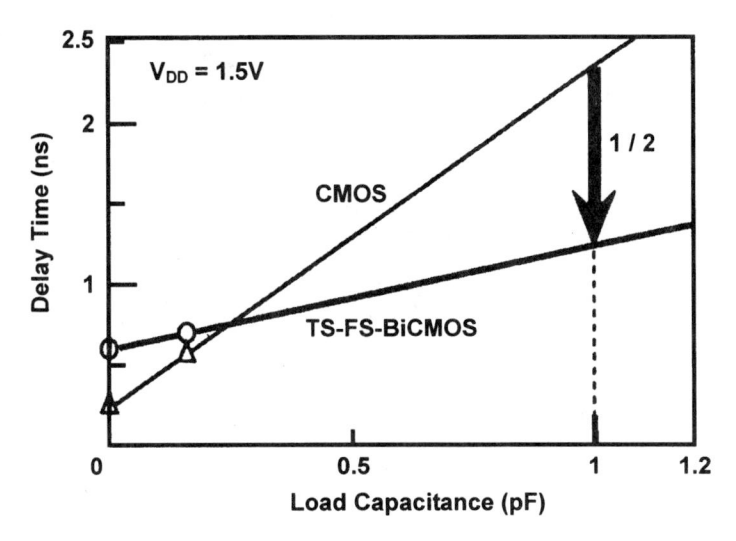

Figure 4-146 Measured dependence of gate delay time on the load capacitance.

4.11.2 1.5 V Bootstrapped BiCMOS Logic Gate (BS-BiCMOS)

The BS-BiCMOS inverter [76], as shown in Figure 4-147, uses the bootstrapping technique to pull up the output voltage to V_{DD} and the transient saturation technique to pull down the output voltage to ground. It operates at a power supply voltage of 1.5 V and gives a transient logic swing of 1.1 V.

4.11.2.1 Circuit Operation

During the input low-to-high transition (pull-down), Pbs turns on causing Cbs to be charged up to V_{DD}. The transistor Pb1 shuts off and subsequently turns transistor T1 off. Meanwhile transistor Pb2 turns on but transistor Nb2 turns off. Since the input is initially low, the high-output voltage causes transistor Pb3 to turn on and transistor Nb3 to turn off. Hence, current flows through transistors Pb2 and Pb3 into the base of transistor T2 causing it to saturate until the output voltage is pulled down to a low value. When this occurs, Pb3 turns off via the feedback inverter, while Nb3 turns on to discharge the base of T2, turning it off and preventing a short circuit current in T1 and T2 during the next pull-up that could cause unnecessary power dissipation.

When the input voltage changes from high to low (pull-up), Pb1 turns on and Nb1 turns off. The base of T1 is bootstrapped to a value above V_{DD} through the bootstrapping capacitor Cbs. The output voltage is subsequently charged up to a value greater than the $(V_{DD} - V_{BE(T1)})$ and pulled up gradually to V_{DD} via Pfb.

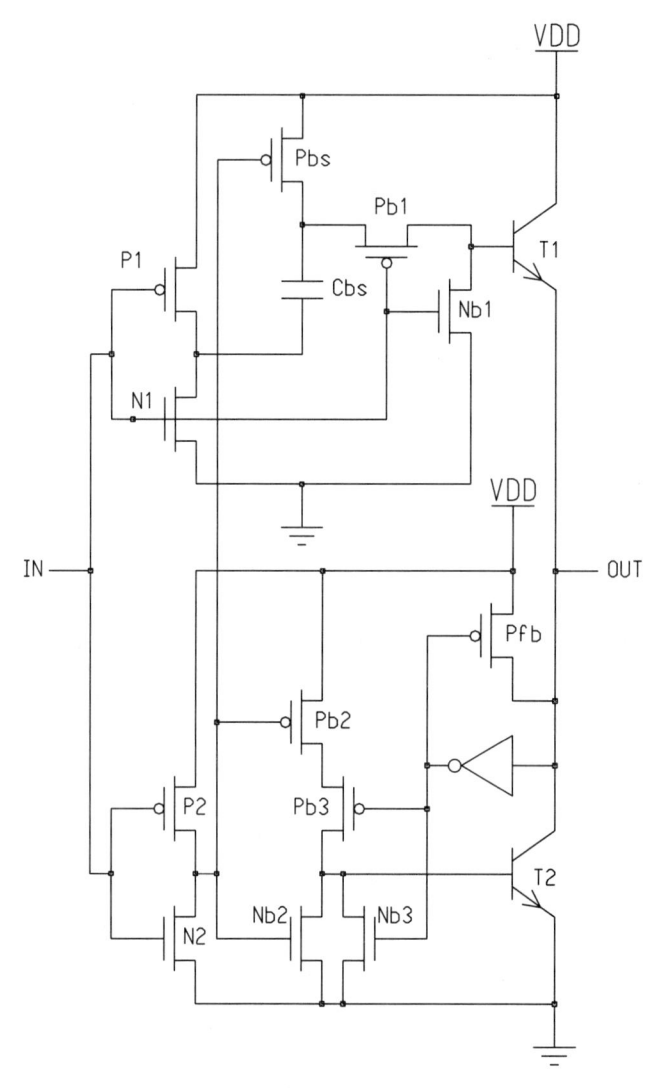

Figure 4–147 BS-BiCMOS inverter.

4.11.2.2 Simulation Results

Figure 4–148 compares the dependence of the inverter delay on different supply voltages for four different BiCMOS configurations. It shows that for sub-3 V supplies and a load capacitance of 1 pF, the BS-BiCMOS is much faster than all other configurations due to the bootstrapping technique employed to enhance both the circuit switching speed and the output voltage swing. The BS-BiCMOS can be realized in a conventional BiCMOS process with only CMOS and npn transistors. It does not require, unlike the MBiCMOS, any special processing steps to alter the MOS threshold voltage to operate at a supply voltage of less than 2 V. However, some of its major disadvantages are the poor bootstrapping feature when the threshold voltage of the MOS devices is low, a large crossover capacitance, and the large count of input transistors, which limits its use for multi-input gates.

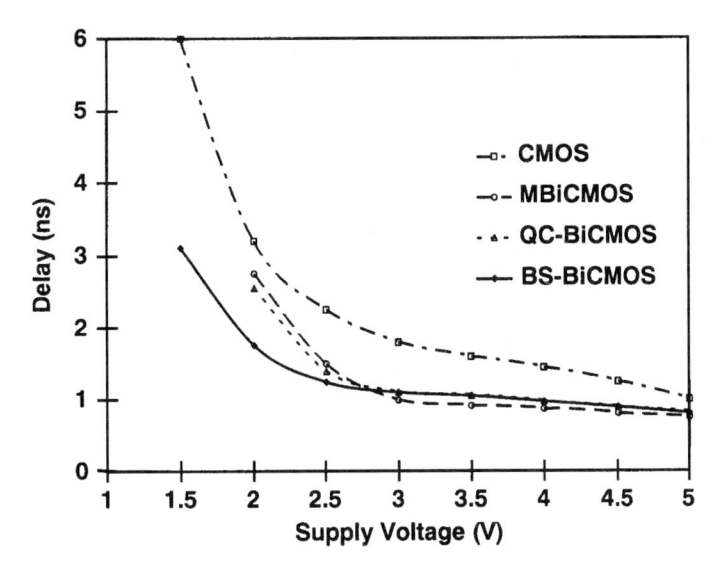

Figure 4–148 Comparison of dependence of the inverter delay on supply voltages for different types of design.

4.11.3 Bootstrapped Full-Swing BiCMOS/BiNMOS Inverter

To overcome the shortcomings of the BS-BiCMOS inverter, two circuit configurations were reported [78], which were designed to operate in the 1.2–3.3 V range. Figure 4–149 illustrates the bootstrapped full-swing BiCMOS inverter (BF-BiCMOS).

4.11.3.1 Circuit Operation

When the input rises from low to high (pull-down), the output voltage is initially high. The transistors MPf and MP2 turn on and subsequently switch the transistor Q2 on, and the output node is pulled down towards ground. But as the output voltage decreases near ground level, the inverter I3 turns MPf off and MNd on. The conduction of MNd depletes the base charge of Q2 and consequently switches Q2 off. Therefore, the output voltage level is maintained by the two inverters I2 and I3 forming the CMOS latch.

During the pull-up phase, the input is initially high and the output voltage is low, the transistor MPd turns on, and the base of transistor Q1 is precharged to V_{T_p} (which is less than V_{BEon} of Q1). At the same time, the transistor MPp turns on to charge the bootstrapped capacitor,

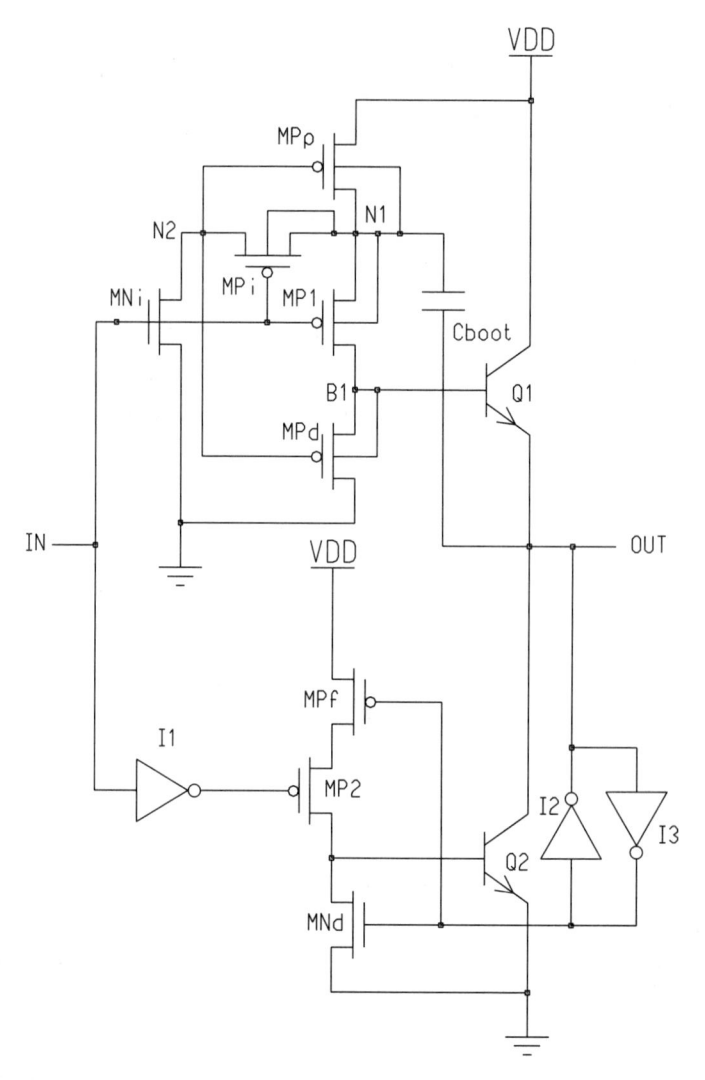

Figure 4–149 Bootstrapped full-swing BiCMOS inverter (BF-BiCMOS).

C_{boot}, to V_{DD}, also known as the precharge cycle. When the input becomes low, MP1 switches on and since the base-emitter junction of transistor Q1 has been precharged earlier near V_{BEon}, it switches on immediately. The output node then starts to charge the load capacitor C_L towards V_{DD}.

When the input is low, the transistor MPp switches off, disconnecting the node N1 from V_{DD}. As the output voltage rises to V_{DD}, the voltage at the node N1 increases to $(V_{DD} + V_{BEon})$. This process is known as the bootstrapping cycle. The bootstrapping cycle prevents the discharge of the bootstrapped node through reverse current from the node N1 to V_{DD}. Also during this cycle, the device MPi turns on and the gate of the precharge transistor MPp is pulled up towards the voltage of the node N1.

In comparison to the BS-BiCMOS, the BF-BiCMOS circuit has several major advantages, namely: (1) the bootstrapped capacitor is driven by the output rather than the input as in the BS-BiCMOS, (2) the gate of the precharge transistor (MP_p) is driven to ($V_{DD} + V_{BEon}$) during the bootstrapping cycle rather than just V_{DD} as in the BS-BiCMOS, and (3) a pMOS transistor MPd is used to deplete the base charge of Q1 to a precharged level of V_{Tp}, resulting in better performance.

4.11.3.2 Design Considerations

4.11.3.2.1 Bootstrapping Condition

An expression can be derived for the bootstrapping capacitor, C_{boot}, which is an important component in the bootstrapping phenomenon. Let the parasitic capacitance on the node N1 be C_p. During the precharge cycle, the total charge on the node N1 consists of charges on the bootstrapped capacitor, $V_{DD} \times C_{boot}$, and those on the parasitic capacitance, $V_{DD} \times C_p$. Hence the total charge on the node N1 during the precharge cycle is:

$$Q_{N1} = V_{DD} C_{boot} + V_{DD} C_P \qquad 4.18$$

During the bootstrapping cycle, the charge on C_p is $(V_{DD} + V_{BEon}) \times C_p$ and that on C_{boot} is $V_{BEon} \times C_{boot}$. Thus the total charge on the node N1 is given by:

$$Q_{N1}' = (V_{DD} + V_{BEon}) C_P + V_{BEon} C_{boot} \qquad 4.19$$

The base charge is given by:

$$Q_b = Q_{N1} - Q_{N1}' \qquad 4.20$$

The base charge can also be written as:

$$Q_b = I_b t_r \qquad 4.21$$

where I_b is the average base current of Q1 and t_r is the output rise time. From Eqs. 4.18 through 4.21, we obtain:

$$C_{boot} = \frac{I_b t_r}{V_{DD} - V_{BEon}} + \frac{V_{BEon} C_P}{V_{DD} - V_{BEon}} \qquad 4.22$$

Eq. 4.22 shows that as the power supply is scaled down, the bootstrapping capacitance, C_{boot}, must be increased. But when device scaling accompanies power supply scaling, the output rise time, t_r, drops and enables C_{boot} to be kept small.

Circuit simulations were performed using the 0.35 μm BiCMOS process, which has been optimized for a power supply voltage of 3.3 V to illustrate the effect of C_{boot} on the output signal at a supply voltage of 1.5 V. The simulation results are shown in Figure 4–150(a). Table 4–18 shows the key device parameters for the 0.35 μm BiCMOS process.

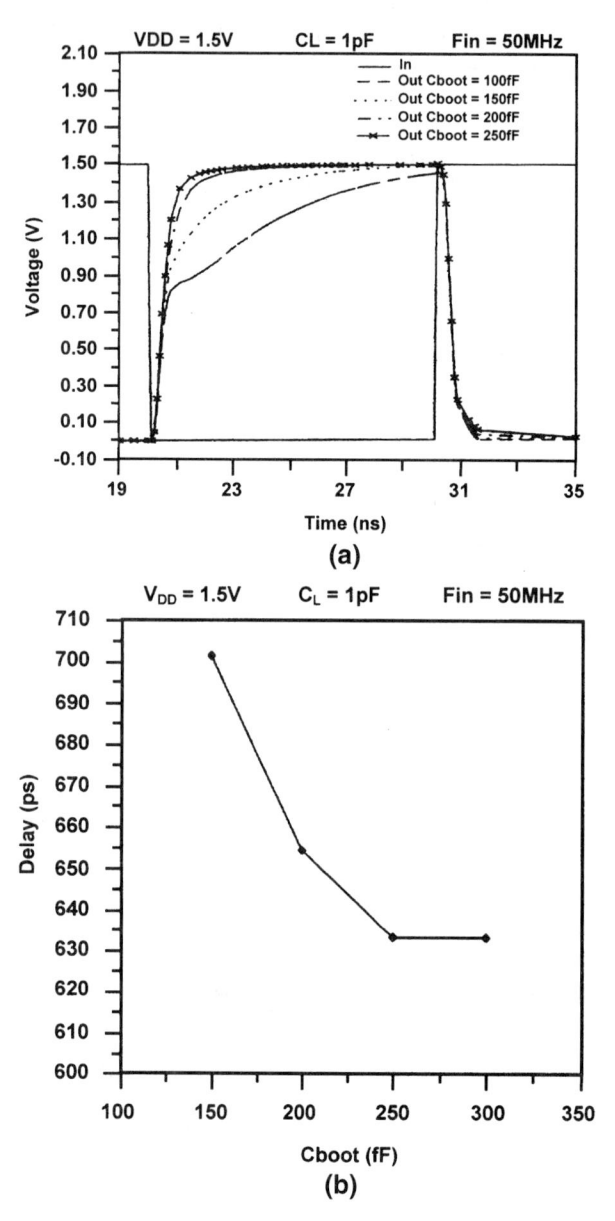

Figure 4–150   (a) Effect of C_{boot} on the output signal (b) Delay of BF-BiCMOS inverter vs. C_{boot}.

Table 4–18 Key device parameters for the 0.35 μm BiCMOS process.

MOS	nMOS	pMOS
L_g	0.35 μm	0.35 μm
L_{eff}	0.23 μm	0.34 μm
I_{DS}	4.9 mA	2.4 mA
	@ $V_{DS} = V_{GS}$ = 3.3 V, W = 10 μm	
T_{OX}	9 nm	9 nm
V_T	0.55 V	−0.65 V
Bipolar	**npn**	**pnp**
A_E	1×5 μm^2	1×4 μm^2
β_f	90	90
τ_f	7 ps	21 ps
C_{JE}	16 fF	28 fF
C_{JC}	17 fF	28 fF
C_{JS}	52 fF	73 fF
R_E	30 Ω	37 Ω
R_C	28 Ω	31 Ω
R_B	265 Ω	260 Ω

From Figure 4–150(a), the optimum C_{boot} that gives the least distortion to the output signal is 250 fF. Figure 4–150(b) shows the delay of four identical BF-BiCMOS inverters with different C_{boot} values. It is clear from Figure 4–150(b) that C_{boot} equals 250 fF gives the lowest delay. Table 4–19 shows the optimum value of C_{boot} for various supply voltages of the BF-BiCMOS.

Table 4–19 Optimum C_{boot} vs. supply voltage of BF-BiCMOS (@ 50 MHz).

V_{DD} (V)	3.3	2.5	2.0	1.5	1.2	1.0
C_{boot} (fF)	100	100	150	250	400	650

4.11.3.2.2 Implementation of the Bootstrapping Capacitance The bootstrapping capacitance can be implemented using a pMOS transistor with its source, drain, and body tied together, as illustrated in Figure 4–151.

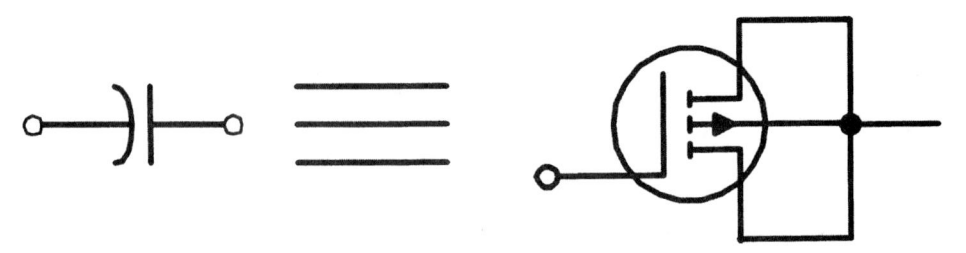

Figure 4–151 Implementation of the bootstrapping capacitor using a pMOS transistor.

Simulations have shown that for the 1.5 V supply voltage, a full output voltage swing can be easily achieved if the width and the channel length of the bootstrapped pMOSFET are equal to 13 μm and 6 μm, respectively.

4.11.3.2.3 Bulk/Well Biasing From Figure 4–149, it can be seen that transistors MPi, MPp, and MP1 share a common n-well, which is connected to the bootstrapped node N1. This has the effect of preventing their source/drain-well junctions from being forward-biased during the bootstrapping cycle, which would otherwise activate the low-impedance parasitic thyristor (SCR) and cause latch-up. Figure 4–152 illustrates the parasitic SCR in relation to latch-up.

4.11.3.2.4 Effect of the Parasitic PNP Transistor The parasitic pnp transistor, as shown in Figure 4–153, is formed by the p-base, the n-collector and the p-substrate. Although this transistor was taken into consideration during simulation, further analysis [77] indicates that its effect on the circuit performance is insignificant. The reasons are first the output would have reached its steady state level by the time the parasitic pnp transistor turns on. Second, due to the low current gain and high forward base transit time of the parasitic pnp transistor, its collector current is negligible.

The BF-BiCMOS gate was simulated with and without the parasitic pnp transistor. The result indicated that there was no substantial impact on the performance of the circuit.

4.11.3.3 BiNMOS Version of the Bootstrapped Circuit

The BF-BiNMOS configuration is illustrated in Figure 4–154. Unlike the BF-BiCMOS configuration, the pull-down section of the Bi-NMOS version uses only one nMOS transistor, MN1.

Another feature of the BF-BiNMOS is that the pull-up section includes a pMOS transistor (MPf). This transistor, together with inverter I3, allows the voltage of nodes N1 and B1 to be set to V_{DD} towards the end of the bootstrapping cycle. Therefore the base-emitter voltage of Q1 is almost equal to zero at the end of bootstrapping. For low-output loads and for an input transition

Bootstrapped BiCMOS Logic Circuits 289

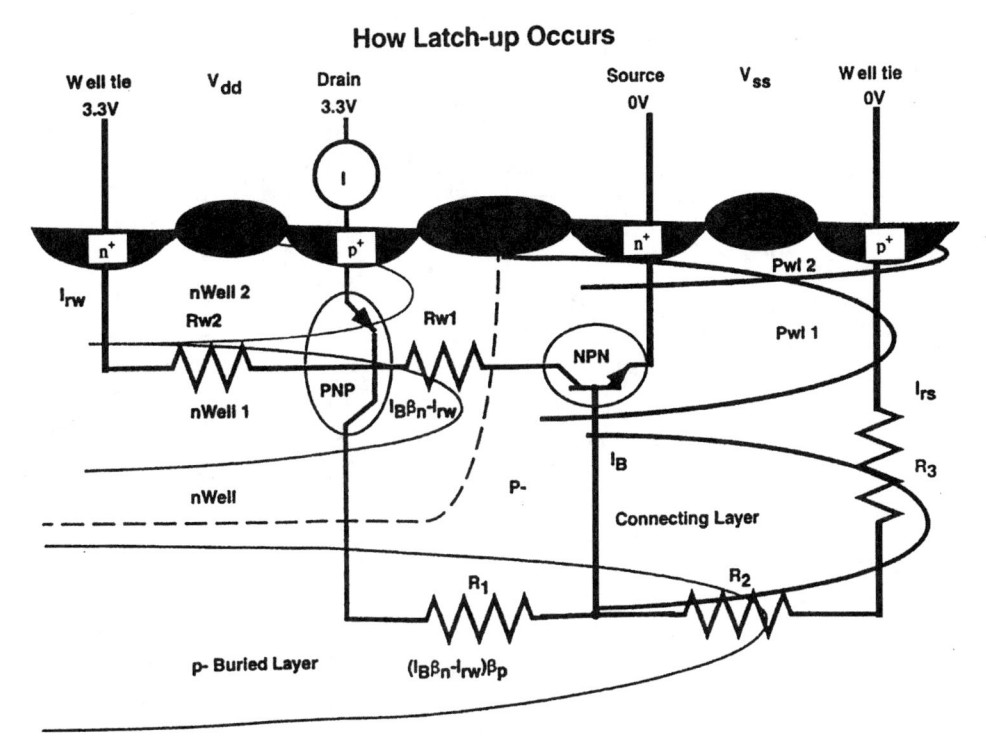

Figure 4–152 Cross-section of CMOS buried/connecting layer architecture indicating current paths during latch-up.

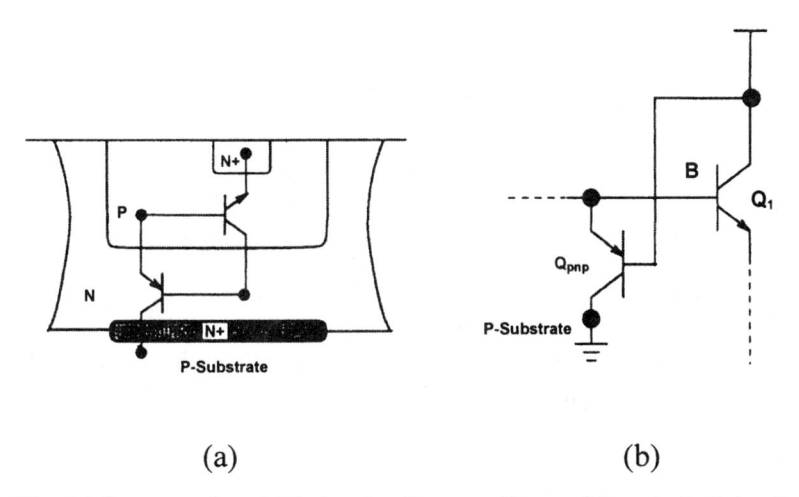

(a)   (b)

Figure 4–153 (a) Cross-section of Q1 showing the parasitic pnp (b) equivalent circuit of Q1 with the parasitic pnp.

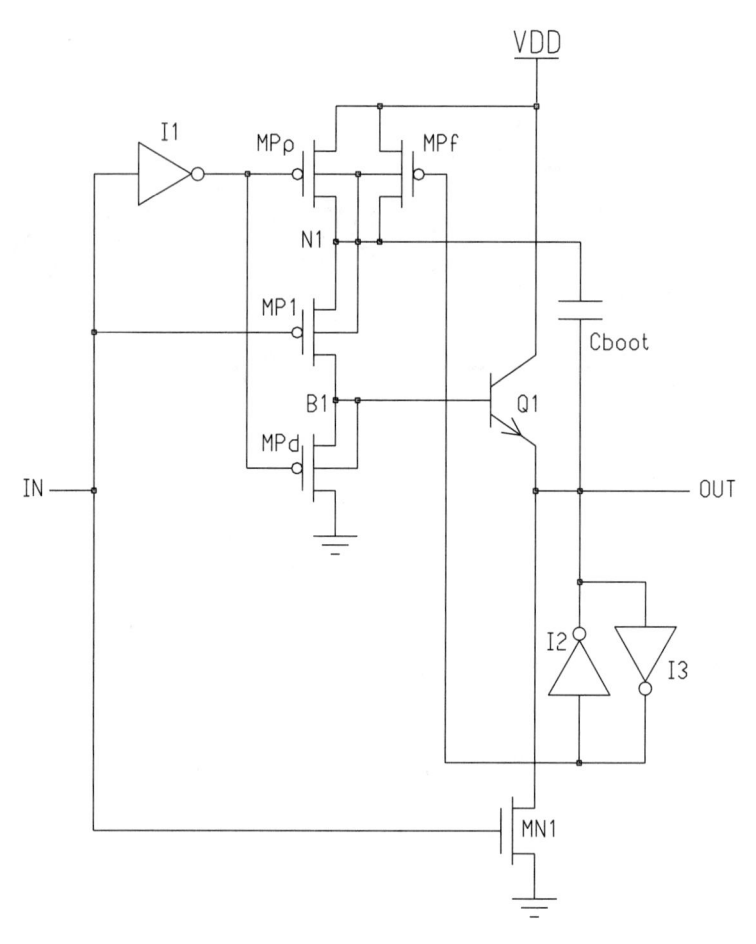

Figure 4–154 Bootstrapped full-swing BiNMOS inverter (BF-BiNMOS).

from low to high, the output node discharges through the pull-down nMOS device MN1 without turning on the transistor Q1. The turning off of Q1 aids in reducing the fall time delay. Figure 4–155 shows the HSPICE simulations of the input voltage, the output voltage, and the base voltage of Q1 for the BF-BiNMOS inverter.

4.11.3.4   Circuit Performance Comparison

HSPICE simulations [44] were used to compare the circuit performance of the BF-BiCMOS and BF-BiNMOS configurations with other members of the logic family, such as the CMOS, BiNMOS, TS-FS-BiCMOS, and BS-BiCMOS. Figure 4–156 shows the two-input NAND gate configurations chosen to evaluate the circuit performance.

Bootstrapped BiCMOS Logic Circuits

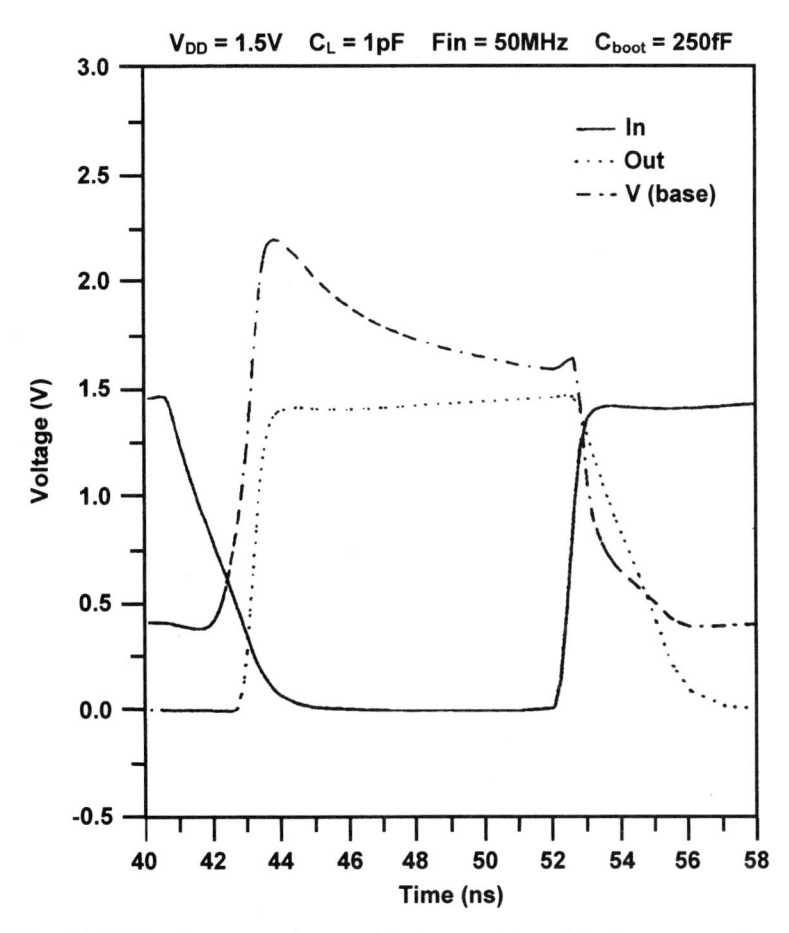

Figure 4–155 HSPICE voltage waveforms of the input voltage(in), the output voltage(out), and the base voltage of Q1 for the BF-BiNMOS inverter.

The simulations were carried out using a chain of four identical gates with the input to the first gate being a signal of 50 MHz frequency. The delay times were extracted from the simulations by measuring the 50% delay time of the third gate in each chain. The 0.35 μm BiCMOS technology (refer to Table 4–18) was used to perform the simulation.

4.11.3.4.1 Delay Circuit delays at supply voltages of 3.3 V and 1.5 V are illustrated in Figure 4–157.

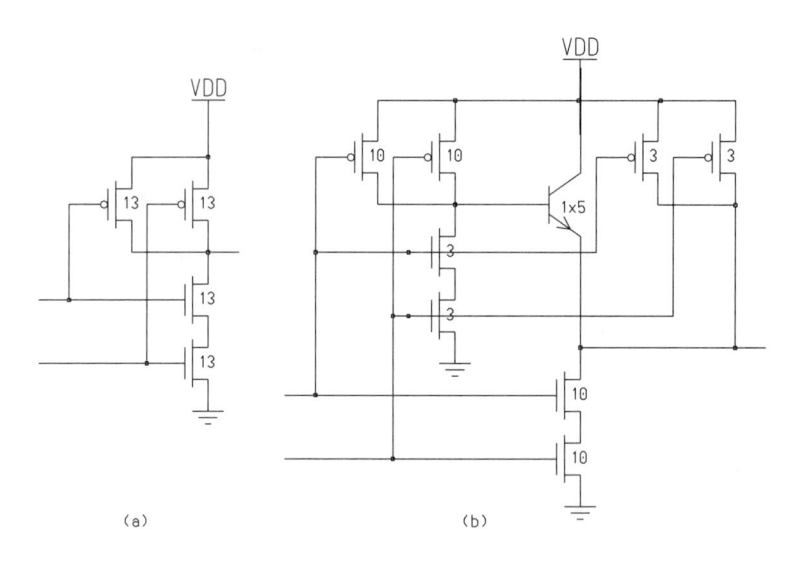

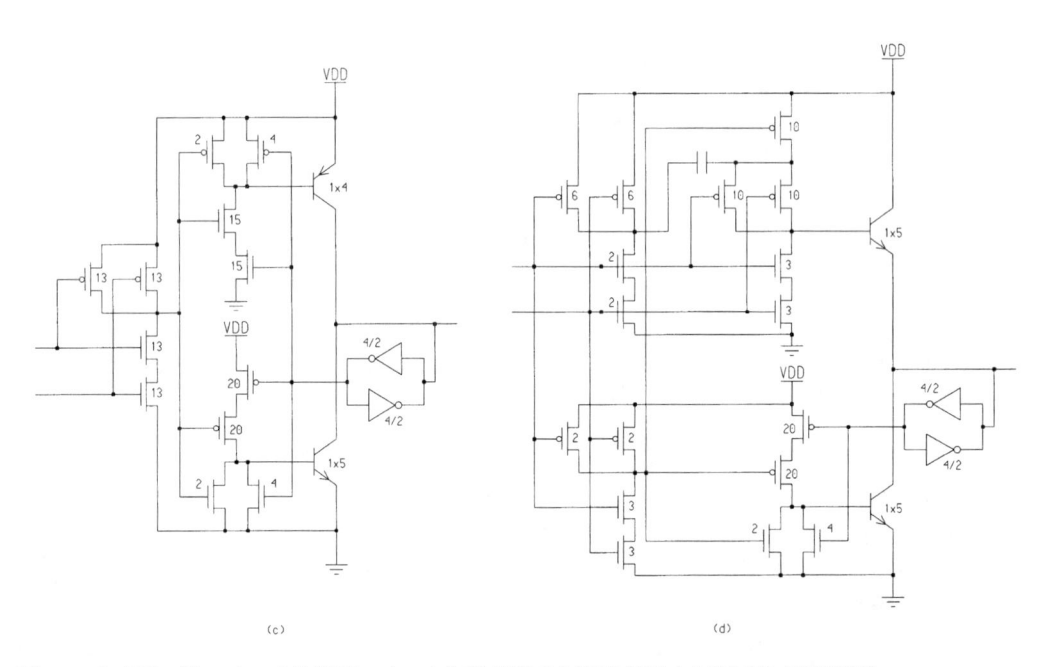

Figure 4–156 Two-input NAND gates (a) CMOS (b) BiNMOS (c) TS-FS-BiCMOS (d) BS-BiCMOS.

Bootstrapped BiCMOS Logic Circuits

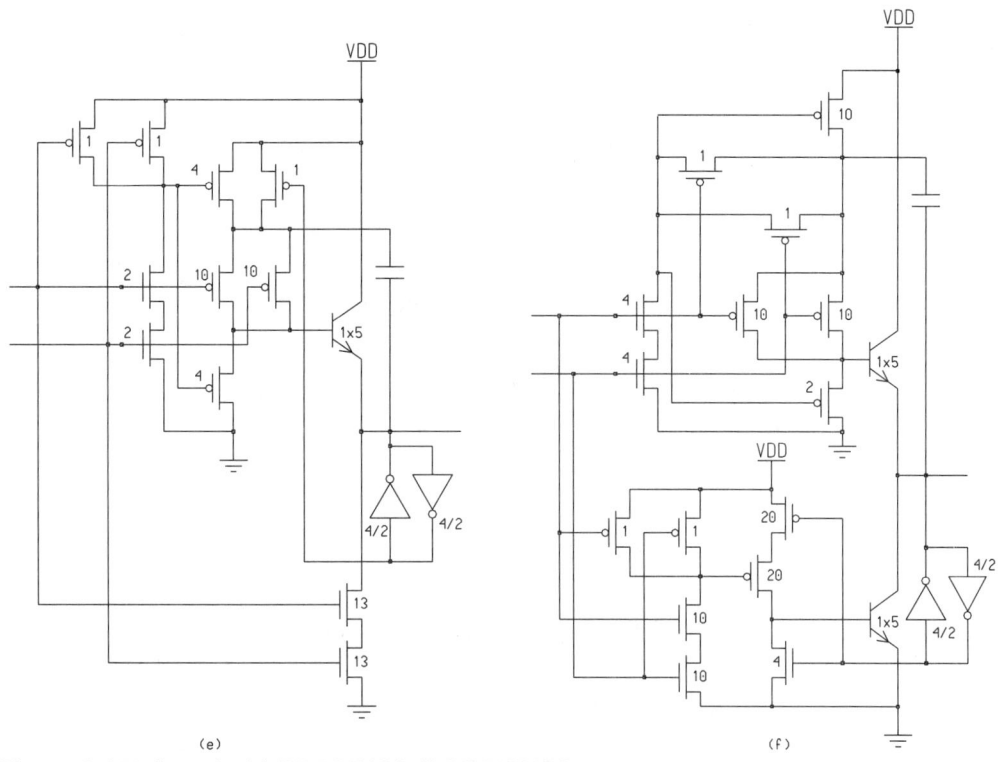

(e)

(f)

Figure 4–156 (cont.) (e) BF-BiNMOS (f) BF-BiCMOS.

Figure 4–157(a) shows that at a load capacitance of 0.3 pF and above, BF-BiCMOS is faster than are all the other gates. For a small load capacitance of 0.2 pF and below, the BiNMOS is the fastest among all the BiCMOS families. Figure 4–157(b) shows that beyond a load capacitance of 0.15 pF, the BF-BiCMOS has the least delay at a supply voltage of 1.5 V. With the CMOS as the reference circuit, bar charts were plotted in Figure 4–158 to illustrate the percentage delay reduction of the other configurations over the CMOS circuit at various supply voltages. Their percentage delay reduction was calculated using the expression:

$$(\frac{\text{Delay}_{\text{CMOS}} - \text{Delay}_{\text{CKT}}}{\text{Delay}_{\text{CMOS}}}) \times 100\%$$

Figure 4–158(a) shows that for $C_L = 0.2$ pF and at a supply voltage of 3.3 V, the BF-BiCMOS and BF-BiNMOS circuits exhibit the most delay reduction. For supply voltages below 3.3 V, only the BF-BiCMOS and BF-BiNMOS circuits have the best delay reduction. The performance of the BiNMOS circuit degrades for supply voltages of 2 V and below. The BS-BiCMOS and TS-FS-BiCMOS circuits exhibit negative delay reductions at all different supplies.

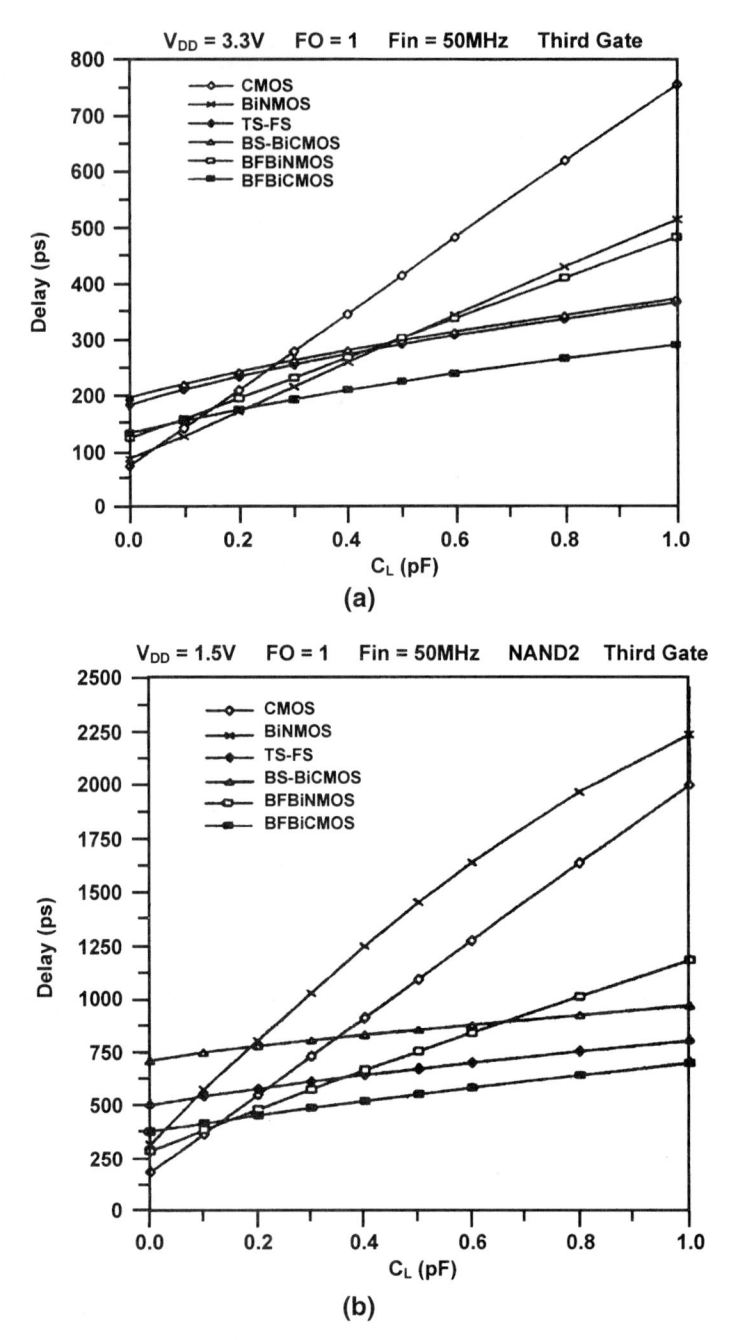

Figure 4–157 Delay of two-input CMOS, BiNMOS, TS-FS-BiCMOS, BS-BiCMOS, BF-BiNMOS and BF-BiCMOS NAND gates at (a) 3.3 V and (b) 1.5 V vs. the load capacitance.

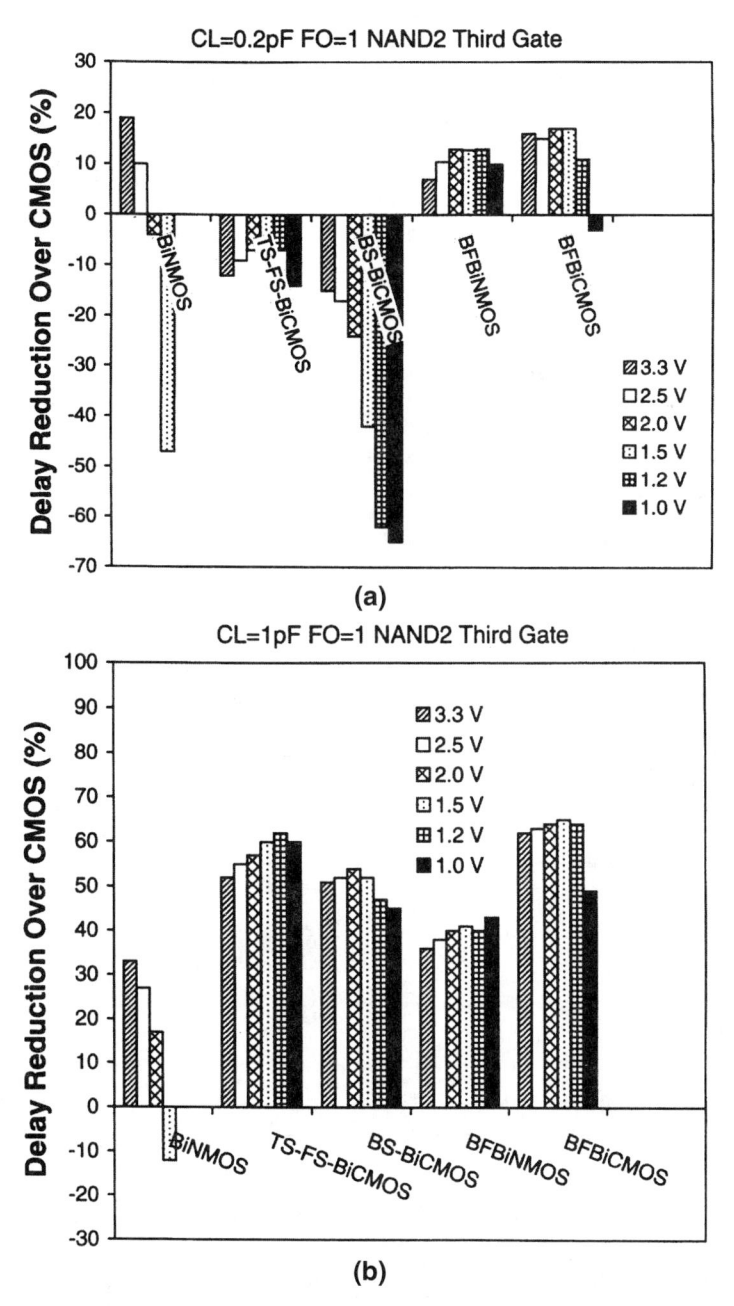

Figure 4–158 Bar charts of the percentage delay reduction over CMOS at 3.3 V, 2.5 V, 2 V, 1.5 V, 1.2 V, and 1 V supply voltage for two different C_L values: (a) 0.2 pF (b) 1 pF.

Figure 4–158(b) shows that for $C_L = 1$ pF, almost all configurations exhibit reasonable performance at various supplies. The BF-BiCMOS has the best delay performance as compared with those of other circuit configurations except at 1 V where the TS-FS-BiCMOS circuit is much faster.

4.11.3.4.2 Area The transistor count and area ratio of the two-input NAND gates shown in Figure 4–156 were compared and shown in Table 4–20. BS-BiCMOS has the largest number of device count, which limits its use for multi-input gates. Although the CMOS gate has the least device count, its poor delay performance does not qualify it for high-speed operation. The area ratios of both the BF-BiCMOS and BF-BiNMOS are reasonable considering their superior delay performance.

Table 4–20 Device count and area ratio of two-input NAND gates.

Circuit Configuration	MOS	Bipolar	Area Ratio	C_{boot} Area Percentage
CMOS	4	–	1	–
BiNMOS	8	1	1.5	–
TS-FS-BiCMOS	16	2	3.5	–
BS-BiCMOS	22	2	4.0	6.5%
BF-BiCMOS	20	2	3.9	6%
BF-BiNMOS	16	1	2.7	15%

4.11.3.4.3 Crossover Capacitance The crossover capacitance is a measure of the load at which the BiCMOS circuits begin to have a speed advantage over the CMOS. Figure 4–159 shows plots of the crossover capacitance as a function of the supply voltage for different circuit configurations.

The TS-FS-BiCMOS and BS-BiCMOS circuits have a substantially higher crossover capacitance than the BF-BiCMOS and BF-BiNMOS have. In the voltage range of 1.2 V to 3.3 V, the BF-BiCMOS and BF-BiNMOS need an equivalent minimum fan-out of 5 in order to outperform the CMOS (since the crossover capacitance is about 125 fF).

4.11.3.4.4 Power Dissipation Figure 4–160 shows a comparison of the BF-BiCMOS and BF-BiNMOS with the other configurations in terms of power dissipation. The TS-FS-BiCMOS circuit has the highest power dissipation and both the BiNMOS and CMOS circuits exhibit the lowest power dissipation. Although the BF-BiCMOS and BF-BiNMOS circuits have reasonably high-power dissipation, this dissipation is compensated by their high-speed performance. At a load capacitance of 1 pF and a power supply voltage of 3.3 V, the BF-BiCMOS

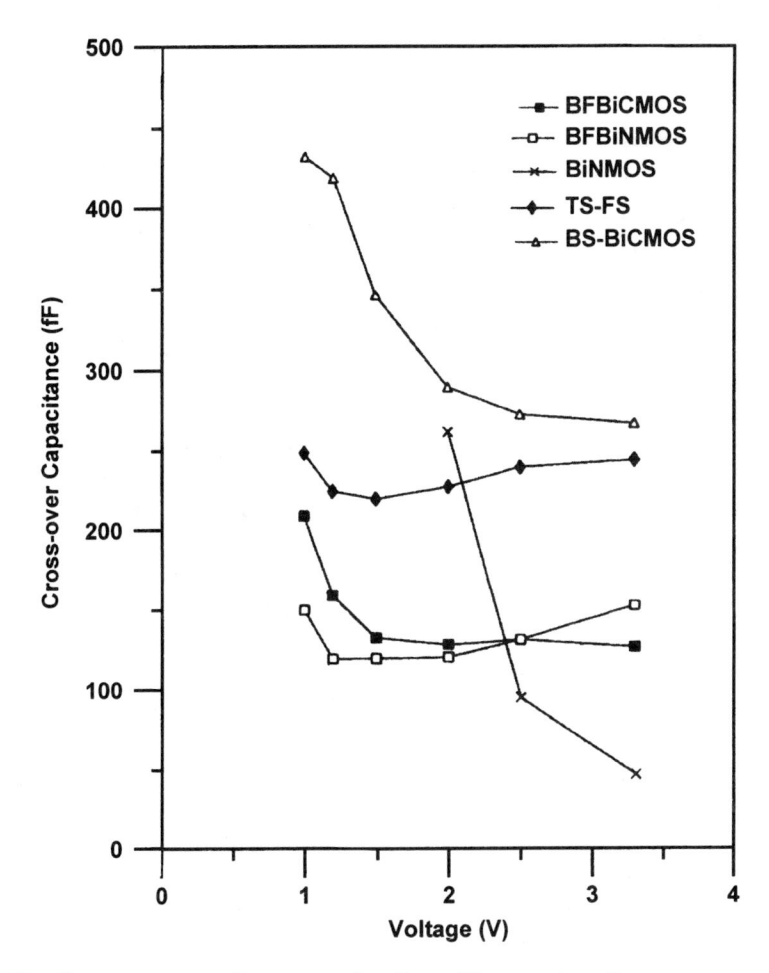

Figure 4–159 Crossover capacitance as a function of the supply voltage.

gate is dissipating twice the amount of power as the CMOS gate, but it is about 2.7 times faster than the CMOS gate. At a supply of 1.5 V and C_L equals to 1 pF, the BF-BiCMOS gate consumes about 110 µW more power than does the CMOS gate, but it is approximately three times faster. Hence, this proves that the BF-BiCMOS and BF-BiNMOS family is able to provide higher speed with reasonable power dissipation as compared to the CMOS family.

4.11.4 Double Bootstrapped BiCMOS Logic Gates (DB-BiCMOS)

The BF-BiCMOS and BF-BiNMOS circuits have been shown to give high speed and full-swing operation at low-supply voltages. However, they use a relatively large number of devices and have a large intrinsic capacitive output loading. In an attempt to sustain and even enhance

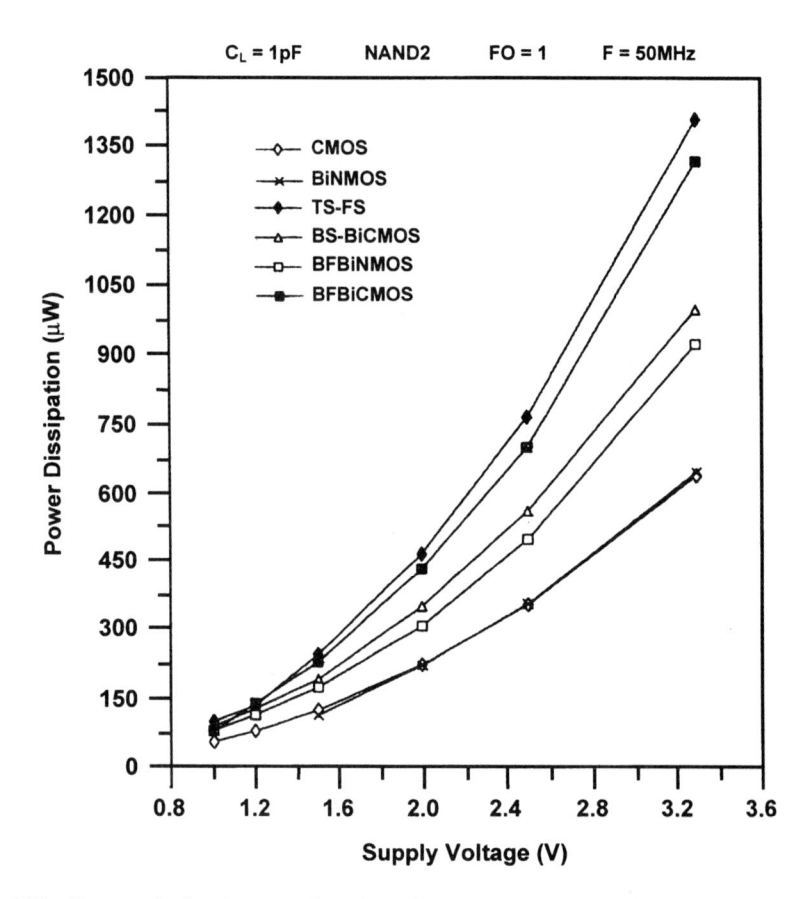

Figure 4–160 Power dissipation as a function of the supply voltage.

the speed and swing of the BiCMOS circuits, this section describes a novel BiCMOS inverter [79] that is able to operate down to a power supply voltage of 1.1 V. The crossover capacitance of this circuit is about 50% lower than that of the BF-BiCMOS and BF-BiNMOS circuits. The proposed BiCMOS inverter, known as DB-BiCMOS, is shown in Figure 4–161.

4.11.4.1 Circuit Operation and Simulation Results

During the pull-up phase, a low-input voltage turns transistor P2 on and bootstraps the base of Q1 above V_{DD} via P3. The transistor P3 is the pMOS implementation of the bootstrapping capacitor. When the base voltage of Q1 increases to a value above V_{DD}, the transistors P1 and P6 turn off. The input low voltage also causes transistor P4 to turn on and it charges up the transistor P5, which acts as a capacitor, to V_{DD}. The transistor P8 is used to maintain the output node at V_{DD} during the pull-up phase.

Bootstrapped BiCMOS Logic Circuits

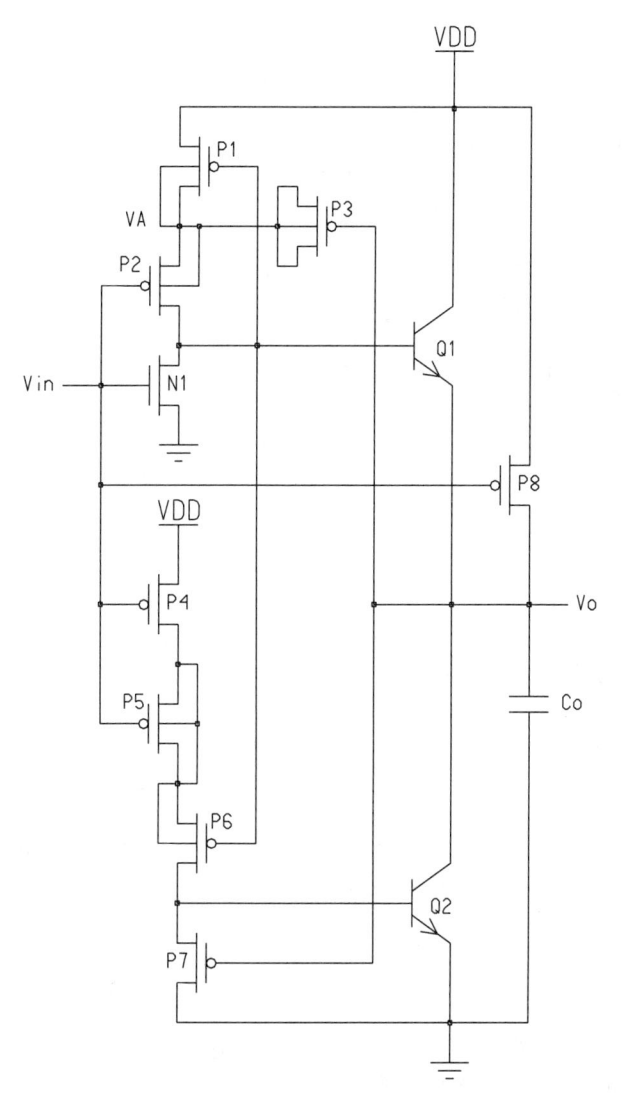

Figure 4–161 DB-BiCMOS inverter.

During the pull-down phase, a high-input voltage turns on the transistor N1 and depletes the base charge of Q1. The falling voltage at the base of Q1 causes P1 to conduct and charge up P3 to V_{DD}. At the same time, P6 turns on causing Q2 to conduct almost immediately. As a result, the turn-on time of Q2 and the crossover capacitance of the circuit are reduced. Figure 4–162 shows the HSPICE simulation results of the DB-BiCMOS inverter during the pull-up and pull-down phases.

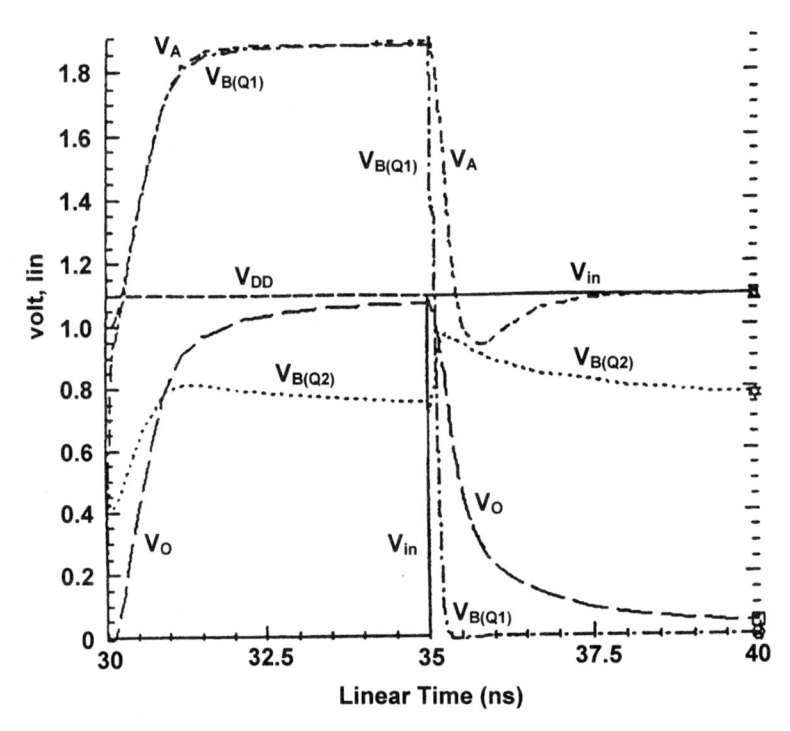

Figure 4–162 Pull-up and pull-down responses of the DB-BiCMOS inverter.

Figure 4–163 shows that for a load capacitance of 1 pF, an output voltage swing of 2.92 Vp-p can be achieved when a square wave of frequency 1 kHz and an amplitude of 3 Vp-p is applied to the inverter input of the DB-BiCMOS circuit.

4.11.4.2 Simulation Results and Performance Comparison

Extensive circuit simulations using HSPICE BSIM model were carried out on three BiCMOS technologies: 3 V/0.5 μm, 2 V/0.35 μm, and 1.1 to 1.5 V/0.25 μm, with threshold voltages of about ± 0.65V, ± 0.42V, and ± 0.35V respectively. Comparisons were made based on an area ratios of 1:1:1:1.2 for the DB-BiCMOS inverter, an optimized two-stage CMOS buffer [27], and the BiCMOS circuits [3,77], respectively. The effect of the output capacitance on the propagation delay, the power-delay product, and the maximum operating frequency are shown in Figures 4–164, 4–165, and 4–166, respectively.

As depicted from Figure 4–164, the DB-BiCMOS inverter is faster than the BBCMOS and CMOS circuits when the output capacitance goes above 0.07 pF, and its crossover capacitance is about 50% lower than that of the BBCMOS circuit [77]. As demonstrated in Figures 4–165 and 4–166, the plots reveal that the DB-BiCMOS circuit has both the least power-delay product as

Bootstrapped BiCMOS Logic Circuits 301

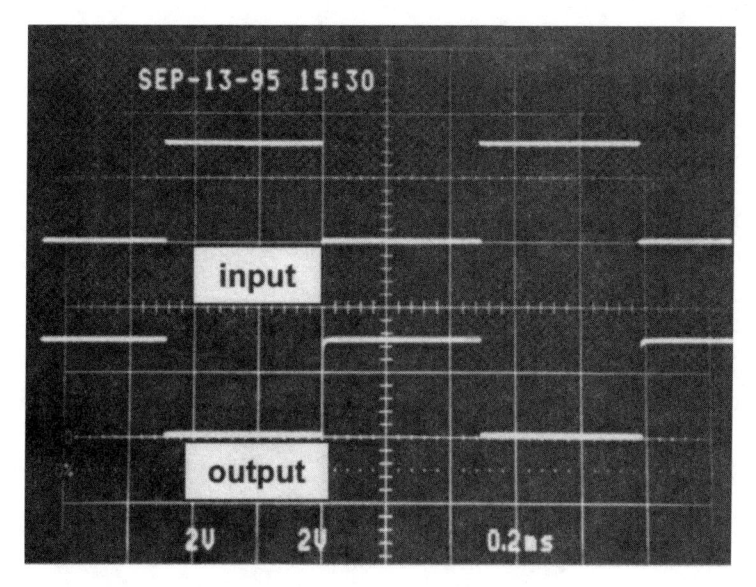

Figure 4–163 Input and output waveforms of the DB-BiCMOS circuit.

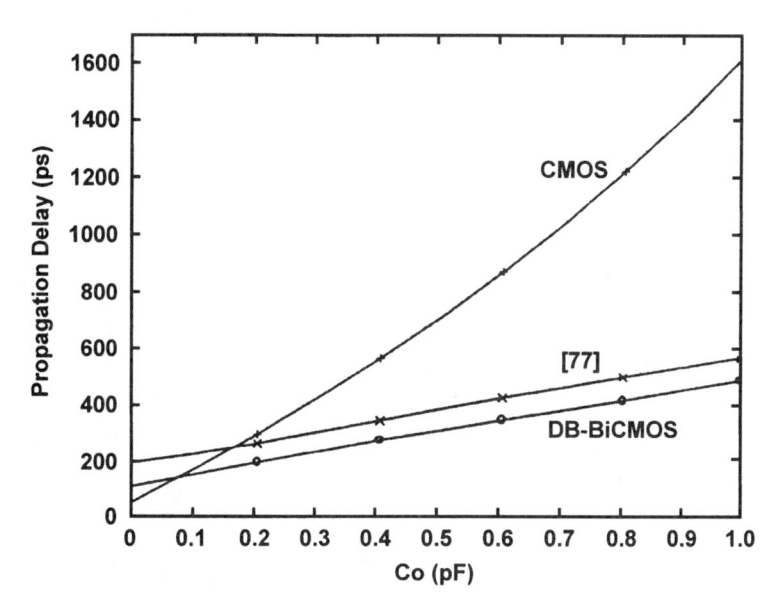

Figure 4–164 Propagation delay of the DB-BiCMOS circuit compared to that of BBCMOS and CMOS at 1.5 V, 0.25 µm technology.

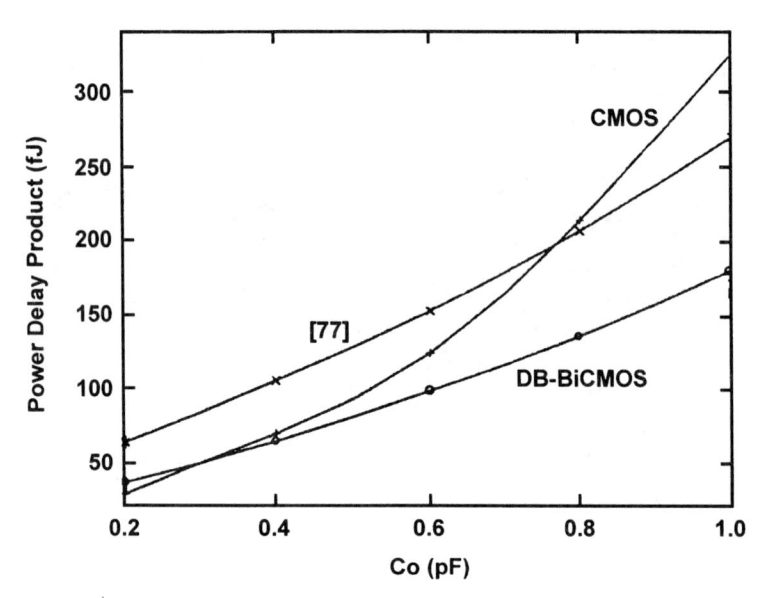

Figure 4–165 Power-delay product of the DB-BiCMOS circuit compared to that of BBCMOS and CMOS at 1.5 V, 0.25 µm technology.

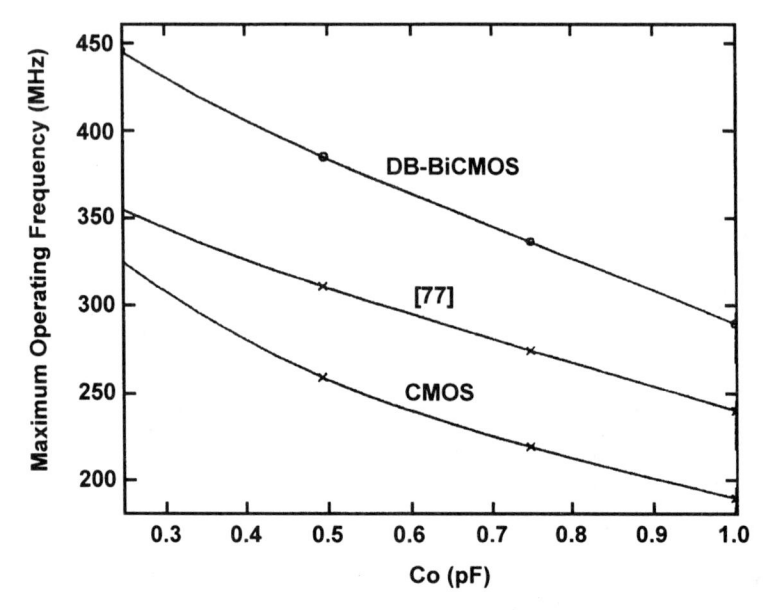

Figure 4–166 Maximum operating frequency of the DB-BiCMOS circuit compared to that of BBCMOS and CMOS at 1.5 V, 0.25 µm technology.

Bootstrapped BiCMOS Logic Circuits

well as the highest operating frequency. Figures 4–167 and 4–168 show the effect of the supply voltage on the speed and average power dissipation. Although the DB-BiCMOS circuit consumes more power, its speed performance is superior to all the other three circuit configurations. Note that the MBiCMOS outperforms the CMOS until the 2 V power supply. Below 2 V, it fails to operate. Figure 4–169 shows the effect of the operating frequency on the propagation delay and the output voltage swing of the DB-BiCMOS and BBCMOS circuits. The BBCMOS fails to function when the operating frequency exceeds 400 MHz, and its output voltage swing deteriorates much faster than the DB-BiCMOS when the operating frequency is increased. This translates to a faster rate of decline in the propagation delay, as shown in Figure 4–169. The DB-BiCMOS circuit can be easily implemented using a standard noncomplementary BiCMOS process. High-speed and full-swing operation are achieved by the bootstrapping technique during the pull-up and pull-down cycles.

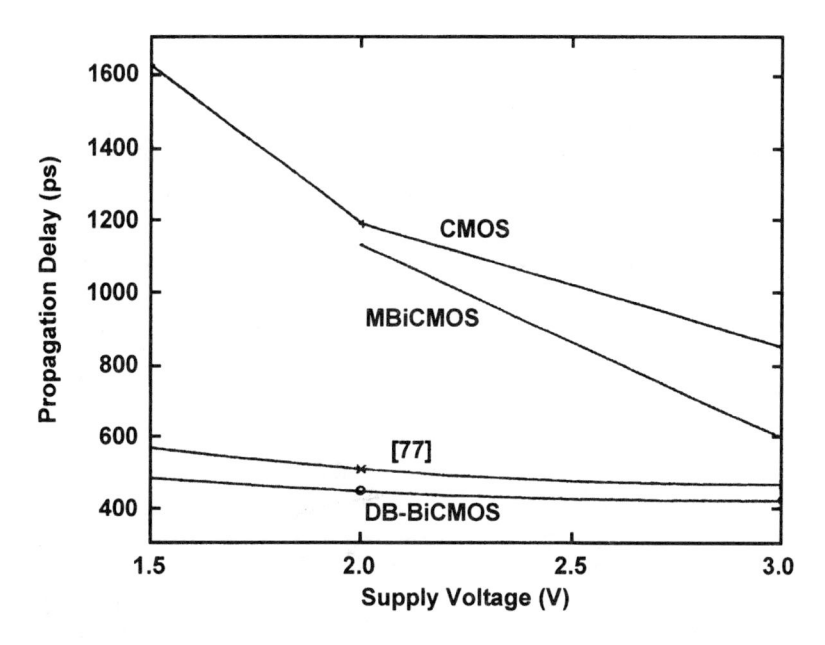

Figure 4–167 Effect of the supply voltage on the propagation delay of the DB-BiCMOS, BBCMOS, MBiCMOS and the CMOS (@C_O = 1 pF).

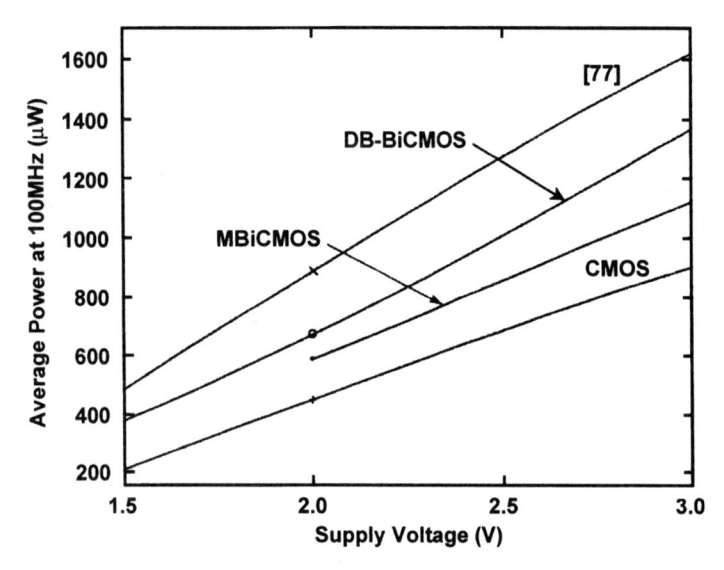

Figure 4–168 Effect of the supply voltage on the average power dissipation of the DB-BiCMOS, BBCMOS, MBiCMOS and the CMOS (@ C_O = 1 pF).

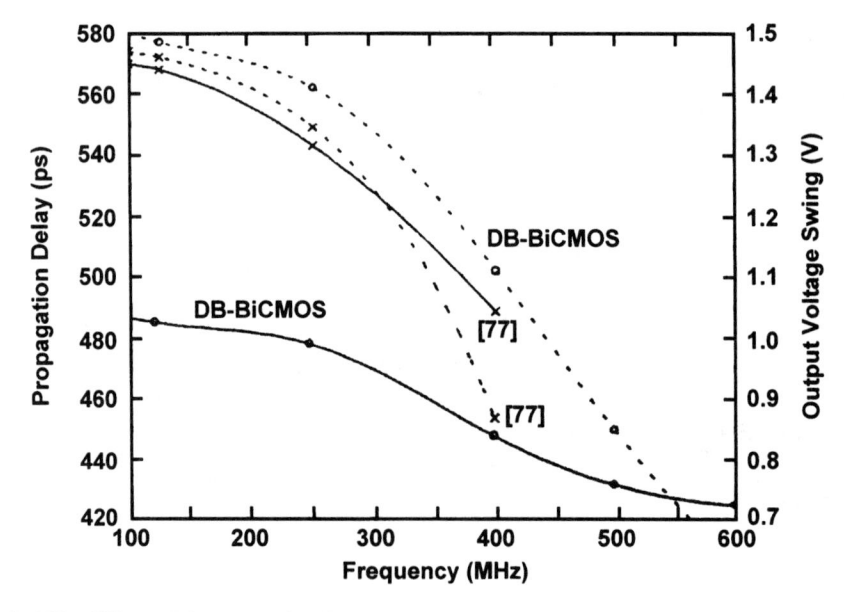

Figure 4–169 Effect of the operating frequency on propagation delay and output voltage swing of the DB-BiCMOS and BBCMOS circuits. (Solid lines represent propagation delay and dotted lines represent output voltage swing.)

4.12 References

[1] Y. G. Paul, P. Bernie, W. S. Shih, and R. Y. John, "Study of BiCMOS Logic Gate Configurations for Improved Low-Voltage Performance," *IEEE J. Solid-State Circuits*, Vol. 28, No. 3, pp. 371–374, Mar. 1993.

[2] O. Takashi and K. Tohru, "A New 0.8 V Logic Swing, 1.6 V Operational High Speed BiCMOS Circuit," *BCTM, Proc.*, pp. 187–190, 1992.

[3] R. B. Ritts, P. A. Raje, J. D. Plummer, K. C. Saraswat, and K. M. Cham, "Merged BiCMOS Logic to Extend the CMOS/BiCMOS Performance Crossover Below 2.5 V Supply," *IEEE J. Solid-State Circuits*, Vol. 26, No. 11, pp. 1606–1614, 1991.

[4] P. R. Gray and R. G. Meyer, *Analysis and Design of Analog Integrated Circuits*, p. 16, Wiley, New York, 1977.

[5] S. M. Sze, *Physics of Semiconductor Devices*, p. 451, 2nd Edition, New York, Wiley, 1981.

[6] H. J. J. De Man, "The Influence of Heavy Doping on the Emitter Efficiency of a Bipolar Transistor," *IEEE Trans. Electron Devices*, Vol ED-18, No. 10, p. 833, 1971.

[7] K. Nakamura, T. Oguri, T. Atsumo, M. Takada, A. Ikemoto, H. Suzaki, T. Nishigori, T. Yamazaki, "A 6 ns 4-Mbit ECL I/O BiCMOS SRAM with LV-TTL Mask Option," *IEEE ISSCC, Tech. Dig.*, pp. 212–213, Feb 1992.

[8] H. J. Shin, "Performance Comparison of Driver Configurations and Full-Swing Techniques for BiCMOS Logic Circuits," *IEEE J. Solid-State Circuits*, Vol. 25, No. 3, pp. 863–865, 1990.

[9] M. Fujishima, K. Asada, and T. Sugano, "Appraised of BiCMOS from Circuit Voltage and Delay Time," *Symp. VLSI Circuits, Tech. Dig.*, pp. 91–92, Jun 1990.

[10] H. J. Shin, "Full-Swing Complementary BiCMOS Logic Circuits," *BCTM, Proc.*, pp. 229–232, Sep 1989.

[11] H. J. Shin, C. L. Chen, E. D. Johnson, Y. Taur, S. Ramaswamy, and G. Boudon, "Full-Swing BiCMOS Logic Circuits with Complementary Emitter-Follower Driver Configuration," *IEEE J. Solid-State Circuits*, Vol. 26, No. 4, pp. 578–584, 1991.

[12] E. D. Johnson, et al., "A 0.5 μm CMOS-Based BiCMOS Technology," *IEEE IEDM, Tech. Dig.*, 1989.

[13] R. B. Ritts and J. D. Plummer, "Merged BiCMOS with Vertical pnp," *Device Res. Conf. Dig. Tech. Papers*, Part II, B-8, 1990.

[14] S. Ogura et al., "Merged Complementary BiCMOS for Logic Applications," *Symp. VLSI Technology, Tech. Dig.*, pp. 81–82, 1990.

[15] T. Hanibuchi, M. Veda, K. Higashitani, M. Hatanaka, and K. Mashiko, "A Bipolar-pMOS Merged Basic Cell for 0.8 μm BiCMOS Sea of Gates," *IEEE CICC, Proc.*, pp. 4.2.1–4.2.4, 1990.

[16] J. B. Kuo, G. P. Rosseel, and R. W. Dutton, "Two-Dimensional Analysis of a Merged BipMOS Device," *IEEE Trans. Computer-Aided Design*, Vol. 8, No. 8, pp. 929–932, 1989.

[17] H. Momose et al., "Characterization of Speed and Stability of BiNMOS Gates with a Bipolar and pMOSFET Merged Structure," *IEDM, Tech. Dig.*, pp. 231–232, 1990.

[18] A. Watanabe, T. Nagano, S. Shukuri and T. Ikeda, "Future BiCMOS Technology for Scaled Supply Voltage," *IEDM, Tech. Dig.*, pp. 429–432, Dec. 1989.

[19] S. C. Lee and D. W. Schucker, "BiCMOS Driver Circuit," *U.S. Patent* 4616146, Oct 7, 1986.

[20] E. W. Greeneich and K. L. Mclaughlin, "Analysis and Characterization of BiCMOS for High-Speed Digital Logic," *IEEE J. Solid-State Circuits*, Vol. 23, No. 4, pp. 558–565, 1988.

[21] G. P. Rosseel and R. W. Dutton, "Influence of Device Parameters on the Switching Speed of BiCMOS Buffers," *IEEE J. Solid-State Circuits*, Vol. 24, No. 2, pp. 90–99, 1989.

[22] J. Shott, C. Knorr, and M. Prisbe, "BiCMOS Technology Overview," Center for Integrated Systems, Stanford, CA, Stanford BiCMOS Project Tech. Rep, Sep 1990.

[23] H. Momose, Y. Unno, and T. Madea, "Supply Voltage Design Tradeoffs Between Speed and nMOSFET Reliability of Half-Micrometer BiCMOS Gates," *IEEE Trans. Electron Devices*, Vol. 38, No. 3, pp. 566–572, 1991.

[24] P. A. Raje, K. C. Saraswat, and K. M. Cham, "Performance-Driven Scaling of BiCMOS Technology," *IEEE Trans. Electron Devices*, Vol. 39, No. 3, pp. 685–694, 1992.

[25] H. J. Shin, "Full-Swing Logic Circuits in a Complementary BiCMOS Technology," *Symp.VLSI Circuits, Tech. Dig.*, pp. 89–90, 1990.

[26] M. Fujishima, K. Asada, and T. Sugano, "Evaluation of Delay-Time Degradation of Low-Voltage BiCMOS Based on a Novel Analytical Delay-Time Modeling," *IEEE J. Solid-State Circuits*, Vol. 26, No. 1, pp. 25–31, 1991.

[27] S. H. K. Embabi, A. Bellaouar, M. I. Elmasry and R. A. Hadaway, "New Full-Voltage Swing BiCMOS Buffers," *IEEE J. Solid-State Circuits*, Vol. 26, No. 2, pp. 150–153, 1991.

[28] K. Yano, M. Hiraki, S. Shukuri, Y. Sawahata, M. Hirao, N. Ohki, T. Nishida, and K. Seki, "Quasi-Complementary BiCMOS for Sub-3 V Digital Circuits," *Symp. VLSI Circuits, Tech. Dig.*, pp. 123–124, 1991.

[29] M. Hiraki, K. Yano, M. Minami, K Satoh, N. Matsuzaki, A. Watanabe, T. Nishida, K. Sasaki, and K. Seki, "A 1.5 V Full-Swing BiCMOS Logic Circuit," *IEEE ISSCC, Tech. Dig.*, pp. 48–49, 1992.

[30] M. I. Elmasry, "Multi-Drain nMOS for VLSI Logic Design," *IEEE J. Solid-State Circuits*, Vol. SC-17, No. 4, pp. 779–781, 1982.

[31] HSPICE User's Manual, Meta-Software, Inc., Campbell, CA, 1988.

[32] M. Elhady, M. H. Elsaid, I. M. Hafez, and H. Haddara, "New Full Voltage Swing Multi-Drain/Multi-Collector Complementary BiCMOS Buffers (M^{2}CBiCMOS)," *Solid-State Electronics*, Vol. 38, No. 1, pp. 211–216, 1995.

[33] V. V. Sophie, S. S. Wong, J. C. S. Woo, and P. K. Ko, "High-Gain Lateral Bipolar Action in a MOSFET Structure," *IEEE Trans. Electron Devices*, Vol. 38, No. 11, pp. 2487–2496, 1991.

[34] S. H. K. Embabi, A. Bellaouar and M. I. Elmasry, *Digital BiCMOS Integrated Circuit Design*, Kluwer Academic, Boston, 1993.

[35] K. Yano, M. Hiraki, S. Shukuri, Y. Onose, M. Hirao, N. Ohki, T. Nishida, K. Seki, and K. Shimohigashi, "Quasi-Complementary BiCMOS for Sub-3 V Digital Circuits," *IEEE J. Solid-State Circuits*, Vol. 26, No. 11, pp. 1708–1718, 1991.

[36] P. R. Gray and R. G. Mayer, *Analysis and Design of Analog Integrated Circuits*, 1st Edition, pp. 311–313, Wiley, New York, 1979.

[37] K. Yano, M. Aoki, and T. Masuhara, "PARAMOST - A New Parasitic Resistance Model for Deep Submicron MOS Transistors," Extended Abstract, *Int. Conf. Solid-State Device and Material*, pp. 85–88, Tokyo, 1986.

[38] M. Suzuki, S. Tachibana, and H. Higuchi, "Drivability of BiCMOS Logic Gate," (in Japanese), *Int. Electron. Commun. Eng. Meeting, Tech. Dig.*, pp. 2–127, 1986.

[39] A. E. Gamal, J. L. Kouloheris, D. How, and M. Morf, "BiNMOS a Basic Cell for BiCMOS Logic Circuits," *IEEE CICC, Proc.*, pp. 8.3.1–8.3.4, May 1989.

[40] H. Hara, et al., "0.5 μm 2M-Transistor BiPNMOS Channelless Gate Array," *IEEE J. Solid-State Circuits*, Vol. 26, No. 11, pp. 1615–1620, 1991.

[41] H. Hara, et al., "0.5 μm 3.3 V BiCMOS Standard Cells with 32-kbit Cache and Ten-Port Register File," *IEEE J. Solid-State Circuits*, Vol. 27, No. 11, pp. 1579–1584, 1992.

[42] A. Bellaouar, I. S. Abu Khater, M. I. Elmasry and A. Chikima, "Full-Swing Schottky BiCMOS/BiNMOS and the Effects of Operating Frequency and Supply Voltage Scaling," *IEEE J. Solid-State Circuits*, Vol. 29, No. 6, pp. 693–699, 1994.

[43]   M. Hiraki, K. Yano, M. Minami, K. Satoh, N. Matsuzaki, A. Watanabe, T. Nishida, K. Sasaki, and K. Seki, "A 1.5V Full-Swing BiCMOS Logic Circuit," *IEEE J. Solid-State Circuits*, Vol. 27, No. 11, pp. 1568–1574, 1992.

[44]   HSPICE Version H92, Meta-Software, Inc., Campbell, CA, 1992.

[45]   K. Yano et al., "3.3 V BiCMOS Circuit Techniques for 250 MHz RISC Arithmetic Modules," *IEEE J. Solid-State Circuits*, Vol. 27, pp. 373–381, 1992.

[46]   S. S. Rofail, "Low-Voltage, Low-Power BiCMOS Digital Circuits," *IEEE J. Solid-State Circuits*, Vol. 29, No. 5, pp. 572–579, 1994.

[47]   K. S. Yeo and S. S. Rofail, "A Full-Swing High Speed CBiCMOS Digital Circuit for Low-Voltage Applications," *IEE Proceeding - Circuits, Devices and Systems*, Vol. 142, No. 1, pp. 8–14, 1995.

[48]   M. S. Elrabaa, M. S. Obrecht, and M. I. Elmasry, "Novel Low-Voltage Low-Power Full-Swing BiCMOS Circuit," *IEEE J. Solid-State Circuits*, Vol. 29, No. 2, pp. 86–93, 1994.

[49]   S. C. Lee et al., "BiCMOS Circuits for High Performance VLSI," *Symp. VLSI Technology, Tech. Dig.*, pp. 46–47, Sep 1984.

[50]   Y. Nishio et al., "A Sub-Nanosecond Low Power Advanced Bipolar CMOS Gate Array," *ICCD, Proc.*, pp. 428–433, Oct 1984.

[51]   I. Masuda et al., "High-Speed Logic Circuits Combining Bipolar and CMOS Technology," *Inst. Electron. Commun. Engineering Trans.*, Japan, Vol. J67-C, No. 12, pp. 999–1005, 1984.

[52]   HSPICE User's Manual, Meta-Software, Inc., Campbell, CA, 1990.

[53]   N. Yoji et al., "A Feedback-Type BiCMOS Logic Gate," *IEEE J. Solid-State Circuits*, Vol. 24, No. 5, pp. 1360–1362, 1989.

[54]   H. Hara et al., "0.5 μm 2M-Transistor BiPNMOS Channelless Gate Array," *IEEE ISSCC, Tech. Dig.*, pp. 148–149, 1991.

[55]   M. S. Obrecht and M. I. Elmasry, "TRASIM - A Compact and Efficient 2-D Transient Simulator for Arbitrary Planar Semiconductor Devices," *IEEE Transaction on Computer-Aided Design of Integrated Circuits and Systems*, Vol. 14, No. 4, pp. 447–458, 1995.

[56]   S. S. Rofail and K. S. Yeo, "New Complementary BiCMOS Digital Gates for Low-Voltage Environments," *Solid-State Electronics*, Vol. 39, No. 5, pp. 681–687, 1996.

[57]   H. Higuchi, A. Anzai, N. Homma, and A. Hayasaka, "High Performance Bipolar VLSIs: Their Present Status and Future," *Symp. VLSI Technology, Tech. Dig.*, pp. 110–113, 1982.

[58] K. Y. Toh, P. K. Ko, and R. G. Meyer, "An Engineering Model for Short-Channel MOS Devices," *IEEE J. Solid-State Circuits*, Vol. 23, No. 4, pp. 950–958, 1988.

[59] P. A. Raje, K. C. Saraswat, and K. M. Cham, "Accurate Delay Models for Digital BiCMOS," *IEEE Trans. Electron Devices*, Vol. 39, No. 6, pp. 1456–1464, 1992.

[60] M. S. Elrabaa and M. I. Elmasry, "Design and Optimization of Buffer Chains and Logic Circuits in a BiCMOS Environment," *IEEE J. Solid-State Circuits*, Vol. 27, No. 5, pp. 792–801, 1992.

[61] J. Yuan and C. Sevensson, "High-Speed CMOS Circuit Techniques," *IEEE J. Solid-State Circuit*, Vol. 24, No. 1, pp. 62–70, 1989.

[62] R. H. Krembec, C. M. Lee, and H. S. Law, "High-Speed Compact Circuits with CMOS," *IEEE J. Solid-State Circuit*, Vol. 17, No. 6, pp. 614–619, 1982.

[63] J. B. Kuo, S. S. Chen, C. S. Chiang, K. W. Su, and J. H. Lou, "A 1.5 V BiCMOS Dynamic Logic Circuit Using a BipMOS Pull-Down Structure for VLSI Implementation of Full Adders," *IEEE Trans. Circuits and Systems*, Vol. 41, No. 4, pp. 329–332, 1994.

[64] J. B. Kuo, H. J. Liao, and H. P. Chen, "A BiCMOS Dynamic Carry Look-Ahead Adder Circuit for VLSI Implementation of High Speed Arithmetic Unit," *IEEE J. Solid-State Circuit*, Vol. 28, No. 3, pp. 375–378, 1993.

[65] J. B. Kuo, H. J. Liao, and H. P. Chen, "A BiCMOS Dynamic Carry Look-Ahead Circuit with Carry Skip for High-Speed Arithmetic Circuits," *IEEE Trans. Circuits and Systems*, Vol. 39, No. 12, pp. 869–871, 1992.

[66] H. P. Chen, H. J. Liao, and J. B. Kuo, "A High-Speed BiCMOS Dynamic Carry Look Ahead Circuit with Carry Skip for CPU VLSI," *IEDM, Tech. Dig.*, pp. 79–80, Nov 1992.

[67] J. B. Kuo, H. J. Liao and H. P. Chen, "A BiCMOS Dynamic Full Adder Circuit for VLSI Implementation of High-Speed Parallel Multipliers using Wallace Tree Reduction Architecture," *BCTM, Proc.*, pp. 191–194, Oct 1992.

[68] J. B. Kuo, J. J. Huang, S. S. Chen, and C. S. Chiang, "A BiCMOS Dynamic Defuzzify Circuit for VLSI Implementation of Large Scale Fuzzy Logic Controllers," *Symp. VLSI Technology, System and Application, Tech. Dig.*, pp. 267–271, May 1993.

[69] J. B. Kuo, H. P. Chen and H. J. Huang, "A Race-Free BiCMOS Dynamic Divider Circuit using a Non-Restoring Iterative Architecture with Carry Look Ahead for VLSI Implementation of CPU," *IEEE Int. Circuit and System Symp.*, pp. 2027–2030, May 1993.

[70] J. B. Kuo, K. W. Su, S. S. Chen, and C. S. Chiang, "A BiCMOS Dynamic Minimum Circuit using a Parallel Comparison Algorithm for VLSI Implementation of Large-Scale Fuzzy Logic Controllers," *Electronics Letters*, Vol. 29, No. 6, pp. 551–553, 1993.

[71] H. Okamura, T. Atsumo, K. Takeda, M. Takada, K. Lmai, and Y. Kinoshita, "A Sub-2 V BiCMOS Logic Circuit with a BiCMOS Charge Pump," *IEEE J. Solid-State Circuit*, Vol. 31, No. 1, pp. 84–90, 1996.

[72] T. Nagano et al., "What Can Replace BiCMOS at Low Supply Voltage Regime?" *IEDM, Tech. Dig.*, pp. 393–395, Dec 1992.

[73] G. Kitoukawa et al., "A 23 ns 1-Mbit BiCMOS DRAM," *IEEE J. Solid-State Circuits*, Vol. 25, No. 5, pp. 1102–1109, 1990.

[74] H. Momose, K. M. Cham, C. I. Drowley, H. R. Grinolds, and H. S. Fu, "0.5 μm BiCMOS Technology," *IEDM, Tech. Dig.*, pp. 838–840, 1987.

[75] E. Takeda, Y. Ohji, and H. Kume, "High Field Effects in MOSFETS," *IEDM, Tech. Dig.*, pp. 60–63, 1985.

[76] R. Y. V. Chik and C. A. T. Salama, "1.5 V Bootstrapped BiCMOS Logic Gate," *Electronics Letters*, Vol. 29, No. 3, pp. 307–309, 1993.

[77] S. H. K. Embabi, A. Bellaouar, and K. Islam, "A Bootstrapped Bipolar CMOS (B^{2}CMOS) Gate for Low-Voltage Applications," *IEEE J. Solid-State Circuits*, Vol. 30, No. 1, pp. 47–53, 1995.

[78] A. Bellaouar, M. I. Elmasry, and S. H. K. Embabi, "Bootstrapped Full-Swing BiCMOS/BiNMOS Logic Circuits for 1.2 - 3.3 V Supply Voltage Regime," *IEEE J. Solid-State Circuits*, Vol. 30, No. 6, pp. 629–636, 1995.

[79] K. S. Yeo and S. S. Rofail, "A 1.1 V Full-Swing Double Bootstrapped BiCMOS Logic Gates," *IEEE Proceedings - Circuits, Devices and Systems*, Vol. 143, No. 1, pp. 41–45, 1996.

CHAPTER 5

Delay Time and Power Dissipation Sensitivity Analyses of Multi-Generation BiCMOS Digital Circuits

5.1 Introduction

In this chapter, the sensitivity of the BiCMOS circuit delay and power dissipation to the key device and circuit parameters are investigated. It will be demonstrated that a few MOS/BJT device parameters affect the BiCMOS circuit speed and power consumption critically. This vulnerability, together with the tolerance associated with each device and process parameter, will either decrease or increase the circuit speed and power. Sensitivity coefficients are defined and generated for the conventional BiCMOS circuit as well as for two low-voltage circuits recently reported [1,2]. These circuits are designed to operate for the 3.3 V supply voltage regime. However, recently, the power dissipation of high-performance digital integrated circuits is gaining importance due to portability and reliability issues. Hence, the supply voltage is further scaled down to the sub-1.5 V regime. In such an environment, it is expected that even a small variation in the process or device parameters can cause an adverse effect on the circuit performance. Therefore, a study of the influence of device parameters' variations on the circuit performance of four different very low-voltage (1.5 V) BiCMOS circuits is presented. Finally, a method to calculate the worst case speed degradation for a given set of device and process parameter tolerances is described.

Delay time and power dissipation sensitivity analyses are especially crucial for small technologies and high-performance BiCMOS circuits, whereby the circuit speed, in particular, is easily affected by adverse effects caused by reduced dimensions and an increase in the device and process parameters tolerances. In the next section, we shall establish a link between key BJT/MOS process parameters and their device/circuit parameters.

5.2 Relationship Between Key BJT/MOS Process and Device/Circuit Parameters

The following relations adopted from Embabi et al. [3] illustrate the linkage between the bipolar process parameters and those device/circuit parameters that affect the BiCMOS delay time. The process parameters include the emitter doping concentration (N_E), the emitter depth (W_E), the emitter length (L_E), the emitter width (B_E), the emitter area (A_E), the peak base doping (N_B), the base width (W_B), the collector doping concentration (N_C), and the collector width (W_C). The relations are as follows:

$$\tau_f = \frac{W_B^2}{2D_{nB}} \qquad 5.1$$

$$\beta_f = \frac{D_{nB} N_E W_E}{D_{pE} N_B W_B} \qquad 5.2$$

$$R_b = \frac{L_E}{q\mu_{pB} N_B W_B B_E} \qquad 5.3$$

$$R_C = \frac{W_C}{q\mu_{nC} N_C A_E} \qquad 5.4$$

$$I_K = q\upsilon_{max} N_C A_E (1 + \frac{2\varepsilon_{si} V_{BC}}{q N_C W_C^2}) \qquad 5.5$$

$$C_{JE} = A_E \sqrt{\frac{q\varepsilon_{si} N_B}{2\phi_J}} \qquad 5.6$$

$$C_{JC} = A_C \sqrt{\frac{q\varepsilon_{si} N_C}{2\phi_J}} \qquad 5.7$$

$$C_{JS} = A_S \sqrt{\frac{q\varepsilon_{si} N_{sub}}{2\phi_J}} \qquad 5.8$$

and from Lie et al. [4], the threshold voltage of a MOS device can be expressed as

$$V_T = V_{TO} - \Delta V_T \qquad 5.9$$

where:

$$V_{TO} = V_{FB} + \frac{qN_{sub} X_{dep} T_{ox}}{\varepsilon_{ox}} + \phi_s \qquad 5.9a$$

Sensitivity Analysis of the Conventional BiCMOS Circuit

$$\Delta V_T = \left\{ \frac{2(V_{bi}-\phi_s)+[V_{DS}+(V_{bi}-\phi_s)](1-e^{-L/l_c})}{4\sinh^2 \frac{L}{2l_c}} + \frac{2\sqrt{(V_{bi}-\phi_s)^2+(V_{bi}-\phi_s)[V_{DS}+(V_{bi}-\phi_s)](e^{-L/l_c}-1)}}{4\sinh^2 \frac{L}{2l_c}} \right\} \quad 5.9b$$

$$\phi_s = \frac{2kT}{q}\ln\left(\frac{N_{sub}}{n_i}\right) \quad 5.9c$$

$$X_{dep} = \sqrt{\frac{2\varepsilon_{si}(\phi_s - V_{BS})}{qN_{sub}}} \quad 5.9d$$

$$l_c = \sqrt{\frac{\varepsilon_{si}T_{ox}X_{dep}}{\varepsilon_{ox}m}} \quad 5.9e$$

where D_{nB} and D_{pE} are the diffusion constants of electrons in the base and holes in the emitter, respectively. μ_{pB} and μ_{nC} are the mobility of holes in the base and electrons in the collector, respectively, υ_{max} is the maximum carrier velocity, V_{BC}, V_{BS}, V_{FB}, and V_{bi} are the base-collector voltage, substrate bias, flat-band voltage, and built-in potential, respectively, ε_{si} is the permittivity of silicon, l_c is the characteristic length [4], and m is a fitting parameter [4]. Table 5–1 tabulates the percentage change in the key bipolar and MOS device/circuit parameters, defined as α_i, based on a ±15% change in the process parameters' values. The results help us to identify the key process parameters affecting the delay sensitivity.

5.3 Sensitivity Analysis of the Conventional BiCMOS Circuit

The delay sensitivity of the conventional BiCMOS circuit, as illustrated in Figure 5–1, will be analyzed and numerically evaluated using HSPICE in the following subsections.

5.3.1 Delay Time Sensitivity to Key Device Parameters

The BiCMOS circuit delay is given approximately as [6]:

$$t_d \cong \tau_f \sqrt{\frac{\beta_f I_K}{I_{Dsat}}} + \frac{V_{os}(C_O + C_1)}{2\sqrt{\beta_f I_K I_{Dsat}}} \quad 5.10$$

where I_{Dsat} is the MOS drain saturation current, V_{os} is the output voltage swing, C_O is the output load capacitance, and C_1 is the sum of the MOS source/drain parasitic capacitance, the output parasitic capacitance of the gate, and the base-collector capacitance at the base of bipolar transistor. In deriving Eq. 5.10, the input pMOS device, P_1, is assumed to operate in the saturation

Table 5–1 Percentage change in the key BJT and MOS device/circuit parameters based on a ±15% change in the process parameters' values. [5]

α_i (%)	τ_f	β_f	I_K	R_C	R_b	C_{JC}	C_{JE}	C_{JS}	V_T
N_E		15 −15							
N_B		−13.04 17.65			−13.04 17.65		7.24 −7.80		
N_C			−0.98 1.32	−13.04 17.65		7.24 −7.80			
N_{SUB}							7.24 −7.80		13.8 −15.43
W_B	32.25 −27.75	−13.04 17.65			−13.04 17.65				
W_E		15 −15							
W_C			−1.83 2.88	15 −15					
A_E			15 −15	−13.04 17.65			15 −15		
A_C						15 −15			
A_S								15 −15	
μ_n				−13.04 17.65					
μ_p					−13.04 17.65				
L_E					15 −15				
L									26.16 −29.05
B_E					−13.04 17.65				
T_{OX}									−3.37 6.00

region. This assumption is true only for submicrometer MOSFETs that have small saturation voltage [7]. At high-level injection, the time required to turn on Q_1 is negligibly small as compared to the time in the other two terms in Eq. 5.10. Therefore, this equation predicts accurately

Sensitivity Analysis of the Conventional BiCMOS Circuit

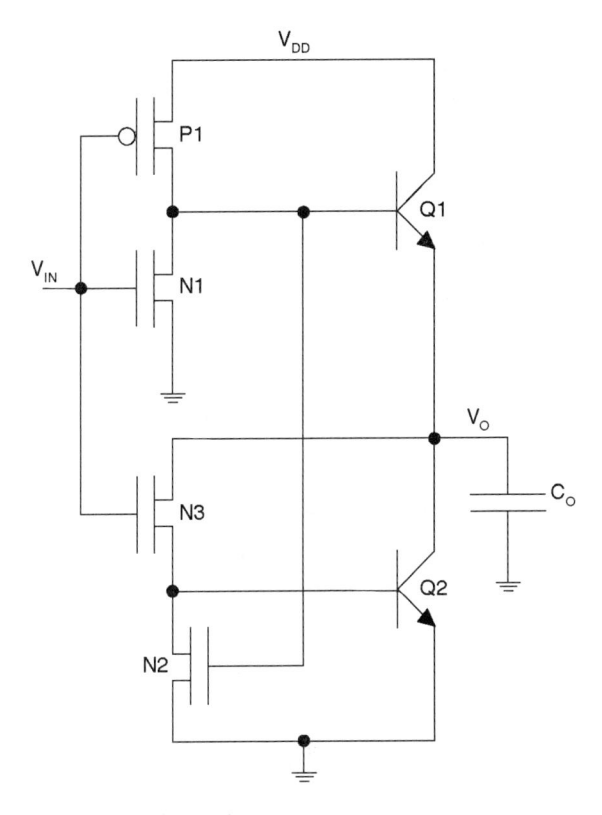

Figure 5–1 Conventional BiCMOS circuit.

the delay time for the output voltage to reach $V_{DD}/2$ at high level injection (a condition that has been evident in the numerical simulations for $C_O \geq 0.8$ pF).

5.3.1.1 The Forward Transit Time τ_f

The transient response of the BiCMOS circuit is influenced by the bipolar transit time [8,9]. Since the gate delay is proportional to the forward transit time, at high-level injection, where τ_f increases substantially with the collector current, the circuit gate delay is severely degraded. The delay sensitivity to τ_f according to Eq. (5.10) is given by:

$$\frac{\Delta t_d}{t_d} = \Psi_{\tau_f} \frac{\Delta \tau_f}{\tau_f} \qquad 5.11a$$

where:

$$\Psi_{\tau_f} = \frac{2\beta_f I_K \tau_f}{2\beta_f I_K \tau_f + V_{os}(C_O + C_1)} \qquad 5.11b$$

5.3.1.2 The Forward Current Gain β_f

It is widely known that the delay time is inversely proportional to β_f. The delay sensitivity to β_f based on Eq. 5.10 is given as:

$$\frac{\Delta t_d}{t_d} = -\Psi_{\beta_f} \frac{\Delta \beta_f}{\beta_f} \qquad 5.12$$

where:

$$\Psi_{\beta_f} = \frac{1}{2}\left[\frac{B - A\beta_f}{B + A\beta_f}\right] \qquad 5.12a$$

$$A = \tau_f \sqrt{\frac{I_K}{I_{Dsat}}} \quad ; \quad B = \frac{V_{os}(C_0 + C_1)}{2\sqrt{I_K I_{Dsat}}} \qquad 5.12b$$

Equation 5.12, as will be shown later in subsections 5.3.2, 5.3.3, and 5.3.4, helps to show how the delay sensitivity to β_f varies with the load capacitance, scaling, and the quality of the BJT.

5.3.1.3 The Knee Current I_K

The BiCMOS delay time decreases with I_K [10]. The delay sensitivity to I_K shows a similar trend to that of β_f and is given by:

$$\frac{\Delta t_d}{t_d} = -\Psi_{I_K} \frac{\Delta I_K}{I_K} \qquad 5.13$$

where:

$$\Psi_{I_K} = \frac{1}{2}\left[\frac{B' - A' I_K}{B' + A' I_K}\right] \qquad 5.13a$$

$$A' = \tau_f \sqrt{\frac{\beta_f}{I_{Dsat}}} \quad ; \quad B' = \frac{V_{os}(C_0 + C_1)}{2\sqrt{\beta_f I_{Dsat}}} \qquad 5.13b$$

5.3.1.4 The Base Resistance R_b, and The Collector Resistance R_C

The effect of the base resistance on the BiCMOS delay is minimal as long as its value is much less than the channel resistance of the pMOS device. On the other hand, the impact of the collector resistance on the BiCMOS delay is significant, especially at high collector currents. A large voltage drop across the collector resistance can drive the BJT into saturation, thereby increasing the circuit delay. Hence, any variable that causes the BJT to move from nonsaturation towards saturation will be expected to increase the delay sensitivity to R_C.

5.3.1.5 The MOS Channel Width W and Length L, Oxide Thickness T_{OX}, and Threshold Voltage V_T

At low level injection, the BiCMOS delay t_d is inversely proportional to W [6]. On the other hand, for high-level injection, from Eq. 5.10, t_d is proportional to $1/\sqrt{W}$. A similar trend is shown for the MOS oxide thickness, where the circuit delay is proportional to T_{OX} at low-level injection, and to $\sqrt{T_{OX}}$ at high-level injection. Therefore, it is expected that the delay sensitivity to W and T_{OX} decreases at high-level injection. The delay sensitivities to the MOSFET's channel length and threshold voltage, for both low level injection (LLI) and high-level injection (HLI), are given by [6,7]:

$$S_L^{t_d}(\text{LLI}) = \frac{\frac{\Delta t_d}{t_d}}{\frac{\Delta L}{L}} = \frac{1}{\left[1 + \frac{V_{GS} - V_T}{E_{crit}L}\right]} \qquad 5.14a$$

$$S_L^{t_d}(\text{HLI}) = \frac{\frac{\Delta t_d}{t_d}}{\frac{\Delta L}{L}} = \frac{1}{2\left[1 + \frac{V_{GS} - V_T}{E_{crit}L}\right]} \qquad 5.14b$$

$$S_{V_T}^{t_d}(\text{LLI}) = \frac{\frac{\Delta t_d}{t_d}}{\frac{\Delta V_T}{V_T}} = \frac{V_T(V_{GS} - V_T + 2E_{crit}L)}{(V_{GS} - V_T + E_{crit}L)(V_{GS} - V_T)} \qquad 5.14c$$

$$S_{V_T}^{t_d}(\text{HLI}) = \frac{\frac{\Delta t_d}{t_d}}{\frac{\Delta V_T}{V_T}} = \frac{V_T(V_{GS} - V_T + 2E_{crit}L)}{2(V_{GS} - V_T + E_{crit}L)(V_{GS} - V_T)} \qquad 5.14d$$

where V_{GS} is the gate-to-source voltage and E_{crit} is the critical electric field. From the above equations, it is clear that high-level injection tends to reduce the delay sensitivity to V_T and L.

5.3.2 The Effects of the Load Capacitance on the Circuit Delay Sensitivity

From Eq. 5.11(a), it is evident that an increase in the load capacitance, C_O, at high-level injection causes the delay sensitivity to τ_f to decrease. From Eq. 5.12(a), it can be seen that C_O also affects the delay sensitivity to β_f through the factor Ψ_{β_f}. Figure 5-2 shows a plot of Ψ_{β_f} against C_O for two sets of values for τ_f and I_K.

According to Figure 5-2, it is clear that an increase in C_O will bring about a corresponding increase in the factor Ψ_{β_f}, and hence an increase in the delay sensitivity to β_f. A similar trend is

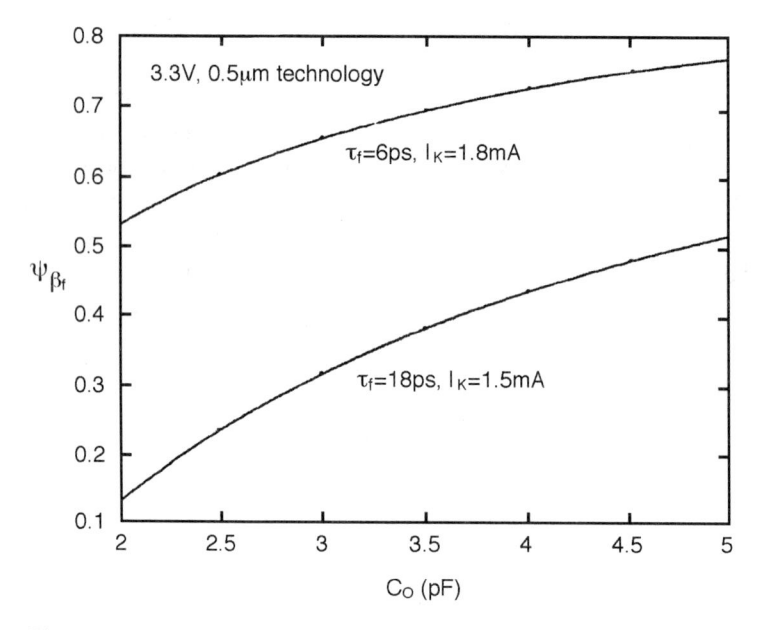

Figure 5–2 Ψ_{β_f} against C_O of the conventional BiCMOS circuit for the (3.3 V, 0.5 μm) technology for two different sets of τ_f and I_K.

expected for Ψ_{I_K} based on Eq. 5.13(a). The delay sensitivity to R_C increases with C_O. This increase is attributed to the increased chances of driving the BJT into saturation. It can be shown that the higher the load capacitance the lower the collector resistance needed to saturate the transistor [6]. Figure 5–3 shows the critical value of R_C (defined as the collector resistance to cause Q_1 to move from nonsaturation to saturation) against the load capacitance for the 0.2 μm and 0.8 μm technologies used in the analysis. The critical value of R_C is obtained by using the analysis in Toh et al. and Fang et al. [7,9]. It is evident that the higher the load capacitance, the lower the collector resistance needed to saturate the transistor.

The critical collector resistance is given by [7,11]:

$$R_C(\text{critical}) = \frac{R_{ch} + R_b}{\beta_f^*} \qquad 5.15$$

where:

$$R_{ch} = \frac{2L}{kW\mu_p C_{ox}(V_{DD} - |V_{TP}|)} \qquad 5.15a$$

$$k = \frac{1}{1 + \dfrac{E_{crit} L_e}{V_{GS} - V_T}} \qquad 5.15b$$

Sensitivity Analysis of the Conventional BiCMOS Circuit

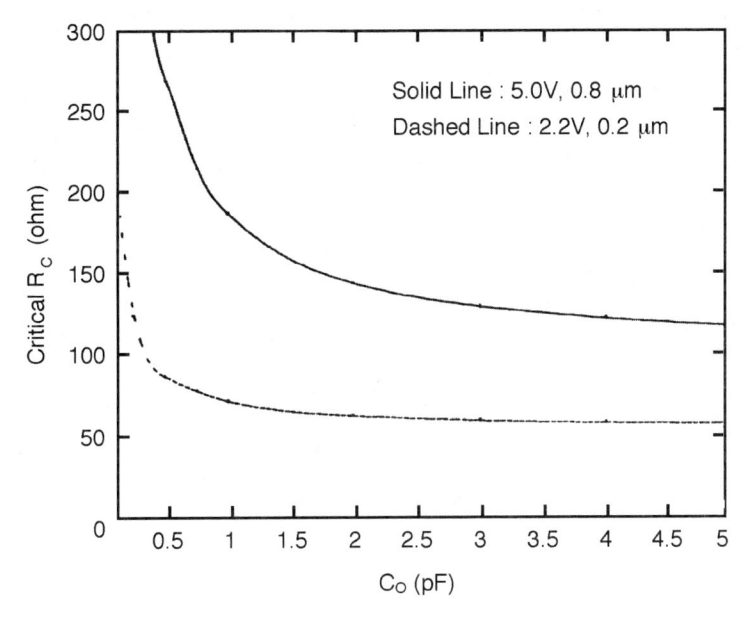

Figure 5–3 R_C (critical) against C_O of the conventional BiCMOS circuit for the 0.2 μm and 0.8 μm technologies.

$$E_{crit} = \frac{2\upsilon_{sat}}{\mu_{eff}} \quad\quad 5.15c$$

L_e is the device electrical channel length, μ_{eff} is the effective carrier mobility, and β_f^* is the effective current gain. Higher load capacitance will also lead to high-level injection, hence causing lower delay sensitivity to W, L, T_{OX} and V_T (as explained in section 5.3.1.5).

5.3.3 The Effects of the Quality of the BJT on the Circuit Delay Sensitivity

The forward transit time, the knee current, and the collector resistance are key parameters indicating the quality of the bipolar device. Table 5.2 gives a comparison of the key bipolar parameters for the high- and low-quality bipolar device.

Table 5–2 Comparison of high- and low-quality bipolar parameters.

Quality of Bipolar	τ_f	I_K	R_C	R_b	C_{JC}	C_{JS}
High	9 ps	2.5 mA	150	500	9.6 fF	25.2 fF
Low	27 ps	2 mA	650	580	8.6 fF	35.5 fF

From the preceding table, it can be deduced that from high quality to low quality, τ_f and R_C increase by factors of approximately 3 and 4, respectively, while I_K decreases between 15% and 20%. According to Eqs. 5.11(a), 5.12(a), and 5.13(a), it can be seen that low quality BJT would cause Ψ_{τ_f} to decrease slightly, and both Ψ_{β_f} and Ψ_{I_K} to decrease sharply, especially at low load capacitance (as illustrated in Figure 5–2). Because of the relatively high value of R_C, the low quality BJT increases the chances of saturation which, in turn, increases the delay sensitivity to R_C, especially at high load capacitance. Figure 5–4 shows the collector current obtained from HSPICE, normalized with respect to I_K, against time for the high- and low-quality BJT in the 0.8 µm technology. It can be seen that the high-quality BJT causes the device to operate deeper in the high level injection mode thereby increasing the delay sensitivities to both β_f and I_K.

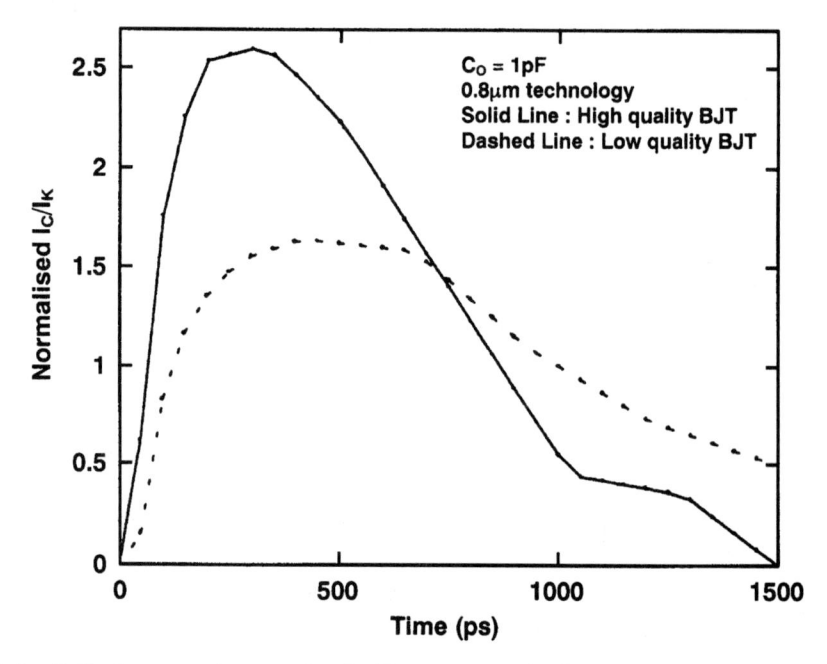

Figure 5–4 Collector current, normalized with respect to the knee current, against time for the conventional BiCMOS circuit with high- and low-quality BJT.

5.3.4 The Effects of Scaling the Technology on the Circuit Delay Sensitivity

The MOS and BJT device parameters are affected by scaling the technology [6]. For instance, scaling the technology from (5 V, 0.8 µm) to (2.2 V, 0.2 µm) reduces I_K from 2.5 to 1.2 mA, τ_f from 9 to 3.5 ps, T_{OX} from 15 to 6 nm, and V_T from 0.8 to 0.25 V. In Eqs. 5.12(b) and 5.13(b), the decrease in τ_f decreases the factors A and A' in Eqs. 5.12(b) and 5.13(b), respectively, which consequently increases the delay sensitivities to β_f and I_K through Ψ_{β_f} and Ψ_{I_K} (as explained in Section 5.3.2).

Sensitivity Analysis of the Conventional BiCMOS Circuit

The delay sensitivity to the changes in the value of R_C is also affected by scaling. In general, an increase in R_C would cause a drop in $I_C \Delta R_C$ in the collector base voltage V_{CB}, which, in turn, reduces the collector current I_C, thereby increasing the delay. In general, scaled technologies have lower values of I_C and this effect will tend to reduce the delay sensitivity to R_C. Figure 5-5 shows the effects of a ±15 % change in R_C on the values of V_{CB} and I_C.

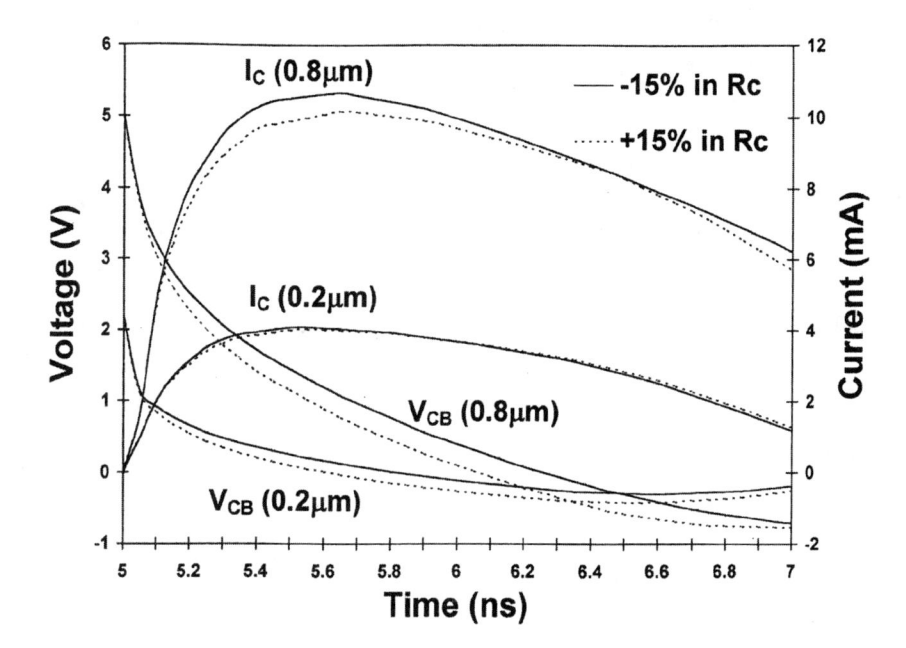

Figure 5-5 $V_{CB}(t)$ and $I_C(t)$ of the pull-up transient response of the conventional BiCMOS circuit for the (5 V, 0.8 μm) and (2.2 V, 0.2 μm) technologies and a change of ±15 % in R_C.

The above figure shows that for the (5 V, 0.8 μm) technology, V_{CB} and I_C have a wider variation as compared to that of the (2.2 V, 0.2 μm) technology. The slope of the V_{CB} plot with time for the 0.8 μm case is also higher than the slope of the 0.2 μm technology. This difference leads to driving the BJT, beyond certain time, deeper into saturation, hence increasing the delay sensitivity to R_C even further. The slope $\frac{\delta V_{CB}}{\delta t}$ is analytically described by:

$$\frac{\delta V_{CB}}{\delta t} \cong -R_{ch}\frac{\delta I_B}{\delta t} - R_b\frac{\delta I_B}{\delta t} + R_C\frac{\delta I_c}{\delta t} \qquad 5.16$$

where R_{ch} is the channel resistance of the pMOS device and R_b is the base resistance. In deriving Eq. 5.16, $\frac{\delta R_{ch}}{\delta t}$ has been assumed to be negligibly small as compared to all the terms in the equation. The numerical results have shown that the first term in the right-hand side of Eq. 5.16 is the dominant one. $R_{ch}\left(=\frac{1}{g_{ds}}\right)$ has been calculated using the model in Toh et al. and Fang et al. [7,11] to give ≈9000 Ω and ≈3000 Ω for the 0.8 μm and 0.2 μm technologies, respectively. This, together with the higher value of $\frac{\delta I_B}{\delta t}$ for the 0.8 μm case, accounts for the steeper slope $\frac{\delta V_{CB}}{\delta t}$ for the 0.8 μm technology than that of the 0.2 μm.

5.3.5 HSPICE Simulations and Analysis

Numerical simulations using HSPICE have been performed on the conventional BiCMOS circuit to evaluate its delay time sensitivity to the key device parameters. The simulations were performed on three technologies (5 V, 0.8 μm), (3.3 V, 0.5 μm) and (2.2 V, 0.2 μm). For each technology, high- and low-quality bipolar devices and three different load capacitances (0.1 pF, 1 pF, and 5 pF) have been considered. The HSPICE model files used in the simulations for the MOS and BJT devices are shown in Tables 5–3 and 5–4, respectively.

Table 5–3 Model files for the MOS device.

MOSFET		(2.2 V, 0.2 μm)		(3.3 V, 0.5 μm)		(5.0 V, 0.8 μm)	
Parameters	Units	nMOS	pMOS	nMOS	pMOS	nMOS	pMOS
V_T	V	0.2594	−0.270	0.6358	−0.643	0.8235	−0.818
T_{OX}	nm	6.0000	6.0000	12.000	12.000	15.000	15.000
R_{shm}	Ω/sq	47.540	90.000	55.540	100.00	75.540	121.00
C_{jw}	pF/m	200.00	245.00	180.00	220.00	150.00	171.70
C_{jm}	μF/m^2	580.00	678.70	500.00	618.70	400.00	498.70
M_{jw}	–	1.0882	0.1416	1.0882	0.1416	1.0882	0.1416
M_{j0}	–	1.0893	0.4831	1.0893	0.4831	1.0893	0.4831
C_{gdom}	pF/m	208.00	208.00	208.00	208.00	208.00	208.00
C_{gsom}	pF/m	208.00	208.00	208.00	208.00	208.00	208.00
C_{gbom}	pF/m	819.00	819.00	819.00	819.00	819.88	819.88

Table 5–4 Model files for the BJT device.

High Quality BJT		(2.2 V, 0.2 μm)		(3.3 V, 0.5 μm)		(5.0 V, 0.8 μm)	
Parameters	Units	npn	pnp	npn	pnp	npn	pnp
β_f	–	80.000	50.000	100.00	60.000	100.00	60.000
I_K	mA	1.2000	0.6660	1.8000	0.9990	2.5000	1.3880
τ_f	ps	3.5000	8.1960	6.0000	14.050	9.0000	21.080

(continued)

Sensitivity Analysis of the Conventional BiCMOS Circuit

Table 5–4 Model files for the BJT device.

High Quality BJT		(2.2 V, 0.2 μm)		(3.3 V, 0.5 μm)		(5.0 V, 0.8 μm)	
Parameters	Units	npn	pnp	npn	pnp	npn	pnp
R_C	Ω	150.00	150.00	150.00	150.00	150.00	150.00
R_E	Ω	30.000	60.000	30.000	60.000	30.000	60.000
R_b	Ω	500.00	111.10	600.00	133.30	500.00	111.10
C_{je}	fF	5.5000	4.3180	9.0000	7.0660	12.100	9.5000
C_{jc}	fF	5.0000	10.000	7.5000	15.000	9.6000	19.200
C_{js}	fF	12.000	23.810	18.200	36.110	25.200	50.000
X_{tf}	–	7.2000	3.4900	7.2000	3.4900	7.2000	3.4900
V_{tf}	–	1.3000	0.0000	1.3000	0.0000	1.3000	0.0000
I_{tf}	mA	1.5000	0.9600	2.0000	1.2800	3.0000	1.9200
I_{kr}	mA	6.0000	3.3600	6.0000	3.3600	9.0000	5.0400

Figure 5–6 shows the percentage change in the T_{plh} (defined as the propagation delay during which the output voltage changes from low to high) for the (5 V, 0.8 μm) technology, based on a ±15% change in the device parameter's value, together with the effects of the load capacitance. Figure 5–7 shows a similar set of data, but for the low-quality bipolar device.

The results show that, in general, high load capacitance yields higher delay sensitivities to β_f, I_K, and R_C. It is understood that employing a low quality BJT gives a relatively poor transient performance. This poor performance occurs because the values of τ_f and R_C are much higher as compared to those of the high-quality BJT; however, the results in Figure 5–7 reveal that its delay sensitivity to key device parameters, except R_C, decreases as the load capacitance increases. The complete set of sensitivity coefficients for the different BiCMOS technologies is shown in Table 5–5. Note that the letters 'H' and 'L' denote high- and low-quality BJT, respectively. Also, the delay sensitivity factor, λ_i, is defined to be the ratio between the normalized change in the delay time over the normalized change in the device parameter's value:

$$\lambda_i = \frac{\frac{\Delta t_d}{t_d}}{\frac{\Delta D_i}{D_i}} \qquad 5.17$$

where D_i is the device parameter under consideration.

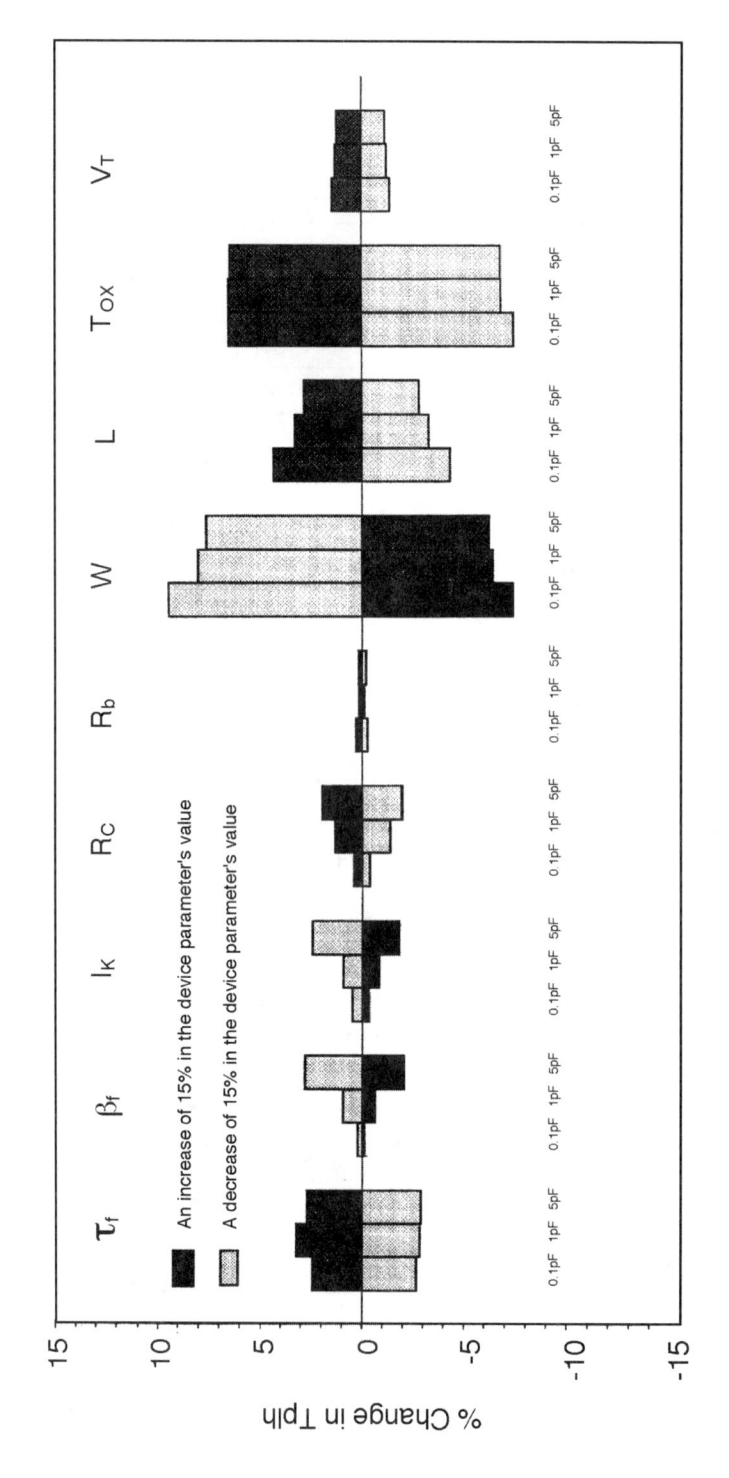

Figure 5–6 Delay sensitivities based on a ±15% change in device parameters together with the effects of the load capacitance, for 5 V, 0.8 μm technology and using high-quality BJT.

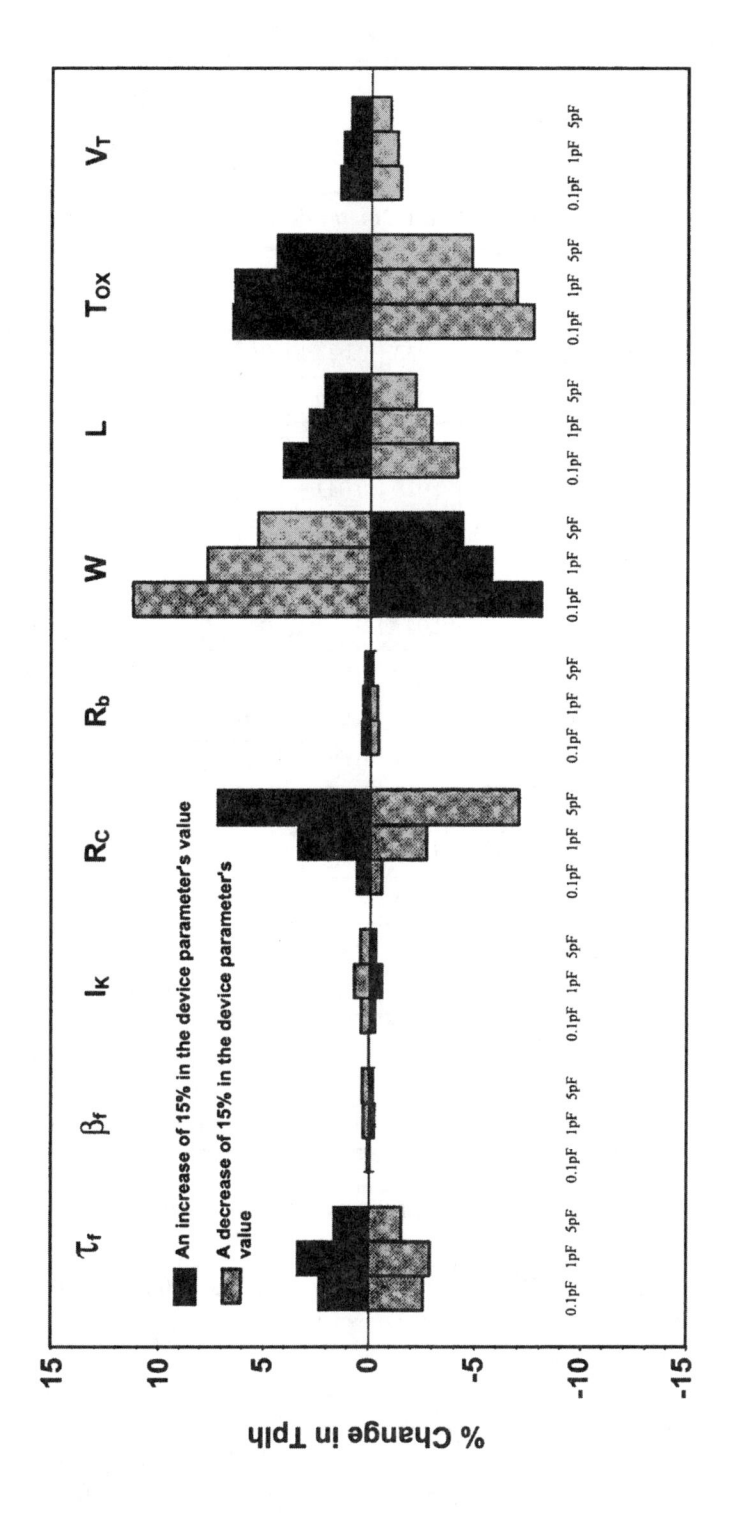

Figure 5–7 Delay sensitivities based on a ±15% change in device parameters together with the effects of the load capacitance, for 5 V, 0.8 μm technology and using low-quality BJT.

Table 5–5 Delay sensitivity factors, λ_i, for the conventional BiCMOS circuit.

Delay Sensitivity Factors, λ_i, for Conventional BiCMOS Circuit											
MOS/BJT Device Parameters			τ_f	β_f	I_K	R_C	R_b	V_T	L	W	T_{OX}
5.0V, 0.8 μm	H	Co = 0.1pF	0.161	0.013	0.032	0.027	0.018	0.094	0.288	0.630	0.431
5.0V, 0.8 μm	H	Co = 1.0pF	0.213	0.063	0.062	0.090	0.008	0.084	0.219	0.535	0.432
5.0V, 0.8 μm	H	Co = 5.0pF	0.179	0.187	0.165	0.131	0.010	0.078	0.189	0.507	0.427
5.0V, 0.8 μm	L	Co = 0.1pF	0.154	0.005	0.028	0.042	0.026	0.096	0.277	0.756	0.438
5.0V, 0.8 μm	L	Co = 1.0pF	0.223	0.019	0.049	0.229	0.023	0.086	0.195	0.519	0.431
5.0V, 0.8 μm	L	Co = 5.0pF	0.107	0.022	0.030	0.488	0.015	0.063	0.145	0.359	0.298
3.3V, 0.5 μm	H	Co = 0.1pF	0.178	0.016	0.047	0.025	0.023	0.123	0.161	0.634	0.474
3.3V, 0.5 μm	H	Co = 1.0pF	0.237	0.081	0.090	0.076	0.014	0.110	0.134	0.565	0.459
3.3V, 0.5 μm	H	Co = 5.0pF	0.166	0.230	0.197	0.084	0.010	0.098	0.122	0.523	0.442
3.3V, 0.5 μm	L	Co = 0.1pF	0.159	0.006	0.034	0.035	0.033	0.125	0.207	0.602	0.451
3.3V, 0.5 μm	L	Co = 1.0pF	0.269	0.027	0.045	0.188	0.021	0.097	0.132	0.541	0.431
3.3V, 0.5 μm	L	Co = 5.0pF	0.176	0.047	0.057	0.349	0.012	0.067	0.091	0.391	0.325
2.2V, 0.2 μm	H	Co = 0.1pF	0.158	0.021	0.053	0.020	0.028	0.042	0.052	0.759	0.456
2.2V, 0.2 μm	H	Co = 1.0pF	0.240	0.153	0.148	0.030	0.022	0.038	0.028	0.592	0.453
2.2V, 0.2 μm	H	Co = 5.0pF	0.081	0.346	0.290	0.057	0.013	0.036	0.019	0.571	0.448
2.2V, 0.2 μm	L	Co = 0.1pF	0.162	0.009	0.034	0.031	0.062	0.037	0.029	0.490	0.491
2.2V, 0.2 μm	L	Co = 1.0pF	0.200	0.044	0.050	0.123	0.028	0.028	0.018	0.450	0.396
2.2V, 0.2 μm	L	Co = 5.0pF	0.137	0.097	0.087	0.385	0.012	0.025	0.007	0.395	0.310

5.4 Sensitivity Evaluation and Comparison of Low-Voltage, Low-Power BiCMOS Circuits

Scaling the technology of BiCMOS circuits tends to accentuate their sensitivity trends in general and to β_f and I_K in particular. It is a well-known fact that a scaled conventional BiCMOS circuit still suffers from two V_{BE} losses, thereby causing degradation to both the circuit speed and its output voltage swing [12,13]. This problem becomes very serious at low-supply voltage that some BiCMOS circuits fail to operate below 2 V supplies. To overcome this problem, new BiCMOS circuits for low-voltage, low-power environments [1,2,14–17] have been developed.

Two comparative studies will be presented below. In the first study, the conventional BiCMOS circuit (refer to Figure 5–1) and circuits (1) and (2) reported in Yeo, Rofail, and Shin

[1,2], respectively (and illustrated in Figures 5–8 and 5–9) have been simulated using HSPICE, and their delay sensitivities based on a ±15% change in the key device parameters' values are obtained. The results are illustrated in Figures 5–10 and 5–11.

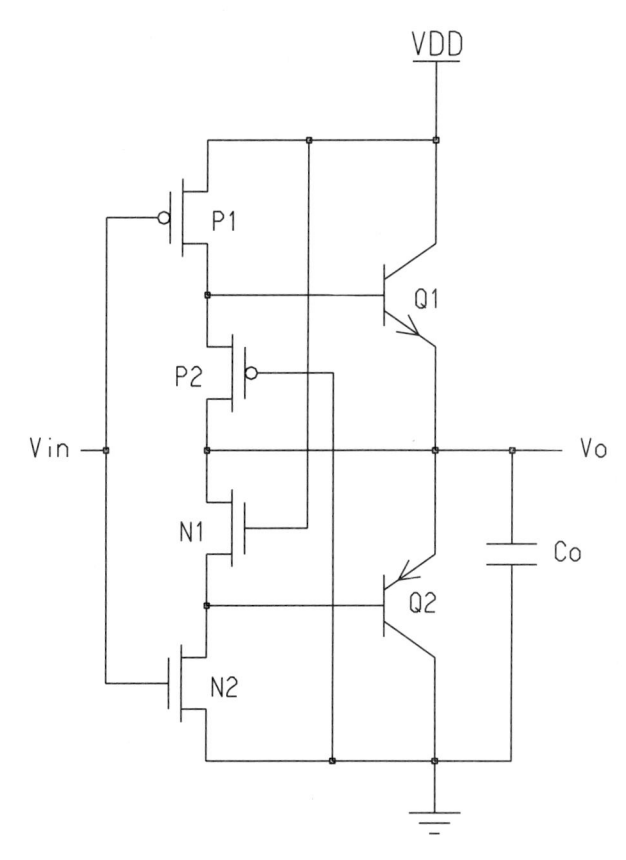

Figure 5–8 Circuit (1). [1]

The results indicate that circuit (1) is less sensitive to changes in the device parameters, except for R_C, than are the other two circuits. The improved sensitivity of circuit (1) is attributed to the BiFET [1] effect and the fact that the drain current is a fraction of the base current driving the BJT [13]. Table 5–6 tabulates the complete sensitivity evaluation and comparison for circuits (1) and (2) employing either high or low quality bipolar devices.

The second comparative study will investigate the delay time and average power dissipation sensitivities of four low-voltage BiCMOS circuits particularly suitable for the 1.5 V power supply voltage regime. These circuits are the feedback complementary BiCMOS (FB-CBiCMOS) [14], transient-saturated full-swing BiCMOS (TS-FS-BiCMOS) [15], bootstrapped

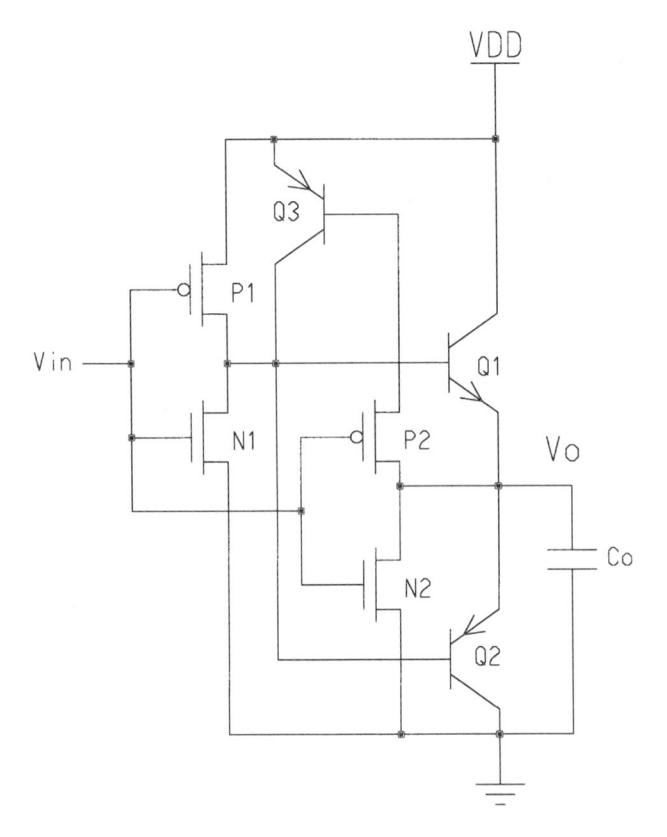

Figure 5–9 Circuit (2). [2]

BiCMOS (BS-BiCMOS) [16] and bootstrapped full-swing BiCMOS (BF-BiCMOS) [17]. They are illustrated in Figure 5–12(a–d), respectively. As mentioned in the Introduction, this study is mandatory as the feature size of today's devices is, in general, less than 0.5 μm. In such a small-geometry, two-dimensional effects become significant and if coupled with the fluctuation in the process or device parameters will affect the circuit performance considerably.

From the sensitivity analysis of these four circuits, the designer is able to identify the right low-voltage circuit to use, as well as the critical parameter(s) affecting the performance of the circuit. HSPICE simulations have been performed for each of the four BiCMOS circuits based on three BiCMOS generations (3.3 V/0.8 μm, 2.2 V/0.5 μm, 1.5 V/0.35 μm), three loading capacitances (0.1 pF, 1 pF and 3 pF), and two different qualities (high and low) of the BJT. For each condition, transient analysis is performed for each parameter, one at a time. The sensitivity coefficient (defined in Eq. 5.17 is obtained for a 1% increase in the device parameter. From the sensitivity coefficients, general trends can be identified for the four circuits under different conditions.

It was found that the MOS parameters and the collector resistance of the bipolar device are, in general, the most sensitive parameters. The MOS parameters affect the switching of the

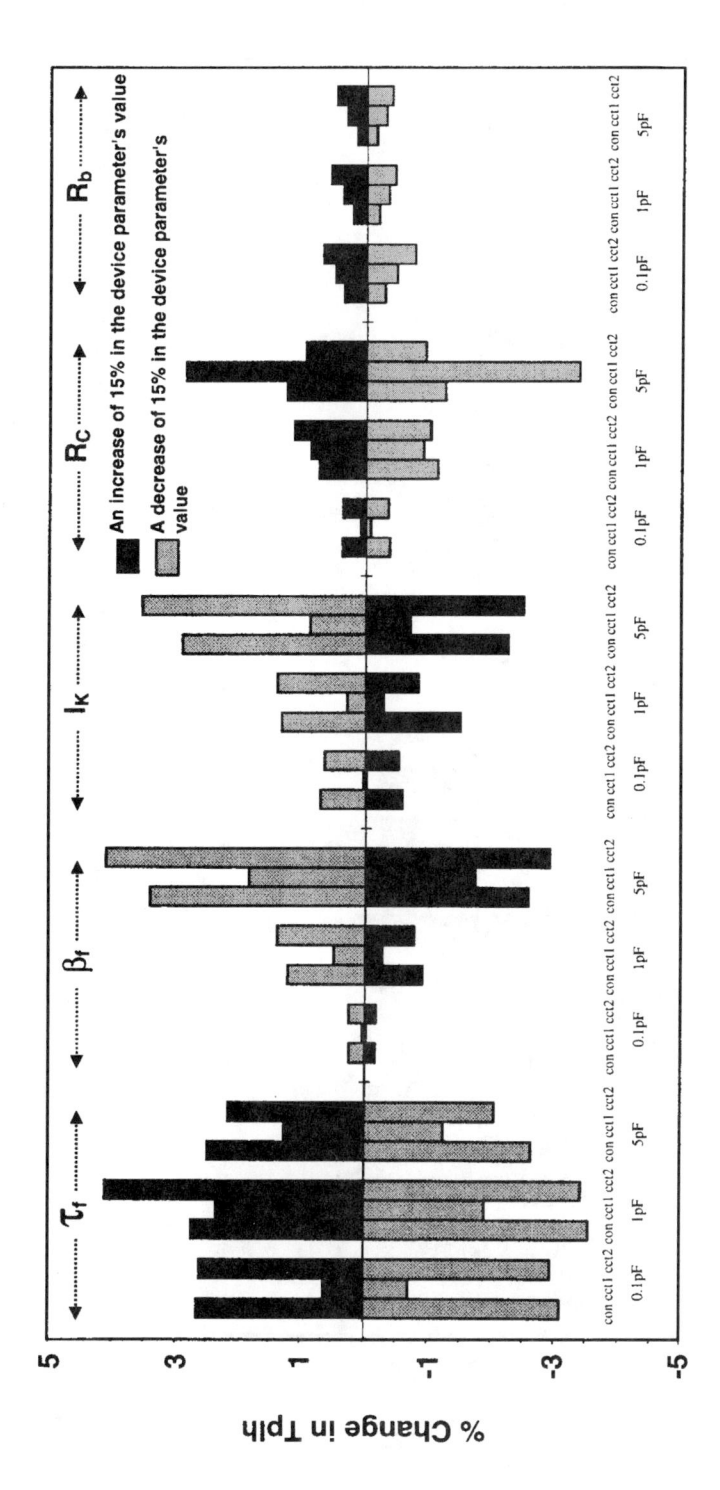

Figure 5–10 Delay sensitivities comparison of the conventional BiCMOS circuit, circuits (1) and (2) based on a ±15% change in the device parameter's value, for the 3.3 V, 0.5 μm technology: τ_f, β_f, I_K, R_C, and R_b.

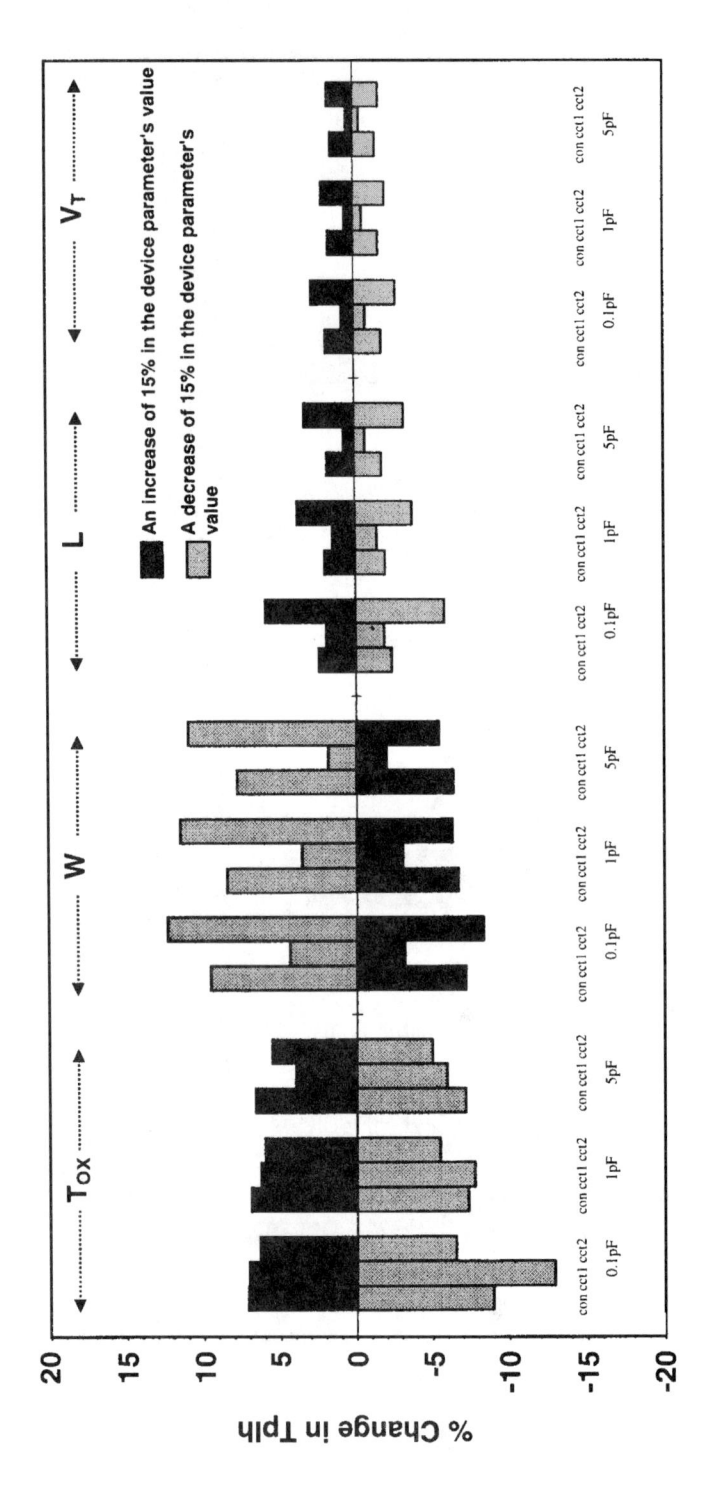

Figure 5-11 Delay sensitivities comparison of the conventional BiCMOS circuit, circuits (1) and (2) based on a ±15% change in the device parameter's value, for the 3.3 V, 0.5 μm technology: T_{ox}, W, L, and V_T.

Table 5–6 Delay sensitivity factors, λ_j, for circuits (1) and (2) based on the 3.3 V, 0.5 μm technology.

MOS/BJT Device Parameters			τ_f	β_f	I_K	R_C	R_b	V_T	L	W	T_{OX}
Cct 1	H	Co = 0.1pF	0.042	0.003	0.002	0.005	0.033	0.054	0.128	0.289	0.470
Cct 1	H	Co = 1.0pF	0.157	0.033	0.019	0.058	0.024	0.040	0.099	0.238	0.418
Cct 1	H	Co = 5.0pF	0.084	0.124	0.098	0.192	0.020	0.028	0.048	0.124	0.265
Cct 1	L	Co = 0.1pF	0.042	0.001	0.005	0.008	0.050	0.051	0.121	0.274	0.513
Cct 1	L	Co = 1.0pF	0.128	0.006	0.004	0.108	0.049	0.040	0.620	0.204	0.406
Cct 1	L	Co = 5.0pF	0.084	0.006	0.001	0.310	0.039	0.023	0.048	0.120	0.310
Cct 2	H	Co = 0.1pF	0.175	0.017	0.043	0.024	0.045	0.186	0.392	0.825	0.424
Cct 2	H	Co = 1.0pF	0.275	0.093	0.094	0.076	0.037	0.139	0.252	0.771	0.400
Cct 2	H	Co = 5.0pF	0.145	0.276	0.239	0.064	0.031	0.112	0.218	0.733	0.368
Cct 2	L	Co = 0.1pF	0.171	0.007	0.034	0.036	0.056	0.125	0.254	0.686	0.433
Cct 2	L	Co = 1.0pF	0.222	0.027	0.052	0.096	0.040	0.106	0.185	0.599	0.388
Cct 2	L	Co = 5.0pF	0.199	0.075	0.073	0.224	0.028	0.104	0.171	0.590	0.378

MOSFET from the linear region to the saturation mode, which will, in turn, influence the charging and discharging current. R_C plays a crucial role because it determines when the BJT enters the saturation mode. If the BJT saturates prematurely, the charging current and discharging current will be degraded. Figure 5–13(a–d) compares the delay sensitivity of the four circuits for the 1.5 V, 0.35 μm technology and Figure 5–14(a–d) compares their power dissipation (Pavg) sensitivity. In these figures, "H" and "L" denote high and low quality BJT respectively.

Based on the results shown in Figures 5–13(a–d) and 5–14(a–d), it is clear that the delay and power dissipation sensitivities are affected by the loading, the quality of the BJT, and the scaling of the technology. The following section discusses these issues.

5.4.1 The Effects of Loading on the Delay and Power Dissipation Sensitivities

With regard to the delay time sensitivity, it can be seen from Figure 5–13(a–d) that the delay time sensitivity to R_C generally increases with the loading. However, the sensitivity to the MOS parameters decreases in general with the load. This decrease occurs because when the loading is low, the time taken for the BJT to charge or discharge the capacitance is very critical, but as the loading increases, the corresponding demand of current would probably cause the BJT to saturate early. Thus, at high loading, R_C becomes very significant.

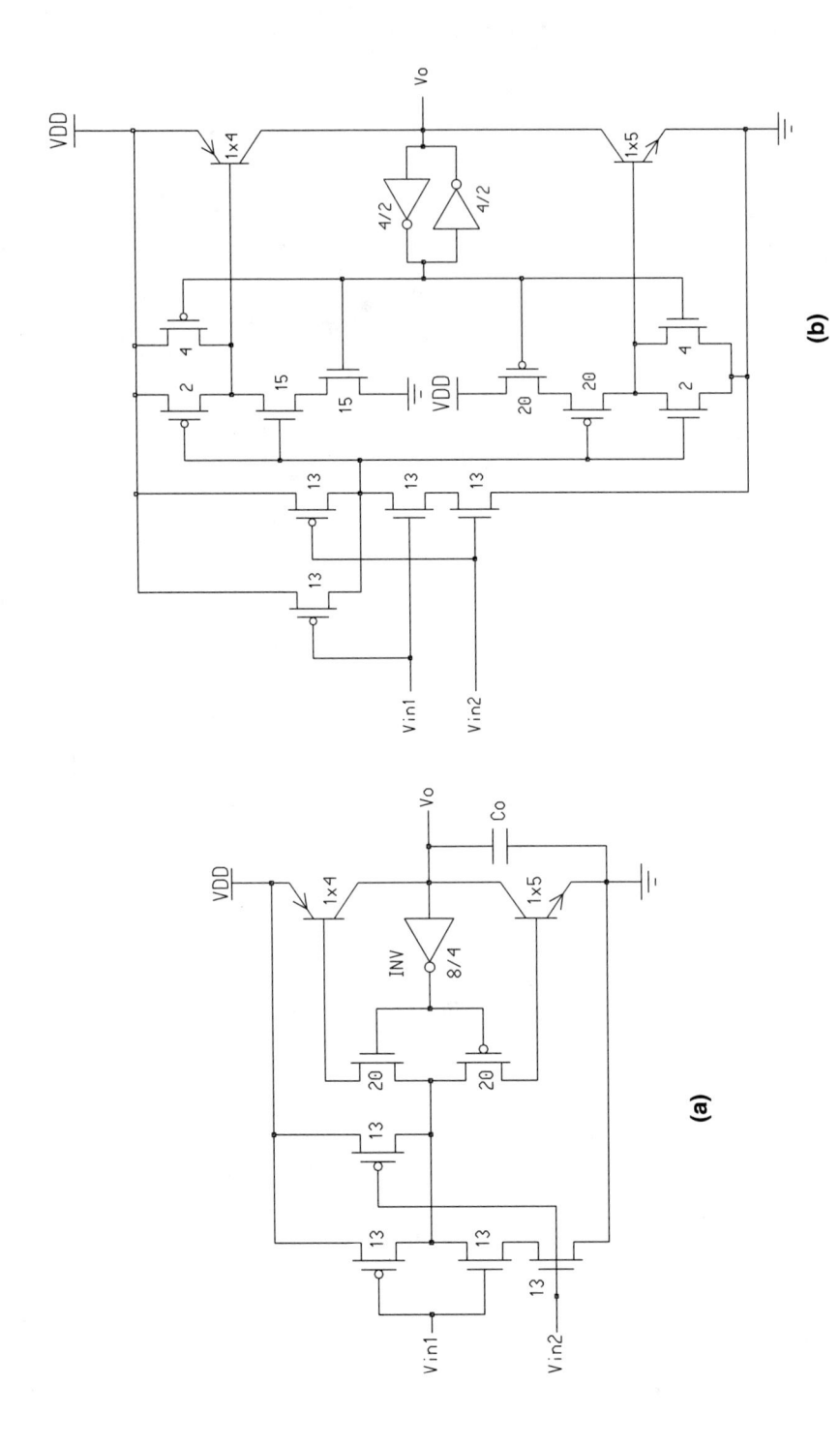

Figure 5–12 (a) FB-CBiCMOS AND gate, (b) TS-FS-BiCMOS NAND gate.

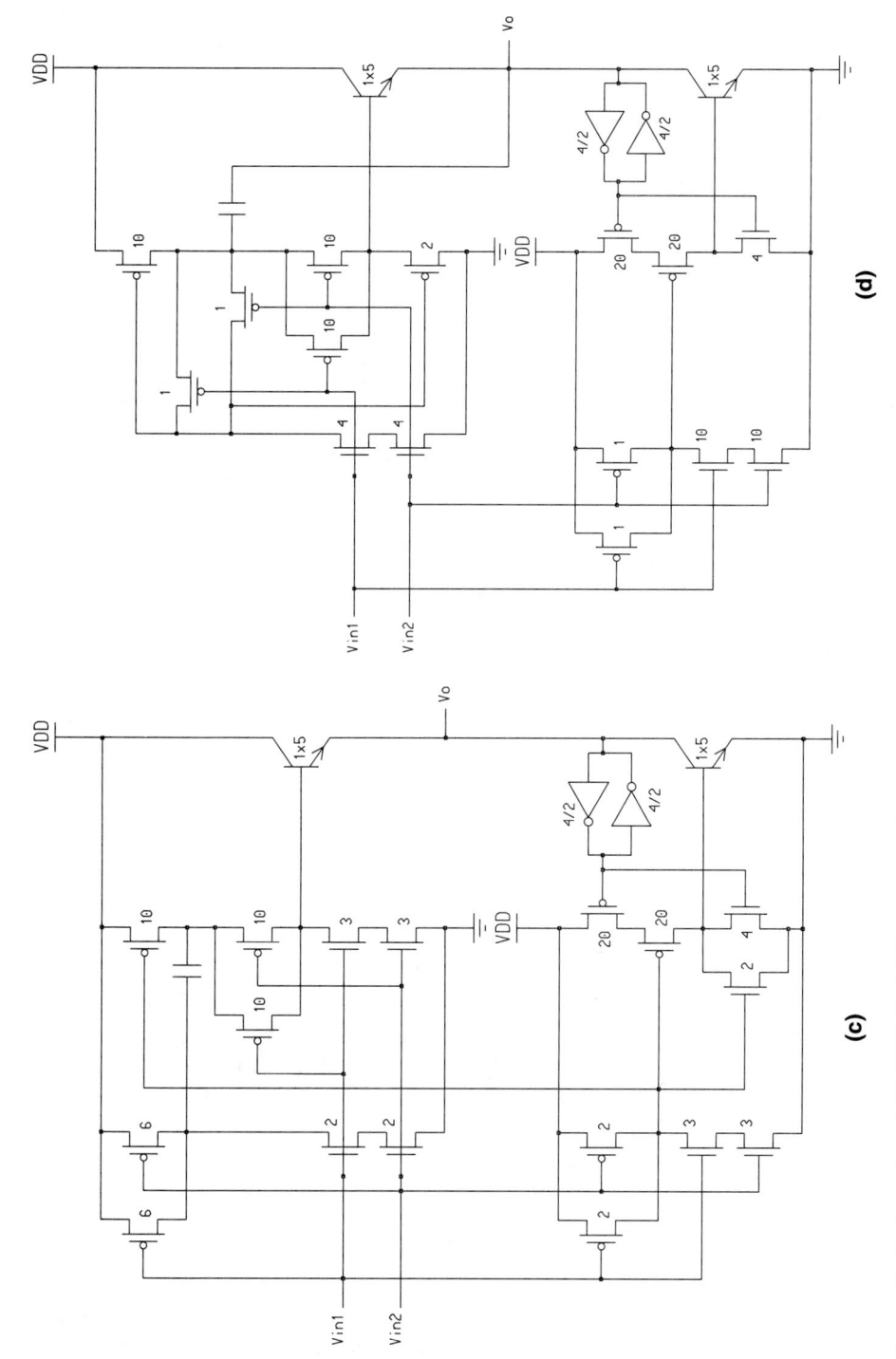

Figure 5–12 (cont.) (c) BS-BiCMOS NAND gate, (d) BF-BiCMOS NAND gate.

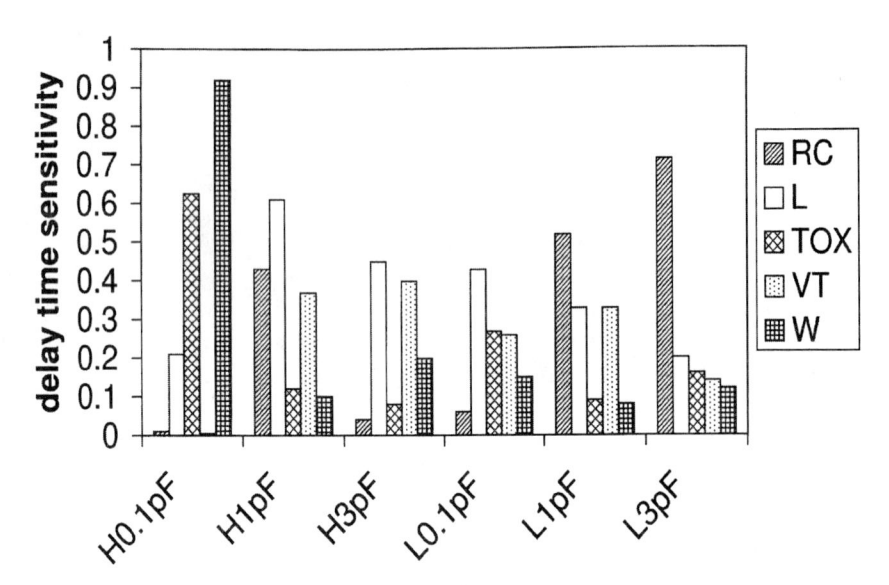

Figure 5–13a Delay sensitivity of FB-CBiCMOS [18].

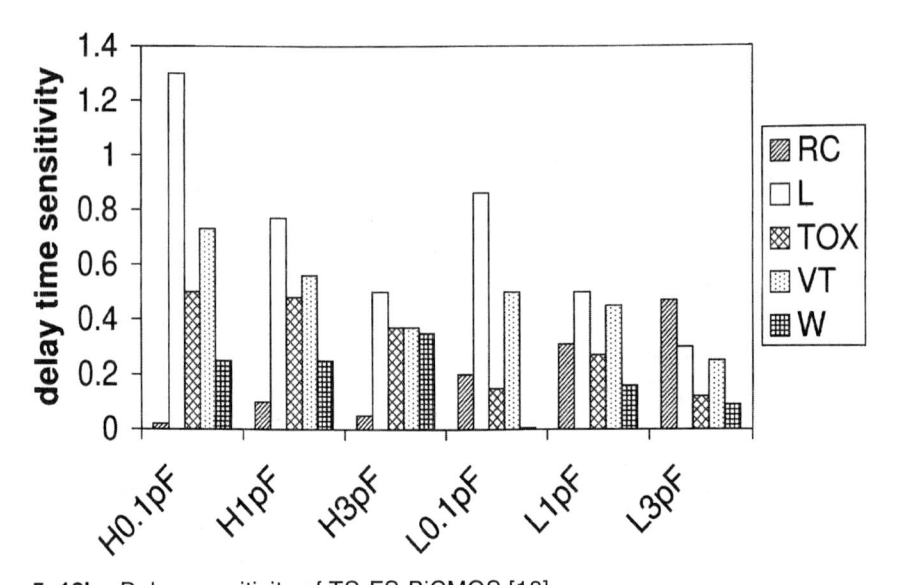

Figure 5–13b Delay sensitivity of TS-FS-BiCMOS [18].

Comparing the results in Figure 5–13(a to d), it has been found that the FB-CBiCMOS is very sensitive to changes in R_C. A possible reason is that the BJT of the FB-CBiCMOS is saturated for a longer period of time than for the other three circuits (note that the other three circuits employ transient saturation technique in their design). Another finding is that among all the MOS parameters, L is the most influential.

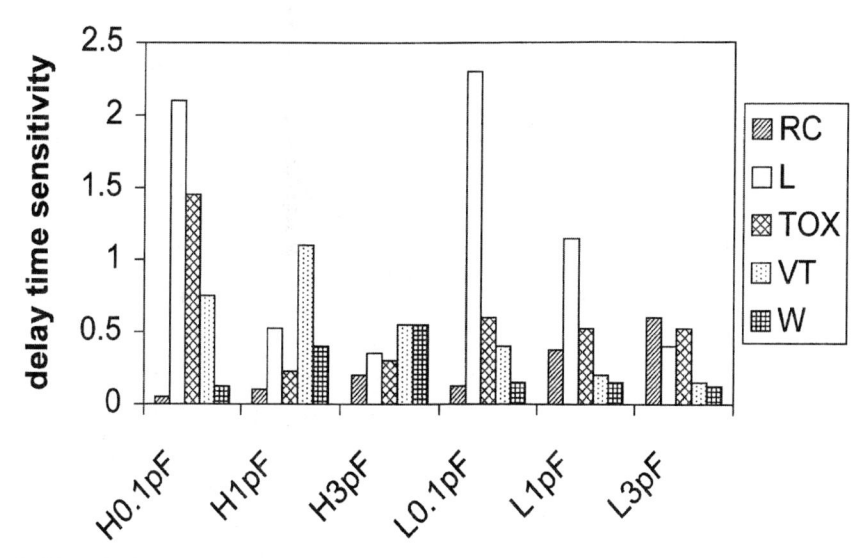

Figure 5-13c Delay sensitivity of BS-BiCMOS [18].

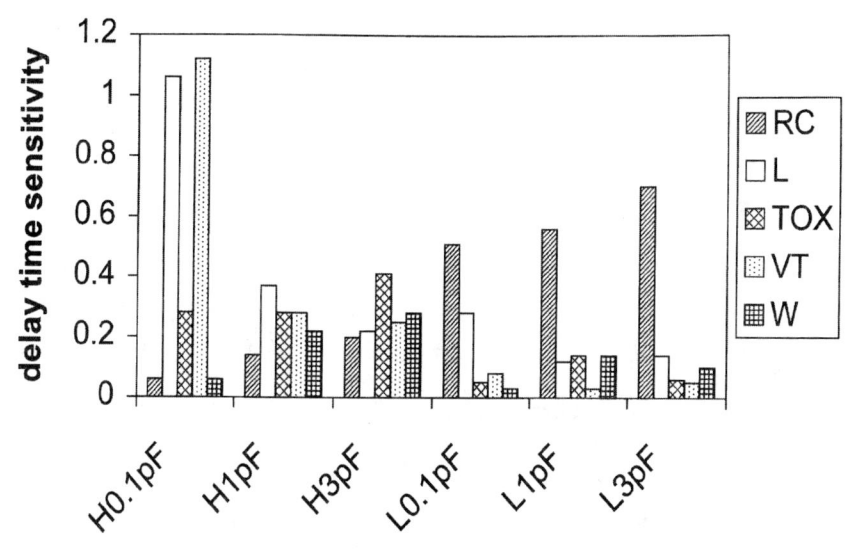

Figure 5-13d Delay sensitivity of BF-BiCMOS [18].

For average power dissipation consideration, from Figure 5–14(a–d), it can be seen that the power consumption sensitivities of all four circuits to the device parameters are the lowest when the loading is high because the dynamic power dissipation of the BiCMOS circuit is generally dominated by the transient component, which depends on both the internal and the load capacitance. Hence, when the loading is high, fluctuation in the internal capacitance will be nullified by the load capacitance.

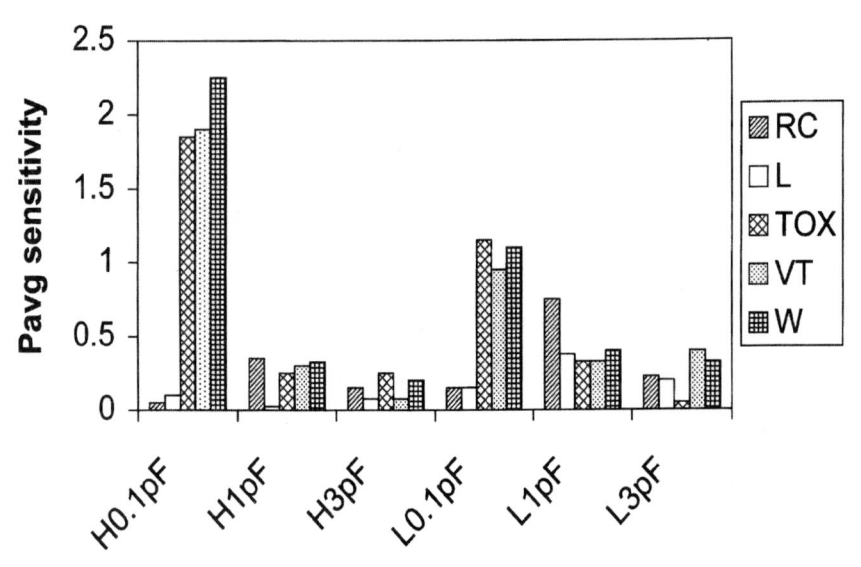

Figure 5-14a (a) Power dissipation sensitivity of FB-CBiCMOS [18].

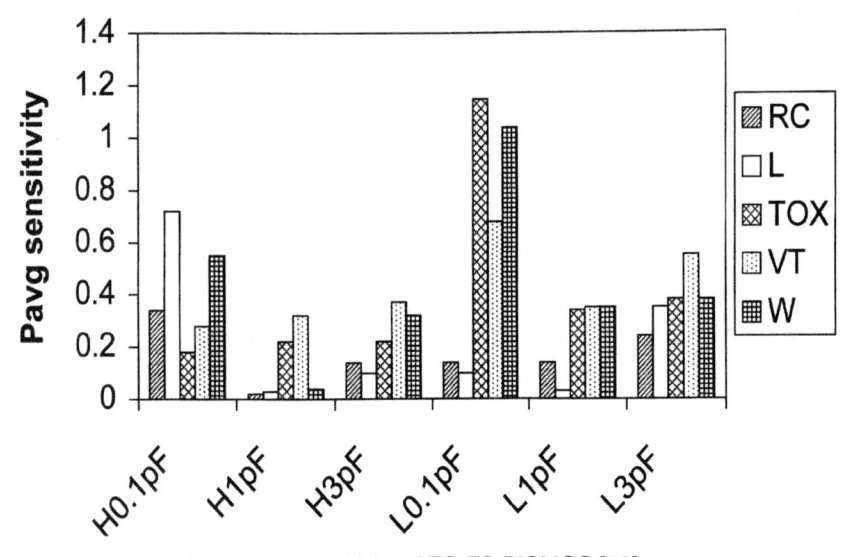

Figure 5-14b Power dissipation sensitivity of TS-FS-BiCMOS [18].

5.4.2 The Effects of the Quality of the BJT on the Delay and Power Dissipation Sensitivities

From Figure 5-13(a-d), it can be observed that the delay sensitivity to R_C is higher for low-quality BJT because low quality BJT has a higher R_C value than the high quality BJT (dominant R_C effect). Hence, its delay time sensitivity shows considerable change with the variation

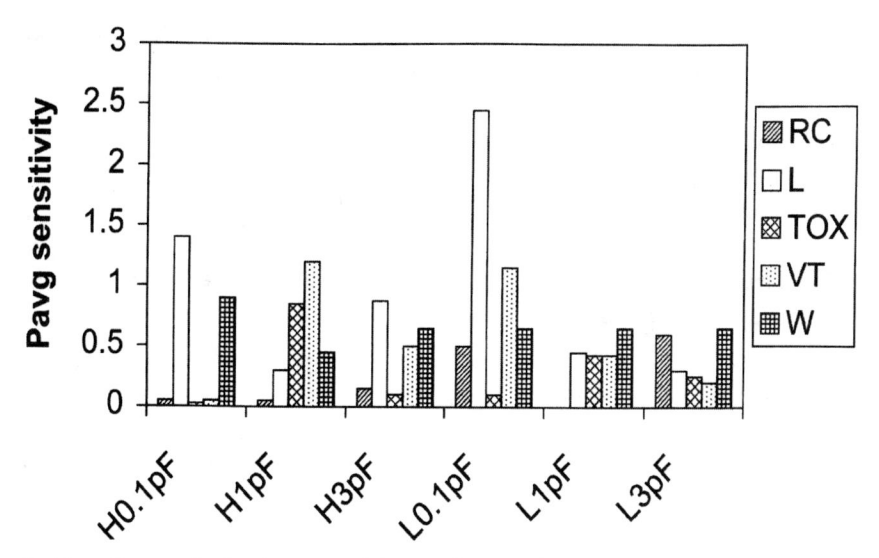

Figure 5–14c Power dissipation sensitivity of BS-BiCMOS [18].

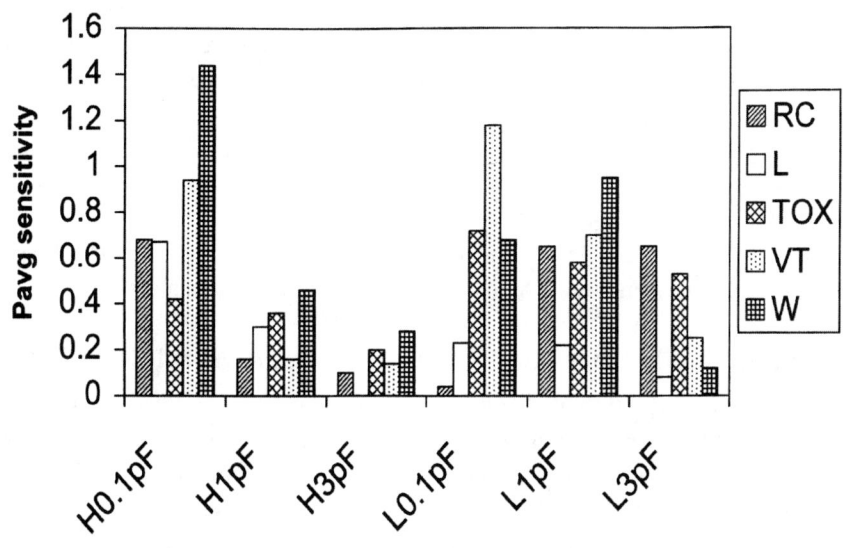

Figure 5–14d Power dissipation sensitivity of BF-BiCMOS [18].

of collector resistance, R_C, especially at high load, where high current flows through the collector. Therefore, due to the large voltage drop across R_C, low quality BJT is more likely to saturate early, thereby causing increased delay sensitivity. The delay time sensitivities of all circuits to other BJT parameters such as τ_f, β_f, R_b, and R_E are unaffected by the quality of the BJT.

The results in Figure 5–14(a–d) show the effect of the quality of the BJT on the power dissipation sensitivity of the four circuits. It is observed that at low loads, high quality BJT for the FB-CBiCMOS and BF-BiCMOS circuits are more sensitive to the device parameters, whereas for the TS-FS-BiCMOS, low quality BJT is more sensitive to the device parameters. For BS-BiCMOS circuit shown in Figure 5–14(c), no clear trend can be seen for changes either in loading or the quality of the BJTs.

5.4.3 The Effects of Scaling the Technology on the Delay and Power Dissipation Sensitivities

It can be seen that scaling the technology has a similar impact on both the power consumption and the delay sensitivities. As the technology changes from 3.3 V, 0.8 μm to 1.5 V, 0.35 μm, both the power dissipation and the delay sensitivities of the BS-BiCMOS, FB-CBiCMOS, and TS-FS-BiCMOS first decrease and then increase with scaling. Only BF-BiCMOS exhibits a different trend, whereby the power consumption and delay sensitivities decrease with scaling. As for the FB-CBiCMOS and BF-BiCMOS circuits, their power dissipation sensitivities decrease as the technology is scaled down, while for the TS-FS-BiCMOS and BS-BiCMOS, their power dissipation sensitivities are more significant at smaller technologies.

In general, we would expect the power consumption sensitivity to the device parameter to decrease as the circuit is scaled down because the load capacitance predominates. The sensitivity comparison reveals that both the BF-BiCMOS and FB-CBiCMOS have the advantage over the BS-BiCMOS and TS-FS-BiCMOS at the 1.5 V, 0.35 μm technology. As such, they are most suitable for low-voltage applications. However, at higher technology (0.8 μm) and supply voltage (3.3 V), BS-BiCMOS and TS-FS-BiCMOS circuits are more suitable. Figure 5–15 shows the average delay and power dissipation sensitivities of all the four circuits (BS-BiCMOS, BF-BiCMOS, FB-CBiCMOS, and TS-FS-BiCMOS which are denoted by BS, BF, FB, and TS, respectively, in the figure) employing high quality BJTs for a load capacitance of 1 pF.

5.5 Delay Sensitivity Upper and Lower Bounds: A Worst Case Scenario

In this section, a method to predict the upper and lower limits of the delay time as related to its sensitivity, to the process, and to the device parameters and the associated tolerances will be presented. The increase in the delay time resulting from a ±15% change in a process parameter P_i, according to this method, is given as:

$$\frac{\Delta t_d}{t_d} = \alpha_i \lambda_i^+ \qquad 5.18$$

where α_i has the same definition as that shown in Table 5–1 and λ_i^+ is the relative increase in t_d normalized with respect to the relative change in the device parameter's value. Note that the positive index in λ_i indicates that the change in the value of the device parameter has taken place in the direction that causes an increase in t_d. To calculate an upper bound for the delay sensitivity, a worst case scenario is adopted. If one process parameter is taken into consideration, it will be the

Conclusions

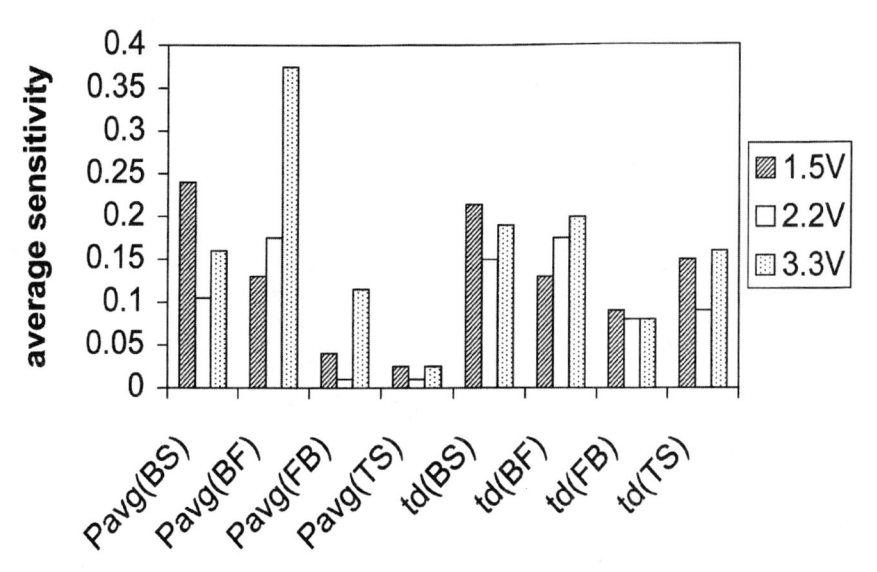

Figure 5–15 Effect of scaling the technology on the delay and power dissipation sensitivities of the four circuits.

one causing the greatest increase in t_d. The influence of more than one process parameter can then be incorporated in Eq. 5.18 according to their α and λ values, with the highest $(\alpha\lambda)$ taken first. Hence, the upper bound for the relative increase in the delay time will be given as:

$$\left(\frac{\Delta t_d}{t_d}\right)_u = \sum_{i=1}^{n} \alpha_i \lambda_i^+ \qquad 5.19$$

where n is the number of process parameters. A similar approach is adopted to calculate the lower bound for the delay sensitivity of circuits (1) and (2) (as described in section 5.4) against the number of process parameters. The results are obtained based on a ±15% tolerance for all values of the process parameters.

Figure 5–16 reveals that the upper and lower bounds for the delay sensitivity of circuit (1) are 5% to 12% and –5% to –10%, respectively, and that of circuit (2) are 10% to 22% and –8% to –19%, respectively. This method of obtaining the upper and lower bounds for the delay sensitivity is applicable to other circuits as well.

5.6 Conclusions

This chapter has presented a detailed sensitivity analysis of a few BiCMOS circuits, starting with the conventional BiCMOS circuit and later extended to cover some recently reported low-voltage BiCMOS circuits. It was found that the BiCMOS delay sensitivity is influenced by key device and process parameters, scaling the technology, load capacitance, the quality of the BJT, and even the circuit configuration. The BiCMOS delay sensitivity to the collector resis-

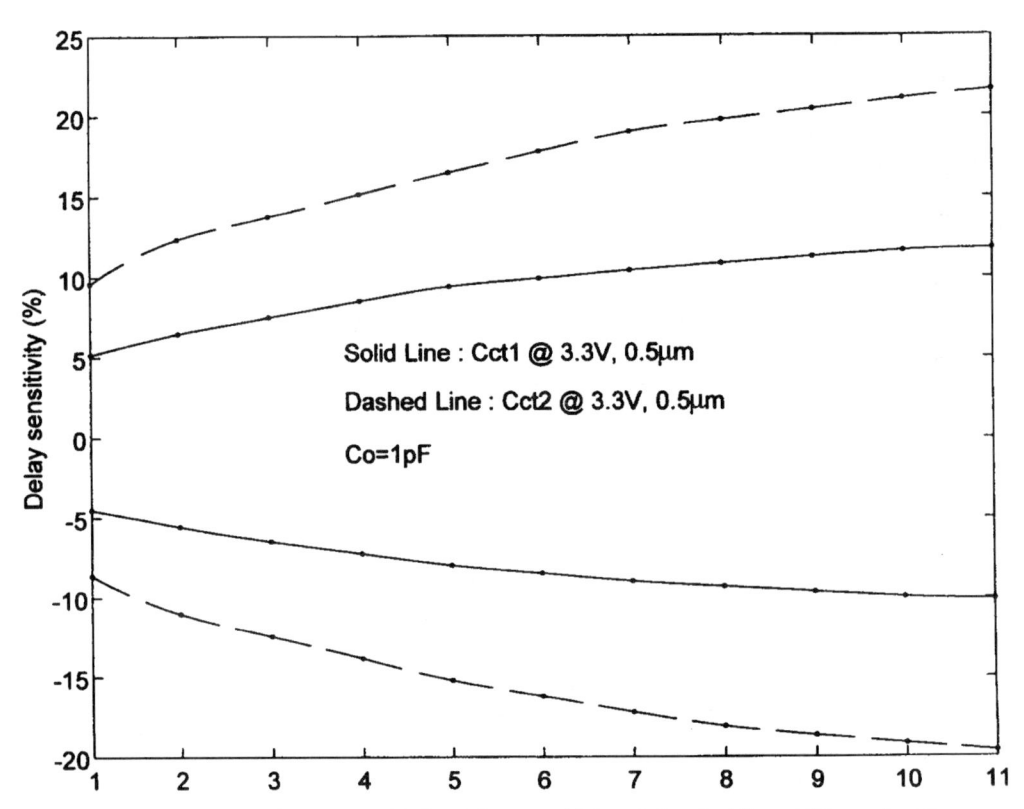

Figure 5–16 Upper and lower bounds for delay sensitivity of circuits (1) and (2) against the number of process parameters at a load of 1 pF and using high quality BJT.

tance increases with any variable that causes the BJT to move from nonsaturation towards saturation. The sensitivity to the MOS channel length, channel width, threshold voltage and oxide thickness is affected by other device or circuit parameters according to their influence on the BJT injection level. It can be concluded that high load capacitance and low quality BJT, in general, lead to increased delay sensitivities to the current gain β_f and the knee current I_K. Furthermore, technology scaling tends to accentuate the trends, especially with respect to changes in β_f and I_K. It has also been shown that the circuit design can affect the circuit's sensitivities. The sensitivity comparison has shown that the BF-BiCMOS and FB-CBiCMOS have low delay sensitivity coefficients. As such, they are suitable for low-voltage applications. A method has been described to calculate, based on a worst case scenario, the upper and lower bounds for the sensitivity as a function of numerically generated coefficients together with the process parameters' tolerances. For a ±15% tolerance, an upper bound of 5% to 22% for the delay sensitivity could result depending on the kind of BiCMOS circuit used. The presented methodology could be used, in general, to evaluate the delay sensitivities of technologies for comparison purposes.

5.7 References

[1] K. S. Yeo and S. S. Rofail, "Full-Swing High Speed CBiCMOS Digital Circuit for Low-Voltage Applications," *IEE Proceedings - Circuits, Devices, and Systems*, Vol. 142, No. 1, pp. 8–14, February 1995.

[2] H. J. Shin, "Performance Comparison of Driver Configurations and Full-Swing Techniques for BiCMOS Logic Circuits," *IEEE J. Solid-State Circuits*, Vol. 25, No. 3, pp. 863–865, June 1990.

[3] S. H. K. Embabi, A. Bellaouar, and M. I. Elmasry, *Digital BiCMOS Integrated Circuit Design*, Kluwer Academic Publishers, 1993.

[4] Liu et al., "Threshold Voltage Model for Deep-Submicrometer MOSFETs," *IEEE Trans. Electron Devices*, Vol. 40, No. 1, pp. 86–94, January 1993.

[5] S. S. Rofail, K. S. Yeo, and Y. S. Sin, "Delay Time Sensitivity Analysis of Multi-Generation BiCMOS Digital Circuits," *IEE Proceedings - Circuits, Devices, and Systems*, Vol. 144, No. 2, pp. 1–8, February 1997.

[6] A. Bellaouar, S. H. K. Embabi, and M. I. Elmasry, "Scaling of Digital BiCMOS Circuits," *IEEE J. Solid-State Circuits*, Vol. 25, No. 4, pp. 932–941, August 1990.

[7] K. Y. Toh, P. K. Ko, and R. G. Meyer, "An Engineering Model for Short-Channel MOS Devices," *IEEE J. Solid-State Circuits*, Vol. 23, No. 4, pp. 950–958, August 1988.

[8] P. A. Raje, K. C. Saraswat, and K. M. Cham, "Accurate Delay Models for Digital BiCMOS," *IEEE Trans. Electron Devices*, Vol. 39, No. 6, pp. 1456–1464, June 1992.

[9] S. Zhang, T. S. Kalkur, S. Lee, and D. Chen, "Analysis of the Switching Speed of BiCMOS Buffer Under High Current," *IEEE J. Solid-State Circuits*, Vol. 29, No. 7, pp. 787–796, July 1994.

[10] G. P. Rosseel and R. W. Dutton, "Influence of Device Parameters on the Switching Speed of BiCMOS Buffers," *IEEE J. Solid-State Circuits*, Vol. 24, No. 1, pp. 90–99, February 1989.

[11] W. Fang, A. Brunnschweiler, and P. Ashburn, "An Accurate Analytical BiCMOS Delay Expression and Its Application to Optimizing High-Speed BiCMOS Circuits," *IEEE J. Solid-State Circuits*, Vol. 27, No. 2, pp. 191–202, February 1992.

[12] M. Fujishima, K. Asada and T. Sugano, "Evaluation of Delay-Time Degradation of Low-Voltage BiCMOS Based on a Novel Analytical Delay-Time Modeling," *IEEE J. Solid-State Circuits*, Vol. 26, No. 1, pp. 25–31, January 1991.

[13] S. S. Rofail, "Low-Voltage Low-Power BiCMOS Digital Circuits," *IEEE J. Solid-State Circuits*, Vol. 29, No. 5, pp. 572–579, May 1994.

[14] S. S. Rofail and K. S. Yeo, "New Complementary BiCMOS Digital Gates for Low-Voltage Environment," *Solid-State Electronics*, Vol. 39, No. 5, pp. 681–687, 1996.

[15] M. Hiraki et al., "A 1.5V Full-Swing BiCMOS Logic Gate," *IEEE J. Solid-State Circuits*, Vol. 27, No. 11, pp. 1568–1573, November 1992.

[16] R. Y. V. Chik, and C. A. T. Salama, "1.5V Bootstrapped BiCMOS Logic Gate," *Electronics Letters*, Vol. 29, No. 3, pp. 307–309, 1993.

[17] A. Bellaouar, M. I. Elmasry, and S. H. K. Embabi, "Bootstrapped Full-Swing BiCMOS/BiNMOS Logic Circuits for 1.2 - 3.3V Supply Voltage Regime," *IEEE J. Solid-State Circuits*, Vol. 30, No. 6, pp. 629–0–636, June 1995.

[18] K. S. Yeo, K. L. Chan, and M. M. Tang, "To Study the Influence of Device Parameters on the Circuit Performance of Low-Voltage BiCMOS Circuits," *Second National Undergraduate Research Programme Congress, Proc.*, Singapore, Vol. 2, pp. 993–998, 1997.

5.8 Appendix

In this appendix, the derivation of the delay sensitivity to τ_f is given. Similarly, the delay sensitivity to β_f and I_K could be derived.

$$t_d \cong \tau_f \sqrt{\frac{\beta_f I_K}{I_{Dsat}}} + \frac{V_{os}(C_O + C_1)}{2\sqrt{\beta_f I_K I_{Dsat}}} \qquad 5.20$$

$$\frac{\partial t_d}{\partial \tau_f} = \sqrt{\frac{\beta_f I_K}{I_{Dsat}}} \qquad 5.21$$

$$\frac{\partial t_d}{\partial \tau_f} = \frac{\Delta t_d}{\Delta \tau_f} \qquad 5.22$$

$$\frac{\Delta t_d}{t_d} = \frac{\partial t_d}{\partial \tau_f} \frac{\Delta \tau_f}{t_d} \qquad 5.23$$

$$\frac{\Delta t_d}{t_d} = \frac{\tau_f \sqrt{\frac{\beta_f I_K}{I_{Dsat}}}}{\tau_f \sqrt{\frac{\beta_f I_K}{I_{Dsat}}} + \frac{V_{os}(C_O + C_1)}{2\sqrt{\beta_f I_K I_{Dsat}}}} \frac{\Delta \tau_f}{\tau_f}$$

$$= \frac{2\beta_f I_K \tau_f}{2\tau_f \beta_f I_K + V_{os}(C_O + C_1)} \frac{\Delta \tau_f}{\tau_f}$$

$$= \Psi_{\tau_f} \frac{\Delta \tau_f}{\tau_f} \qquad 5.24$$

The Authors

Samir S. Rofail received a B.Sc. degree in electrical engineering from Cairo University, Egypt, and a B.Sc. degree in mathematics from Ain Shams University, Egypt, in 1971 and 1973, respectively. He received M.A.Sc. and Ph.D. degrees, both in electrical engineering, from the University of Waterloo, Waterloo, Ontario, Canada, in 1975 and 1978, respectively. He has been involved in teaching, research, and consulting in the areas of semiconductor devices and integrated circuits for the last twenty years. He worked for Cairo University from 1971 to 1973 and for the University of Waterloo as a teaching and research assistant from 1973 to 1978. In addition, Professor Rofail held faculty positions in the United States, Canada, Saudi Arabia, and Singapore.

From 1990 to 1992, Professor Rofail was with the VLSI group at the University of Waterloo working on the modeling and design of BiCMOS digital circuits. He has been with Nanyang Technological University (NTU) in Singapore since 1992, where he is currently the coordinator of the IC-Design group. At NTU, he led an intensive research activity with a primary focus on low-voltage, low-power BiCMOS/CMOS circuits. Professor Rofail's research interests also include the modeling of deep sub-micrometer MOSFETs and high performance VLSI circuits. He serves as a technical reviewer for several international journals and has published more than 40 journal papers, several of which are on IC-Design for low-power applications. Professor Rofail is listed in Who's Who in the World.

Kiat-Seng Yeo has been with Nanyang Technological University (NTU) in Singapore since 1993, where he is currently a member of the faculty at the School of Electrical and Electronic Engineering. Previously he was with Toshiba and STMicroelectronics. Professor Yeo has many years of teaching and research experience in integrated circuit design and holds B.Eng. (Hons.) and Ph.D. degrees in electrical engineering from Nanyang Technological University. In addition, he was awarded a book prize in 1992 for excellent academic achievements.

Currently, Professor Yeo provides consulting to statutory boards and multinational corporations in the areas of semiconductor devices and electronic circuit design. He has been extensively involved in the modeling and fabrication of small MOS/Bipolar integrated technologies for the last six years. His research interests also include the design of new circuits and systems (based on scaled technologies) for low-voltage, low-power applications; integrated circuit design of BiCMOS/CMOS multiple-valued logic circuit, domino logic, and memories; and device characterization of deep sub-micrometer MOSFETs. He holds one patent and has published numerous articles on CMOS/BiCMOS technology and integrated circuit design in leading technical journals worldwide, and is a technical reviewer for several international journals.

Index

A

Active device isolation, 43-44
Adder:
    full, 254
    half, 254
Analog/digital BiCMOS process flow, 60-64
Analytical transient model, FB-BiCMOS, 245-47
Applications, BiCMOS, 7-10
    Digital Signal Processing (DSP), 9
    random access memories (RAMs), 7-9
    telecommunications applications, 9-10
Autodoping, and epitaxial layer, 42

B

Base design techniques, 36-37
Base resistance, 195
BF-BiCMOS:
    delay of two-input gates, 294
    power dissipation, 296-97
    transistor count/area count, 296
    two-input NAND gates, 293

BF-BiNMOS, 288-90
    delay of two-input gates, 294
    power dissipation, 296-97
    transistor count/area count, 296
    two-input NAND gates, 293
BiCMOS, 140, 142
    applications of, 7-10
        Digital Signal Processing (DSP), 9
        random access memories (RAMs), 7-9
        telecommunications applications, 9-10
BiCMOS dynamic circuit, 256-58
BiNMOS logic circuits, 208-23
bootstrapped, 279-304
    bootstrapped BiCMOS logic gate (BS-BiCMOS) (1.5 V), 281-83
    bootstrapped full-swing BiCMOS/BiNMOS inverter, 283-97
    double bootstrapped BiCMOS logic gates (DB-BiCMOS), 297-304
conventional, structure of, 179
Darlington-type BiCMOS inverter, 190

delay analysis, 201-4
Digital Signal Processing (DSP), 9
double bootstrapped, 297-304
DRAMS, 9
driver configuration, 147-49
future directions, 17-20
   decrease in power supply voltage, 17-18
   toward BiCMOS technology, 18-19
   universal VDD concept, 19-20
Integrated Circuits (ICs), 50-52
interconnect issues, 44-50
   metallization, 48-49
   planarization, 48-49
   silicidation, 44-48
key parameters used in circuit simulations, 195
measured unloaded delay, 157
process technology, 27-82
pull-down operation of, 171
QC-BiCMOS logic circuits, 189-208
   circuit concept, advantages of, 190-93
   circuit performance comparison and analysis, 193-201
   delay analysis, 201-4
   design issues, 204
   experimental results, 204-6
R + N type, 224-29
   experimental results, 228-29
   simulation results, 226-27
shunting techniques, 160
SPICE simulations on, 140
totem-pole gate, 1-2
two-input NAND gates, 213, 221
BiCMOS charge pump with HB-BiCMOS, 264-68
   simulated output voltage of, 267
BiCMOS DRAMS, 9
BiCMOS driver configuration, 147-49

BiCMOS dynamic circuit, 256-58
   speed advantages, 258
BiCMOS dynamic logic circuit (1.5 V), 252-58
   circuit operation, 254
   circuit performance comparison/discussion, 254-58
   dynamic full adder, 254
BiCMOS full-swing circuits using positive capacitively coupled feedback technique, 230-44
   circuit performance comparison, 235-41
      AND gates, 238-41
      simple buffers, 235-38
   design issues regarding feedback circuit, 241-44
BiCMOS Integrated Circuits (ICs), 50-52
   starting material for IC fabrication, 52
BiCMOS process (0.35 mm), key device parameters for, 287
BiCMOS process (0.5 mm), key device parameters for, 216
BiCMOS process technology, 4-6, 27-82
   advantages of, 6
   analog/digital BiCMOS process flow, 60-64
   as "AND" function of high-speed and high-density, 7
   base design techniques, 36-37
   BiCMOS fabrication, 6
   charge pump with HB-BiCMOS, 264-68
   DRAMS, 9
   driver configuration, 147-49
   dynamic circuit, 256-58
   dynamic logic circuit (1.5 V), 252-58
   full-swing circuits using positive capacitively coupled feedback technique, 230-44

BiCMOS interconnect issues, 44-50
    metallization, 48-49
    planarization, 48-49
    silicidation, 44-48
BiCMOS isolation issues, 39-44
    active device isolation, 43-44
    buried layers, 43
    epitaxial layer, 42
    latch-up phenomenon, 39
    trench isolation, 40-42
bipolar process techniques, 33-36
classification of BiCMOS technologies, 50-56
    high-performance 0.5 mm digital BiCMOS process changes, 56-60
    high-performance 0.8 mm BiCMOS process flow, 50-56
as CMOS speed enchancer/bipolar power miser, 4
CMOS technology compared to, 4-6
collector design techniques, 37
comparison of new process flow with, 65
convergence of bipolar and CMOS processes, 28-30
and design of high-speed mega-byte DRAMs, 9
device parameters of, 232
and devices with reduced power supply environment, 6
disadvantages of, 6
emitter design techniques, 38-39
future trends in, 68-73
    bipolar device structure improvements, 69-71
    silicon-on-insulator (SOI) technology, 72-73
and latch-up, 6
low-power ultra low-capacitance BiCMOS process, 65-68

process description, 66
shallow trench isolation (STI), 66-67
main disadvantage of, 68
manufacturing considerations, 68
and pure bipolar technology, 7
and pure CMOS technology, 7
SRAMs, 7-9
totem-pole gate, 1-2
twin-well BiCMOS process: CMOS processing issues of, 30-33
    source/drain structures and channel profiles, 30-33
    threshold voltage issues, 33
BiCMOS SRAMs, 7-9
    development trends in memory cell sizes/chip sizes, 7-8
    future of, 9
BiCMOS totem-pole gate, 1-2
BiNMOS, 140, 142, 259
    delay of two-input gates, 294
    performance comparison, 146
    power dissipation, 296
    SPICE simulations on, 140
    transistor count/area count, 296
    two-input NAND gate, 196, 213, 221, 292
    voltage waveforms of, 143
Bipolar circuits, power dissipation/gate vs. clock frequency for, 5
Bipolar compatible CMOS (BiCMOS), 1
Bipolar device structure improvements, 69-71
Bipolar process techniques, 33-36
Bipolar transistors, 5
Bipolar turn-on voltage technique, quasi-reduction of, 136-37
BJT clamping diode, FS-CMBL with, 158-59
Bootstrapped BiCMOS logic circuits, 279-304
    BF-BiNMOS, 288-90

bootstrapped BiCMOS logic gate
  (BS-BiCMOS) (1.5 V), 281-83
    circuit operation, 291
    simulation results, 282-83
bootstrapped full-swing
  BiCMOS/BiNMOS inverter, 283-97
    bootstrapping condition, 285-87
    bulk/well biasing, 288
    circuit operation, 283-85
    design considerations, 285-88
    implementation of the bootstrapping
      capacitance, 288
    parasitic PNP transistor, 288
  circuit performance comparison,
    290-97
    area, 296
    crossover capacitance, 296
    delay, 291-96
    power dissipation, 296-97
  double bootstrapped BiCMOS logic
    gates (DB-BiCMOS), 297-304
    circuit operation/simulation results,
      298-300
    input/output waveforms of, 301
    simulation results/performance
      comparison, 300-304
Boron, 97
Borophosphosilicate glass (BPSG), 49
BP-BiNMOS, 261
BP-CBiCMOS, 261
BS-BiCMOS:
  crossover capacitance, 296
  delay of two-input gates, 294
  transistor count/area count, 296
  two-input NAND gates, 292
BSIM (Berkeley Short-Channel IGFET
  Model) model, 85, 125
    BSIM3v3, 125
      model features, 92-93
      features of, 92

Buffers, 235-38
Buried-contact structure, cross-sectional view
  of, 48
Buried layers, 43
Bus architecture, 17
  and power dissipation reduction, 17

C

Capacitance:
  crossover, 222-23
  parasitic, 159
CBiCMOS, 178-85, 196
  CBiCMOS-A circuit, 184, 186
  CBiCMOS-B circuit, 184-85, 187-88
  CBiCMOS-C circuit, 185, 187-89
  simulation results and discussion,
    185-89
  structure of, 179
C-BiCMOS, two-input NAND gates, 196,
  215, 222
Channel length, 317
  dependence of delay time on, 201
  scaling, 121
Channel width, 317
Charge pump, 264-68
Chemical Vapour deposition (CVD)-nitride
  layers, 43
Circuit variations, FS-CMBL, 156-57
Clock frequency, for bipolar/CMOS circuits,
  5
CMOS, 1, 27, 140, 142, 218
  BiCMOS technology vs., 4-5
  bipolar technology compared to, 4-6
  delay of two-input gates, 294
  measured delay, 157
  measured unloaded delay, 157
  moderate-depth trench isolation for, 42
  performance comparison, 146
  power dissipation/gate vs. clock
    frequency for, 5

Index 349

power required by, 4-5
SPICE simulations on, 140
SRAMs, development trends in
   memory cell sizes/chip sizes, 7-8
transistor count/area count, 296
transitory process changes, 58
two-input NAND gates, 196, 213, 292
   device count/area for, 221
voltage waveforms of, 143
CMOS-B, 172-73
CMOS low-voltage BiCMOS digital circuits, 131-310
CMOS-M, 172
CMOS SRAMs, development trends in
   memory cell sizes/chip sizes, 7-8
Collector design techniques, 37
Collector-emitter (CE) shunt, 160-62
   FS-CMBL, 160
Collector resistance, and epitaxial layer
   thickness, 37
Common-emitter (CE) driver configuration, 140, 147
Complementary bipolar (CB) technology, 56-57
Complementary Bipolar IC U (CBIC-U), 57
Complementary Bipolar IC V (CBIC-V), 57
Complementary feedback BiCMOS
   (FB-BiCMOS) logic gates (1.2 V), 244-52
   analytical transient model, 245-47
   performance evaluation, 247-51
   two-input FB-BiCMOS AND gate, 252
Complementary-MOS:
   full-swing BiCMOS logic circuits with
      complementary emitter-follower
      driver configuration, 140-66
   pMOS/NPN pull-down technique with
      and without source-well tie, 132-36
   quasi-reduction of bipolar turn-on
      voltage technique, 136-40

circuit operation, 137-40
simulation results and analysis, 140
source-well and quasi-reduction of
   bipolar turn-on voltage techniques, 132-40
Conventional BiCMOS converter, 1-2
Crossover capacitance, 222-23
Cutoff frequency, 195
CVD oxide, deep trench isolation regions
   filled with, 73
CVD-oxide layer, 54-55

D

Darlington-type BiCMOS inverter, 190, 196
DB-BiCMOS, See Double bootstrapped
   BiCMOS logic gates (DB-BiCMOS)
DEC Alpha 21164, power dissipation, 4
Deep collector regions, 53
Deep n+ polysilicon plug contact, 37-38
Deep submicron MOSFET, parasitic
   resistance for, 16
Delay analysis:
   BiCMOS, 201-4
   QC-BiCMOS, 197-98, 201-4
      low-voltage limitation, 203-4
      pull-down delay, 201-3
      pull-up delay, 203
Delay time:
   dependence on channel length, 201
   dependence on fan-out load, 200
Device currents model:
   sub-half micron DC model formulation, 111-20
      bipolar currents, 114-19
      extraction of device currents, 111
      MOS currents, 119-20
      space-charge currents, 112-14
Device parameters:
   sub-half micron DC model formulation, 120-24

350 Index

device transconductance, 121-22
  output conductance, 122-24
Device transconductance, 121-22
Digital Signal Processing (DSP), 9
Double bootstrapped BiCMOS logic gates (DB-BiCMOS), 297-304
  circuit operation/simulation results, 298-300
  input/output waveforms of, 301
  simulation results/performance comparison, 300-304
Double diffused graded drain structure, 31-32
Drain current, 195
Drain-Induced Barrier Lowering (DIBL) effect, 84, 108
Drain-substrate capacitance, 195
Drain-to-source voltage, and Gradual-Channel Approximation (GCA), 87
Dual-implanted polysilicon gate approach, 33

E

Electrical parameter shift, 3
Electromigration, 3
Electron devices, switching energy of, 18-19
Emitter-Coupled Logic (ECL) circuits, 5
Emitter-follower (EF) driver configuration, 140
Epitaxial layer:
  and autodoping, 42
  thickness, and collector resistance, 37
Etching, 49

F

Fan-out load, dependence of delay time on, 200
Feedback, FS-CMBL with/without, 153-54
Feedback BiCMOS (FB-BiCMOS) inverting circuit, 244-52

Feedback-type BiCMOS logic circuits, 223-58
  BiCMOS dynamic logic circuit (1.5 V), 252-58
    circuit operation, 254
    circuit performance comparison/discussion, 254-58
    dynamic full adder, 254
  BiCMOS full-swing circuits using positive capacitively coupled feedback technique, 230-44
    circuit performance comparison, 235-41
    design issues regarding feedback circuit, 241-44
  bootstrapped BiCMOS logic circuits, 279-304
    bootstrapped BiCMOS logic gate (BS-BiCMOS) (1.5 V), 281-83
    bootstrapped full-swing BiCMOS/BiNMOS inverter, 283-97
    double bootstrapped BiCMOS logic gates (DB-BiCMOS), 297-304
  complementary feedback BiCMOS (FB- BiCMOS) logic gates (1.2 V), 244-52
    analytical transient model, 245-47
    performance evaluation, 247-51
    two-input FB-BiCMOS AND gate, 252
  conventional BiCMOS circuits, 223-24
  high-beta BiCMOS (HB-BiCMOS) logic circuits, 258-68
    BiCMOS charge pump with, 264-68
    circuit operation, 261-62
    circuit simulations/discussion, 262-64
    defined, 258-61
  power dissipation, 229

propagation delay, 228-29
R + N type and feedback-type (FB)
   BiCMOS logic gate, 224-29
      experimental results, 228-29
      simulation results, 226-27
   ring oscillator pattern, 228
   transiently saturated full-swing
      BiCMOS (TS-FS-BiCMOS) logic
      circuits, 215, 269-79
         circuit concept/operation, 269-72
         circuit performance
            comparison/analysis, 272-76
         design issues, 276-78
         experimental results and analysis,
            278-79
Forward source-body effect factor, 108
FS-CMBL (full-swing complementary
   MOS/bipolar logic) circuits, 153-55, 177
      with BJT clamping diode, 158-59
      circuit variations, 156-57
      collector-emitter (CE) shunt, 160-62
      experimental results and analysis,
         155-56, 160-66
      with feedback, 153-54
      with low-parasitic capacitance (Cp)
         clamping diode, 159
      with low-threshold clamping diode, 158
      measured delay, 157
      measured unloaded delay, 157
      without feedback, 153
      without feedback control and clamping,
         159
Full adder, 254
Full-swing BiCMOS logic circuits with
   complementary emitter-follower driver
   configuration, 140-66
      BiCMOS driver configuration, 147-49
         common-emitter driver
            configuration, 147

emitter-follower EF driver
   configuration, 149
full-swing techniques, 149-50
gated-diode driver configuration,
   148
comparison between driver
   configurations, 150-52
feedback-type BiCMOS logic circuits,
   223-58
      BiCMOS dynamic logic circuit
         (1.5V), 252-58
      BiCMOS full-swing circuits using
         positive capacitively coupled
         feedback technique, 230-44
      complementary feedback BiCMOS
         logic gates (1.2 V), 244-52
      conventional BiCMOS circuits,
         223-24
      power dissipation, 229
      propagation delay, 228-29
      R + N type and feedback-type (FB)
         BiCMOS logic gate, 224-29
      ring oscillator pattern, 228
FS-CMBL (full-swing complementary
   MOS/bipolar logic) circuits, 153-55
      with BJT clamping diode, 158-59
      circuit variations, 156-57
      collector-emitter (CE) shunt, 160-62
      experimental results and analysis,
         155-56, 160-66
      with feedback, 153-54
      with low-parasitic capacitance (Cp)
         clamping diode, 159
      with low-threshold clamping diode,
         158
      without feedback, 153
      without feedback control and
         clamping, 159
full-swing multi-drain/multi-collector
   BiCMOS buffers, 183-89

circuit implementation/operation, 184-85
simulation results/discussion, 185-89
full-swing Schottky BiCMOS/BiNMOS logic circuits, 208-23
  area comparison, 221-22
  basic concept, 208-10
  circuit implementations, 210-12
  circuit simulation and discussion, 212-16
  crossover capacitance, 222-23
  delay dependence on the frequency of operation, 218-19
  operation of, 208-10
  power dissipation, 219-21
  scaling the supply voltage, 217-18
full voltage swing multi-drain/multi-collector complementary BiCMOS buffers, 177-84
  conventional BiCMOS and CBiCMOS buffers, 178-79
  multi-drain/multi-collector BiCMOS buffers, 179-81
  simulation results and discussion, 181-83
merged BiCMOS (MBiCMOS) logic gates, 131, 166-77
  circuit performance and comparison, 169-71
  experimental tests and analysis, 172-75
  full-swing MBiCMOS gate, 177
  MBiCMOS gate, 167-69
  simulation results and analysis, 175-77

quasi-complementary BiCMOS (QC-BICMOS) logic circuits, 189-208
  circuit concept, advantages of, 190-93
  circuit performance comparison and analysis, 193-201
  delay analysis, 197-98, 201-4
  design issues, 204
  experimental results, 204-6
  pull-down waveform of, 198
Full-swing complementary MOS/bipolar logic (FS-CMBL) circuits, 146
Full-swing M-BiCMOS gate, 177
Full-swing multi-drain/multi-collector BiCMOS buffers, 183-89
  circuit implementation/operation, 184-85
  first implementation, 184
  second implementation, 184-85
  third implementation, 185
  simulation results/discussion, 185-89
Full-swing Schottky BiCMOS/BiNMOS logic circuits, 208-23
  area comparison, 221-22
  basic concept, 208-10
  circuit implementations, 210-12
  circuit simulation and discussion, 212-16
  crossover capacitance, 222-23
  delay dependence on the frequency of operation, 218-19
  operation of, 208-10
  power dissipation, 219-21
  supply voltage, scaling, 217-18
Full voltage swing multi-drain/multi-collector complementary BiCMOS buffers, 177-84

Index 353

conventional BiCMOS and CBiCMOS
    buffers, 178-79
multi-drain/multi-collecxtor BiCMOS
    buffers, 179-81
simulation results and discussion,
    181-83

G

Gated-diode (GD) driver configuration, 140, 148
Gate delay time of CMOS inverter vs.
    threshold voltage/power supply voltage, 12
Gate oxide thickeness, 195
Gate resizing, 15-16
    and power dissipation reduction, 15-16
Graded drain approach, to hot-carrier
    generation, 31
Gradual-Channel Approximation (GCA), 87

H

Half adder, 254
High-beta BiCMOS (HB-BiCMOS) logic
    circuits, 258-68
    BiCMOS charge pump with, 264-68
    circuit operation, 261-62
    circuit simulations/discussion, 262-64
    defined, 258-61
High-performance 0.5 mm digital BiCMOS
    process changes, 56-60
    CMOS transistor process changes, 58
    complementary bipolar (CB)
        technology, 56-57
    NPN profile optimization, 59-60
High-performance 0.8 mm BiCMOS process
    flow, 50-56
High-performance BiCMOS technology,
    comparison of CMOS and BiCMOS
    process flows for, 52

Hole injection barrier, in hybrid-mode
    devices, 117
Hot-carrier generation, drain structures to
    reduce, 31
HSPICE, 181, 245, 290

I

Input/output (I/O), and power dissipation
    reduction, 17
Inter-Level Dielectric (ILD), 62
Intrinsic-doped epitaxial layer, 42
Inverter time vs. supply voltage, 11
Isolation schemes, base-collector edge
    capacitance for, 68

J

Junction capacitance, 195
Junction fatigue, 3

K

Knee current, 195, 316

L

Latch-up, 39, 168
    and BiCMOS technology, 6
Lateral autodoping, 42
Lateral bipolar gain, 118-19
Lightly doped drain (LDD), 30-33, 85
Load capacitance, 16
    measured gate delay as function of, 156
    and power dissipation reduction, 16
    propagation as a function of, 188
Loaded gate delays, load capacitance vs., 165
Loaded power consumption, supply voltage
    vs., 166
Local Interconnect (LI) technology, 47
LOCal Oxidation of Silicon (LOCOS), 43-44, 68

Low-parasitic capacitance clamping diode, FS-CMBL with, 159
Low-power dissipation in stand-by mode, 17
Low-threshold clamping diode, FS-CMBL with, 158
Low-voltage low-power BiCMOS circuits:
  loading, effects on delay and power dissipation sensitivities, 331-36
  quality of BJT, effects on delay and power dissipation sensitivities, 336-38
  sensitivity evaluation/comparison of, 326-38
Low-voltage low-power design, 10-20
  BiCMOS, future directions, 17-20
  design limitations, 10-13
    power supply voltage, 10
    scaling, 13
    threshold voltage, 11-12
  power dissipation:
    of LSIs, 14-15
    of a single logic gate, 13-14
  power dissipation reduction techniques, 15-17
    amount of computation/algorithm, 16
    bus architecture, 17
    gate resizing, 15-16
    input/output (I/O), 17
    load capacitance, 16
    parasitic effects, 16
    power dissipation in gates, 15
    power dissipation in the standby mode, 17
    power supply voltage, 15
Low-voltage, low-power operation, 2-4
Low-voltage operational bipolar FET (LV-BiFET), 137
LSIs, power dissipation of, 14-15
LV-BiFET, 137-42
  performance comparison, 146
  pull-up/pull-down operation of, 138
  retaining operation of, 139-40
  SPICE simulations on, 140
  voltage waveforms of, 143

M

MBiCMOS, 167-69, 177
  circuit design, 167-68
  circuit operation and analysis, 168
  pull-down operation of, 171
Measured delay:
  load capacitance vs., 175
  power supply voiltage vs., 174
  supply voltage for, 174
Measured gate delays:
  as function of load capacitance, 156
  load capacitance vs., 163-64
Measured power-delay product, power supply voltage vs., 176
Merged BiCMOS (MBiCMOS) logic gates, 131, 166-77
  circuit performance and comparison, 169-71
  experimental tests and analysis, 172-75
  full-swing MBiCMOS gate, 177
  MBiCMOS gate, 167-69
  simulation results and analysis, 175-77
Merged BipMOS device, 230-35
  dimensions of, 231
Metallization, 48-49
Metal Oxide Semiconductor (MOS), 1
Microprocessors, power dissipation in, 4
MIPS R10000, power dissipation, 4
Moderate-depth trench isolation for CMOS, 42
MOS-C, 87
MOS current model, 119-20
MOSFET current models, 87-96
MOSFET in a hybrid mode environment, 96-124

device fabrication, 100
model parameters extraction, 100-102
sub-half micron DC model formulation, 102-24
surface p-channel for sub-half micron devices, 97-100
See also Sub-half micron DC model formulation

Motorola:
  SRAM technology using triple-level polysilicon, 39
  triple-poly BiCMOS device, detailed cross-section of, 41

Multi-generation BiCMOS digital circuits, 311-42
  delay and power dissipation sensitivities:
    effects of loading on, 331-36
    effects of quality of BJT on, 336-38
    effects of scaling the technology on, 338
  delay sensitivity upper/lower bounds, 338-39
  key BJT/MOS process and device/circuit parameters, relationship between, 312-13
  low-voltage low-power BiCMOS circuits:
    loading, effects on delay and power dissipation sensitivities, 331-36
    quality of BJT, effects on delay and power dissipation sensitivities, 336-38
    scaling the technology, effects on delay and power dissipation sensitivities, 338
    sensitivity evaluation/comparison of, 326-38
  sensitivity analysis of the conventional BiCMOS circuit, 313-26

base resistance/collector resistance, 316
delay time sensitivity to key device parameters, 313-17
forward current gain, 316
forward transit time, 315
HSPICE simulations/analysis, 322-26
knee current, 316
load capacitance, effects on circuit delay sensitivity, 317-19
MOS channel width/length, 317
oxide thickness, 317
quality of BJT, effects on circuit delay sensitivity, 319-20
scaling the technology, effects on circuit delay sensitivity, 320-22
threshold voltage, 317

N

New process flow, comparison with BiCMOS process, 65-66
Nickel Cadmium (NiCd) batteries, 3
Nickel-Metal Hydride (Ni-MH), 3
NMOS LDD section, 30
Non-LDD MOSFETs, 85
npn profile organization, 59-60
NTL circuits, 140, 143
  performance comparison, 146
  SPICE simulations on, 140
  voltage waveforms of, 143

O

Output conductance, 122-24
Oxide-Nitride-Oxide (ONO), 62
Oxide spacer technology, 30-31

P

Package-related failure, 3

Parasitic bipolar transistors, 39
Parasitic effects, and power dissipation reduction, 16
Performance improvement, and power dissipation, 3
Phosphosilicate glass (PSG), 49
Planarization, 48-49
    of inter-level oxides, 51
Plug, 37
PMOSFET, 96-97
    space-charge current of, 95
pnpn path:
    elimination methods, 39MOS device modeling, 83-130
    elimination methods, 39
    MOSFET current models, 87-96
    MOSFET in a hybrid mode environment, 96-124
        device fabrication, 100
        model parameters extraction, 100-102
        sub-half micron DC model formulation, 102-24
        surface p-channel for sub-half micron devices, 97-100
    threshold voltage models, 83-86
    See also Sub-half micron DC model formulation
Poly emiter, 38
Poly-ridge emitter transistor (PRET), 65-67
Polysilicon buffer layer, 43
Polysilicon buffer LOCOS process, key steps in, 44
Portable systems, 3
Power dissipation:
    in gates, 15
    of microprocessors, 4
    and performance improvement, 3
    in standby mode, 15

Power dissipation reduction techniques, 15-17
    amount of computation/algorithm, 16
    bus architecture, 17
    gate resizing, 15-16
    input/output (I/O), 17
    load capacitance, 16
    parasitic effects, 16
    power dissipation in gates, 15
    power dissipation in the standby mode, 17
    power supply voltage, 15
PowerPC 620, power dissipation, 4
Power supply voltage:
    decrease ian, 17-18
    and low-voltage low-power design, 10
    and power dissipation reduction, 15
    scaling, 3
Process complexity, comparison of technologies in terms of, 51
Pull-down delay, 201-3
Pull-up delay, 203

Q

QC-BiCMOS, 218
    two-input NAND gates, 214, 221
Quasi-complementary BiCMOS (QC-BICMOS) logic circuits, 189-208
    circuit concept, advantages of, 190-93
    circuit performance comparison and analysis, 193-201
Quasi-pnp, 191

R

R + N type and feedback-type (FB) BiCMOS logic gate, 224-29
    experimental results, 228-29
    simulation results, 226-27
Random access memories (RAMs), 7-9

Index

Refractory metal-silicides, and LI technology, 47

S

SBiCMOS, two-input NAND gates, 214, 221
SBiNMOS, 218
    two-input NAND gates, 214, 222
Scaling:
    benefits of, 13
    of channel length, 121
    and low-voltage low-power design, 13
Scanning Electron Microscopy (SEM), 100-101
Schottky BiNMOS (SBiNMOS) gate:
    circuit concept, advantages of, 190-93
    circuit performance comparison and analysis, 193-201
    delay analysis, 197-98, 201-4
        low-voltage limitation, 203-4
        pull-down delay, 201-3
        pull-down waveform of, 198
        pull-up delay, 203
    design issues, 204
    experimental results, 204-6
    See also Full-swing Schottky BiCMOS/BiNMOS logic circuits
Self-alignment schemes, with single level polysilicon/double level polysilicon, 39
Semiconductor Industry Association (SIA), 96, 101
Semi-empirical short-channel model (SPICE LEVEL 3), 84
Sensitivity:
    circuit delay, 317-20
    power dissipation, 326-38
Shallow trench isolation (STI), 66-67
SideWAll-Masked Isolation (SWAMI) technique, 43, 45
Silicidation, 44-48, 55
Silicon interconnect fatigue, 3

Silicon-on-insulator (SOI) technology, 72-73
Simulated delay, supply voltage vs., 176
Simulated delay time:
    number of fan-out at power supply voltage, 144
    supply voltage vs., 144
Simulated power, delay at power supply voltage vs., 146
Simulated power consumption:
    number of fan-out at power supply voltage vs., 145
    supply voltage vs., 145
Single logic gate, power dissipation of, 13-14
Sinker, 37
SOI technology, 72-73
Source-well tie:
    pMOS/NPN pull-down technique with/without, 132-36
    and quasi-reduction of bipolar turn-on voltage techniques, 132-40
SPICE:
    MOSFET model levels for circuit simulation, 88-89
        BSIPM model, 92-93
        Level-1 MOSFET model, 88
        Level-2 MOSFET model, 88-90
        Level-3 MOSFET model, 90-92
SPICE simulations:
    device parameters used for, 274
    key BJT device parameters used in, 186
    key CMOS device parameters used in, 186
    on Lv-BiFET, 140
SSBiCMOS, 218
    two-input NAND gates, 215, 222
Sub-half micron DC model formulation, 102-24
    device currents model, 111-20
        bipolar currents, 114-19
        extraction of device currents, 111

MOS currents, 119-20
space-charge currents, 112-14
device parameters, 120-24
device transconductance, 121-22
output conductance, 122-24
threshold voltage model, 102-11
experimental threshold voltage, extraction of, 102-5
impact of short-channel effects on classical threshold voltage model, 15-106
temperature depends and hybrid-mode device threshold voltage model, 106-11
Sub-half micron devices, surface p-channel for, 97-100
Surface p-channel for sub-half micron devices, 97-100
SWAMI, 43, 45

T

Telecommunication chips, DSP portions of, 10
Telecommunications applications, and BiCMOS, 9-10
Tetraethylorthosilicate (TEOS), 30, 100
Threshold voltage, and low-voltage low-power design, 11-12
Threshold voltage models, 83-86
sub-half micron DC model formulation, 102-11
experimental threshold voltage, extraction of, 102-5
impact of short-channel effects on classical threshold voltage model, 15-106
temperature depends and hybrid-mode device threshold voltage model, 106-11
Titanium silicide (TiSi2) process, 45-49

TMA TSUMPREM-4 simulation, 101
Transconductance peak method, 102-4
Transiently saturated full-swing BiCMOS (TS-FS-BiCMOS) logic circuits, 218, 269-79
circuit concept/operation, 269-72
circuit performance comparison/analysis, 272-76
crossover capacitance, 296
delay of two-input gates, 294
design issues, 276-78
experimental results and analysis, 278-79
power dissipation, 296
transistor count/area count, 296
two-input NAND gates, 215, 222, 269, 292
voltage swing of parasitic capacitance in, 278
Transistor formation, 100
Transport of Ions in Matter (TRIM), 101-2
TRASIM, 232
Trench isolation, 40-42
TS-FS-BiCMOS, See Transiently saturated full-swing BiCMOS (TS-FS-BiCMOS) logic circuits
Two-dimensional transient device simulator (TRANSIM), 232
Two-input FB-BiCMOS AND gate, 252
Two-input gates, device count and gate area of, 277
Two-input NAND gates, device count/area of, 221-22
Two-terminal MOS capacitor (MOS-C), 87

U

ULSI (Ultra Large Scale Integration), 9
UltraSparc, power dissipation, 4
Unloaded gate delays, supply voltage vs., 165

Unloaded power consumption, supply voltage vs., 166

V

Vertical gate field, and Gradual-Channel Approximation (GCA), 87

Video Signal Parallel Pipeline Processor (VSP3), 9
VLSI (Very Large Scale Integration) circuits, 1, 4, 6, 27
VSP3 (300 MHz Video Signal Parallel Pipeline Processor), 9

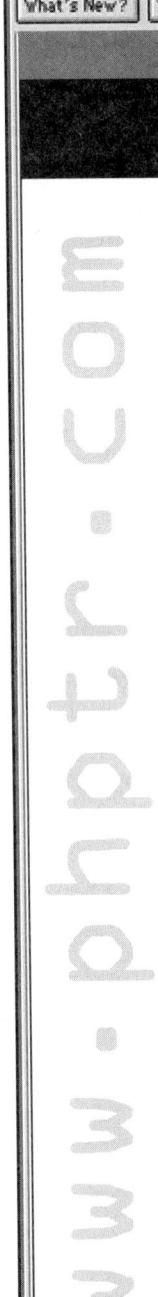

Keep Up-to-Date with
PH PTR Online!

We strive to stay on the cutting-edge of what's happening in professional computer science and engineering. Here's a bit of what you'll find when you stop by **www.phptr.com**:

@ Special interest areas offering our latest books, book series, software, features of the month, related links and other useful information to help you get the job done.

Deals, deals, deals! Come to our promotions section for the latest bargains offered to you exclusively from our retailers.

$ Need to find a bookstore? Chances are, there's a bookseller near you that carries a broad selection of PTR titles. Locate a Magnet bookstore near you at www.phptr.com.

What's New at PH PTR? We don't just publish books for the professional community, we're a part of it. Check out our convention schedule, join an author chat, get the latest reviews and press releases on topics of interest to you.

Subscribe Today! **Join PH PTR's monthly email newsletter!**

Want to be kept up-to-date on your area of interest? Choose a targeted category on our website, and we'll keep you informed of the latest PH PTR products, author events, reviews and conferences in your interest area.

Visit our mailroom to subscribe today! **http://www.phptr.com/mail_lists**